Quantum Mechanics and Electrodynamics

Jaroslav Zamastil • Jakub Benda

Translated with the assistance of Tereza Uhlířová

Quantum Mechanics and Electrodynamics

Springer

Jaroslav Zamastil
Charles University in Prague
Prague, Czech Republic

Jakub Benda
Charles University in Prague
Prague, Czech Republic

This is an English-language translation of the Czech-language original publication: Kvantová mechanika a elektrodynamika by Jaroslav Zamastil and Jakub Benda; Karolinum Press;

ISBN 978-3-319-65779-0 ISBN 978-3-319-65780-6 (eBook)
DOI 10.1007/978-3-319-65780-6

Library of Congress Control Number: 2017949677

Printed on acid-free paper

This Springer imprint is published by Springer Nature
The registered company is Springer International Publishing AG
The registered company address is: Gewerbestrasse 11, 6330 Cham, Switzerland

Preface

> We have not redefined quantum theory; we carry it to its logical conclusion. (...) We learned it second or third hand, as an established discipline whose rules and techniques we came to feel as intuitive and natural, not as a peculiar displacement of classical: we found and find it almost painful to do 19th century physics. The great Bohr-Einstein philosophical debates which fascinate historians and the philosophers are to us a bit wrong-headed (...) [1]

A Few Words of Explanation

Nowadays, when there is a large number of books on quantum mechanics available, some of them are indeed of high quality and rightfully regarded as the "classics" (e.g. [4, 8] for the uninitiated, [7, 19, 24] for intermediate and [15] for advanced readers), when other remarkable expositions appear (e.g. [25]), when there is a large number of books on quantum electrodynamics and quantum field theory, some of them are indeed excellent (ranging widely from easier pieces such as [3, 12, 18, 23, 26] to more difficult ones such as [2, 5, 13, 21, 27]), and when almost any textbook ever written can be found on the Internet, downloaded and printed, we feel a need to start with an excuse, or at least with a few words of explanation, for coming up with yet another book on quantum mechanics and electrodynamics.

1. Symmetries of a problem at hand (such as parity, angular momentum etc.) and their application in the solutions of quantum mechanical problems (e.g. the Wigner-Eckart theorem) are usually subject to such an abstract exposition that an uninitiated reader, even if he understands the concept, does not see their usefulness. For this reason, we have devoted considerable attention to finding the solutions to the simplest, physically interesting problems that cannot be solved exactly, such as the anharmonic oscillator and the helium atom. In contrast to other textbooks, we carry out the solutions to their very end and provide more than an outline of them. Today, when routines for the diagonalization of matrices form a standard part of libraries of advanced programming languages such as

Maple, Matlab, Mathematica, Octave, Scilab, Maxima etc., we want to show the reader how such problems can be solved alone using a laptop.

Furthermore, we want to emphasize yet another aspect of the use of symmetry in quantum mechanics: namely the advantage of an algebraic over an analytical approach to both precisely solvable and insolvable problems. The algebraic approach provides an elegant solution to the very few physically interesting problems that can be solved exactly, such as the harmonic oscillator, angular momentum, composition of angular momenta and hydrogen-like atoms. It enables a complete solution to be constructed, including the pertinent wave functions without knowledge of various orthogonal polynomials etc. Its advantage becomes even more obvious when dealing with problems that cannot be solved exactly; namely, the calculation of the Hamiltonian matrix elements for variational calculation of anharmonic oscillators, helium and more complex atoms, the inclusion of the continuous part of the spectra in the case of complex atoms etc.

It is needless to emphasize that the problem of symmetries and their applications in quantum mechanics is not a minor one, nor is it at all complicated. Once the basic mathematical concepts have been absorbed that are needed to understand quantum mechanics, such as the concept of an operator, a commutator, eigenstates and eigenvalues, one is able to understand everything about symmetries in addition to their applications, without any recourse to the mathematical niceties of group theory.

2. We wanted to explain relativistic quantum electrodynamics with minimal necessary formalism and with an emphasis on its physical content and its applications to atomic physics.

   Modern expositions of quantum electrodynamics, such as [18, 21, 27],[1] heavily rely on either the Dyson-Wick expansion or path integrals, both of which are undoubtedly very powerful tools. However, for an uninitiated reader, both the physical content of the formalism and its connection to the usual formalism of quantum mechanics are rather, if not completely, obscured. Moreover, the applications are almost exclusively restricted to the high-energy scattering processes.

   Two conclusions could thus be drawn. First, it may appear that the "low-energy" physicists do not need to possess any knowledge of quantum field theory. Second, the usual formalism of quantum mechanics would have to be dismissed at the very beginning and a completely new formalism of the quantum field theory would have to be learnt. However, this task is clearly too difficult and bearing in mind the first point, it is not worth the effort. Furthermore, when dealing with the most important low-energy quantum electrodynamic processes, such as spontaneous emission and atom-photon scattering, it suffices to consider the non-relativistic theory, which is much more comprehensible.

[1] We find them excellent and strongly recommend them to the reader.

We do not share this view for a variety of reasons. First and foremost, even experimenters could know the theoretical value contained in precise measurements. In addition, once beyond the first approximation, even within the non-relativistic theory, the so-called virtual processes are encountered. However, these processes cannot be fully described within non-relativistic theory. Therefore, this theory is clearly incomplete. Moreover, methods of quantum field theory have proved to be of enormous use far beyond the application for which they were primarily invented. For instance, they find their applications in both classical and quantum many-body problems, see for example [16, 17, 27]. Finally, we believe that quantum field theory—like the general theory of relativity—is part of the basic education of every physicist.

When dealing with relativistic quantum electrodynamics, we first proceed within the framework of the ordinary quantum-mechanical formalism and only then show how the mathematical description is simplified by adopting the Feynman view of positrons as electrons running backwards in time. Although this derivation is neither the shortest nor the most formally satisfying, it enables one to determine the key steps in the transition from the non-relativistic to the relativistic theory, to see where in the Feynman diagrams and rules the ordinary perturbation method and Coulomb law are hidden etc. The approach adopted in this book lies closest to that in the book [23].

We made a considerable effort to minimize the discussion of purely formal aspects of problems on the one hand and to completely avoid the phrase "it can be shown that" on the other hand. This means that we tried to carefully motivate and derive everything within a physicist's level of rigor. This also means that a number of (important) topics are deliberately treated in a manner that can be justifiably considered superficial. In particular, this applies to our treatment of the spinors, adequate formalism of relativistic quantum field theory, renormalization, Lie algebras etc. Even the discussion of Lorentz and gauge invariance does not go into much depth. There are necessarily omissions in every book; a part of the process of writing is to decide what should be omitted and what should be included. The choice of the themes and their exposition in our work was dictated by our strong conviction about what should be learnt in the first place. The purpose of every physical theory worthy of that name is to "get the numbers out". Despite the knowledge of deep connections having its value, we believe that one should learn how the theory works in the first place, i.e. how numbers can be obtained that experimenters can measure.

Regarding the level of rigor, we want to stress that the mathematical level of rigor is completely alien to us. The art of approximation forms an integral part of "the art of physics" and we strongly feel that concepts should be explained to others in a way that is as close as possible to how we think about them ourselves.

If the reader is not satisfied either with our choice of the topics or with our method of exposition, he is encouraged to consult other textbooks, for instance, those mentioned at the beginning of this Preface. The books [15] and [2] are of an encyclopaedic nature.

Regarding the literature, we refer to a few basic books we consider to be generally known and available. There is such a large number of books on quantum mechanics that even a complete list of them exceeds our abilities. In the first three chapters, where we introduce the fundamentals of quantum mechanics, we do not list our inspirations. However, we try to do so in the advanced parts of this book. Note, though, that the absence of a citation does not necessarily imply that our exposition is original.

There are a number of exercises throughout the text. They are intended for the reader to examine how much he understood. The exercises are, according to the difficulty, denoted by a number of ▲. If the reader is able to complete only one exercise, he is surely able to recognize whether a greater number of ▲ marks a more or less difficult exercise.

Finally, in this book we focus on the exposition of the "classics" within quantum mechanics and electrodynamics. Note, though, that the word "classics" does not imply that these topics ceased to be of interest. For instance, the Nobel prize of 2012 was awarded for experiments that manipulated the individual mesoscopic objects and thus probed the boundary between the quantum and classical behaviour, see for example [6]. In 2005, the Nobel prize was awarded for extremely accurate spectroscopic measurements. When combined with the theoretical predictions based on quantum electrodynamics, these measurements enable us to deduce the nuclear properties, such as the proton size, with substantially better accuracy than by any other means. This in turn leads to surprising findings, see for example [22]. Thus, what follows are the "classics"; nevertheless, the classics are still alive! [2]

Prerequisites

The following is mainly for autodidacts, who have our sympathies. Should this book bring joy and not frustration, the following should hold.

1. The reader has a good command of single-variable calculus, complex numbers and basic linear algebra.
2. The reader possesses some knowledge of multi-variable calculus and vector analysis. The level of exposition in for example, Feynman's freshman lectures [9] should suffice.
3. For parts concerning quantum electrodynamics, namely Chapters 6 and 7, it is necessary to know the complex analysis at the level of for example [10].

[2] We do not go as far as to try to explain the latest developments, such as the two mentioned above. We believe that by mastering the content of this book, the reader will be able to catch up on them on his own.

4. The reader should possess some knowledge of classical mechanics, electrodynamics and special relativity. Again, knowledge at the level of [9] should suffice. Some knowledge of the Hamilton formulation of classical mechanics would be useful, though not crucial. The "classic" textbooks are [11, 14].

Acknowledgments

We would like to express our gratitude

- To Jakub Zahumenský, Lada Vybulková and Tomáš Javůrek for their help when creating the very first version of the manuscript.
- To Roman Čurík and Tomáš Mančal for a number of suggestions and improvements.
- To our dear mentor, Prof. Lubomír Skála, for his support of the whole project in its initial phase. Unfortunately, he passed away before completion.

Errors

An up-to-date list of errors we have not discovered in time will be kept at

`http://quantum.karlov.mff.cuni.cz/~jzamastil/`

Prague, Czech Republic

Jaroslav Zamastil
Jakub Benda

Contents

List of Exercises

Notation, Convention, Units, and Experimental Data

Notation

a	Scalar quantity, components of vectors and their magnitudes
a^*	Complex conjugation
$\mathbf{A}$	Three-dimensional vector
$\boldsymbol{A}$	Four-dimensional vector ("four-vector")
$\mathsf{A}, \mathsf{A}^\top, \mathsf{A}^+$	Matrix, its transposition and Hermitian conjugation
1	Unit matrix
$\mathrm{diag}\,\{a_1, \ldots, a_n\}$	Diagonal matrix determined by its eigenvalues
$\hat{\mathsf{A}}$	Scalar operator, component of a vector operator
$\hat{\boldsymbol{\mathsf{A}}}$	Three-dimensional vector operator
$\hat{\boldsymbol{A}}$	Four-dimensional vector operator
$[\hat{\mathsf{a}}, \hat{\mathsf{b}}] = \hat{\mathsf{a}}\hat{\mathsf{b}} - \hat{\mathsf{b}}\hat{\mathsf{a}}$	Commutator
$\{\hat{\mathsf{a}}, \hat{\mathsf{b}}\} = \hat{\mathsf{a}}\hat{\mathsf{b}} + \hat{\mathsf{b}}\hat{\mathsf{a}}$	Anticommutator
$\|+\rangle$	Spin state $\left\|\frac{1}{2}, +\frac{1}{2}\right\rangle$
$\|-\rangle$	Spin state $\left\|\frac{1}{2}, -\frac{1}{2}\right\rangle$
$\nabla_i = \frac{\partial}{\partial x_i}$	Differential vector operator
$\hat{\mathsf{V}}_\pm = \hat{\mathsf{V}}_1 \pm \mathrm{i}\hat{\mathsf{V}}_2$	Often-used combination of components of vector operator
$\Re$	Real part of a complex number
$\Im$	Imaginary part of a complex number
$\dot{q} = \frac{\mathrm{d}q}{\mathrm{d}t}$	Time derivative
$\mathrm{d}\Omega = \mathrm{d}\varphi\mathrm{d}\vartheta \sin\vartheta$	Differential of solid angle

The Summation Convention

- The scalar product is in both three- and four-dimensional spaces denoted by a centered dot, i.e., $\mathbf{a} \cdot \mathbf{b}$ and $\boldsymbol{a} \cdot \boldsymbol{b}$.
- The components of three-vectors carry a Latin index $(i, j, k, \ldots)$; the components of four-vectors are distinguished using Greek indices $(\mu, \nu, \ldots)$.
- The Einstein summation convention is used throughout the book: if two same indices appear, they are summed over; for instance,

$$a_i b_i = a_1 b_1 + a_2 b_2 + a_3 b_3 ,$$

or in the case of four-vectors, we always use the metric $(+1, -1, -1, -1)$ with indices being always subscripts

$$a_\mu b_\mu = a_0 b_0 - a_1 b_1 - a_2 b_2 - a_3 b_3 .$$

The Component Formalism

The scalar product may be written in components by means of the abovementioned Einstein summation convention as

$$\mathbf{A} \cdot \mathbf{B} = A_j B_j = \delta_{ij} A_i B_j , \qquad \nabla \cdot \mathbf{A} = \frac{\partial A_i}{\partial x_i} , \qquad \nabla^2 = \frac{\partial^2}{\partial x_i \partial x_i} ,$$

where the Kronecker symbol

$$\delta_{ij} = \begin{cases} 1 & (i = j) , \\ 0 & (i \neq j) . \end{cases}$$

The vector product may be written in components as

$$(\mathbf{A} \times \mathbf{B})_i = \varepsilon_{ijk} A_j B_k , \qquad (\nabla \times \mathbf{A})_i = \varepsilon_{ijk} \frac{\partial A_k}{\partial x_j} ,$$

where the Levi-Civita symbol

$$\varepsilon_{123} = \varepsilon_{231} = \varepsilon_{312} = 1 ,$$

$$\varepsilon_{213} = \varepsilon_{132} = \varepsilon_{321} = -1,$$

and

$$\varepsilon_{ijk} = 0\,,$$

when any two indices ij, jk, or ik take the same value.

All identities of vector algebra and analysis used in the text may be derived from the identity

$$\varepsilon_{ijk}\varepsilon_{ipq} = \delta_{jp}\delta_{kq} - \delta_{jq}\delta_{kp}\,.$$

The simplest way to prove it is by direct substitution of specific values.

Units

Unless stated otherwise, we use the so-called natural units where $\hbar = c = \varepsilon_0 = 1$. The dimensions of a few basic physical quantities in this system of units are displayed in Table 1.

We will encounter the transitions between the SI and natural system of units only in very few cases. In Eq. (3.11), Sect. 3.2.3, we show the relation between a frequency given in hertz and the corresponding energy difference in electronvolts, and the other transitions for cross section, time, electric intensity, and magnetic induction follow.

When converting units in Eq. (6.156), we proceed as follows. In natural units, the electron mass equals [20]

$$m_e = 0.5109989\,\mathrm{MeV}\,, \tag{1}$$

and the reduced Planck constant $\hbar$ and light velocity c are equal to one,

$$\hbar = 1 = \{\hbar\}\,\mathrm{J\,s}\,, \quad \{\hbar\} = 1.0545717 \times 10^{-34} \tag{2}$$

Table 1 System of units used in the text

		Dimension in the system of units	
Quantity	Symbol	SI	Natural
Energy	E	J	eV
Mass	m	kg	eV
Velocity	v	m/s	1
Time	t	s	eV^{-1}
Cross section	σ	m^2	eV^{-2}
Electric intensity	E	V/m	eV^2
Magnetic induction	B	T	eV^2

and

$$c = 1 = \{c\}\,\mathrm{m\,s^{-1}}\,, \quad \{c\} = 2.99792458 \times 10^{8}\,, \tag{3}$$

where the numbers in the curly brackets are values in SI units [20]. One electronvolt equals[3]

$$1\,\mathrm{eV} = 1e \times 1\,\mathrm{V} = \{e\}\,\mathrm{J}\,, \quad \{e\} = 1.6021766 \times 10^{-19}\,. \tag{4}$$

Combining the last three equations, we find

$$(1\,\mathrm{eV})^{-1} = \left\{\frac{\hbar c}{e}\right\}(1\,\mathrm{m})\,, \quad \left\{\frac{\hbar c}{e}\right\} = 1.9732696 \times 10^{-7}\,. \tag{5}$$

Substituting Eq. (1) for m and Eq. (5) into Eq. (6.156), we arrive at the displayed value.

In Sects. 4.4.7 and 4.5.3, we need to transform the intensities of magnetic and electric fields from the SI units to atomic units. For intensity of the electric field, we have

$$1\,\frac{\mathrm{eV}}{\mathrm{m}} = \left\{\frac{\hbar c}{e}\right\}(\mathrm{eV})^2\,;$$

hence for γ_{E}, Eq. (4.73), we obtain

$$\gamma_{\mathrm{E}} = \frac{eE}{m_e^2\alpha^3} = \frac{\{E\}(\mathrm{eV})^2}{m_e^2\alpha^3}\left\{\frac{\hbar c}{e}\right\} \simeq 1.945\{E\} \times 10^{-12}\,. \tag{6}$$

Likewise, for the intensity of the magnetic field, we have

$$1e \times 1\mathrm{T} = 1\frac{\mathrm{eV}}{\mathrm{m}}\frac{\mathrm{s}}{\mathrm{m}} = \left\{\frac{\hbar c^2}{e}\right\}(\mathrm{eV})^2\,;$$

hence for γ_{B}, Eq. (4.57), we find

$$\gamma_{\mathrm{B}} = \frac{eB}{m_e^2\alpha^2} = \frac{\{B\}(\mathrm{eV})^2}{m_e^2\alpha^2}\left\{\frac{\hbar c^2}{e}\right\} \simeq 4.254\{B\} \times 10^{-6}\,. \tag{7}$$

Finally, in Sect. 6.2.6, we need to convert the lifetime of the excited states from the natural to SI units. The combination of Eqs. (2) and (4) yields

$$1(\mathrm{eV})^{-1} = \frac{1}{\{e\}1J} = \left\{\frac{\hbar}{e}\right\} s\,.$$

[3] In this book, the elementary charge is denoted by "e," the Euler number by "e," and one electronvolt by "eV."

Thus, to express the lifetime in seconds, we use Eq. (6.99)

$$\tau_I = \frac{1}{m}\left\{\frac{\hbar}{e}\right\}\frac{1}{\frac{\sum w_{I\to F}}{m}}, \tag{8}$$

where we evaluate the dimensionless quantity $\frac{\sum w_{I\to F}}{m}$ from Eq. (6.97) and for m we substitute m_e from Eq. (1).

Fundamental Constants

Basic physical constants that must be taken from experiment are introduced in the following sections:

- The fine structure constant α in 3.2.3
- The Rydberg constant multiplied by the speed of light $R_\infty c$ in 3.2.3
- The ratios of electron to proton, electron to muon, and electron to deuteron masses in 3.2.6

Experimental Data

Quantitative experimental data that will be compared with the theoretical predictions may be found in the following sections:

- The transition $2s - 1s$ in hydrogen, deuterium, and muonium in 3.2.6
- The transition $1^3s - 1^1s$ in hydrogen and muonium in 3.3
- The transition $2p_{3/2} - 2p_{1/2}$ in hydrogen in 3.5.2
- The transitions $2^3s - 2^3p_0$, $2^3s - 2^3p_1$, and $2^3s - 2^3p_2$ in positronium in 4.4.9
- The transition $2\,{}^1S - 1\,{}^1S$ in helium in 5.3.7
- The transition $1^3s - 1^1s$ in positronium and the lifetime of the state 1^1s of positronium in 7.3.2
- The electron gyromagnetic ratio g_e in 7.5.7
- The transition $2p_{1/2} - 2s$ in hydrogen, muonium, and muonic hydrogen in 7.6.6.
- The transition $2^3s - 1^3s$ in positronium in 7.8.5

References

1. P.W. Anderson, *More and Different: Notes from a Thoughtful Curmudgeon* (World Scientific, Singapore, 2011)
2. V.B. Beresteckij, E.M. Lifshitz, L.P. Pitaevskij, *Quantum Electrodynamics* (Butterworth-Heinemann, London, 1982)
3. J.D. Bjorken, S.D. Drell, *Relativistic Quantum Mechanics* (McGraw-Hill, New York, 1965)
4. D.I. Blokhintsev, *Quantum Mechanics* (Reidel, Dordrecht, 1964)
5. L.S. Brown, *Quantum Field Theory* (Cambridge University Press, Cambridge, 1994)
6. M. Brune et al., Phys. Rev. Lett. **77**, 4887 (1996); S. Haroche, Rev. Mod. Phys. **85**, 1083 (2013); S. Haroche, M. Brune, J.M. Raimond, Phys. Today **66**, 27 (2013); S. Haroche, J.M. Raimond, *Exploring the Quantum, Atoms, Cavities, Photons* (Oxford University Press, Oxford, 2006)
7. P.A.M. Dirac, *Principles of Quantum Mechanics* (Oxford University Press, Oxford, 1947)
8. R.P. Feynman, R.B. Leighton, M. Sands, *Feynman Lectures on Physics 3* (Addison-Wesley, Reading, 1971)
9. R.P. Feynman, R.B. Leighton, M. Sands, *Feynman Lectures on Physics 1, 2* (Addison-Wesley, Reading, 1977)
10. B.A. Fuchs, B.V. Shabat, *Functions of a Complex Variable and Some of Their Applications*, vol. 1 (Pergamon Press, Oxford, 1964)
11. H. Goldstein, C.P. Jr., Poole, J.L. Safko, *Classical Mechanics* (Addison-Wesley, Reading, 2001)
12. W. Greiner, J. Reinhardt, *Quantum Electrodynamics* (Springer, Berlin, 2008)
13. J.M. Jauch, F. Rohrlich. *The Theory of Photons and Electrons: The Relativistic Quantum Field Theory of Charged Particles With Spin One-Half* (Addisson Wesley, Reading, 1955)
14. L.D. Landau, E.M. Lifshitz, *Mechanics* (Elsevier/Butterworth-Heinemann, Amsterdam/London, 1976)
15. L.D. Landau, E.M. Lifshitz, *Quantum Mechanics, Non-relativistic Theory* (Elsevier/Butterworth-Heinemann, Amsterdam/London, 1976)
16. I. Lindgren, J. Morrison, *Atomic Many-Body Theory* (Springer, Berlin, 1982)
17. S.-K. Ma, *Modern Theory of Critical Phenomena* (Westview Press, Boulder, 2000)
18. F. Mandl, G. Shaw, *Quantum Field Theory* (Wiley, New York, 1993)
19. A. Messiah, *Quantum Mechanics* (North Holland, Amsterdam, 1961)
20. P.J. Mohr, B.N. Taylor, D.B. Newell, Rev. Mod. Phys. **84**, 1527 (2012)
21. M. Peskin, D. Schroeder, *An Introduction to Quantum Field Theory* (Westview Press, Boulder, 1995)
22. R. Pohl et al., Nature **466**, 213 (2010)
23. J.J. Sakurai, *Advanced Quantum Mechanics* (Addison Wesley, Reading, 1967)
24. L. Schiff, *Quantum Mechanics* (McGraw Hill College, New York, 1968)
25. J. Schwinger, *Quantum Mechanics: Symbolism of Atomic Measurement* (Springer, Berlin, 2003)
26. M. Veltman, *Diagrammatica: The Path to Feynman Diagrams* (Cambridge University Press, Cambridge, 1994)
27. A. Zee, *Quantum Field Theory in a Nutshell*, 2nd edn. (Princeton University Press, Princeton, 2010)

Chapter 1
Foundations of Quantum Mechanics

In this chapter, we introduce the fundamental principles of quantum mechanics. We commence by discussing the famous Stern-Gerlach experiments for a particle with the spin $1/2$ as several key quantum mechanical phenomena may be well understood thereof. Using these very experiments as an example, we then illustrate how the basic principles are incorporated within the mathematical formalism of quantum mechanics. Subsequently, we generalize this mathematical scheme for more complicated systems. Finally, focusing on the harmonic oscillator as an example, we show the relation between an abstract and a specific approach to the formalism.

1.1 Basic Principles

There is nothing mystical about quantum mechanics provided we are willing to accept the following two principles. They comprise all of the unusual aspects of the behavior of microscopic particles which, whether based on everyday experience or from the point of view of classical physics, one may consider rather odd. These principles cannot be logically derived; fathers of the quantum mechanics[1] arrived at them from experimental results. On the other hand, supposing we accept these principles, basically anything within quantum mechanics follows logically therefrom. Shall quantum mechanics be altered one day, however improbable that seems to the authors, alteration of the two following principles would be inevitable.

1. *We are capable of only predicting the probability of processes.*
   Physicists encountered this fact for the first time during analysis of radioactive processes and atomic radiation. As early as in 1905, Rutherford stated that the

[1]Those are W. Heisenberg, E. Schrödinger, P. Dirac, and M. Born, with key contributions from M. Planck, A. Einstein, N. Bohr, L. De Broglie, and W. Pauli.

J. Zamastil, J. Benda, *Quantum Mechanics and Electrodynamics*,
DOI 10.1007/978-3-319-65780-6_1

amount of nuclei $N(t)$ which undergo radioactive decay within time period t is given by the formula $N(t) = N(t = 0)e^{-\gamma t}$, where γ represents a constant for the particular nucleus (i.e., whether we are considering uranium, radium, or any other nucleus). Only as the years passed, physicists came to the conclusion that nothing better than a statistical law valid for great N can be derived. The reason is, we do not know when a particular nucleus undergoes the decay. We can merely predict the probability of decay of a typical nucleus at any moment, hence only the mean number of particles decaying during a given time period.

In 1917, Einstein analyzed black-body radiation and reached the conclusion that one may describe spontaneous emission in the same manner as the above-discussed radioactive decay. An atom in an excited state emits a photon, after a lapse of time, and thus relaxes into the ground state. However, for each individual atom, we are able to determine the time lapse merely by statistical means. The irony is that Einstein, who was most likely the very first person to realize the necessity of probabilistic approach, in fact never accepted it.

The strict determinism is thus disrupted when it comes to the microscopic world. That is, the same cause (such as an excited atom) leads to various results (the atom emits a photon and relaxes to the ground state at different times, e.g., sometimes after 2×10^{-9} s, other time after 5×10^{-9} s, etc.).

What makes the behavior of microscopic particles so unusual is the way we calculate the probabilities.

2. *The probability of an event P is given by the square of the probability amplitude A, A being generally a complex number,* $P = |A|^2 = AA^*$. *Amplitudes of independent processes multiply. In case a system can evolve from the initial to the final state via two in principle indistinguishable ways, the respective amplitudes add to each other.*

This second principle is generally known as the *principle of superposition*. The well-known double-slit experiment (Fig. 1.1) serves well for the illustration of these principles. The probability that a particle, such as an electron, emitted from the source Z is detected by the detector D is given—in accordance with the above-stated principle—as

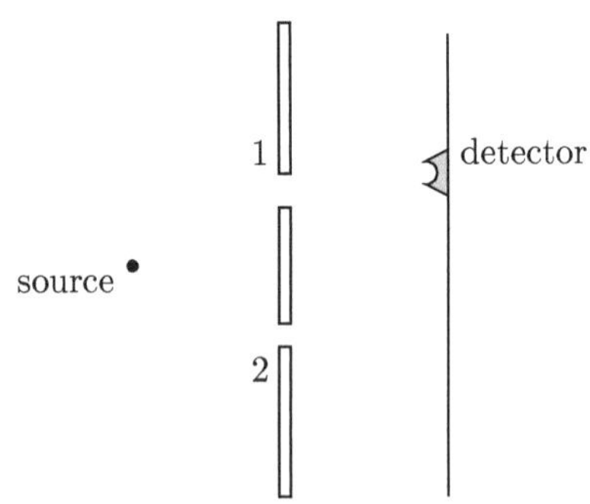

Fig. 1.1 Double-slit experiment

$$\begin{aligned}P(Z \to D) &= |A(Z \to 1)A(1 \to D) + A(Z \to 2)A(2 \to D)|^2 \\ &= |A(Z \to 1)A(1 \to D)|^2 + |A(Z \to 2)A(2 \to D)|^2 \\ &\quad + 2\Re\left(A(Z \to 1)A(1 \to D)A^*(Z \to 2)A^*(2 \to D)\right).\end{aligned}$$

The first term on the right-hand side (rhs) of the equation gives the probability of a particle passing through the first slit, while the second term represents the probability of a particle passing through the other slit. These probabilities are given as products of the probabilities that a particle emitted from the source Z reaches one of the slits and that the particle arrives at the detector D from the slit. Since these two processes are independent of each other, the respective probabilities multiply, which follows from the principle that amplitudes of independent processes multiply. Such results answer to the common sense. However, what turns out differently from what one could expect is the total probability of a particle starting at the source and reaching the detector, as it does *not* equal the sum of a particle passing through each of the slits. This is due to a third, so-called interference, term. Considering a situation where particles leave the source one at a time,[2] pass through the apparatus and are detected, and inspecting the final distribution of the particles at the plane of the detector (such as by using a photographic plate), we obtain a so-called interference pattern. It consists of alternating bright and dark stripes as there is a large number of impacting particles in case of the former and very few in case of the latter.

It is much easier to grasp the interference pattern if we perform the entire experiment with waves, instead of particles. Waves exit the source and reach the two slits which we can then consider as new sources, and waves from them subsequently reach the plane of the detector. However, the individual waves differ in phase which in turn depends on the place of impact. Some waves arrive at the detector with the same phase, i.e., their "peaks" and "valleys" are identical, while—at other locations—the waves meet with opposite phases, i.e., "peaks" of one wave match "valleys" of the other and vice versa. The former then corresponds to the bright stripes with many impacting particles, while the latter results in dark stripes with none.

Note though that we always detect electrons, protons, neutrons, etc. as particles. That is, the process of detection is precisely located in space and time (a particle hits the detector "here and now").

Wave-particle duality is thus incorporated into quantum mechanics via the two above-stated principles. We always determine the probability amplitudes according to the second principle in order to accurately describe the behavior of electrons in terms of waves, such as observed in the case of the double-slit experiment. However, the fact that electron is detected as a particle then leads to probabilistic interpretation of the square of the amplitudes. It shall be noted that the double-slit experiment was relatively recently accomplished [14].

[2]It certainly is experimentally possible to arrange that only one particle passes through the slit.

Another example of quantum mechanical interference is scattering of two indistinguishable particles. In his famous experiment, Ernst Rutherford bombarded heavy atoms with α-particles. This experiment lead Rutherford to an atomic model which still holds—at least roughly—to present days. Due to significantly greater mass of α-particles in comparison to that of electrons, it is possible to neglect the influence of electrons on the motion of α-particles. The scattering probability of an α-particle into solid angle $\mathrm{d}\Omega = \sin\vartheta \mathrm{d}\vartheta \mathrm{d}\varphi$ per unit time divided by area density of the incident α-particles is given according to the first principle as

$$\frac{\mathrm{d}\sigma}{\mathrm{d}\Omega} = |f(\vartheta)|^2 , \tag{1.1}$$

where σ is called the cross-section and $f(\vartheta)$ the scattering amplitude. The axis from which the angle ϑ is determined matches the initial direction of the impacting α-particles, as depicted in Fig. 1.2. Assuming the mutual interaction of the α-particles and the nucleus to be purely electrostatic, Rutherford succeeded in deriving a formula for the amplitude $f(\vartheta)$ (e.g., [7]) which agreed with the experimental results. From a historical point of view, Rutherford was extremely lucky to have bombarded heavy nuclei, since the electrostatic effect does not allow the α-particles to come to close proximity of the nuclei. In case of lighter nuclei, the effects of nuclear forces come into play. Rutherford was even more lucky that he did not study scattering of α-particles by α-particles. Had he done so, he would have obtained results incomprehensible at that time (1912). The problem is, it is impossible to tell apart the process of an α-particle being scattered by the angle ϑ and the process with the deflection angle equal to $\pi - \vartheta$ (Fig. 1.3) since the particles are *in principle indistinguishable*. In this case, the angular distribution of deflected α-particles equals

$$\frac{\mathrm{d}\sigma}{\mathrm{d}\Omega} = |f(\vartheta) + f(\pi - \vartheta)|^2 . \tag{1.2}$$

Fig. 1.2 Scattering of an α-particle on a nucleus

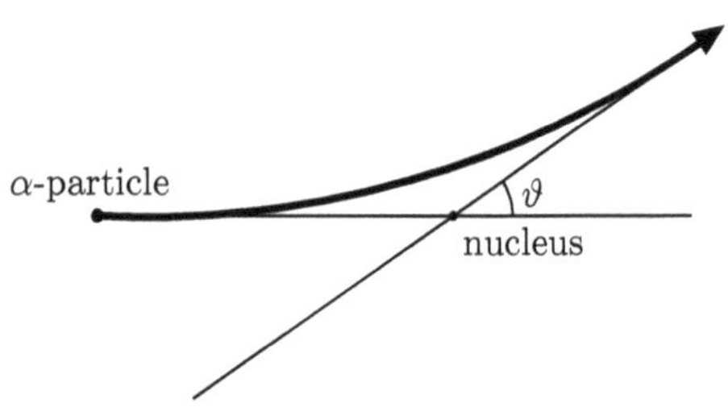

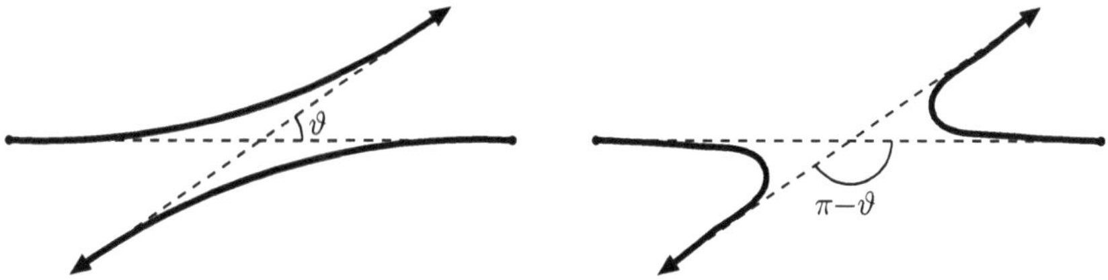

Fig. 1.3 Scattering of two indistinguishable particles

Fig. 1.4 Stern-Gerlach apparatus

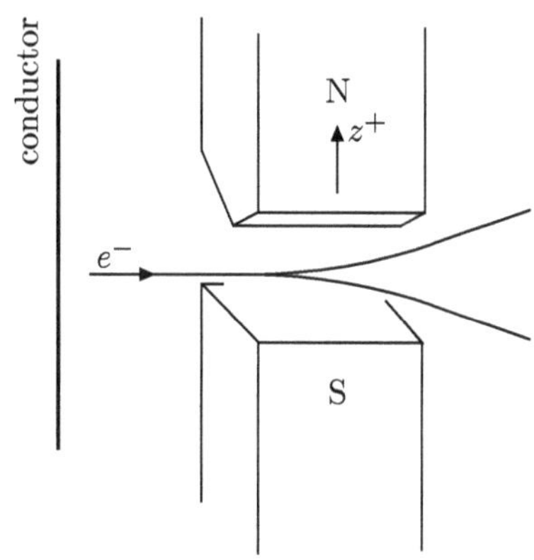

1.2 Mathematical Scheme of the Quantum Theory

1.2.1 Stern-Gerlach Experiments

In 1921, Otto Stern designed the first of the series of Stern-Gerlach experiments and successfully realized it in collaboration with Walter Gerlach 1 year later. This series of experiments focuses on measurements of the inner degree of freedom of an electron—the spin, projection of which into any direction, i.e., the "state of the system," may acquire only two values. Mathematical apparatus of quantum mechanics can be easily explained using this simple example, and subsequent generalization for more complicated systems is then quite straightforward.

Electrons (in the original experiment silver atoms) are heated up in a cavity with a tiny exit hole. A collimated beam constituted by the emitted electrons enters inhomogeneous magnetic field, which is maintained by two opposite magnetic poles, one of them having a sharp tip-shape, as depicted in Fig. 1.4. In an external electromagnetic field, an electron behaves like an electric monopole and a magnetic dipole. The magnitude of the *magnetic dipole moment* $\boldsymbol{\mu}$ is directly proportional to the intrinsic mechanical angular momentum, the *spin* $\mathbf{s}$,

$$\boldsymbol{\mu} = K\mathbf{s}\,.$$

The coefficient of proportionality is given as

$$K = g_e e/m_e\,,$$

where g_e is called the *gyromagnetic ratio* and equals approximately 1 for an electron, see Eq. (3.51).[3] The classical energy of a magnetic dipole of the magnitude $\boldsymbol{\mu}$ in a field with magnetic induction $\mathbf{B} = (0, 0, B)$ reads

$$E = -\boldsymbol{\mu} \cdot \mathbf{B} = -K\mathbf{s} \cdot \mathbf{B} = -Ks_z B\,. \tag{1.3}$$

The force the magnetic field exerts on electrons is given as

$$F_z = -\frac{\partial E}{\partial z} = Ks_z \frac{\partial B}{\partial z}\,.$$

We thus see that it is the field inhomogenity that plays the crucial role in splitting of the beam on the basis of angular momentum, and not the magnitude of the magnetic field.

It follows thereof that the force has a "downward" direction when acting on electrons with $s_z < 0$ and "upward" in case of electrons with $s_z > 0$, and therefore splitting of the electron beam occurs according to the values of s_z. In the framework of classical physics, there should be no prevailing value in the distribution of s_z of the detected particles and the distribution shall be continuous since the orientation of the electrons leaving the heated cavity is utterly random. However, the reality differs. There are merely two distinguishable electron beams coming from the apparatus—that is, only two values $s_z = \pm 1/2$ occur. It seems natural to ask: is there something that makes the z axis special? The answer is no, of course. Projection of the electron spin along any direction acquires only two values.

We now consider a sequence of three pairs of magnets (a "modified" SG experiment) which are placed as shown in Fig. 1.5. The electron beam enters the first magnetic field and splits into two according to the spin orientation of the individual electrons, where the probabilities of the spin being $1/2$ or $-1/2$ are equal (we restrict ourselves to the z-component of the spin). Both beams subsequently pass to the second magnetic field. However, there is a screen blocking one of the beams with a particular spin projection (along the z-axis), while the other beam proceeds to the third magnetic field, which bends it into its original direction.

Figures 1.6 and 1.7 depict schematics for the modified Stern-Gerlach apparatus that will be employed throughout the reading.

We will now describe several typical arrangements one may encounter in such experiments.

[3] One may also encounter a different definition of gyromagnetic ratio $K = (g_e/2)(e/m_e)$.

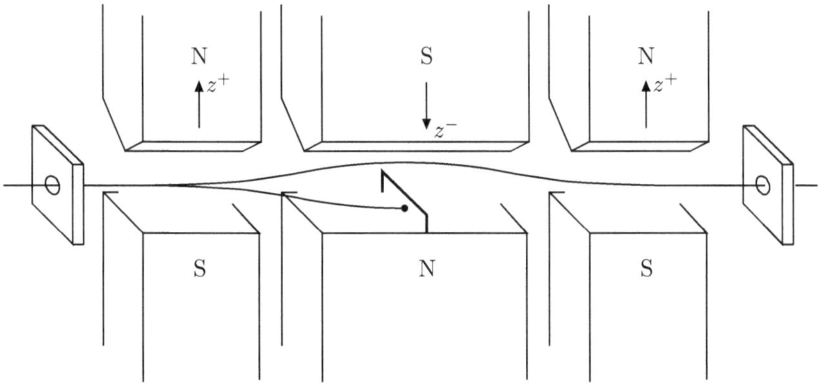

Fig. 1.5 A "modified" Stern-Gerlach apparatus with a shield allowing only one of the beams to pass

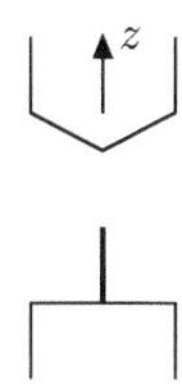

Fig. 1.6 Schematics for modified Stern-Gerlach apparatus selecting the electrons with the spin pointing "upward," $s_z = +\frac{1}{2}$

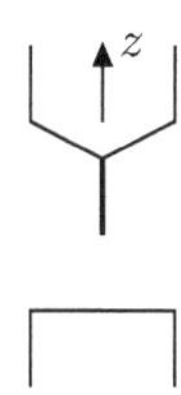

Fig. 1.7 The same as in Fig. 1.6, but now the electrons leaving the apparatus have spin pointing "downward," $s_z = -\frac{1}{2}$

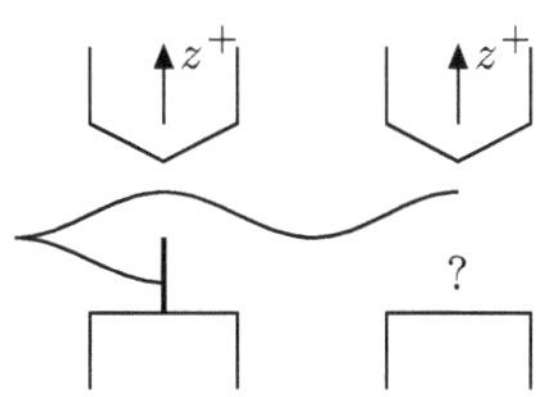

Fig. 1.8 Two successive modified Stern-Gerlach apparatuses selecting the electrons with the spin projections onto the same axis

1. (Fig. 1.8) Spin of all electrons exiting the first apparatus points "upwards." What is the probability of detecting an electron with its spins oriented "downwards"? The answer is zero. This surely makes sense as we allowed only electrons with the "upward" spin to leave the first apparatus.
2. (Fig. 1.9) The spin of all electrons exiting the first apparatus has an "upward" orientation. What is the probability of detecting an electron with a spin oriented to the "right" at the second apparatus, which was rotated by 90° with respect to the first one? One half of the electrons leaving the first apparatus. However, for an

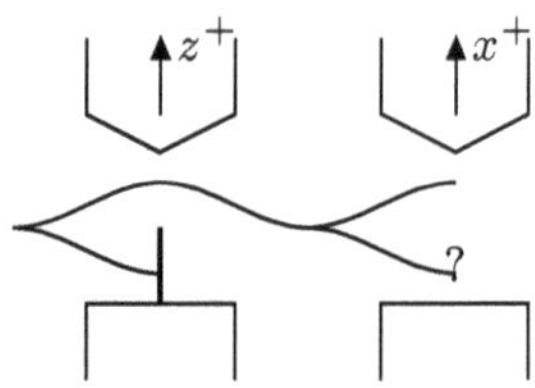

Fig. 1.9 The same as in Fig. 1.8, but now the second apparatus is rotated about the angle $\pi/2$ with respect to the first one

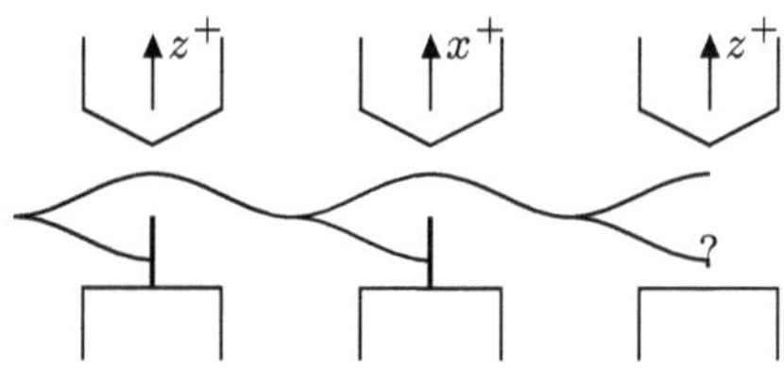

Fig. 1.10 Three successive modified Stern-Gerlach apparatuses. The second apparatus is rotated about the angle $\pi/2$ with respect to the first and the third

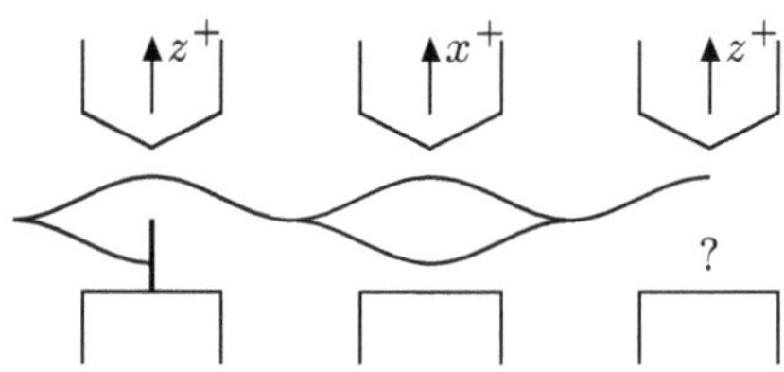

Fig. 1.11 The same as in Fig. 1.10, but now with no selection made in the second apparatus

individual electron, we cannot decide where it is deflected "left" or "right." This experiment thus illustrates the first principle, that is, we are capable of predicting only the probability of individual processes.

3. (Fig. 1.10) The spin of all electrons leaving the first apparatus is again oriented "upwards." After passing the second apparatus, the spin points "left." What is the probability of detecting an electron with a "downward" spin at the third apparatus? One fourth of the electrons exiting the first apparatus. At first sight, this may seem rather incomprehensible, as we allowed only electrons with an "upward" spin to exit the first apparatus! Apparently, the information whether a spin points "left" or "right" destroys the information of the spin pointing "upwards" or "downwards." This experiment well illustrates the multiplication principle of independent processes. That is, the probability of an "upward" and subsequently "left" spin equals one half, and likewise the probability of a "left" spin then pointing "downwards" equals one half. It follows thereof, the probability of a spin pointing "upwards, then left, and finally downwards" is one fourth.
4. (Fig. 1.11) The spin of all electrons leaving the first apparatus points "upwards." The second apparatus now allows also all electrons with a "right" spin to pass (unlike the previous case where these electrons were eliminated). What is the probability of detecting an electron with a "downward" spin at the third apparatus? Zero. Once again, this may appear rather odd. When we block one of the two possible ways, one fourth of the electrons reaches the detector in the third apparatus. Yet obstructing neither of the two beams results in no particles impacting on the detector. This experiment likewise demonstrates the quantum

mechanical interference, i.e., the second principle: we add the probability amplitudes of indistinguishable process, not the probabilities themselves. This fourth case apparently matches the first one since the second apparatus does not change the electron beam at all. That is, we do not conduct any measurements, hence do not affect the state of the system.

We make use of these four experimental setups to introduce mathematical methods of describing physical phenomena. *We describe the state of a system by a vector from an abstract space of states.*[4] We use a *ket*-vector to describe a system entering the measuring apparatus

$$|\psi\rangle = \begin{pmatrix} a \\ b \end{pmatrix} \tag{1.4}$$

and a *bra*-vector for a system exiting the apparatus[5]

$$\langle\psi| = \begin{pmatrix} a^* & b^* \end{pmatrix}, \tag{1.5}$$

where a and b are generally complex numbers and their meaning will become clear from the further text. The transition from a ket-vector to a bra-vector $\langle\psi| = (|\psi\rangle)^+$ is called *Hermitian conjugation* and corresponds, as we can see, to transposition and complex conjugation. Scalar product of a vector $\langle\phi|$,

$$\langle\phi| = \begin{pmatrix} c^* & d^* \end{pmatrix}, \tag{1.6}$$

and a vector $|\psi\rangle$ equals

$$\langle\phi|\psi\rangle = c^*a + d^*b, \tag{1.7}$$

and one can easily prove that

$$\langle\phi|\psi\rangle = \langle\psi|\phi\rangle^*. \tag{1.8}$$

The probability amplitude of a transition from one state into another is then given by the scalar product of the corresponding state vectors. The following text will clarify the necessity to identify scalar products with probability amplitudes and not the probabilities themselves.

We now illustrate these general expressions on the SG experiments. The probability of a particle passing through a specific SG setup equals the square of the amplitude magnitude. For the first case (Fig. 1.8), the experiments dictate the relations

[4]Mathematicians call it the Hilbert space.

[5]These symbols, as well as the entire notation, comes from English physicist Paul Dirac. Scalar product $\langle\phi|\psi\rangle$ is a “bra-c-ket” in English, meaning $\langle\phi|$ is a “bra” and $|\psi\rangle$ is a “ket.”

$$P(\pm z, \pm z) = |\langle \pm z | \pm z \rangle|^2 = 1\,, \qquad P(\pm z, \mp z) = |\langle \pm z | \mp z \rangle|^2 = 0\,. \tag{1.9}$$

The simplest plausible choice of vectors $|+z\rangle$, $|-z\rangle$, $\langle +z|$, and $\langle -z|$ meeting these requirements reads

$$|+z\rangle = \begin{pmatrix} 1 \\ 0 \end{pmatrix}, \qquad |-z\rangle = \begin{pmatrix} 0 \\ 1 \end{pmatrix}, \qquad \langle +z| = (1\ 0)\,, \qquad \langle -z| = (0\ 1)\,. \tag{1.10}$$

With this choice, the components a and b of a general vector $|\psi\rangle$ given by Eq. (1.4) satisfy

$$a = \langle +z | \psi \rangle\,, \qquad b = \langle -z | \psi \rangle\,,$$

which means a and b are the probability amplitudes of a general state $|\psi\rangle$ transforming into the states $|+z\rangle$ and $|-z\rangle$, respectively.

For the second case (Fig. 1.9), we have

$$P(+z, -x) = |\langle +z | -x \rangle|^2\,, \qquad P(\pm z, \pm x) = P(\pm z, \mp x) = 1/2\,.$$

Most certainly the following must also hold (since nothing makes the z-axis special!)

$$P(\pm x, \pm x) = |\langle \pm x | \pm x \rangle|^2 = 1\,, \qquad P(\pm x, \mp x) = |\langle \pm x | \mp x \rangle|^2 = 0\,. \tag{1.11}$$

These conditions are satisfied for example by the following choice of vectors

$$|+x\rangle = \frac{1}{\sqrt{2}} \begin{pmatrix} 1 \\ 1 \end{pmatrix}, \qquad |-x\rangle = \frac{1}{\sqrt{2}} \begin{pmatrix} 1 \\ -1 \end{pmatrix}, \tag{1.12a}$$

$$\langle +x| = \frac{1}{\sqrt{2}} (1\ 1)\,, \qquad \langle -x| = \frac{1}{\sqrt{2}} (1\ -1)\,. \tag{1.12b}$$

Having found possible forms of the vectors $|\pm x\rangle$, we can readily calculate the probabilities for the third case (Fig. 1.10),[6]

$$P(+z, +x, +x, -z) = |\langle +z | +x \rangle \langle +x | -z \rangle|^2 = \left| (1\ 0) \frac{1}{\sqrt{2}} \begin{pmatrix} 1 \\ 1 \end{pmatrix} \frac{1}{\sqrt{2}} (1\ 1) \begin{pmatrix} 0 \\ 1 \end{pmatrix} \right|^2 = \frac{1}{4}\,,$$

and also for the fourth experimental setup. For this instance, however, we have to pay attention to add the probability amplitudes of the individual processes, not the probabilities themselves, such as for the double-slit experiment:

[6] Taking into account $\langle \phi | \psi \rangle = \langle \psi | \phi \rangle^*$, see (1.8), hence $|\langle \phi | \psi \rangle|^2 = |\langle \psi | \phi \rangle|^2$, it obviously makes no difference whether we write the order of events from left to write, $P = |\langle +z | +x \rangle \langle +x | -z \rangle|^2$, or from right to left, $P = |\langle -z | +x \rangle \langle +x | +z \rangle|^2$.

$$
\begin{aligned}
P(+z, \pm x, \pm x, -z) &= |\langle +z|+x\rangle\langle +x|-z\rangle + \langle +z|-x\rangle\langle -x|-z\rangle|^2 \\
&= \left|\frac{1}{2}(1\ 0)\begin{pmatrix}1\\1\end{pmatrix}(1\ 1)\begin{pmatrix}0\\1\end{pmatrix} + \frac{1}{2}(1\ 0)\begin{pmatrix}1\\-1\end{pmatrix}(1\ -1)\begin{pmatrix}0\\1\end{pmatrix}\right|^2 \\
&= 0\,.
\end{aligned}
$$

The hereby built-up formalism allows us to elucidate the meaning of the statement "measuring affects the state of the system." To do so, we now introduce a *tensor product*

$$
|\psi\rangle\langle\phi| = \begin{pmatrix}a\\b\end{pmatrix}\begin{pmatrix}c^* & d^*\end{pmatrix} = \begin{pmatrix}ac^* & ad^*\\ bc^* & bd^*\end{pmatrix}.
$$

We can thus regard the effect of the measuring device on the state of the system as a projection operator (a matrix) acting on a vector. As an example, we take the third SG experiment (Fig. 1.10). The state after the first filtration reads $\langle +z|$. After passing the second SG filter, all particles are in the state $\langle +z|+x\rangle\langle +x|$, which we can understand as the original state $\langle +z|$ modified by the operator

$$
\hat{1}_{+x} = |+x\rangle\langle +x| = \frac{1}{2}\begin{pmatrix}1 & 1\\ 1 & 1\end{pmatrix}.
$$

The third experimental setup shows that the spin projection of the electrons that passed through the second apparatus along the x-axis equals one half. This means we have conducted a measurement, and thus altered the state of the system as the electrons find themselves in different states prior to and after traveling through the apparatus,

$$
\langle +z|\hat{1}_{+x} = \langle +z|+x\rangle\langle +x| = (1\ 0)\,\frac{1}{2}\begin{pmatrix}1 & 1\\ 1 & 1\end{pmatrix} = \frac{1}{2}(1\ 1) = \frac{1}{\sqrt{2}}\langle +x|\,.
$$

On the other hand, the state of the system does not change for the fourth arrangement (neither of the ways is blocked at the second apparatus). This agrees with no measuring being performed:

$$
\begin{aligned}
\langle +z|(|+x\rangle\langle +x| + |-x\rangle\langle -x|) &= (1\ 0)\left[\frac{1}{2}\begin{pmatrix}1 & 1\\ 1 & 1\end{pmatrix} + \frac{1}{2}\begin{pmatrix}1 & -1\\ -1 & 1\end{pmatrix}\right] \\
&= (1\ 0)\,\mathbf{1} = (1\ 0) = \langle +z|\,.
\end{aligned}
$$

It follows that a crucial operator equality

$$
|+x\rangle\langle +x| + |-x\rangle\langle -x| = \mathbf{1}\,,
$$

that mathematicians call the *completeness relation* for the basis $\{|+x\rangle, |-x\rangle\}$ bears a clear physical meaning: placing a device into the path of the studied system but making no measurements leads to the same final state as placing there no measuring device at all. In mathematical description, this corresponds to multiplication by a unit matrix.

The physical interpretation of the other important relation for bases, the *orthonormality relation*, see for example (1.9) or (1.11), should be clear too. Placing the very same type of device into the path of the studied system does not produce any new information. The reason is, the system maintains the selfsame state, as demonstrated by the first experiment, Fig. 1.8.

From a practical point of view, the relations of completeness and orthonormality show to be extremely useful, as they allow for calculations of transition amplitudes. A scalar product of the completeness relation

$$|+z\rangle\langle+z| + |-z\rangle\langle-z| = \mathbf{1}$$

with the vector $|\psi\rangle$ leads to decomposition of the vector $|\psi\rangle$ into the basis of eigenstates of the spin projection on the z-axis,

$$|\psi\rangle = \mathbf{1}|\psi\rangle = |+z\rangle\langle+z|\,\psi\rangle + |-z\rangle\langle-z|\,\psi\rangle = \begin{pmatrix}\langle+z|\,\psi\rangle\\ \langle-z|\,\psi\rangle\end{pmatrix},$$

where the last equation follows from Eq. (1.10). We can thus rewrite the scalar product of two vectors as

$$\langle\phi|\,\psi\rangle = \langle\phi|+z\rangle\langle+z|\,\psi\rangle + \langle\phi|-z\rangle\langle-z|\,\psi\rangle = \left(\langle\phi|+z\rangle\ \langle\phi|-z\rangle\right)\begin{pmatrix}\langle+z|\,\psi\rangle\\ \langle-z|\,\psi\rangle\end{pmatrix}.$$

The left-hand side (lhs) of the equation comprises the vectors $\langle\phi|$ and $|\psi\rangle$ in an abstract form, while on the rhs they acquire a specific form in the basis of eigenstates of the spin projection along the z-axis.

If you deem such decompositions slightly odd (you should not), consider the following analogy of the last three equations in an ordinary two-dimensional Euclidean space:

$$\begin{aligned}\mathbf{1} &= \mathbf{e}_x\mathbf{e}_x + \mathbf{e}_y\mathbf{e}_y\,,\\ \mathbf{F} &= \mathbf{F}\cdot\mathbf{e}_x\mathbf{e}_x + \mathbf{F}\cdot\mathbf{e}_y\mathbf{e}_y = F_x\mathbf{e}_x + F_y\mathbf{e}_y\,,\\ \mathbf{F}\cdot\mathbf{v} &= \mathbf{F}\cdot\mathbf{e}_x\mathbf{e}_x\cdot\mathbf{v} + \mathbf{F}\cdot\mathbf{e}_y\mathbf{e}_y\cdot\mathbf{v} = F_xv_x + F_yv_y\,.\end{aligned}$$

The only difference is that there is no need to distinguish between a ket- and a bra-vector in the Euclidean space with a real and orthogonal basis.

Note, though, that while vectors commonly used in classical mechanics (such as a position vector, etc.) are vectors of the three-dimensional Euclidean space, which is a direct abstraction of the space of our everyday experience, vectors employed in quantum mechanics belong to the abstract space which has no natural connection to our three-dimensional space.

The projection of the spin on the y-axis naturally complies with the same rules as the projection on the z and x axes (no coordinate may be preferred over the others). This leads to the requirements:

$$|\langle\pm z|\pm y\rangle|^2 = |\langle\pm z|\mp y\rangle|^2 = |\langle\pm x|\pm y\rangle|^2 = |\langle\pm x|\mp y\rangle|^2 = 1/2\,,$$
$$|\langle\pm y|\pm y\rangle|^2 = 1\,, \qquad |\langle\pm y|\mp y\rangle|^2 = 0\,.$$

For the given choice of vectors $|\pm z\rangle$ and $|\pm x\rangle$, Eqs. (1.10) and (1.12), these conditions cannot be fulfilled for purely real components. However, considering also complex numbers, we can easily prove that the following choice of vectors meets the previously given requirements (mind the complex conjugation upon transition from bra- to ket-vectors, see (1.5))

$$|+y\rangle = \frac{1}{\sqrt{2}}\begin{pmatrix}1\\ \mathrm{i}\end{pmatrix}, \qquad |-y\rangle = \frac{1}{\sqrt{2}}\begin{pmatrix}1\\ -\mathrm{i}\end{pmatrix}, \tag{1.13a}$$

$$\langle +y| = \frac{1}{\sqrt{2}}\left(1\ -\mathrm{i}\right), \qquad \langle -y| = \frac{1}{\sqrt{2}}\left(1\ \mathrm{i}\right). \tag{1.13b}$$

1.2.2 Operators

For the purposes of the following chapters, we now introduce the notion of operators. We have already encountered them in connection with the tensor product of two vectors. A general operator $\hat{\mathsf{A}}$ projects a vector $|\varphi\rangle$ onto a vector $|\psi\rangle$,

$$\hat{\mathsf{A}}|\varphi\rangle = |\psi\rangle\,.$$

By Hermitian conjugation, we obtain

$$\langle\psi| = \langle\varphi|\hat{\mathsf{A}}^+\,,$$

where $\hat{\mathsf{A}}^+$ is the Hermitian conjugate to $\hat{\mathsf{A}}$. For a two-dimensional space (a physical system with two possible states), the particular forms for $|\psi\rangle = \begin{pmatrix}a_1\\ a_2\end{pmatrix}$, $|\varphi\rangle = \begin{pmatrix}b_1\\ b_2\end{pmatrix}$ and $\hat{\mathsf{A}} = \begin{pmatrix}A_{11} & A_{12}\\ A_{21} & A_{22}\end{pmatrix}$ read

$$\begin{pmatrix}A_{11} & A_{12}\\ A_{21} & A_{22}\end{pmatrix}\begin{pmatrix}b_1\\ b_2\end{pmatrix} = \begin{pmatrix}a_1\\ a_2\end{pmatrix}, \qquad \left(a_1^*\ a_2^*\right) = \left(b_1^*\ b_2^*\right)\begin{pmatrix}A_{11}^+ & A_{12}^+\\ A_{21}^+ & A_{22}^+\end{pmatrix}.$$

By writing the complex conjugate of the first equation and comparing it with the other, we obtain

$$A^+_{ij} = A^*_{ji} \,,$$

that is, we apply transposition and complex conjugation to find the Hermitian conjugated matrix $\hat{\mathsf{A}}^+$ to the original matrix $\hat{\mathsf{A}}$.

1.2.3 Time Evolution in Quantum Theory

So far, we only dealt with a description of a state when a system either enters or exits the apparatus. However, time elapses between these two events, which thus begs the question of the time evolution of the system.

We denote two physical (experimentally feasible) states of a system at time t_0 by the state vectors $\langle\varphi(t_0)|$ and $|\psi(t_0)\rangle$. At any moment later in time t, these systems are described by the state vectors

$$|\psi(t)\rangle = \hat{\mathsf{U}}(t,t_0)|\psi(t_0)\rangle \qquad \text{and} \qquad \langle\varphi(t)| = \left(\hat{\mathsf{U}}(t,t_0)|\varphi(t_0)\rangle\right)^+ = \langle\varphi(t_0)|\hat{\mathsf{U}}^+(t,t_0)\,, \tag{1.14}$$

where $\hat{\mathsf{U}}(t,t_0)$ is called the *evolution operator* or the *propagator*. Provided the dynamics of the system is invariant with respect to time displacement, the probability of a system changing from the state $\langle\varphi(t_0)|$ into the state $|\psi(t_0)\rangle$ at the time t_0 must necessarily equal the probability of the system changing from the state $\langle\varphi(t)|$ into the state $|\psi(t)\rangle$ at the time t,

$$|\langle\varphi(t_0)|\,\psi(t_0)\rangle|^2 = |\langle\varphi(t)|\,\psi(t)\rangle|^2 = |\langle\varphi(t_0)|\hat{\mathsf{U}}^+(t,t_0)\hat{\mathsf{U}}(t,t_0)|\psi(t_0)\rangle|^2 \,.$$

This implies the $\hat{\mathsf{U}}$ be *unitary*

$$\hat{\mathsf{U}}^+(t,t_0)\hat{\mathsf{U}}(t,t_0) = \mathbf{1}\,,$$

or equivalently

$$\hat{\mathsf{U}}^+(t,t_0) = \hat{\mathsf{U}}^{-1}(t,t_0) = \hat{\mathsf{U}}(t_0,t)\,. \tag{1.15}$$

Furthermore, the evolution operator must meet the decomposition requirement

$$\hat{\mathsf{U}}(t,t_0) = \hat{\mathsf{U}}(t,t')\hat{\mathsf{U}}(t',t_0)\,.$$

That is, evolution from the initial time t_0 to t must yield the same final state as starting at t_0 and reaching t', and subsequently moving from t' to t. Finally, we demand

$$\hat{\mathsf{U}}(t,t) = \mathbf{1}\,.$$

These requirements are met by the choice

$$\hat{\mathsf{U}}(t, t_0) = \exp(-\mathrm{i}\hat{\mathsf{H}}(t - t_0)), \tag{1.16}$$

where after substitution from Eq. (1.16) into Eq. (1.15) we obtain

$$\hat{\mathsf{H}} = \hat{\mathsf{H}}^+ ;$$

it follows thereof that $\hat{\mathsf{H}}$ is a *Hermitian* (also self-adjoint) operator. Matrix elements of a Hermitian operator then evidently comply with

$$H_{ij} = H_{ji}^* .$$

Mathematical properties and physical meaning of this Hermitian operator will be discussed in more detail later in this section. Its significance follows from the so-called *time-dependent Schrödinger equation*, named after Austrian physicist Erwin Schrödinger. It describes the time evolution of a system

$$\mathrm{i}\frac{\mathrm{d}|\psi(t)\rangle}{\mathrm{d}t} = \mathrm{i}\frac{\mathrm{d}}{\mathrm{d}t}\hat{\mathsf{U}}(t, t_0)|\psi(t_0)\rangle = \hat{\mathsf{H}}\hat{\mathsf{U}}(t, t_0)|\psi_0\rangle = \hat{\mathsf{H}}|\psi(t)\rangle , \tag{1.17}$$

which we obtain by differentiation of Eq. (1.14) with respect to time.

1.2.4 Stationary States

To find the physical meaning of the operator $\hat{\mathsf{H}}$, we focus on cases when its action on a vector reduces to a mere multiplication by a number, i.e.,

$$|\psi_n(t)\rangle = \hat{\mathsf{U}}(t, t_0)|\psi_n(t_0)\rangle = \mathrm{e}^{-\mathrm{i}\hat{\mathsf{H}}(t-t_0)}|\psi_n(t_0)\rangle = \mathrm{e}^{-\mathrm{i}E_n(t-t_0)}|\psi_n(t_0)\rangle . \tag{1.18}$$

The reason is that these states $|\psi_n\rangle$ differ from the others in one crucial property: if a system is in such a state at time t_0, it stays so and we surely find it in this state at any later time t:

$$|\langle\psi_n(t_0)|\,\psi_n(t)\rangle|^2 = 1 . \tag{1.19}$$

Such states are called *stationary*. As there are many of such states $|\psi_n\rangle$ and corresponding numbers E_n for the operator $\hat{\mathsf{H}}$, we label them with a subscript n.

It follows from Eqs. (1.18) and (1.19) that the numbers E_n describe *a property of a system which does not change with time*, i.e., it is conserved throughout time. From classical mechanics we know that such a quantity that is conserved due to the system dynamics being invariant with respect to time displacement is *energy*. For example, we have for a one-dimensional motion in a potential field

$$m\frac{\mathrm{d}^2x}{\mathrm{d}t^2} = F = -\frac{\mathrm{d}V}{\mathrm{d}x}\,.$$

Provided the dynamics is invariant with respect to time displacement, i.e., the potential V is independent of time, $V = V(x)$, we may multiply the equation by $\frac{\mathrm{d}x}{\mathrm{d}t}$, rearrange it and finally obtain the well-known law of conservation of energy

$$\frac{\mathrm{d}}{\mathrm{d}t}\left(\frac{m}{2}\left(\frac{\mathrm{d}x}{\mathrm{d}t}\right)^2 + V\right) = 0\,.$$

Therefore, we interpret the numbers E_n as possible energy values of the system. Inserting Eq. (1.18) into the Schrödinger equation (1.17) leads to

$$\hat{\mathsf{H}}|\psi_n\rangle = E_n|\psi_n\rangle\,, \tag{1.20}$$

we thus see that E_n are the *eigenvalues* and $|\psi_n\rangle$ the *eigenstates* or *eigenvectors* of the operator $\hat{\mathsf{H}}$. We call this operator the *Hamilton operator* or simply the "Hamiltonian," and Eq. (1.20) is known as the *time-independent Schrödinger equation*, or as an *eigenproblem* for the Hamilton operator.

We now return to the SG experiments to elucidate the time evolution and stationary states in quantum theory. It follows from the discussion of the outcomes of the experiment depicted in Fig. 1.8 that *for a magnetic field oriented along the z axis, the states $|\pm z\rangle$ are stationary, hence eigenstates of the Hamiltonian.* Equation (1.3), $E = -KBs_z$, shows that measuring energy corresponds to measuring the spin. Accordingly, the Hamiltonian operator matches—but for a constant—the projection operator of spin along the z axis. Thus the conserved property of the states $|\pm z\rangle$ is the value of the spin projection along the z axis, $s_z = \pm 1/2$. *Therefore, the eigenvalues of the Hamiltonian operator corresponding to the eigenvectors $|\pm z\rangle$ equal, again but for the constant* $(-KB)$, $\pm 1/2$. Multiplying then the completeness relation

$$|+z\rangle\langle +z| + |-z\rangle\langle -z| = \mathbf{1}$$

by the operator $\hat{\mathsf{H}}/(-KB)$ yields, see Eq. (1.10),

$$\begin{aligned}\hat{\mathsf{H}}/(-KB) &= \hat{\mathsf{S}}_z = \hat{\mathsf{S}}_z|+z\rangle\langle +z| + \hat{\mathsf{S}}_z|-z\rangle\langle -z| \\ &= \frac{1}{2}|+z\rangle\langle +z| - \frac{1}{2}|-z\rangle\langle -z| = \frac{1}{2}\begin{pmatrix}1 & 0\\ 0 & -1\end{pmatrix}.\end{aligned} \tag{1.21}$$

Equation $\hat{\mathsf{S}}_z = \frac{1}{2}|+z\rangle\langle +z| - \frac{1}{2}|-z\rangle\langle -z|$ is called the *spectral decomposition of the operator $\hat{\mathsf{S}}_z$.*

Analogously, we find that the states $|\pm x\rangle$ ($|\pm y\rangle$) are stationary for a magnetic field oriented along the x (y) axis and the conserved observable is the spin projection along the respective axis with the value $\pm 1/2$. In these two cases, the operator $\hat{\mathsf{H}}$ acquires the form, see Eqs. (1.12) and (1.13),

$$\hat{\mathsf{H}}/(-KB) = \hat{\mathsf{S}}_x = \hat{\mathsf{S}}_x|+x\rangle\langle+x| + \hat{\mathsf{S}}_x|-x\rangle\langle-x|$$
$$= \frac{1}{2}|+x\rangle\langle+x| - \frac{1}{2}|-x\rangle\langle-x| = \frac{1}{2}\begin{pmatrix}0 & 1\\ 1 & 0\end{pmatrix} \tag{1.22}$$

and

$$\hat{\mathsf{H}}/(-KB) = \hat{\mathsf{S}}_y = \hat{\mathsf{S}}_y|+y\rangle\langle+y| + \hat{\mathsf{S}}_y|-y\rangle\langle-y|$$
$$= \frac{1}{2}|+y\rangle\langle+y| - \frac{1}{2}|-y\rangle\langle-y| = \frac{1}{2}\begin{pmatrix}0 & -\mathrm{i}\\ \mathrm{i} & 0\end{pmatrix}. \tag{1.23}$$

The spin operators introduced by these equations are usually written in the form

$$\hat{\boldsymbol{S}} = \frac{1}{2}\boldsymbol{\sigma}\,, \tag{1.24}$$

where the matrices $\boldsymbol{\sigma}$ are called, after the Austrian physicists Wolfgang Pauli, *Pauli spin matrices* and satisfy

$$\sigma_i\sigma_j = \delta_{ij} + \mathrm{i}\varepsilon_{ijk}\sigma_k\,, \tag{1.25}$$

which may be proved best by direct substitution. It follows immediately from this equation that the spin projection operators along each axis do not commute mutually

$$[\hat{\mathsf{S}}_i, \hat{\mathsf{S}}_j] = \mathrm{i}\varepsilon_{ijk}\hat{\mathsf{S}}_k\,, \tag{1.26}$$

where *commutator* of any two operators is defined as

$$[\hat{\mathsf{A}}, \hat{\mathsf{B}}] = \hat{\mathsf{A}}\hat{\mathsf{B}} - \hat{\mathsf{B}}\hat{\mathsf{A}}\,.$$

This example illustrates a more general principle: observables that we cannot measure simultaneously correspond to non-commuting operators within the mathematical scheme of quantum mechanics.

1.2.5 Properties of Hermitian Operators

All Hermitian operators comply with the following statements (and we will often take advantage thereof throughout this book).

1. *Eigenvalues of a Hermitian operator are real.*

   Let $|n\rangle$ be an eigenvector of a Hermitian operator $\hat{\mathsf{H}}$ corresponding to the eigenvalue E_n:

$$\hat{\mathsf{H}}|n\rangle = E_n|n\rangle \,. \tag{1.27}$$

   After Hermitian conjugation of Eq. (1.27) we obtain

$$\langle n|\hat{\mathsf{H}} = \langle n|E_n^* \,, \tag{1.28}$$

   which shows that $\langle n|$ is likewise an eigenvector of the operator $\hat{\mathsf{H}}$ and corresponds to the eigenvalue E_n^*. We now show that these two eigenvalues equal. Multiplying Eq. (1.27) with a bra-vector $\langle n|$ from left and Eq. (1.28) with a ket-vector $|n\rangle$ from right, and comparing these two equations with each other then yields

$$E_n = E_n^* \,, \tag{1.29}$$

   hence $E_n \in \mathbb{R}$. In contrast, eigenvalues of non-Hermitian operators may be both complex and real. The following two operators are examples of non-Hermitian operators with complex and real eigenvalues, respectively:

$$\hat{\mathsf{A}} = \begin{pmatrix} 0 & -1 \\ 1 & 0 \end{pmatrix}, \hat{\mathsf{B}} = \begin{pmatrix} 1 & -\mathrm{i}\alpha \\ -\mathrm{i}\alpha & -1 \end{pmatrix},$$

   where $\alpha \in (0, 1)$.

2. *Eigenvectors corresponding to two different eigenvalues are mutually orthogonal.*

   It follows from Eqs. (1.27), (1.28), and (1.29) that

$$\hat{\mathsf{H}}|n\rangle = E_n|n\rangle \,,$$
$$\langle m|\hat{\mathsf{H}} = E_m\langle m| \,.$$

   Multiplying the former with a bra-vector $\langle m|$ and the latter with a ket-vector $|n\rangle$, and taking their difference results in

$$0 = (E_n - E_m)\langle m|n\rangle \,.$$

   If $E_n \neq E_m$,

$$\langle m|n\rangle = 0 \,. \tag{1.30}$$

   Note (without proof herein) that in case of a degeneracy, i.e., there exist several eigenvectors corresponding to the same eigenvalue, it is always possible to choose the eigenvectors so that they are mutually orthogonal.

3. *The basis of eigenvectors* $|n\rangle$ *is complete within the space upon which the Hermitian operator* $\hat{\mathsf{H}}$ *acts.* Meaning

$$\sum_{n=1}^{N} |n\rangle\langle n| = \mathbf{1}, \tag{1.31}$$

where N represents the number of degrees of freedom of the studied system, i.e., the dimension of the abstract vector space upon which the operator $\hat{\mathsf{H}}$ acts. We will not prove this statement herein in its most general sense. Note, though, that aside from mathematical nuances related to the transition to infinite-dimensional spaces, this statement is almost trivial. Every N-by-N-matrix has N mutually orthogonal eigenvectors (as shown previously), therefore they must form a complete basis in an N-dimensional space.

Despite the Hamiltonian being a Hermitian operator, it is often very convenient to work with non-Hermitian operators. For instance, consider the eigenstates of the spin projection along the z axis,

$$\hat{\mathsf{S}}_z|\pm\rangle = \pm\frac{1}{2}|\pm\rangle, \tag{1.32}$$

which we previously denoted as $|\pm z\rangle$ and for the purpose of the following we will henceforth denote merely as $|\pm\rangle$. Introducing the so-called ladder operators

$$\hat{\mathsf{S}}_\pm = \hat{\mathsf{S}}_x \pm \mathrm{i}\hat{\mathsf{S}}_y,$$

and considering Eqs. (1.10), (1.22), and (1.23), we find for their acting on the states $|\pm\rangle$ that

$$\hat{\mathsf{S}}_+|+\rangle = 0, \quad \hat{\mathsf{S}}_+|-\rangle = |+\rangle, \tag{1.33}$$

$$\hat{\mathsf{S}}_-|+\rangle = |-\rangle, \quad \hat{\mathsf{S}}_-|-\rangle = 0. \tag{1.34}$$

These equations show that the operator $\hat{\mathsf{S}}_+$ "raises" the spin projection along the z axis and the operator $\hat{\mathsf{S}}_-$ "lowers" the projection. This thus demonstrates that acting of the non-Hermitian operators $\hat{\mathsf{S}}_\pm$ on the eigenvectors of the operator $\hat{\mathsf{S}}_z$ is simpler than acting of the Hermitian operators $\hat{\mathsf{S}}_x$ and $\hat{\mathsf{S}}_y$. Owing to the operator identity

$$\hat{\boldsymbol{A}}\cdot\hat{\boldsymbol{B}} = \frac{1}{2}(\hat{\mathsf{A}}_+\hat{\mathsf{B}}_- + \hat{\mathsf{A}}_-\hat{\mathsf{B}}_+) + \hat{\mathsf{A}}_z\hat{\mathsf{B}}_z, \tag{1.35}$$

which holds for any two vector operators $\hat{\boldsymbol{A}}$ and $\hat{\boldsymbol{B}}$, we may easily always replace $\hat{\mathsf{A}}_x$ and $\hat{\mathsf{A}}_y$ with $\hat{\mathsf{A}}_+$ and $\hat{\mathsf{A}}_-$.

1.2.6 Ambiguity in the Determination of States

In the previous section, we derived the form of an operator knowing its eigenvectors and eigenvalues, see Eqs. (1.21), (1.22) and (1.23). The usual case is the opposite, though: we need to find the eigenvalues and eigenvectors of a given operator. However, Eq. (1.27) does not define the eigenvectors unambiguously. For instance, eigenvectors and eigenvalues of the operator $\hat{\mathsf{S}}_z$ are given by the equation

$$\hat{\mathsf{S}}_z|S_z\rangle = S_z|S_z\rangle\,.$$

Considering Eqs. (1.4) and (1.21), we obtain

$$\begin{pmatrix} 1 & 0 \\ 0 & -1 \end{pmatrix}\begin{pmatrix} a \\ b \end{pmatrix} = 2S_z\begin{pmatrix} a \\ b \end{pmatrix}, \tag{1.36}$$

which is a system of two equations for two unknowns a and b with a parameter S_z. Depending on the value of the parameter S_z, this system has either no solution at all or infinitely many. Specifically for $S_z = \pm 1/2$, the system (1.36) determines only one of the two unknowns a and b. We obtain $b = 0$ and arbitrary a for $S_z = 1/2$, and $a = 0$ and arbitrary b for $S_z = -1/2$. Similarly, writing equation for eigenvalues and eigenvectors of the operator $\hat{\mathsf{S}}_x$ yields

$$\begin{pmatrix} 0 & 1 \\ 1 & 0 \end{pmatrix}\begin{pmatrix} a \\ b \end{pmatrix} = 2S_x\begin{pmatrix} a \\ b \end{pmatrix}. \tag{1.37}$$

We obtain for $S_x = 1/2$ that

$$b = a \tag{1.38}$$

and for $S_x = -1/2$

$$b = -a\,, \tag{1.39}$$

however, a remains still completely undetermined. We may remove this freedom by imposing a normalization requirement:

$$\langle\psi|\,\psi\rangle = 1 \Rightarrow |a|^2 + |b|^2 = 1\,. \tag{1.40}$$

Nevertheless, neither this condition defines the eigenvector unambiguously, as Eqs. (1.36) and (1.40), or alternatively (1.37) and (1.40), still hold when we multiply a and b with the *same* phase factor $\mathrm{e}^{\mathrm{i}\varphi}$. Furthermore, the probability of transition from one state into another given as $|\langle\phi|\,\psi\rangle|^2$ obviously maintains its validity upon the substitution $|\psi\rangle \to |\psi\rangle\mathrm{e}^{\mathrm{i}\varphi_1}$, $\langle\phi| \to \langle\phi|\mathrm{e}^{-\mathrm{i}\varphi_2}$, where generally $\varphi_1 \neq \varphi_2$.

Thus we see that the herein developed formalism contains freedom in terms of the *total* phase factor of the state vectors. The *relative* phase factors, on the other hand, are well defined by the eigenproblem, see, e.g., Eqs. (1.38) and (1.39).

For a real Hamiltonian (which may not always be the case, see (1.23)), we usually choose the total phase so that the corresponding eigenvectors are real as well.

Exercise 1: General Projection of the Spin $S = 1/2$

Consider an SG experiment where only electrons with their spin projection +1/2 leave the apparatus. What is the probability of detecting the projection of +1/2 at the second apparatus if it is rotated by an arbitrary angle with respect to the first one? Hint: Using Eq. (1.35), find the projection operator of spin along a general direction $\hat{S}_{\mathbf{n}} = \hat{\mathbf{S}} \cdot \mathbf{n}$, *where*

$$\mathbf{n} = (\sin\vartheta\cos\varphi, \sin\vartheta\sin\varphi, \cos\vartheta)\,.$$

Find a vector satisfying the equation $\hat{S}_{\mathbf{n}}|+\mathbf{n}\rangle = \frac{1}{2}|+\mathbf{n}\rangle$. *The desired probability is then given as* $|\langle +z|+\mathbf{n}\rangle|^2$.

Note that this exercise is not listed here for the only purpose of practice. Vectors $|\pm\mathbf{n}\rangle$ are called *helicity spinors* and play a crucial role at the frontier of modern quantum field theory, see, e.g., [12, 13].

1.2.7 Rabi Method of Magnetic Moments

Isidor Rabi came with the brilliant idea of placing a coil with variable current at the second magnet of the modified SG apparatus shown in Fig. 1.5. This current, as it passes through the coil, creates a variable magnetic field in the xy plane which is perpendicular to the permanent magnetic field. At a certain frequency ω of the current, there is a very high probability of the electron spin flipping to the opposite direction. In such a case, the third magnet further deflects the trajectory of the electron, instead of unbending it back into the original direction, and thus the electron disappears from the electron beam.

This method is applicable to any particle with spin 1/2 and a nonzero magnetic moment, and allows for very accurate measurements of the constant K in Eq. (1.3). A more detailed description of the Rabi method may be found in, e.g., [4, 5]. However, a kind reader who comprehended everything so far is most certainly capable of deriving the necessary theory for this experiment with the help of the following notes.

1. For the Hamiltonian operator in the Schrödinger equation (1.17), we substitute from Eq. (1.3) where we substituted the vector **S** with a vector operator $\hat{\mathbf{S}}$:

$$\mathrm{i}\frac{\mathrm{d}|\psi\rangle}{\mathrm{d}t} = -K\hat{\mathbf{S}}\cdot\mathbf{B}|\psi\rangle\,. \tag{1.41}$$

The vector of magnetic induction reads

$$\mathbf{B} = (B_1\cos\omega t, B_1\sin\omega t, B_0)\,,$$

where obviously B_0 represents the intensity of the permanent (in time) field along the z axis, and B_1 stands for the intensity of the variable field.

2. We write the state vector $|\psi\rangle$ in the basis of the eigenstates of the spin projection along the z axis:

$$|\psi(t)\rangle = c_+(t)|+\rangle + c_-(t)|-\rangle\,. \tag{1.42}$$

After its insertion into Eq. (1.41), we multiply this equation from left with $\langle+|$ and subsequently with $\langle-|$. We thus obtain a system of two differential equations for two unknown functions $c_\pm(t)$:

$$\mathrm{i}\dot{c}_+ = H_{++}c_+ + H_{+-}c_-\,,\quad \mathrm{i}\dot{c}_- = H_{-+}c_+ + H_{--}c_-\,, \tag{1.43}$$

where $H_{\pm\pm}$ and $H_{\pm\mp}$ are the matrix elements of the Hamiltonian in the mentioned basis

$$H_{++} = (-K)\langle+|\hat{\mathbf{S}}\cdot\mathbf{B}|+\rangle = (-KB_z)\langle+|\hat{\mathsf{S}}_z|+\rangle = -KB_0/2\,,$$
$$H_{+-} = (-K)\langle+|\hat{\mathbf{S}}\cdot\mathbf{B}|-\rangle = (-KB_-/2)\langle+|\hat{\mathsf{S}}_+|-\rangle = -(KB_1/2)\mathrm{e}^{-\mathrm{i}\omega t}\,,\ \text{etc.};$$

that is

$$\hat{\mathsf{H}} = \begin{pmatrix} H_{++} & H_{+-} \\ H_{-+} & H_{--} \end{pmatrix} = -\frac{K}{2}\begin{pmatrix} B_0 & B_1\mathrm{e}^{-\mathrm{i}\omega t} \\ B_1\mathrm{e}^{+\mathrm{i}\omega t} & -B_0 \end{pmatrix}.$$

3. Assuming the system is in the state $|+\rangle$ at time $t = 0$, we have the initial condition for Eq. (1.43):

$$c_+(0) = 1,\quad c_-(0) = 0\,. \tag{1.44}$$

The probability of the system being in the state $|-\rangle$ at any later time t is given as

$$P_{+\to-} = |\langle-|\psi(t)\rangle|^2 = |c_-(t)|^2\,, \tag{1.45}$$

where we substituted from Eq. (1.42) and used the orthonormality equation, Eq. (1.9).

This way, we transformed the problem into a mathematical one the reader shall be able to solve on his own. All left to do is to solve Eq. (1.43) with the initial conditions (1.44) and then insert the results into (1.45).

Exercise 2: Rabi Oscillations I

Find the time-dependence of the probability (1.45) and also its dependence on the parameters K, B_0, B_1, ω.

1.3 Systems with More Degrees of Freedom

In the previous sections, we developed mathematical formalism of quantum mechanics: a system is described by a state vector from the abstract (Hilbert) space of states, and the probability of its transition from one state into another is given by the scalar product of the corresponding state vectors. Time evolution of the system is determined by the Hamiltonian operator $\hat{\mathsf{H}}$ via the Schrödinger equation. Eigenvalues of this operator correspond to the possible energy values of the system that we can measure.[7]

It clearly follows from this brief summary that knowing the Hamiltonian of the system allows for—at least in principle—prediction of anything that can be discovered about the system. In case of the Stern-Gerlach experiments, we obtained the Hamiltonian from analysis of the experiments. However, this method is generally inapplicable to systems with an infinite number of degrees of freedom. Moreover, shall quantum mechanics predict experimental results, such as what energy values we will measure, we must know the Hamiltonian "in advance." In the following sections, we show how to find the Hamiltonian when the behavior of the system is well known within classical physics.

1.3.1 Expected Values of Operators and Their Time Evolution

As we move from the macroscopic towards the microscopic world, i.e., to objects of the size 10^{-10} m or less, description within the framework of classical mechanics

[7] To be more precise, the differences of the eigenvalues of the operator.

fails. Quantum mechanics, which in contrast provides a correct description on this scale, should nevertheless contain classical mechanics somehow. To elucidate the connection, it is advantageous to introduce the notion of expectation value.

The expectation value of an observable described by the operator $\hat{\mathsf{A}}$ upon repeated measurements of the state $|\psi\rangle$ is given by the expression

$$\langle \hat{\mathsf{A}} \rangle = \sum_{j=1}^{N} a_j p(a_j) = \sum_{j=1}^{N} \langle \psi | a_j \rangle a_j \langle a_j | \psi \rangle = \langle \psi | \hat{\mathsf{A}} | \psi \rangle \, . \tag{1.46}$$

We used the definition of the mean value in the first equality, and the probability of the system's transition from the state $\langle \psi |$ into the state $|a_j\rangle$, given as $p(a_j) = |\langle \psi | a_j \rangle|^2$, in the second equality. Since Eq. (1.8) holds, we obtain $p(a_j) = \langle \psi | a_j \rangle \langle a_j | \psi \rangle$. Finally, the third step involved spectral decomposition of the operator $\hat{\mathsf{A}}$

$$\hat{\mathsf{A}} = \sum_{j=1}^{N} a_j |a_j\rangle\langle a_j| \, ,$$

where N represents the number of degrees of freedom of the studied system. Specifically, if inner degrees of freedom of a particle with the spin 1/2 are of interest, $N = 2$.

We now turn our attention to the time evolution of the expectation value $\langle \hat{\mathsf{A}} \rangle$, i.e., we examine the dependence

$$\langle \hat{\mathsf{A}}(t) \rangle = \langle \psi(t) | \hat{\mathsf{A}} | \psi(t) \rangle \, .$$

It follows from Eq. (1.14) that

$$\langle \hat{\mathsf{A}}(t) \rangle = \langle \psi | \hat{\mathsf{U}}^{+}(t, t_0) \hat{\mathsf{A}} \hat{\mathsf{U}}(t, t_0) | \psi \rangle \, .$$

We may regard this equation also in another way: time evolution of the expectation value of the observable A is given by the expectation value of a time-dependent operator

$$\hat{\mathsf{A}}(t) = \hat{\mathsf{U}}^{+}(t, t_0) \hat{\mathsf{A}} \hat{\mathsf{U}}(t, t_0) \tag{1.47}$$

in a time-independent state $|\psi\rangle$.

We now adopt the latter approach, called the Heisenberg picture or also the *Heisenberg representation*, after German physicist Werner Heisenberg, and inspect the time evolution of the operator (1.47). Using the Schrödinger equation (1.16) and definition (1.47), we obtain

$$\frac{d\hat{\mathsf{A}}(t)}{dt} = \frac{d\hat{\mathsf{U}}^{+}(t, t_0)}{dt} \hat{\mathsf{A}} \hat{\mathsf{U}}(t, t_0) + \hat{\mathsf{U}}^{+}(t, t_0) \hat{\mathsf{A}} \frac{d\hat{\mathsf{U}}(t, t_0)}{dt}$$

$$= -\mathrm{i}\hat{\mathsf{U}}^{+}(t,t_0)[\hat{\mathsf{A}},\hat{\mathsf{H}}]\hat{\mathsf{U}}(t,t_0) = -\mathrm{i}[\hat{\mathsf{A}}(t),\hat{\mathsf{H}}(t)]\,, \tag{1.48}$$

where we inserted $\mathbf{1} = \hat{\mathsf{U}}(t,t_0)\hat{\mathsf{U}}^{+}(t,t_0)$ between $\hat{\mathsf{A}}$ and $\hat{\mathsf{H}}$ in the third equality. The other approach, called the *Schrödinger representation*, places the time evolution into the state vectors and regards operators representing observables as time-independent. In most cases, we will employ this latter picture.

We now use Eq. (1.48) to describe a motion of a particle with the spin 1/2 in a magnetic field, $\hat{\mathsf{A}}(t) = \hat{\mathsf{S}}_i(t)$, $\hat{\mathsf{H}}(t) = -K\hat{\mathsf{S}}_j(t)B_j$,

$$\frac{\mathrm{d}\hat{\mathsf{S}}_i(t)}{\mathrm{d}t} = -\mathrm{i}KB_j[\hat{\mathsf{S}}_i(t),\hat{\mathsf{S}}_j(t)] = -K\varepsilon_{ijk}B_j\hat{\mathsf{S}}_k(t)\,,$$

where we used Eq. (1.26) in the second equality. Rewriting the last equation in a vector notation yields

$$\frac{\mathrm{d}\hat{\mathbf{S}}(t)}{\mathrm{d}t} = -K\mathbf{B}\times\hat{\mathbf{S}}(t)\,,$$

which is the classical equation for the motion of a magnetic dipole in a magnetic field. Its classical derivation can be found in, e.g., [6].

1.3.2 *Canonical Quantization*

In classical mechanics, the Hamiltonian function is a function of canonical coordinates and momenta of particles. However, there are projections of the position and momentum vectors onto the coordinate axes that we cannot measure simultaneously.[8] This situation quite resembles that of the spin projections onto coordinate axes, and we have seen that within the mathematical formalism of quantum mechanics, this fact is represented by the noncommutativity of the corresponding operators. Furthermore, we have seen an analogy between the classical and quantum-mechanical equations in Heisenberg representation for the motion of a particle with the spin 1/2. Therefore, it is reasonable to demand that equations determining the time evolution of the coordinates and momenta operators have the same forms as the corresponding equations in classical theory. The exact

[8] We do not want to extensively discuss this matter here, as everyone has surely encountered it ample times. Nevertheless we mention Heisenberg's intuitive argument. We know from the Compton experiment (see, e.g., [1]) that a photon with the wavelength λ has the momentum $p = 2\pi/\lambda$. The more precisely we wish to locate a particle, the shorter the wavelength of the photon must be, hence higher momenta. Thus reducing the uncertainty in the particle's position results in higher uncertainty of its momentum.

forms of the commutation relations between the coordinates and momenta operators then follow thereof.

We now consider a particle moving in a static potential given by the function $V(\mathbf{r})$. Its classical Hamiltonian reads

$$H = \frac{\mathbf{p} \cdot \mathbf{p}}{2m} + V(\mathbf{r}), \tag{1.49}$$

and we will look for a quantum Hamiltonian of the form

$$\hat{\mathsf{H}} = \frac{\hat{\boldsymbol{p}} \cdot \hat{\boldsymbol{p}}}{2m} + V(\hat{\boldsymbol{r}}). \tag{1.50}$$

Substituting $\hat{\mathsf{A}}$ in (1.48) for the position or momentum operator, respectively, yields

$$\frac{\mathrm{d}\hat{\mathsf{x}}_i(t)}{\mathrm{d}t} = -\mathrm{i}\hat{\mathsf{U}}^+(t,t_0)[\hat{\mathsf{x}}_i, \hat{\mathsf{H}}]\hat{\mathsf{U}}(t,t_0), \quad \frac{\mathrm{d}\hat{\mathsf{p}}_i(t)}{\mathrm{d}t} = -\mathrm{i}\hat{\mathsf{U}}^+(t,t_0)[\hat{\mathsf{p}}_i, \hat{\mathsf{H}}]\hat{\mathsf{U}}(t,t_0). \tag{1.51}$$

The commutator of $\hat{\mathsf{x}}_i$ and $\hat{\mathsf{H}}$ reads

$$[\hat{\mathsf{x}}_i, \hat{\mathsf{H}}] = \frac{1}{2m}[\hat{\mathsf{x}}_i, \hat{\mathsf{p}}^2] + [\hat{\mathsf{x}}_i, V(\hat{\mathsf{x}})] = \frac{1}{2m}\left([\hat{\mathsf{x}}_i, \hat{\mathsf{p}}_j]\hat{\mathsf{p}}_j + \hat{\mathsf{p}}_j[\hat{\mathsf{x}}_i, \hat{\mathsf{p}}_j]\right) + 0, \tag{1.52}$$

since $\hat{\mathsf{x}}_i$ commutes with any function of itself.[9] We now demand that Eq. (1.51) acquires the same form as their analogs in classical mechanics. The first equation then reads

$$\frac{\mathrm{d}\hat{\mathsf{x}}_i(t)}{\mathrm{d}t} = \frac{\hat{\mathsf{p}}_i(t)}{m}, \tag{1.53}$$

hence

$$[\hat{\mathsf{x}}_i, \hat{\mathsf{p}}_j] = \mathrm{i}\delta_{ij}. \tag{1.54}$$

Similarly, the commutator of $\hat{\mathsf{p}}_i$ and $\hat{\mathsf{H}}$ is

$$[\hat{\mathsf{p}}_i, \hat{\mathsf{H}}] = \frac{1}{2m}[\hat{\mathsf{p}}_i, \hat{\mathsf{p}}^2] + [\hat{\mathsf{p}}_i, V(\hat{\boldsymbol{r}})] = [\hat{\mathsf{p}}_i, V(\hat{\boldsymbol{r}})].$$

[9] The reason is, we can expand any "reasonable" function of $\hat{\mathsf{x}}$ into a series in $\hat{\mathsf{x}}$. That $\hat{\mathsf{x}}$ commutes with any of its powers is obvious.

Furthermore, as we will prove later, see Eq. (1.84),[10]

$$[\hat{\mathsf{p}}_i, V(\hat{\boldsymbol{r}})] = -\mathrm{i}\left.\frac{\partial V(\mathbf{r})}{\partial x_i}\right|_{\mathbf{r}=\hat{\boldsymbol{r}}}, \tag{1.55}$$

hence

$$\frac{\mathrm{d}\hat{\mathsf{p}}_i(t)}{\mathrm{d}t} = -\left.\frac{\partial V(\mathbf{r})}{\partial x_i}\right|_{\mathbf{r}=\hat{\boldsymbol{r}}(t)}. \tag{1.56}$$

Equations (1.53) and (1.56) are the quantum-mechanical analogs of the Newton equations of classical theory.

We can readily generalize our previous steps, and thus obtain a universal process of transformation from classical to quantum theory called the *canonical quantization*. First, we identify the Hamiltonian and canonically conjugated coordinates and momenta, i.e., quantities satisfying the Hamilton equations

$$\dot{x}_j = \frac{\partial H}{\partial p_j}, \quad \dot{p}_j = -\frac{\partial H}{\partial x_j}.$$

For example, we find for the Hamiltonian (1.49) that $\dot{x}_j = \frac{\partial H}{\partial p_j} = \frac{p_j}{m}$ and $\dot{p}_j = -\frac{\partial H}{\partial x_j} = -\frac{\partial V}{\partial x_j}$, which is obviously equivalent to the Newton equation $m\frac{\mathrm{d}^2 x_j}{\mathrm{d}t^2} = -\frac{\partial V}{\partial x_j}$. Next, we postulate the so-called canonical commutation relations (1.54) for the canonically conjugated coordinates and momenta, and substitute the classical quantities for corresponding operators. This Hamiltonian, such as (1.50), is then the correct one within quantum mechanics.

We thus see that dynamics is within the quantum and classical theory very much alike. Kinematics, i.e., representation of states and of their changes, on the other hand, acquires an absolutely new form. However, if the distances between individual energy levels are small enough, see the discussion in the Sect. 6.2.7, quantum mechanics blends with the classical. Note, though, that the classical limit of quantum theory is not entirely trivial, see, e.g., [8, 9, 11].

1.3.3 Harmonic Oscillator

We now illustrate this general method on one of the most fundamental systems—on the harmonic oscillator. Its importance stems from the fact that it belongs to the

---

[10]We briefly show that this relation holds for one-dimensional motion. We expand $V(x)$ in a Taylor series $V(x) = V_0 + V_1 x + V_2 x^2 + \ldots$ and using Eq. (1.54) we obtain $[\hat{\mathsf{p}}, V(\hat{\mathsf{x}})] = V_1[\hat{\mathsf{p}}, \hat{\mathsf{x}}] + V_2[\hat{\mathsf{p}}, \hat{\mathsf{x}}^2] + \ldots = -\mathrm{i}(V_1 + 2V_2\hat{\mathsf{x}} + \ldots) = -\mathrm{i}\left.\frac{\mathrm{d}V(x)}{\mathrm{d}x}\right|_{x=\hat{\mathsf{x}}}$.

few physically interesting problems we are able to solve exactly, and that the free electromagnetic field behaves as a set of mutually independent harmonic oscillators, as we will see later in Sect. 6.1.6.

We can expand any potential energy $V(\hat{\mathsf{x}})$ of a one-dimensional one-particle Hamiltonian into a series in $\hat{\mathsf{x}}$

$$V(\hat{\mathsf{x}}) - V(x_0) = \left.\frac{\partial V}{\partial x}\right|_{x_0} (\hat{\mathsf{x}} - x_0) + \frac{1}{2!}\left.\frac{\partial^2 V}{\partial x^2}\right|_{x_0} (\hat{\mathsf{x}} - x_0)^2 + \frac{1}{3!}\left.\frac{\partial^3 V}{\partial x^3}\right|_{x_0} (\hat{\mathsf{x}} - x_0)^3 + \dots \tag{1.57}$$

For a minimum of the potential at x_0, the first derivative equals zero. Furthermore, we neglect third and higher derivatives, i.e., keep only the second, place the minimum of the potential to $x_0 = 0$, and take the energy from the value $V(x_0)$. Finally, following the procedure of canonical quantization, we obtain the Hamiltonian of the harmonic oscillator

$$\hat{\mathsf{H}} = \frac{\hat{\mathsf{p}}^2}{2m} + \frac{1}{2} m\omega^2 \hat{\mathsf{x}}^2 , \qquad \text{where} \qquad [\hat{\mathsf{p}}, \hat{\mathsf{x}}] = -\mathrm{i} .$$

Had we included other terms in the expansion (1.57), the behavior of the system would be much more complex and usually analytically insolvable. We show how to deal with such cases in Chap. 2. It is generally desirable to reduce the number of constants, therefore we will henceforth use dimensionless coordinates and momentum. We accomplish that by the following scaling transformation[11]

$$\hat{\mathsf{x}} \to \lambda \hat{\mathsf{x}}, \quad \hat{\mathsf{p}} \to \frac{\hat{\mathsf{p}}}{\lambda},$$

which we insert into the Schrödinger equation

$$\left[\frac{\hat{\mathsf{p}}^2}{2m\lambda^2} + \frac{1}{2} m\omega^2 \lambda^2 \hat{\mathsf{x}}^2\right] |\psi\rangle = \mathscr{E} |\psi\rangle ,$$

multiply with $m\lambda^2$ and finally set $m^2\omega^2\lambda^4 = 1$; we thus obtain

$$\left[\frac{\hat{\mathsf{p}}^2}{2} + \frac{\hat{\mathsf{x}}^2}{2}\right] |\psi\rangle = m\lambda^2 \mathscr{E} |\psi\rangle = \frac{\mathscr{E}}{\omega} |\psi\rangle = E |\psi\rangle , \tag{1.58}$$

where E is a dimensionless energy.

[11] Upon this particular transformation, the canonical commutation relation $[\hat{\mathsf{p}}, \hat{\mathsf{x}}] = -\mathrm{i}$ maintains its form.

1.3.4 Abstract Solution

Inspired by the formula $a^2 + b^2 = (a + ib)(a - ib)$, it seems reasonable to attempt to decompose the operator on the lhs of Eq. (1.58) to the product. We therefore introduce non-Hermitian operators

$$\hat{a} = \frac{1}{\sqrt{2}}(\hat{x} + i\hat{p}), \quad \hat{a}^+ = \frac{1}{\sqrt{2}}(\hat{x} - i\hat{p}).$$

We can easily prove that the Hamiltonian $\hat{H} = \frac{\hat{p}^2}{2} + \frac{\hat{x}^2}{2}$ acquires quite a simple form

$$\hat{H} = \hat{a}^+\hat{a} + \frac{1}{2} \tag{1.59}$$

when expressed in terms of the operators $\hat{a}$ and $\hat{a}^+$. These two operators do not mutually commute, precisely speaking their commutator equals one,

$$[\hat{a}, \hat{a}^+] = \frac{1}{2}[\hat{x} + i\hat{p}, \hat{x} - i\hat{p}] = -\frac{i}{2}[\hat{x}, \hat{p}] + \frac{i}{2}[\hat{p}, \hat{x}] = 1.$$

An important property of the operators $\hat{a}$ and $\hat{a}^+$ follows from the last two equations: together with the Hamiltonian $\hat{H}$, they form a algebra of three operators which is closed with respect to the commutation operation,

$$[\hat{H}, \hat{a}] = [\hat{a}^+\hat{a}, \hat{a}] = [\hat{a}^+, \hat{a}]\hat{a} = -\hat{a}, \tag{1.60}$$

$$[\hat{H}, \hat{a}^+] = [\hat{a}^+\hat{a}, \hat{a}^+] = \hat{a}^+[\hat{a}, \hat{a}^+] = \hat{a}^+. \tag{1.61}$$

As we will shortly see, this property allows us to determine the spectrum of eigenvalues of the Hamiltonian. Note also that we could have obtained the second equation from the first one via Hermitian conjugation. We now write the equation for eigenvalues and eigenvectors of the Hamiltonian

$$\hat{H}|n\rangle = E_n|n\rangle, \tag{1.62}$$

where n labels the individual eigenvalues, and $n = 0$ stands for the ground state, $n = 1$ the first excited state, etc. Equations (1.60) and (1.61) are operator equalities, meaning they hold whatever vector we act on with both sides of these equations. We thus act with them on the eigenvectors of the Hamiltonian:

$$[\hat{H}, \hat{a}]|n\rangle = -\hat{a}|n\rangle \Rightarrow \hat{H}\left(\hat{a}|n\rangle\right) = \hat{a}\left(\hat{H}|n\rangle - |n\rangle\right) = (E_n - 1)\left(\hat{a}|n\rangle\right), \tag{1.63}$$

$$[\hat{H}, \hat{a}^+]|n\rangle = \hat{a}^+|n\rangle \Rightarrow \hat{H}\left(\hat{a}^+|n\rangle\right) = \hat{a}^+\left(\hat{H}|n\rangle + |n\rangle\right) = (E_n + 1)\left(\hat{a}^+|n\rangle\right), \tag{1.64}$$

where we expanded the commutator, used Eq. (1.62), and also made use of the fact that the number E_n commutes with any operator, hence also with the operators $\hat{a}$ and $\hat{a}^+$.

We see that the state $\hat{a}|n\rangle$ is an eigenstate of the Hamiltonian with the energy $E_n - 1$, and so is the state $\hat{a}^+|n\rangle$ an eigenstate of the same Hamiltonian with the energy $E_n + 1$, compare Eq. (1.62) with Eqs. (1.63) and (1.64). Action of the operator $\hat{a}$ on an eigenstate of the Hamiltonian thus yields an eigenstate with the corresponding eigenvalue smaller by 1, and similarly the action of the operator $\hat{a}^+$ results in an eigenstate with the eigenvalue greater by 1. However, a ground state of the system must exist,[12] so must thus be a state for which

$$\hat{a}|0\rangle = 0\,.$$

This equation holds also when acting with the operator $\hat{a}^+$,

$$\hat{a}^+\hat{a}|0\rangle = \left(\hat{H} - \frac{1}{2}\right)|0\rangle = \left(E_0 - \frac{1}{2}\right)|0\rangle = 0\,,$$

hence $E_0 = 1/2$. Acting with the operator $\hat{a}^+$ on the state $|0\rangle$, we obtain a state with the eigenvalue $E_1 = 1/2 + 1 = 3/2$, etc. The spectrum of the eigenvalues of the Hamiltonian thus reads

$$E_n = n + 1/2\,. \tag{1.65}$$

Note that there are no other eigenvalues with corresponding eigenvectors of the Hamiltonian, i.e., the basis $\{|n\rangle\}$ is complete. Were there another eigenvector between the n-th and $(n+1)$-th state with the corresponding eigenvalue E', we could repetitively apply the operator $\hat{a}$ on this eigenvector and we would obtain successively eigenvectors with eigenvalues $E' - 1$, $E' - 2$, etc. Upon sufficient number of repetitions we would finally reach an eigenstate corresponding to the eigenvalue smaller than $1/2$, which contradicts the requirement $1/2$ be the ground state energy.

It follows from Eq. (1.65) that $E_n \pm 1 = E_{n\pm1}$. Equation (1.62) for the eigenvalue corresponding to the state $|n \pm 1\rangle$ reads then

$$\hat{H}|n \pm 1\rangle = (E_n \pm 1)|n \pm 1\rangle\,.$$

[12] From a purely mathematical point of view, this requirement is excessive. The reason is that the Hamiltonian has a below bounded spectrum since the operator $\hat{a}^+\hat{a}$ is positive-definite as $\langle\psi|\hat{a}^+\hat{a}|\psi\rangle = \langle\chi|\chi\rangle \geq 0$, where $|\chi\rangle = \hat{a}|\psi\rangle$ and the inequality follows from (1.7). As physicists, we engage only with Hamiltonians with a below bounded spectrum; see the interpretation of the Dirac equation in Chap. 7.

Comparison of this equation with Eqs. (1.63) and (1.64) leads to

$$\hat{\mathsf{a}}|n\rangle = \alpha_-(n)|n-1\rangle\,, \quad \hat{\mathsf{a}}^+|n\rangle = \alpha_+(n)|n+1\rangle\,, \tag{1.66}$$

where $\alpha_\pm(n)$ are constants we can usually choose to be real owing to the freedom in the total phase of the eigenvectors. We now determine their values, which will later prove to be indeed useful. We find the Hermitian conjugate of the first equation and multiply it from right by the vector $|n-1\rangle$

$$\langle n|\hat{\mathsf{a}}^+|n-1\rangle = \alpha_-(n)\langle n-1|n-1\rangle\,.$$

In the second of Eq. (1.66), we substitute n for $n-1$ and multiply both sides from left by the vector $\langle n|$,

$$\langle n|\hat{\mathsf{a}}^+|n-1\rangle = \alpha_+(n-1)\langle n|n\rangle\,.$$

Under the normalization requirement for the eigenvectors, the constants must satisfy

$$\alpha_+(n-1) = \alpha_-(n)\,.$$

Furthermore, it must hold, see Eq. (1.66),

$$\hat{\mathsf{a}}^+\hat{\mathsf{a}}|n\rangle = \hat{\mathsf{a}}^+\alpha_-(n)|n-1\rangle = \alpha_+(n-1)\alpha_-(n)|n\rangle\,,$$

and also, see Eq. (1.59),

$$\hat{\mathsf{a}}^+\hat{\mathsf{a}}|n\rangle = \left(\hat{\mathsf{H}} - \frac{1}{2}\right)|n\rangle = n|n\rangle\,. \tag{1.67}$$

From the last three equations, we finally obtain

$$\alpha_-(n) = \sqrt{n}\,, \quad \alpha_+(n) = \sqrt{n+1}\,. \tag{1.68}$$

1.3.5 *Matrix Representation*

We can easily give a specific form to the above-derived abstract solution by choosing a particular representation of the vectors $|n\rangle$. For instance,

$$|0\rangle = \begin{pmatrix} 1 \\ 0 \\ 0 \\ 0 \\ \vdots \end{pmatrix}, \quad |1\rangle = \begin{pmatrix} 0 \\ 1 \\ 0 \\ 0 \\ \vdots \end{pmatrix}, \quad |2\rangle = \begin{pmatrix} 0 \\ 0 \\ 1 \\ 0 \\ \vdots \end{pmatrix}, \quad \dots$$

By Hermitian conjugation we obtain

$$\begin{aligned} \langle 0| &= \left(1\,0\,0\,0\,\dots\right), \\ \langle 1| &= \left(0\,1\,0\,0\,\dots\right), \\ \langle 2| &= \left(0\,0\,1\,0\,\dots\right), \\ &\dots \end{aligned}$$

Acting on the completeness relation

$$\sum_{n=0}^{\infty} |n\rangle\langle n| = \mathbf{1} \tag{1.69}$$

then leads to the matrix expression of the operators $\hat{\mathsf{H}}$, $\hat{\mathsf{a}}$, and $\hat{\mathsf{a}}^{+}$:

$$\hat{\mathsf{H}} = \hat{\mathsf{H}}\mathbf{1} = \sum_{n=0}^{\infty} E_n |n\rangle\langle n| = \operatorname{diag}\left\{\frac{1}{2}, \frac{3}{2}, \frac{5}{2}, \dots\right\},$$

$$\hat{\mathsf{a}} = \hat{\mathsf{a}}\mathbf{1} = \sum_{n=0}^{\infty} \alpha_{-}(n)|n-1\rangle\langle n| = \begin{pmatrix} 0 & 1 & 0 & 0 & \dots \\ 0 & 0 & \sqrt{2} & 0 & \dots \\ 0 & 0 & 0 & \sqrt{3} & \dots \\ 0 & 0 & 0 & 0 & \ddots \\ \vdots & \vdots & \vdots & \ddots & \ddots \end{pmatrix},$$

$$\hat{\mathsf{a}}^{+} = \hat{\mathsf{a}}^{+}\mathbf{1} = \sum_{n=0}^{\infty} \alpha_{+}(n)|n+1\rangle\langle n| = \begin{pmatrix} 0 & 0 & 0 & 0 & \dots \\ 1 & 0 & 0 & 0 & \dots \\ 0 & \sqrt{2} & 0 & 0 & \dots \\ 0 & 0 & \sqrt{3} & 0 & \ddots \\ \vdots & \vdots & \vdots & \ddots & \ddots \end{pmatrix}.$$

We see that the Hamiltonian, as any operator in the basis of its eigenvectors, is diagonal with eigenvalues placed along this diagonal. We could easily verify that in the matrix representation

$$\hat{\mathsf{x}} = \frac{1}{\sqrt{2}}(\hat{\mathsf{a}} + \hat{\mathsf{a}}^+) \tag{1.70}$$

and also

$$\hat{\mathsf{p}} = \frac{1}{\mathrm{i}\sqrt{2}}(\hat{\mathsf{a}} - \hat{\mathsf{a}}^+) \tag{1.71}$$

are—unlike $\hat{\mathsf{a}}$ and $\hat{\mathsf{a}}^+$—Hermitian operators and that the relation $[\hat{\mathsf{x}}, \hat{\mathsf{p}}] = \mathrm{i}$ holds.

1.3.6 Dirac δ-Function

In the next section, we will need to operate with the *Dirac δ-function*. From a physical point of view, it is the best to regard the δ-function as a limit of functions, for instance:

$$\delta(x) = \lim_{\varepsilon \to 0+} \frac{1}{\pi} \frac{\varepsilon}{x^2 + \varepsilon^2}, \tag{1.72}$$

that is

$$\delta(x \neq 0) = 0, \qquad \delta(x = 0) = \infty.$$

From Eq. (1.72), we easily obtain the integral of the δ-function along the entire line[13]

$$\int_{-\infty}^{\infty} \delta(x)\,\mathrm{d}x = \frac{1}{\pi} \lim_{\varepsilon \to 0+} \int_{-\infty}^{\infty} \frac{\varepsilon}{x^2 + \varepsilon^2} \mathrm{d}x = \lim_{\varepsilon \to 0+} 1 = 1, \tag{1.73}$$

and also with the Fourier transform

$$\begin{aligned} \delta(x) &= \lim_{\varepsilon \to 0+} \frac{1}{\pi} \Im \frac{1}{x - \mathrm{i}\varepsilon} = \lim_{\varepsilon \to 0+} \frac{1}{2\mathrm{i}\pi} \left[\frac{1}{x - \mathrm{i}\varepsilon} - \frac{1}{x + \mathrm{i}\varepsilon} \right] \\ &= \lim_{\varepsilon \to 0+} \frac{1}{2\pi} \left[\int_0^{\infty} \mathrm{e}^{-\mathrm{i}k(x - \mathrm{i}\varepsilon)}\,\mathrm{d}k + \int_0^{\infty} \mathrm{e}^{\mathrm{i}k(x + \mathrm{i}\varepsilon)}\,\mathrm{d}k \right] = \lim_{\varepsilon \to 0+} \frac{1}{2\pi} \int_{-\infty}^{\infty} \mathrm{e}^{-\mathrm{i}kx - |k|\varepsilon}\,\mathrm{d}k . \end{aligned} \tag{1.74}$$

The integral of a product of the δ-function and an "ordinary" function complies with

$$\int_{-\infty}^{\infty} f(x)\delta(x - a)\,\mathrm{d}x = f(a). \tag{1.75}$$

---

[13] When solving such problems, we start with the definition of the δ-function (1.72) and leave the limit $\varepsilon \to 0+$ to the very end of our calculation.

We can easily obtain this equation from the Taylor expansion of the function $f(x)$

$$f(x) = f(a) + (x-a)f'(a) + \frac{1}{2!}(x-a)^2 f''(a) + \dots,$$

from the normalization condition (1.73) and from the equation

$$\int x^n \delta(x) \mathrm{d}x = 0, \quad n > 0,$$

that one can easily prove. We calculate the derivations of the δ-function by employing the integration by parts

$$\int_{-\infty}^{\infty} f(x)\delta'(x-a)\,\mathrm{d}x = -f'(a)\,. \tag{1.76}$$

1.3.7 Coordinate Representation

Were we able to solve all quantum-mechanical problems exactly, we could now easily move from the harmonic oscillator to another problem. However, that is not the case, and we will thus discuss additional specific realizations of the abstract solution. Among the infinite amount of possible bases of the Hilbert space of states, there is one that stands out from the others—the basis of the eigenstates of the position operator,

$$\hat{\mathsf{x}}|x\rangle = x|x\rangle\,. \tag{1.77}$$

As there exist uncountably many places where a particle may be located, the spectrum of the position operator is uncountable. The completeness and orthonormality relations then acquire the form

$$\int_{-\infty}^{\infty} |x\rangle\langle x|\,\mathrm{d}x = \mathbf{1}\,, \tag{1.78}$$

$$\langle x|x'\rangle = \delta(x-x')\,. \tag{1.79}$$

The scalar product of two general states and the action of an operator $\hat{\mathsf{A}}$ on a state are then given as

$$\langle\psi_1|\psi_2\rangle = \langle\psi_1|\mathbf{1}|\psi_2\rangle = \langle\psi_1|\left(\int_{-\infty}^{\infty} |x\rangle\langle x|\,\mathrm{d}x\right)|\psi_2\rangle = \int_{-\infty}^{\infty} \langle\psi_1|x\rangle\langle x|\psi_2\rangle\,\mathrm{d}x\,, \tag{1.80}$$

$$\langle x|\hat{\mathsf{A}}|\psi\rangle = \int_{-\infty}^{\infty} \langle x|\hat{\mathsf{A}}|x'\rangle\langle x'|\,\psi\rangle\,\mathrm{d}x'\,. \tag{1.81}$$

The projection of a general state $|\psi\rangle$ onto the coordinate basis $\langle x|\,\psi\rangle$ corresponds to a vector with uncountably many components; we call such vectors functions. For historical reasons, we refer to the projection $\langle x|\,\psi\rangle = \psi(x)$ as *a wave function*. The square of the absolute value of this function gives the probability density of the particle to be found at the point x in the state $|\psi\rangle$, as follows from the general physical interpretation of scalar products.

Once we know the matrix elements of the coordinate and momentum operators between the vectors of coordinate basis, we can easily express any operator in this basis. The coordinate operator acquires a diagonal form in its own basis, i.e., we find from Eq. (1.77)

$$\langle x|\hat{\mathsf{x}}|\psi\rangle = x\langle x|\,\psi\rangle = x\psi(x)\,. \tag{1.82}$$

We obtain the matrix elements of the momentum operator from the canonical commutation relation

$$[\hat{\mathsf{p}},\hat{\mathsf{x}}] = -\mathrm{i}$$

by multiplying it with the coordinate eigenstates from both left and right

$$\langle x|[\hat{\mathsf{p}},\hat{\mathsf{x}}]\,|x'\rangle = -\mathrm{i}\langle x|\,x'\rangle\,.$$

We now expand the commutator on the lhs and use Eq. (1.77)

$$\langle x|[\hat{\mathsf{p}},\hat{\mathsf{x}}]\,|x'\rangle = \langle x|\hat{\mathsf{p}}|x'\rangle(x'-x)\,,$$

and rearrange the rhs using Eq. (1.79). Finally, we multiply both sides of the equation with $\langle x'|\,\psi\rangle$ and integrate with respect to x',

$$\int_{-\infty}^{\infty} \mathrm{d}x'\,\langle x|\hat{\mathsf{p}}|x'\rangle(x'-x)\psi(x') = -\mathrm{i}\psi(x)\,,$$

where we also used Eq. (1.75) on the rhs. The last equation holds if, see Eq. (1.76),

$$\langle x|\hat{\mathsf{p}}|x'\rangle = \mathrm{i}\frac{\mathrm{d}}{\mathrm{d}x'}\delta(x'-x)\,,$$

hence

$$\langle x|\hat{\mathsf{p}}|\psi\rangle = \int_{-\infty}^{\infty} \mathrm{d}x'\,\langle x|\hat{\mathsf{p}}|x'\rangle\langle x'|\,\psi\rangle = -\mathrm{i}\frac{\mathrm{d}}{\mathrm{d}x}\psi(x)\,. \tag{1.83}$$

The commutator of $\hat{\mathsf{p}}$ and any function $f(\hat{\mathsf{x}})$ satisfies

$$\langle x|[\hat{\mathsf{p}},f(\hat{\mathsf{x}})]|\psi\rangle = \int_{-\infty}^{\infty} \mathrm{d}x' \, \langle x|[\hat{\mathsf{p}},f(\hat{\mathsf{x}})]|x'\rangle\psi(x') = \int_{-\infty}^{\infty} \mathrm{d}x' \, (f(x') - f(x))\langle x|\hat{\mathsf{p}}|x'\rangle\psi(x')$$

$$= -\mathrm{i}\frac{\mathrm{d}}{\mathrm{d}x}(f(x)\psi(x)) + \mathrm{i}f(x)\frac{\mathrm{d}\psi(x)}{\mathrm{d}x} = -\mathrm{i}\frac{\mathrm{d}f(x)}{\mathrm{d}x}\psi(x) = -\mathrm{i}\langle x|\frac{\mathrm{d}f(x)}{\mathrm{d}x}|\psi\rangle\,.$$

This must hold for any state $|\psi\rangle$ and any coordinate state $\langle x|$, therefore

$$[\hat{\mathsf{p}},f(\hat{\mathsf{x}})] = -\mathrm{i}\left.\frac{\mathrm{d}f(x)}{\mathrm{d}x}\right|_{x=\hat{\mathsf{x}}}, \tag{1.84}$$

or equivalently

$$\left[\frac{\mathrm{d}}{\mathrm{d}x},f(x)\right] = \frac{\mathrm{d}f(x)}{\mathrm{d}x}\,. \tag{1.85}$$

Generalization for three dimensions then leads to Eq. (1.55).

Henceforth, we will often employ the following common notation: we replace the $\psi(x)$ with merely ψ, and Eqs. (1.82) and (1.83) with

$$\hat{\mathsf{x}}\psi(x) = x\psi(x)$$

and

$$\hat{\mathsf{p}}\psi(x) = -\mathrm{i}\frac{\mathrm{d}}{\mathrm{d}x}\psi(x)\,.$$

In other words, we will not distinguish between the abstract and coordinate representation of vectors and operators.

In the case of the harmonic oscillator, the Schrödinger equation (1.58) acquires the form of a second-order differential equation in the coordinate representation

$$\left(-\frac{1}{2}\frac{\mathrm{d}^2}{\mathrm{d}x^2} + \frac{1}{2}x^2\right)\psi(x) = E\psi(x). \tag{1.86}$$

However, we do not have to directly solve this equation in order to determine the forms of the wave functions. We rather project Eq. (1.66) onto the coordinate basis. Introducing the notation $\langle x|n\rangle = \psi_n(x)$ and using these equations for $n = 0$, $n = 1$, etc. successively leads to

$$\langle x|\hat{\mathsf{a}}|0\rangle = \frac{1}{\sqrt{2}}\langle x|\hat{\mathsf{x}} + \mathrm{i}\hat{\mathsf{p}}|0\rangle = 0 \Rightarrow \frac{1}{\sqrt{2}}\left(x + \frac{\mathrm{d}}{\mathrm{d}x}\right)\psi_0(x) = 0$$

$$\Rightarrow \psi_0(x) = A\mathrm{e}^{-x^2/2}\,, \tag{1.87}$$

$$\langle x|\hat{\mathsf{a}}^+|0\rangle = \frac{1}{\sqrt{2}}\langle x|\hat{\mathsf{x}} - \mathrm{i}\hat{\mathsf{p}}|0\rangle = \langle x|\,1\rangle$$

$$\Rightarrow \psi_1(x) = \frac{1}{\sqrt{2}}\left(x - \frac{\mathrm{d}}{\mathrm{d}x}\right)\psi_0(x) = A\sqrt{2}x\mathrm{e}^{-x^2/2}\,, \tag{1.88}$$

$$\langle x|\hat{\mathsf{a}}^+|1\rangle = \frac{1}{\sqrt{2}}\langle x|\hat{\mathsf{x}} - \mathrm{i}\hat{\mathsf{p}}|1\rangle = \sqrt{2}\langle x|\,2\rangle$$

$$\Rightarrow \psi_2(x) = \frac{1}{2}\left(x - \frac{\mathrm{d}}{\mathrm{d}x}\right)\psi_1(x) = \frac{A}{\sqrt{2}}(2x^2 - 1)\mathrm{e}^{-x^2/2}\,, \tag{1.89}$$

etc. We determine the constant A from the normalization requirement

$$\langle 0|\,0\rangle = \int_{-\infty}^{\infty} \mathrm{d}x\,\langle 0|\,x\rangle\langle x|\,0\rangle = \int_{-\infty}^{\infty} \mathrm{d}x\,|\psi_0(x)|^2 =$$
$$= \int_{-\infty}^{\infty} \mathrm{d}x\,\mathrm{e}^{-x^2}A^2 = 1 \Rightarrow A = \pi^{-1/4}\,.$$

All the other functions then automatically comply with the correct normalization, as one can verify. The kind reader easily verifies that the functions $\psi_0(x)$, $\psi_1(x)$, etc. indeed obey the Schrödinger equation (1.86).

The set of the functions $\{\psi_n(x)\}$ is orthonormal, see Eq. (1.30),

$$\langle m|\,n\rangle = \int_{-\infty}^{\infty} \langle m|\,x\rangle\langle x|\,n\rangle\,\mathrm{d}x = \int_{-\infty}^{\infty} \psi_m(x)^*\psi_n(x)\,\mathrm{d}x = \delta_{mn}\,, \tag{1.90}$$

and complete, see Eq. (1.69),

$$\sum_{n=0}^{\infty}\langle x|\,n\rangle\langle n|\,x'\rangle = \sum_{n=0}^{\infty}\psi_n(x)\psi_n(x')^* = \langle x|\,x'\rangle = \delta(x - x')\,.$$

1.3.8 Momentum Representation

In many cases, such as we will see later in Sects. 7.3.2, 7.6, 7.7, and 7.8, it is advantageous to operate within the momentum representation. In this representation, the basis of the Hilbert space is given by the eigenvectors of the momentum operator,

$$\hat{\mathsf{p}}|p\rangle = p|p\rangle\,. \tag{1.91}$$

The completeness and orthonormality relations for this basis read

$$\int_{-\infty}^{\infty} |p\rangle\langle p|\,\mathrm{d}p = \mathbf{1}\,, \tag{1.92}$$

$$\langle p|p'\rangle = \delta(p-p')\,. \tag{1.93}$$

Starting from the canonical commutation relations, we have derived the action of the operators $\hat{\mathsf{x}}$ and $\hat{\mathsf{p}}$ on any state $|\psi\rangle$ in the coordinate representation, namely Eqs. (1.82) and (1.83). In an analogous way, we can easily find that these two operators $\hat{\mathsf{x}}$ and $\hat{\mathsf{p}}$ act on any state $|\psi\rangle$ in the momentum representation as:

$$\langle p|\hat{\mathsf{p}}|\psi\rangle = p\psi(p)\,, \qquad \langle p|\hat{\mathsf{x}}|\psi\rangle = \mathrm{i}\frac{\mathrm{d}}{\mathrm{d}p}\psi(p)\,,$$

where $\psi(p) = \langle p|\psi\rangle$.

At this point we could ask for the connection between the wave functions in each of the representations. Using the unity decomposition, Eqs. (1.78) and (1.92), we obtain

$$\langle x|\psi\rangle = \int_{-\infty}^{\infty} \langle x|p\rangle\langle p|\psi\rangle\,\mathrm{d}p\,, \qquad \langle p|\psi\rangle = \int_{-\infty}^{\infty} \langle p|x\rangle\langle x|\psi\rangle\,\mathrm{d}x\,. \tag{1.94}$$

We obviously need to find the scalar product of the eigenstates $\langle x|p\rangle$. Therefore, we project Eq. (1.91) onto the coordinate eigenstates

$$\langle x|\hat{\mathsf{p}}|p\rangle = p\langle x|p\rangle \Rightarrow -\mathrm{i}\frac{\mathrm{d}\langle x|p\rangle}{\mathrm{d}x} = p\langle x|p\rangle\,,$$

where we used Eq. (1.83). Inserting the unity decomposition (1.78) into the orthonormality relation (1.93) yields

$$\int_{-\infty}^{\infty} \langle p|x\rangle\langle x|p'\rangle\,\mathrm{d}x = \delta(p-p').$$

The solution to the last two equations then reads

$$\langle x|p\rangle = \frac{1}{\sqrt{2\pi}}\mathrm{e}^{\mathrm{i}px}\,, \tag{1.95}$$

where we used the relation for the Fourier transform of the δ-function, Eq. (1.74).

We now return again to the harmonic oscillator, this time to illustrate the method of calculation of the ground state wave function in the momentum representation. We can start from Eqs. (1.94) and (1.95) and readily find that

$$\langle p|0\rangle = \frac{1}{\sqrt{2\pi}}\int_{-\infty}^{\infty} \mathrm{e}^{-\mathrm{i}px}\langle x|0\rangle\,\mathrm{d}x = A\mathrm{e}^{-p^2/2}\,,$$

where we considered the formula for the integral from the Gaussian function

$$\int_{-\infty}^{\infty} \mathrm{e}^{-au^2-bu}\mathrm{d}u = \mathrm{e}^{b^2/(4a)}\sqrt{\frac{\pi}{a}}\,. \tag{1.96}$$

and where $\langle x|\,0\rangle$ is given by Eq. (1.87). Alternatively, we can obtain the ground state wave function directly by projecting Eq. (1.66) for $n = 0$ onto the momentum basis:

$$\langle p|\hat{\mathsf{a}}|0\rangle = \frac{1}{\sqrt{2}}\langle p|\hat{\mathsf{x}} + \mathrm{i}\hat{\mathsf{p}}|0\rangle = 0 \Rightarrow \left(\frac{\mathrm{d}}{\mathrm{d}p} + p\right)\psi_0(p) = 0 \Rightarrow \psi_0(p) = A\mathrm{e}^{-p^2/2}\,. \tag{1.97}$$

Notice that Eq. (1.95) describes a plane wave. According to quantum mechanics then, a particle with the momentum p is assigned wavelength $\lambda = 2\pi/p$, usually called the *de Broglie wavelength*. Generally speaking, we describe particles in terms of quantum mechanics when the de Broglie wavelength is of large values, i.e., the particle is light in mass and moves slowly, hence behaves like a wave. On the other hand, classical approach suffices—at least roughly—when the particle is heavy or moves quickly, i.e., the de Broglie wavelength acquires small values, and the particle behaves like a tiny ball.

Finally, notice also that it follows from Eqs. (1.94) and (1.95) that the wave function in the momentum representation is a Fourier transform of the wave function in the coordinate representation,

$$\psi(p) = \frac{1}{\sqrt{2\pi}}\int_{-\infty}^{\infty} \psi(x)\mathrm{e}^{-\mathrm{i}px}\mathrm{d}x\,, \tag{1.98}$$

which will come in hand in the next section. By multiplying the coordinate x by a constant a and employing the last equation, we obtain for the wave function in the momentum representation

$$\frac{1}{\sqrt{2\pi}}\int_{-\infty}^{\infty} a^{1/2}\psi(ax)\mathrm{e}^{-\mathrm{i}px}\mathrm{d}x = \frac{1}{\sqrt{2\pi a}}\int_{-\infty}^{\infty} \psi(x)\mathrm{e}^{-\mathrm{i}\frac{p}{a}x}\mathrm{d}x = \frac{1}{a^{1/2}}\psi\left(\frac{p}{a}\right), \tag{1.99}$$

where $\psi(p)$ is given by the previous Eq. (1.98). The factor $a^{1/2}$ ensures correct normalization of the wave function

$$\int_{-\infty}^{\infty} \mathrm{d}x|a^{1/2}\psi(ax)|^2 = \int_{-\infty}^{\infty} \mathrm{d}x|\psi(x)|^2 = 1\,.$$

1.3.9 Gaussian Packet and the Uncertainty Principle

There are several quantities which characterize the statistical distribution of measured values of an observable A: in particular, the expectation value $\langle\hat{\mathsf{A}}\rangle$ we have already encountered, see Eq. (1.46), and the square deviation

$$\left\langle (\Delta \hat{\mathsf{A}})^2 \right\rangle = \left\langle \left(\hat{\mathsf{A}} - \langle \hat{\mathsf{A}} \rangle \right)^2 \right\rangle = \left\langle \hat{\mathsf{A}}^2 - 2\langle \hat{\mathsf{A}} \rangle \hat{\mathsf{A}} + \langle \hat{\mathsf{A}} \rangle^2 \right\rangle = \langle \hat{\mathsf{A}}^2 \rangle - \langle \hat{\mathsf{A}} \rangle^2 ,$$

where in the last step we used the fact that the expectation value is a number, thus we can take it out from the scalar product. A question may arise: why to use square deviation instead of its linear analog $\langle \hat{\mathsf{A}} - \langle \hat{\mathsf{A}} \rangle \rangle$? The reason is, the latter always equals zero. Square deviation, on the other hand, indicates the "spread" of the expectation value, meaning to what extent the result of measuring the observable A will differ from the expectation value $\langle \hat{\mathsf{A}} \rangle$. In other words, the square deviation estimates the uncertainty of measuring the expectation value. We now consider the Gaussian packet

$$\psi(x) = \sqrt{\frac{a}{\sqrt{\pi}}} \mathrm{e}^{-(ax)^2/2} , \tag{1.100}$$

which differs from the ground-state wave function (1.87) merely in the substitution $x \to xa$. One can easily show that for this case, see Eqs. (1.46), (1.77), and (1.78),

$$\langle \hat{\mathsf{x}} \rangle = \langle \psi | \hat{\mathsf{x}} | \psi \rangle = \int_{-\infty}^{\infty} \langle \psi | x \rangle \langle x | \hat{\mathsf{x}} | \psi \rangle \, \mathrm{d}x = \int_{-\infty}^{\infty} x |\psi(x)|^2 \, \mathrm{d}x = 0 , \tag{1.101a}$$

$$\langle \hat{\mathsf{x}}^2 \rangle = \int_{-\infty}^{\infty} x^2 |\psi(x)|^2 \, \mathrm{d}x = \frac{1}{2a^2} . \tag{1.101b}$$

Using Eqs. (1.97), (1.99), and (1.100), one can also find that

$$\psi(p) = \sqrt{\frac{1}{a\sqrt{\pi}}} \mathrm{e}^{-(p/a)^2/2} .$$

It follows then from Eqs. (1.46), (1.91), and (1.92) that

$$\langle \hat{\mathsf{p}} \rangle = \int_{-\infty}^{\infty} p |\psi(p)|^2 \, \mathrm{d}p = 0 , \qquad \langle \hat{\mathsf{p}}^2 \rangle = \int_{-\infty}^{\infty} p^2 |\psi(p)|^2 \, \mathrm{d}p = \frac{a^2}{2} . \tag{1.102}$$

These results illustrate the famous[14] Heisenberg uncertainty principle. The uncertainty in the expectation value of the coordinate or momentum depends on the magnitude of a, but we can reduce each of them to any size. However, as we reduce one of them, the other necessarily grows, and the product of these two uncertainties remains constant

$$\left\langle (\Delta \hat{\mathsf{x}})^2 \right\rangle \left\langle (\Delta \hat{\mathsf{p}})^2 \right\rangle = \frac{1}{4} .$$

[14] In this case, the glory indeed stretches well beyond the boundaries of the physicists' world.

In fact, we observe such behavior only for the ideal case. In general we find that

$$\langle(\Delta\hat{\mathsf{x}})^2\rangle\langle(\Delta\hat{\mathsf{p}})^2\rangle \geq \frac{1}{4}\,.$$

Instead of a general proof which does not bring much of understanding anyway, we rather illustrate this fact on an example. Envisage a particle (for instance an electron) in a magnetic trap. In this trap, the particle is well described by the wavefunction (1.100). We suppose that we switch off the magnetic field at the time $t = 0$, i.e., the particle henceforth moves absolutely free. What is the time evolution of its expectation values? Since the particle is free and we may choose suitable units so that its mass is unity, the Hamiltonian reads $\hat{\mathsf{H}} = \frac{1}{2}\hat{\mathsf{p}}^2$ and obviously we have $[\hat{\mathsf{p}}, \hat{\mathsf{H}}] = 0$. It follows then from Eq. (1.56) that

$$\frac{\mathrm{d}\hat{\mathsf{p}}(t)}{\mathrm{d}t} = 0 \Rightarrow \hat{\mathsf{p}}(t) = \hat{\mathsf{p}}\,, \qquad [\hat{\mathsf{p}}(t)]^2 = \hat{\mathsf{p}}^2$$

and from Eq. (1.53) that

$$\frac{\mathrm{d}\hat{\mathsf{x}}(t)}{\mathrm{d}t} = \hat{\mathsf{p}}(t) \Rightarrow \hat{\mathsf{x}}(t) = \hat{\mathsf{p}}t + \hat{\mathsf{x}}\,, \qquad [\hat{\mathsf{x}}(t)]^2 = \hat{\mathsf{p}}^2t^2 + (\hat{\mathsf{p}}\hat{\mathsf{x}} + \hat{\mathsf{x}}\hat{\mathsf{p}})t + \hat{\mathsf{x}}^2\,.$$

We thus find for the time evolution of the expectation values

$$\langle\hat{\mathsf{p}}(t)\rangle = \langle\hat{\mathsf{p}}\rangle = 0\,, \quad \langle[\hat{\mathsf{p}}(t)]^2\rangle = \langle\hat{\mathsf{p}}^2\rangle = \frac{a^2}{2}\,,$$

$$\langle\hat{\mathsf{x}}(t)\rangle = \langle\hat{\mathsf{p}}t + \hat{\mathsf{x}}\rangle = 0\,, \quad \langle[\hat{\mathsf{x}}(t)]^2\rangle = \langle\hat{\mathsf{p}}^2t^2 + (\hat{\mathsf{p}}\hat{\mathsf{x}} + \hat{\mathsf{x}}\hat{\mathsf{p}})t + \hat{\mathsf{x}}^2\rangle = \frac{t^2a^2}{2} + \frac{1}{2a^2}\,,$$

where we used Eqs. (1.101) and (1.102). One can easily prove that we have $\langle\hat{\mathsf{x}}\hat{\mathsf{p}} + \hat{\mathsf{p}}\hat{\mathsf{x}}\rangle = 0$ for the state (1.100). Finally, we obtain for the product of the uncertainties of the coordinate and momentum

$$\langle(\Delta\hat{\mathsf{x}}(t))^2\rangle\langle(\Delta\hat{\mathsf{p}}(t))^2\rangle = \frac{1}{4}(1 + a^4t^2)\,.$$

We thus see there is only one sole condition for the product of the uncertainties to grow: the particle stays free and time elapses.

This uncertainty principle played a crucial role in the understanding of quantum mechanics. It allows us to comprehend why and how the classical physics must necessarily fail when it comes to the microscopic scales. Moreover, it allows us to calculate the failure.

Note though, that not everything is uncertain in quantum mechanics. For example, quantum mechanics predicts exact values for energies of stationary states. The probabilistic approach is a necessity only when studying transitions between these states. Nevertheless, even this uncertainty is often smaller than the uncertainty in the actual realization of experiments, such as the uncertainty in the initial conditions, and thus is not crucial.

1.4 Final Notes

Note that we did not discuss the "question of measuring" at all. This "problem," roughly speaking, comprises the question why we describe the measuring apparatus in terms of classical mechanics, while the measured system with quantum mechanics, and where the line between these two approaches lies. This question goes hand in hand with the fact that we do not observe quantum-mechanical interference in our everyday lives. Recently, physicists successfully performed remarkable experiments that illustrate the disruption of the quantum-mechanical interference resulting from an "uncontrollable" interaction between meso- and macroscopic systems with their surroundings. One may find a more detailed description and elucidation of these experiments in [2], a motivation in [15], and necessary theory in [3]. These experiments furthermore show that under special circumstances when we manage to control this "uncontrollable" interaction, we can actually observe the quantum-mechanical interference for mesoscopic systems. However, as our control of the interaction gradually weakens, the interference signal likewise fades away. As long as the studied system requires a quantum-mechanical description, no difficulties but one arise—the system cannot serve as a measuring system. This means, if we are to acquire experimental data, the measuring system must be of a "sufficient size" in order to describe it within the framework of classical mechanics. An interesting if very unorthodox view of the "question of measuring" may be found in [10]. Although the authors do not share this view, it is definitely worth learning about it.

References

1. D.I. Blokhintsev, *Quantum Mechanics* (Reidel, Dordrecht, 1964)
2. M. Brune et al., Phys. Rev. Lett. **77**, 4887 (1996); S. Haroche, Rev. Mod. Phys. **85**, 1083 (2013); S. Haroche, M. Brune, J.M. Raimond, Phys. Today **66**, 27 (2013); S. Haroche, J.M. Raimond, *Exploring the quantum, Atoms, Cavities, Photons* (Oxford University Press, Oxford, 2006)
3. C. Cohen-Tannoudji, J. Dupont-Roc, G. Grynberg, *Atom-Photon Inter actions: Basic Processes and Applications*, Wiley Science Paperback Series (Wiley, London, 1992)
4. R.P. Feynman, R.B. Leighton, M. Sands, *Feynman Lectures on Physics 1, 2* (Addison-Wesley, Reading, 1977)
5. C.J. Foot, *Atomic Physics*, Oxford Master Series in Physics (Oxford University Press, Oxford, 2007)
6. L.D. Landau, E.M. Lifshitz, *The Classical Theory of Fields* (Elsevier/Butterworth-Heinemann, Amsterdam/London, 1975)
7. L.D. Landau, E.M. Lifshitz, *Mechanics* (Elsevier/Butterworth-Heinemann, Amsterdam/London, 1976)

8. L.D. Landau, E.M. Lifshitz, *Quantum Mechanics, Non-relativistic Theory* (Elsevier/Butterworth-Heinemann, Amsterdam/London, 1976)
9. A. Messiah, *Quantum Mechanics* (North Holland, Amsterdam, 1961)
10. R. Penrose, *The Road to Reality: Complete Guide to the Laws of the Universe* (Vintage, New York, 2007)
11. J. Schwinger, *Quantum Mechanics: Symbolism of Atomic Measurement* (Springer, Berlin, 2003)
12. A. Zee, *Quantum Field Theory in a Nutshell*, 2nd edn. (Princeton University Press, Princeton, 2010)
13. A. Zee, *Einstein Gravity in a Nutshell* (Princeton University Press, Princeton, 2013)
14. A. Zeilinger et al., Rev. Mod. Phys. **60**, 1067 (1988)
15. W. Zurek, Phys. Today **44**, 36 (1991); W. Zurek, Phys. Today **67**, 44 (2014); M. Schlosshauer, *Decoherence and the Quantum-to-Classical Transition* (Springer, Berlin, 2007)

Chapter 2
Approximate Methods in Quantum Mechanics

However light and elegant the mathematical apparatus of quantum mechanics appears, we can solve exactly only very few physically interesting problems with it. Therefore, we need to opt for appropriate approximations when facing the remaining vast majority of quantum-mechanical problems. In this chapter, we will introduce two basic approaches—the *variational* and *perturbation* methods. Naturally, many others exist (for example the semi-classical approximations). However, those usually focus on a specific class of problems, while we can employ the variational and perturbative methods when facing almost any problem. We will illustrate both methods on the simplest problem one cannot solve analytically—the anharmonic oscillator.

We now return to the potential expansion (1.57) and consider more terms this time. We cannot simply add the next—third-order—term, though. Depending on the sign of the third derivative of the potential at the point where we carry out the expansion, the asymptotic behavior of the third-order polynomial would be either $V(x \to -\infty) \to -\infty$, or $V(x \to +\infty) \to -\infty$, in both cases without a global minimum. This would result in the absence of any bound states; that is, all particles, although initially contained within a local minimum of the potential, would eventually tunnel through the barrier and escape for good, see Fig. 2.1. For detailed discussion of tunneling see, e.g., [2]. It follows thereof the highest power of potential must be even. Therefore, we include all terms up to the fourth order in x,

$$V(\hat{\mathsf{x}}) = \frac{1}{2!} V^{(2)}(0)\, \hat{\mathsf{x}}^2 + \frac{1}{3!} V^{(3)}(0)\, \hat{\mathsf{x}}^3 + \frac{1}{4!} V^{(4)}(0)\, \hat{\mathsf{x}}^4 .$$

As we aim to merely illustrate the method, we further simplify our problem by setting $V^{(3)}(0) = 0$. The form of the resulting Schrödinger equation (after the removal of constants via a scale transform) reads

$$\left[\frac{\hat{\mathsf{p}}^2}{2} + \frac{\hat{\mathsf{x}}^2}{2} + \delta \hat{\mathsf{x}}^4 \right] \psi(x) = E \psi(x) . \tag{2.1}$$

J. Zamastil, J. Benda, *Quantum Mechanics and Electrodynamics*,
DOI 10.1007/978-3-319-65780-6_2

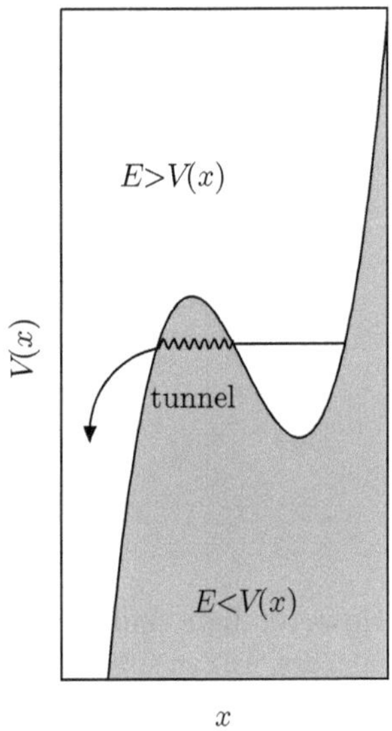

Fig. 2.1 A polynomial potential with the highest nonzero odd power does not support the bound states

2.1 Variational Method

2.1.1 The Ritz Variational Principle

The variational method exploits an interesting observation we will now derive. We have for an eigenstate of the Hamiltonian

$$\hat{\mathsf{H}}|\phi_n\rangle = E_n|\phi_n\rangle \Rightarrow E_n = \frac{\langle\phi_n|\hat{\mathsf{H}}|\phi_n\rangle}{\langle\phi_n|\phi_n\rangle}\,.$$

We take an arbitrary, "test," vector $|\psi\rangle$ and calculate

$$E_{\text{var}}[\psi] = \frac{\langle\psi|\hat{\mathsf{H}}|\psi\rangle}{\langle\psi|\psi\rangle}\,, \tag{2.2}$$

where we call E_{var} the *variational* energy. Obviously, it is not the energy of the state $|\psi\rangle$ since the energy of that state is not well defined as it is generally not an eigenstate of the Hamiltonian. Were we to interpret this number, we would call it the mean value of the energy after many repetitive measurements on the test state. We now subtract the ground state energy from the variational energy,

$$E_{\text{var}}[\psi] - E_0 = \frac{\langle\psi|\hat{\mathsf{H}}|\psi\rangle}{\langle\psi|\psi\rangle} - E_0 = \frac{\langle\psi|(\hat{\mathsf{H}} - E_0)|\psi\rangle}{\langle\psi|\psi\rangle}\,.$$

We use the spectral decomposition of the Hamiltonian and write E_0 as a unity operator expressed in the basis of the Hamiltonian eigenstates,

$$\begin{aligned} E_{\text{var}}[\psi] - E_0 &= \frac{1}{\langle\psi|\psi\rangle}\left(\sum_{n=0}^{\infty} E_n \langle\psi|\phi_n\rangle\langle\phi_n|\psi\rangle - E_0 \sum_{n=0}^{\infty} \langle\psi|\phi_n\rangle\langle\phi_n|\psi\rangle\right) \\ &= \frac{1}{\langle\psi|\psi\rangle}\sum_{n=0}^{\infty}(E_n - E_0)\,|\langle\psi|\phi_n\rangle|^2 \geq 0\,. \end{aligned}$$

The last equation represents the so-called *Ritz variational principle*. One can readily see—for both the energy difference to the right and the squares of scalar products are positive—that the difference between the variational energy and the energy of the ground state is positive as well. Thus, the variational energy always acquires larger values than the energy of the ground state! Owing to this fact, we are able to calculate the ground state of a Hamiltonian and its energy using the variational calculus, namely by searching for the extremum of the functional (2.2). For instance, we can parametrize the state $|\psi\rangle$ using a few variables and search for the extremum of the functional (2.2) as a function of several variables.

2.1.2 *Optimization of Nonlinear Parameters*

We now demonstrate the above-presented approach on the case of the anharmonic oscillator (2.1). One would expect that for small δ, it behaves very much like the harmonic oscillator. For this reason we consider the test function of the ground state in the form

$$\psi(x;\alpha) = \mathrm{e}^{-\alpha x^2/2}\,,$$

where α represents a real parameter. The variational energy is then exactly as in Eq. (2.2)[1]

$$E_{\text{var}} = \frac{\int_{-\infty}^{\infty}\mathrm{e}^{-\alpha x^2/2}\left(-\frac{1}{2}\frac{\mathrm{d}^2}{\mathrm{d}x^2} + \frac{x^2}{2} + \delta x^4\right)\mathrm{e}^{-\alpha x^2/2}\mathrm{d}x}{\int_{-\infty}^{\infty}\mathrm{e}^{-\alpha x^2}\mathrm{d}x} = \frac{1}{4\alpha} + \frac{\alpha}{4} + \frac{3}{4}\delta\frac{1}{\alpha^2}\,. \tag{2.3}$$

We now try to find the minimal energy such a guessed state allows for. The required condition for a minimum of the function states the first derivative be equal to zero:

[1]Here we used Eq. (1.96) for $b = 0$. One can obtain the remaining integrals by differentiating the integrals with respect to a parameter.

$$0 = \frac{\partial E_{\text{var}}}{\partial \alpha} = -\frac{1}{4\alpha^2} + \frac{1}{4} - \frac{3}{2}\frac{\delta}{\alpha^3}\,.$$

After reordering it to $\alpha^3 - \alpha - 6\delta = 0$, we readily see that there is a single real root to this equation. We then obtain approximate energy of the ground state by its substitution into Eq. (2.3).

2.1.3 Optimization of Linear Parameters

The previous section made use of the so-called *nonlinear parameters*. However, another possibility exists: *linear* parametrization. In this case, we express the sought ground state as a finite sum of reference states, for example of the stationary states of the harmonic oscillator,

$$|\psi\rangle = \sum_{j=1}^{N} c_j |j\rangle\,. \tag{2.4}$$

We can certainly use any other basis as well. The only restriction is the basis set $\{|j\rangle\}$ be complete in the limit $N \to \infty$, so that we could express any vector $|\psi\rangle$ as a linear combination of the reference states. After substitution into Eq. (2.2) we arrive at

$$E_{\text{var}} = \frac{\sum_{i,j=1}^{N} c_i c_j \langle i|\hat{\mathsf{H}}|j\rangle}{\sum_{i,j=1}^{N} c_i c_j \langle i|j\rangle} = \frac{\sum_{i,j=1}^{N} c_i c_j (\hat{\mathsf{H}})_{ij}}{\sum_{i,j=1}^{N} c_i c_j S_{ij}}\,. \tag{2.5}$$

The numbers S_{ij} are called the *overlap matrix* elements. As before, we differentiate Eq. (2.5) with respect to all parameters c_k and set these derivatives equal to zero, that is $\frac{\partial E_{\text{var}}}{\partial c_k} = 0$; we thus have

$$0 = \frac{\left(\sum\limits_{i,j=1}^{N}\left(\frac{\partial c_i}{\partial c_k}c_j + c_i\frac{\partial c_j}{\partial c_k}\right)(\hat{\mathsf{H}})_{ij}\right)}{\left(\sum\limits_{i,j=1}^{N} c_i c_j S_{ij}\right)} - \frac{\left(\sum\limits_{i,j=1}^{N} c_i c_j (\hat{\mathsf{H}})_{ij}\right)\left(\sum\limits_{i,j=1}^{N}\left(\frac{\partial c_i}{\partial c_k}c_j + c_i\frac{\partial c_j}{\partial c_k}\right)S_{ij}\right)}{\left(\sum\limits_{i,j=1}^{N} c_i c_j S_{ij}\right)^2}\,.$$

It follows from the independence of the parameters c_k that $\frac{\partial c_i}{\partial c_k} = \delta_{ik}$. Therefore the following equation must hold, bearing in mind Eq. (2.5),

$$\sum_{i,j=1}^{N}\left(\delta_{ik}c_j + c_i\delta_{jk}\right)(\hat{\mathsf{H}})_{ij} = E_{\text{var}}\sum_{i,j=1}^{N}\left(\delta_{ik}c_j + c_i\delta_{jk}\right)S_{ij}\,.$$

If we use real basis, both the Hamiltonian and overlap matrices are symmetric (for they are Hermitian and real); hence

$$\sum_{j=1}^{N} c_j(\hat{\mathsf{H}})_{ij} = E_{\text{var}} \sum_{j=1}^{N} c_j S_{ij}\,,$$

or in matrix notation

$$\mathsf{Hc} = E_{\text{var}}\mathsf{Sc}\,. \tag{2.6}$$

The optimization of the linear parameters thus results in searching for a solution to a *generalized eigenproblem* (2.6).

Once again, we use the anharmonic oscillator (2.1) to illustrate this method. We choose the eigenstates of the harmonic oscillator as basis states, the overlap matrix elements S_{ij} then being equal to δ_{ij}. We thus need to evaluate only the matrix elements of the Hamiltonian matrix, $(\hat{\mathsf{H}})_{ij} = \langle i|\hat{\mathsf{H}}|j\rangle$. After substituting for the Hamiltonian, we obtain

$$\langle i|\hat{\mathsf{H}}|j\rangle = \langle i|\left(\frac{\hat{\mathsf{p}}^2}{2} + \frac{\hat{\mathsf{x}}^2}{2} + \delta\hat{\mathsf{x}}^4\right)|j\rangle = \langle i|(E_j + \delta\hat{\mathsf{x}}^4)|j\rangle = \delta_{ij}\left(j + \frac{1}{2}\right) + \delta\langle i|\hat{\mathsf{x}}^4|j\rangle\,. \tag{2.7}$$

There are two possible ways of finding the numbers $\langle i|\hat{\mathsf{x}}^4|j\rangle$. The first one, straightforward though arduous and tedious, leads via the coordinate representation. For instance, the first element is

$$\langle 0|\hat{\mathsf{x}}^4|0\rangle = \int_{-\infty}^{\infty} \langle 0|x\rangle x^4 \langle x|0\rangle\,\mathrm{d}x = \int_{-\infty}^{\infty} \frac{1}{\sqrt[4]{\pi}}\mathrm{e}^{-x^2/2} x^4 \frac{1}{\sqrt[4]{\pi}}\mathrm{e}^{-x^2/2}\,\mathrm{d}x = \frac{3}{4}\,,$$

where we used Eq. (1.87).

The other one is based on expressing $\hat{\mathsf{x}}^4$ in terms of the ladder operators $\hat{\mathsf{a}}$ and $\hat{\mathsf{a}}^+$ which "act nicely" on the eigenstates of the harmonic oscillator. We can thus find the action of the Hamiltonian very easily. From Eq. (1.70) we have

$$\begin{aligned}\hat{\mathsf{x}}^4 = \left(\frac{1}{\sqrt{2}}(\hat{\mathsf{a}} + \hat{\mathsf{a}}^+)\right)^4 &= \frac{1}{4}\left[\hat{\mathsf{a}}^4 + (\hat{\mathsf{a}}^2\hat{\mathsf{a}}^+\hat{\mathsf{a}} + \hat{\mathsf{a}}^+\hat{\mathsf{a}}^3 + \hat{\mathsf{a}}^3\hat{\mathsf{a}}^+ + \hat{\mathsf{a}}\hat{\mathsf{a}}^+\hat{\mathsf{a}}^2)\right.\\ &+ (\hat{\mathsf{a}}^2(\hat{\mathsf{a}}^+)^2 + (\hat{\mathsf{a}}^+)^2\hat{\mathsf{a}}^2 + \hat{\mathsf{a}}^+\hat{\mathsf{a}}\hat{\mathsf{a}}^+\hat{\mathsf{a}} + \hat{\mathsf{a}}\hat{\mathsf{a}}^+\hat{\mathsf{a}}\hat{\mathsf{a}}^+ + \hat{\mathsf{a}}^+\hat{\mathsf{a}}\hat{\mathsf{a}}\hat{\mathsf{a}}^+ + \hat{\mathsf{a}}\hat{\mathsf{a}}^+\hat{\mathsf{a}}^+\hat{\mathsf{a}})\\ &+ \left.((\hat{\mathsf{a}}^+)^3\hat{\mathsf{a}} + \hat{\mathsf{a}}^+\hat{\mathsf{a}}(\hat{\mathsf{a}}^+)^2 + (\hat{\mathsf{a}}^+)^2\hat{\mathsf{a}}\hat{\mathsf{a}}^+ + \hat{\mathsf{a}}(\hat{\mathsf{a}}^+)^3) + (\hat{\mathsf{a}}^+)^4\right]\,.\end{aligned}$$

Recalling that the operators $\hat{\mathsf{a}}$ and $\hat{\mathsf{a}}^+$ act according to Eqs. (1.66) and (1.68), $\hat{\mathsf{a}}|j\rangle = \sqrt{j}|j-1\rangle$, $\hat{\mathsf{a}}^+|j\rangle = \sqrt{j+1}|j+1\rangle$, and that for example

$$\langle i|\hat{\mathsf{a}}^4|j\rangle = \sqrt{j(j-1)(j-2)(j-3)}\langle i|j-4\rangle = \sqrt{j(j-1)(j-2)(j-3)}\,\delta_{i,j-4}\,,$$

where in the second equality we used the orthogonality equations for the eigenstates, Eq. (1.30), we see that the terms in the expansion of $\hat{x}^4$ can shift the state $|j\rangle$ only to $|j \pm 4\rangle$ (if only one of the two operators $\hat{a}$ and $\hat{a}^+$ is present), to $|j \pm 2\rangle$ (one of the operators occurs three times, the other only once) and to $|j\rangle$ (both operators occur twice). To further simplify our task, we substitute $i \to i-2$ and $i \to i-4$ in the calculated elements $\langle i|\hat{x}^4|i+2\rangle$ and $\langle i|\hat{x}^4|i+4\rangle$, respectively, and obtain the results also for $\langle i-2|\hat{x}^4|i\rangle$ and $\langle i-4|\hat{x}^4|i\rangle$. Since the operator $\hat{x}^4$ is Hermitian, we have $\langle i|\hat{x}^4|j\rangle = \langle j|\hat{x}^4|i\rangle^* = \langle j|\hat{x}^4|i\rangle$ where the last equality holds because eigenstates of the harmonic oscillator form a real basis, see, e.g., Eqs. (1.87)–(1.89). For this reason, we have $\langle i-2|\hat{x}^4|i\rangle = \langle i|\hat{x}^4|i-2\rangle$ and $\langle i-4|\hat{x}^4|i\rangle = \langle i|\hat{x}^4|i-4\rangle$. Thus we see that instead of a tiresome calculation of numerous integrals it suffices to calculate three numbers algebraically,

$$\begin{aligned}
\langle i|\hat{x}^4|i+4\rangle &= \frac{1}{4}\langle i|\hat{a}^4|i+4\rangle = \frac{1}{4}\sqrt{(i+1)(i+2)(i+3)(i+4)}\,,\\
\langle i|\hat{x}^4|i+2\rangle &= \frac{1}{4}\langle i|\left(\hat{a}^2\hat{a}^+\hat{a} + \hat{a}^+\hat{a}^3 + \hat{a}^3\hat{a}^+ + \hat{a}\hat{a}^+\hat{a}^2\right)|i+2\rangle\\
&= \frac{1}{4}\Big(\sqrt{(i+1)(i+2)^3} + \sqrt{i^2(i+1)(i+2)}\\
&\quad + \sqrt{(i+1)(i+2)(i+3)^2} + \sqrt{(i+1)^3(i+2)}\Big)\\
&= \frac{1}{4}(4i+6)\sqrt{(i+2)(i+1)}\,,\\
\langle i|\hat{x}^4|i\rangle &= \frac{1}{4}\langle i|\left(\hat{a}\hat{a}\hat{a}^+\hat{a}^+ + \hat{a}^+\hat{a}^+\hat{a}\hat{a} + \hat{a}^+\hat{a}\hat{a}^+\hat{a}\right.\\
&\quad \left.+ \hat{a}\hat{a}^+\hat{a}\hat{a}^+ + \hat{a}^+\hat{a}\hat{a}\hat{a}^+ + \hat{a}\hat{a}^+\hat{a}^+\hat{a}\right)|i\rangle\\
&= \frac{1}{4}\Big(\sqrt{(i+1)(i+2)(i+2)(i+1)} + \sqrt{i(i-1)^2 i} + \sqrt{i^4}\\
&\quad + \sqrt{(i+1)^4} + \sqrt{i^2(i+1)^2} + \sqrt{(i+1)^2 i^2}\Big)\\
&= \frac{1}{4}(6i^2 + 6i + 3)\,.
\end{aligned}$$

It follows from our previous considerations that

$$\begin{aligned}
\langle i|\hat{x}^4|i-2\rangle &= \frac{1}{4}(4i-2)\sqrt{i(i-1)}\,,\\
\langle i|\hat{x}^4|i-4\rangle &= \frac{1}{4}\sqrt{i(i-1)(i-2)(i-3)}\,.
\end{aligned}$$

Owing to the symmetry mentioned earlier, we have thus all needed matrix elements, and nothing hinders us now from substituting them into Eq. (2.7) and from diagonalizing the Hamiltonian matrix.

It is worthwhile, though, to inspect its structure. It follows from Eq. (2.7) and from the above calculated results that the Hamiltonian does not mix odd and even states. That is, $\langle i|\hat{\mathsf{H}}|j\rangle$ equals zero unless the states $|i\rangle$ and $|j\rangle$ are both even or both odd. This is a deeper result associated with the notion of parity of a state,[2] and is more obvious if we use the coordinate representation. We readily see that the Hamiltonian $\hat{\mathsf{H}}$ in Eq. (2.1) is invariant with respect to interchange $\hat{\mathsf{x}} \to -\hat{\mathsf{x}}$. Therefore, if $\psi_e(x)$ is an even function and $\psi_o(x)$ an odd one, the matrix element of $\hat{\mathsf{H}}$ between these two states is zero,

$$\langle\psi_e|\hat{\mathsf{H}}|\psi_o\rangle = \int_{-\infty}^{\infty} \psi_e^*(x)\hat{\mathsf{H}}\psi_o(x) = 0\,,$$

for an integral of an odd function over an interval that is symmetrical around the origin always equals zero. We thus see that the class of odd states and that of even states are completely separated worlds for our system. We can exploit this finding when diagonalizing the Hamiltonian and reorder the matrix into a block-diagonal form by renumbering the basis

$$\hat{\mathsf{H}} = \begin{array}{c} \\ 0 \\ 2 \\ 4 \\ \cdot \\ 1 \\ 3 \\ 5 \\ \cdot \end{array} \begin{array}{c} \begin{array}{cccccccc} 0 & 2 & 4 & \cdot & 1 & 3 & 5 & \cdot \end{array} \\ \left(\begin{array}{cccccccc} * & * & * & * & & & & \\ * & * & * & * & & & & \\ * & * & * & * & & & & \\ * & * & * & * & & & & \\ & & & & * & * & * & * \\ & & & & * & * & * & * \\ & & & & * & * & * & * \\ & & & & * & * & * & * \end{array} \right) \end{array} .$$

This approach is often called the *adaptation of the basis to the symmetry* of the system. The asterisks mark generally nonzero entries while the empty positions are zero. For example, if we are interested in the ground state energy, we can remove all rows and columns corresponding to odd states as the ground state is an even state.

Table 2.1 lists the results we would obtain considering N even states and setting $\delta = 0.1$, as the reader can (and should) easily verify.

Table 2.1 Variational energy of the anharmonic oscillator

N	1	2	3	4	5
$2E_{\mathrm{var}}(N)$	1.15	1.1191	1.1188	1.1183	1.118293

[2]We will use it again later on, in more complex situations, see Sect. 5.3.8.

Table 2.2 Comparison of results of the nonlinear and linear method

2δ	$2E_{\text{var}}^{\text{nnolin}}(\alpha)$	$2E_{\text{var}}^{\text{lin}}(N=9)$	$2E_0$
10^{-2}	1.0073908	1.0073737	1.0073737
0.2	1.1206	1.11829	1.11829
1	1.4033	1.392355	1.392351
100	5.10	5.28	5.00
2000	13.66	44.79	13.39
40,000	37.0	830	36.3

One can also notice from this table how we usually, though non-rigorously, estimate the error of the variational method. We consider the numbers that do not change with an increasing basis set—are *stabilized*—as final and accurate.

The next table, Table 2.2, summarizes and compares the results obtained employing both variational (linear and nonlinear) methods. One can readily see from the table that the linear method is more accurate for smaller deviations from harmonicity. The accuracy of the nonlinear method in the case $\delta \approx 10{,}000$ is surprising as such a system has little in common with the harmonic oscillator. The explanation lies in the following. If we opt for the linear method, we have N fixed functions to compose the best fit to the correct function. In case of the nonlinear method, on the other hand, we are able to obtain the parameter α directly as a function of the parameter δ. This means that at least for small $|x|$ we are able to adjust the wave function of the harmonic oscillator to the real ground state wave function. The difference between the harmonic oscillator function and the exact ground state function for *large* $|x|$ does not cause much error. The reason is, both functions decrease rapidly for large $|x|$ and the area in turn contributes substantially less to the integral (2.2) than the area of small $|x|$. In practice, one usually uses a combination of the both, linear and nonlinear, methods, see Sect. 5.3.

Finally, we remark without a proof that also the energy estimates for higher excited states calculated by the diagonalization of the Hamiltonian in a finite basis lie above the exact values of these energies, as in the case of the ground state.

2.2 Perturbation Method

2.2.1 *Isolated Levels*

We consider, once again, the Hamiltonian of the anharmonic oscillator,

$$\hat{\mathsf{H}} = \frac{1}{2}\left(\hat{\mathsf{p}}^2 + \hat{\mathsf{x}}^2\right) + \delta\hat{\mathsf{x}}^4 = \hat{\mathsf{H}}_0 + \delta\hat{\mathsf{H}}_1\,, \tag{2.8}$$

this time as a sum of the Hamiltonian of the harmonic oscillator (eigenfunctions and eigenvalues of which we know) and a small perturbation of the order δ. We write the sought eigenstate of the Hamiltonian (2.8) as a sum of the harmonic oscillator eigenstate and a sequence of perturbations of increasing order in δ, and we perform a similar expansion for the eigenvalue:

$$|\psi\rangle = |\psi^{(0)}\rangle + \delta|\psi^{(1)}\rangle + \delta^2|\psi^{(2)}\rangle + \dots$$
$$E = E^{(0)} + \delta E^{(1)} + \delta^2 E^{(2)} + \dots \tag{2.9}$$

We substitute both these expansions into the Schrödinger equation $\hat{\mathsf{H}}|\psi\rangle = E|\psi\rangle$ and collect terms with the same power of δ:

$$(\hat{\mathsf{H}}_0 - E^{(0)})|\psi^{(0)}\rangle = 0\,, \tag{2.10}$$
$$(\hat{\mathsf{H}}_0 - E^{(0)})|\psi^{(1)}\rangle = -(\hat{\mathsf{H}}_1 - E^{(1)})|\psi^{(0)}\rangle\,, \tag{2.11}$$
$$(\hat{\mathsf{H}}_0 - E^{(0)})|\psi^{(2)}\rangle = -(\hat{\mathsf{H}}_1 - E^{(1)})|\psi^{(1)}\rangle + E^{(2)}|\psi^{(0)}\rangle\,, \tag{2.12}$$

and similarly for higher orders.

In the following text, one needs to carefully distinguish between the unrelated upper and lower indices: the former stands for the order of the perturbation, while the latter for the excitation level of the state under consideration. The equation for the eigenstates and eigenvalues of $\hat{\mathsf{H}}_0$ reads:

$$(\hat{\mathsf{H}}_0 - E_n^{(0)})|n\rangle = 0\,, \qquad n = 0, \dots, \infty\,. \tag{2.13}$$

By comparison with Eq. (2.10) we obtain $E^{(0)} = E_N^{(0)}$ and $|\psi^{(0)}\rangle = |N\rangle$, where $|N\rangle$ is the so-called reference state— the N-th excited state of the unperturbed system. Our task now is to find the effect of the perturbation on this particular reference state.

Equations (2.10)–(2.12) do not determine the perturbation functions $|\psi^{(j)}\rangle$ unambiguously. One can easily verify that they are given but for an addition of an arbitrary multiple of the reference state: $|\psi^{(j)}\rangle \to |\psi^{(j)}\rangle + \gamma|\psi^{(0)}\rangle$. This freedom is related to the normalization of the exact solution $|\psi\rangle$. A common choice of the normalization, called the *intermediate normalization*, is

$$\langle\psi^{(0)}|\psi^{(j)}\rangle = \delta_{j0} \quad \Rightarrow \quad \langle\psi^{(0)}|\psi\rangle = 1\,. \tag{2.14}$$

The difficulty while solving the eigenproblem stems from the fact that we need to simultaneously determine the eigenvalues and eigenvectors of the pertinent operator. The advantage of the perturbation method lies in that we are able to “disconnect” this problem. When we multiply Eqs. (2.11) and (2.12) by a bra-vector $\langle\psi^{(0)}|$, we obtain

$$E^{(1)} = \langle\psi^{(0)}|\hat{\mathsf{H}}_1|\psi^{(0)}\rangle\,, \tag{2.15}$$
$$E^{(2)} = \langle\psi^{(0)}|(\hat{\mathsf{H}}_1 - E^{(1)})|\psi^{(1)}\rangle = \langle\psi^{(0)}|\hat{\mathsf{H}}_1|\psi^{(1)}\rangle\,.$$

This shows that we can always find the next energy perturbation with the mere knowledge of the previous perturbed wave function. We can express the first energy perturbation, $E^{(1)}$, using the known reference state $\psi^{(0)}$. If we then solve Eq. (2.11) for $|\psi^{(1)}\rangle$ we are able to calculate $E^{(2)}$, etc.

However, how does one solve Eq. (2.11)? For we generally know nothing about the action of $\hat{\mathsf{H}}_0$ on the state $|\psi^{(1)}\rangle$ we expand the state $|\psi^{(1)}\rangle$ on the basis of the eigenstates of the operator $\hat{\mathsf{H}}_0$,

$$|\psi^{(1)}\rangle = \sum_{n \neq N} c_n^{(1)} |n\rangle .$$

Substituting from Eq. (2.13) and multiplying Eq. (2.11) by a bra-vector $\langle m|$ yield

$$\begin{aligned}
\langle m| \sum_{n \neq N} c_n^{(1)} (E_n^{(0)} - E_N^{(0)}) |n\rangle &= \langle m| (E^{(1)} - \hat{\mathsf{H}}_1) |N\rangle \\
\Rightarrow c_m^{(1)} (E_m^{(0)} - E_N^{(0)}) &= E^{(1)} \delta_{mN} - \langle m|\hat{\mathsf{H}}_1|N\rangle \\
\Rightarrow c_m^{(1)} &= \frac{-1}{E_m^{(0)} - E_N^{(0)}} \langle m|\hat{\mathsf{H}}_1|N\rangle , \quad m \neq N , \quad c_N^{(1)} = 0 ,
\end{aligned}$$

where the last implication follows from Eq. (2.14). From the above expressions for $E^{(2)}$, $|\psi^{(1)}\rangle$ and $c_n^{(1)}$ we finally obtain

$$E^{(2)} = - \sum_{n \neq N} \frac{\left| \langle N|\hat{\mathsf{H}}_1|n\rangle \right|^2}{E_n^{(0)} - E_N^{(0)}} . \tag{2.16}$$

The estimate of the energy up to the second order of perturbation is then

$$E \simeq E^{(0)} + \delta E^{(1)} + \delta^2 E^{(2)} .$$

Exercise 3: Perturbation Method

Using the perturbation method of the second order, estimate the ground state energy of the system with the Hamiltonian $\hat{\mathsf{H}} = \frac{\hat{\mathsf{p}}^2}{2} + \frac{\hat{\mathsf{x}}^2}{2} + \delta \hat{\mathsf{x}}^4$ *for* $2\delta = 10^{-2}$. *Compare the result with the variational estimate.*

2.2.2 Degenerate Levels

In none of the one-dimensional systems examined up to now did we encounter *degeneracy*, namely an existence of several (at least two) distinct levels with the same energy. In fact, there is a rule stating that degeneracy occurs only in more than one dimension. We now apply the perturbation method to a system that does contain degeneracy, namely to two weakly coupled harmonic oscillators. The Hamiltonian of such a system is of the form

$$\hat{\mathsf{H}} = \frac{1}{2}(\hat{\mathsf{p}}_x^2 + \hat{\mathsf{x}}^2) + \frac{1}{2}(\hat{\mathsf{p}}_y^2 + \hat{\mathsf{y}}^2) + \delta\hat{\mathsf{x}}^2\hat{\mathsf{y}}^2 = \hat{\mathsf{H}}_0^{[x]} + \hat{\mathsf{H}}_0^{[y]} + \delta\hat{\mathsf{H}}_1 = \hat{\mathsf{H}}_0 + \delta\hat{\mathsf{H}}_1 \,.$$

The constant δ governs the magnitude of the interaction between the two oscillators. Note that we know the eigenstates of the unperturbed problem given by Eq. (2.13); we can arrive at them by a direct product of the eigenstates of the x-dependent and y-dependent parts of the full Hamiltonian, denoted as $\hat{\mathsf{H}}_0^{[x]}$ and $\hat{\mathsf{H}}_0^{[y]}$, respectively:

$$|ij\rangle = |i(x)\rangle|j(y)\rangle \,.$$

The corresponding eigenvalues are then

$$\hat{\mathsf{H}}_0|ij\rangle = \left(\hat{\mathsf{H}}_0^{[x]} + \hat{\mathsf{H}}_0^{[y]}\right)|i(x)\rangle|j(y)\rangle = \left(i + \frac{1}{2} + j + \frac{1}{2}\right)|i(x)\rangle|j(y)\rangle = E_{ij}^0|ij\rangle$$
$$\Rightarrow E_{ij}^0 = i + j + 1 \,.$$

By the notation $\hat{\mathsf{H}}_0^{[x]} + \hat{\mathsf{H}}_0^{[y]}$ we mean $\hat{\mathsf{H}}_{\mathrm{osc}} \otimes \mathbf{1} + \mathbf{1} \otimes \hat{\mathsf{H}}_{\mathrm{osc}}$, where $\hat{\mathsf{H}}_{\mathrm{osc}}$ is the Hamiltonian of the one-dimensional linear harmonic oscillator. The action of such an operator on a state $|i(x)\rangle|j(y)\rangle$ can be then written as $(\hat{\mathsf{H}}_{\mathrm{osc}} \otimes \mathbf{1} + \mathbf{1} \otimes \hat{\mathsf{H}}_{\mathrm{osc}})|i(x)\rangle|j(y)\rangle = (\hat{\mathsf{H}}_{\mathrm{osc}}|i(x)\rangle)(\mathbf{1}|j(y)\rangle) + (\mathbf{1}|i(x)\rangle)(\hat{\mathsf{H}}_{\mathrm{osc}}|j(y)\rangle)$. As long as we keep δ equal to zero, the energy levels of the system are *degenerate*. It means that the states $|10\rangle$ and $|01\rangle$ have the same energy $E^{(0)} = 2$, the states $|20\rangle$, $|11\rangle$ and $|02\rangle$ have the same energy $E^{(0)} = 3$, and so on.

The *exact* Hamiltonian $\hat{\mathsf{H}}$ is invariant with respect to the interchanges $x \to -x$ and $y \to -y$. This implies that there are four classes of states, i and j even, i and j odd, i odd and j even, and i even and j odd, which are not mixed across by the exact Hamiltonian. This means that in the perturbative as well as in the variational calculation, we can restrict ourselves only to a single class of these states. For clarity, Fig. 2.2 displays the lowest levels combined from even states of the oscillators only.

In the following we focus on searching how the energy of the states $|02\rangle = |0\rangle|2\rangle$ and $|20\rangle = |2\rangle|0\rangle$ changes when perturbation is taken into account.

The application of the perturbation method is slightly more difficult for the degenerate levels than for the isolated ones. The problem is we do not a priori know

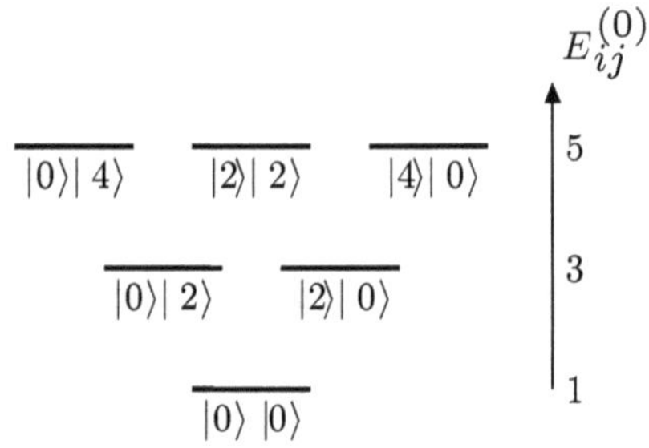

Fig. 2.2 Energy of the degenerate levels

which unperturbed wave function to choose at the very beginning. For instance, for $E^{(0)} = 3$ every linear combination of the form

$$|\psi^{(0)}\rangle = c_1|02\rangle + c_2|20\rangle\,, \tag{2.17}$$

where $c_1 = \langle 02|\psi^{(0)}\rangle$ and $c_2 = \langle 20|\psi^{(0)}\rangle$ obeys Eq. (2.10).

The coefficients c_1 and c_2, together with the first correction to energy $E^{(1)}$ are determined as follows. We multiply Eq. (2.11), $(\hat{\mathsf{H}}_0 - E^{(0)})|\psi^{(1)}\rangle = -(\hat{\mathsf{H}}_1 - E^{(1)})|\psi^{(0)}\rangle$, from left first by $\langle 02|$ and then by $\langle 20|$. We obtain

$$\langle 02|\hat{\mathsf{H}}_1|\psi^{(0)}\rangle = E^{(1)}\langle 02|\psi^{(0)}\rangle\,,$$
$$\langle 20|\hat{\mathsf{H}}_1|\psi^{(0)}\rangle = E^{(1)}\langle 20|\psi^{(0)}\rangle\,.$$

By substitution of the expansion (2.17) into the last two equations, and using the orthonormality of the states $|02\rangle$ and $|20\rangle$,[3] we find

$$c_1E^{(1)} = c_1(\hat{\mathsf{H}}_1)_{11} + c_2(\hat{\mathsf{H}}_1)_{12} \quad \text{and} \quad c_2E^{(1)} = c_1(\hat{\mathsf{H}}_1)_{21} + c_2(\hat{\mathsf{H}}_1)_{22}\,.$$

This can be written in a matrix notation

$$E^{(1)}\begin{pmatrix} c_1 \\ c_2 \end{pmatrix} = \begin{pmatrix} (\hat{\mathsf{H}}_1)_{11} & (\hat{\mathsf{H}}_1)_{12} \\ (\hat{\mathsf{H}}_1)_{21} & (\hat{\mathsf{H}}_1)_{22} \end{pmatrix}\begin{pmatrix} c_1 \\ c_2 \end{pmatrix}\,.$$

One can easily recognize that the rank of the matrix always equals the degree of the degeneracy of the inspected state. We now need to calculate the matrix elements of the operator $\hat{\mathsf{H}}_1$[4]:

---

[3] $\langle 02|\,20\rangle = \langle 0(x)|\,2(x)\rangle\langle 2(y)|\,0(y)\rangle = 0$, see Eqs. (1.30) and (1.90).

[4] The last step in this equation is best understandable in the coordinate representation:

$$\langle 0(x)|\langle 2(y)|\hat{\mathsf{x}}^2\hat{\mathsf{y}}^2|2(y)\rangle|0(x)\rangle = \int \mathrm{d}x \int \mathrm{d}y\, \psi_0(x)\psi_2(y)x^2y^2\psi_2(y)\psi_0(x)$$
$$= \int \mathrm{d}x\psi_0(x)x^2\psi_0(x) \int \mathrm{d}y\psi_2(y)y^2\psi_2(y) = \langle 0|\hat{\mathsf{x}}^2|0\rangle\langle 2|\hat{\mathsf{x}}^2|2\rangle\,.$$

$$
\begin{aligned}
(\hat{\mathsf{H}}_1)_{11} &= \langle 02|\hat{\mathsf{H}}_1|02\rangle = \langle 0(x)|\langle 2(y)|\hat{\mathsf{x}}^2\hat{\mathsf{y}}^2|2(y)\rangle|0(x)\rangle \\
&= \langle 0(x)|\hat{\mathsf{x}}^2|0(x)\rangle\langle 2(y)|\hat{\mathsf{y}}^2|2(y)\rangle = \langle 0|\hat{\mathsf{x}}^2|0\rangle\langle 2|\hat{\mathsf{x}}^2|2\rangle\,.
\end{aligned}
$$

We can rewrite these expressions, once again, in terms of the operators $\hat{\mathsf{a}}$ and $\hat{\mathsf{a}}^+$,

$$
\begin{aligned}
\langle 0|\hat{\mathsf{x}}^2|0\rangle &= \frac{1}{2}\langle 0|\left(\hat{\mathsf{a}}\hat{\mathsf{a}} + \hat{\mathsf{a}}\hat{\mathsf{a}}^+ + \hat{\mathsf{a}}^+\hat{\mathsf{a}} + \hat{\mathsf{a}}^+\hat{\mathsf{a}}^+\right)|0\rangle = \frac{1}{2}\,, \\
\langle 2|\hat{\mathsf{x}}^2|2\rangle &= \frac{1}{2}\langle 2|\left(\hat{\mathsf{a}}\hat{\mathsf{a}} + \hat{\mathsf{a}}\hat{\mathsf{a}}^+ + \hat{\mathsf{a}}^+\hat{\mathsf{a}} + \hat{\mathsf{a}}^+\hat{\mathsf{a}}^+\right)|2\rangle = \frac{5}{2}\,,
\end{aligned}
$$

hence $(\hat{\mathsf{H}}_1)_{11} = \frac{5}{4}$ and similarly for the additional elements. We obtain the equation

$$
E^{(1)}\begin{pmatrix} c_1 \\ c_2 \end{pmatrix} = \frac{1}{4}\begin{pmatrix} 5 & 2 \\ 2 & 5 \end{pmatrix}\begin{pmatrix} c_1 \\ c_2 \end{pmatrix},
$$

which has two solutions, $E^{(1)} \in \{\frac{3}{4}, \frac{7}{4}\}$. Therefore the sought energy of the first excited state composed of even functions only, $E \simeq (i^{[x]}+j^{[y]}+1)+\delta E^{(1)} = 3+\delta E^{(1)}$, splits into two sublevels,

$$
E = \begin{cases} 3 + \frac{7}{4}\delta\,, \\ 3 + \frac{3}{4}\delta\,. \end{cases}
$$

Exercise 4: Degenerate Perturbation Method

In the first order of the perturbation method, determine the splitting of the level $E^{(0)} = 2$ *in a system with the Hamiltonian*

$$
\hat{\mathsf{H}} = \frac{1}{2}(\hat{\mathsf{p}}_x^2 + \hat{\mathsf{x}}^2) + \frac{1}{2}(\hat{\mathsf{p}}_y^2 + \hat{\mathsf{y}}^2) + \lambda\hat{\mathsf{x}}^3\hat{\mathsf{y}}\,,
$$

where $\lambda = \frac{1}{100}$.

2.2.3 Note on the Error of the Perturbation Method

With the exception of very simple systems, such as those we discussed, it is practically impossible to calculate a large number of perturbation coefficients. This leads to the question how large an error we create when we truncate the perturbation series (2.9) after N terms, taking $E^{(0)}$ as the zeroth term. For small magnitude of the parameter δ, we can estimate that the inaccuracy will be of the order δ^{N+1}. For example, in the case of the anharmonic oscillator the series (2.9), terminated after the second term, yields $2E = 1.0073687$; the error being of the order of $(2\delta)^3 = 10^{-6}$. The exact value reads $2E = 1.0073737$ and the second-order estimate is off by 5×10^{-6}, which agrees with the estimate of the error. For $2\delta = 0.2$, we obtain from the second order perturbation method $2E = 1.0975$. Comparing this with the exact value $2E = 1.11829$ confirms the error of 2×10^{-2}, which is, once again, very close to the estimate $(2\delta)^3 \simeq 10^{-2}$. Mathematicians call these series *asymptotic*, which have, alas, little in common with the *convergent* series. A reader willing to discover more on this fascinating topic within mathematical physics should be satisfied by the captivating book [1].

References

1. C.M. Bender, S.A Orszag, *Advanced Mathematical Methods for Scientists and Engineers* (McGraw-Hill, New York, 1978)
2. L.D. Landau, E.M. Lifshitz, *Quantum Mechanics, Non-relativistic Theory* (Elsevier/Butterworth-Heinemann, Amsterdam/London, 1976)

Chapter 3
The Hydrogen Atom and Structure of Its Spectral Lines

In this chapter, we focus on hydrogen-like atoms[1] and their spectral structure in great depth. We show that the spectrum consists of a gross structure resulting from the electrostatic interaction between an electron and the nucleus, a fine structure arising from the spin-orbit interaction, and a hyperfine structure stemming from the spin-spin interaction. These structures are not specific solely for the hydrogen-like atoms though, we encounter them in any system which we can describe in the first approximation within the framework of nonrelativistic quantum mechanics. As we will shortly see, the spin-orbit interaction is an effect of relativistic kinematics and the spin-spin interaction is nothing more than a quantum mechanical analogy of the interaction of two magnetic dipoles. In the case of such systems, the effects of relativistic kinematics are minor and likewise the magnetic interaction is substantially smaller in comparison to the electrostatic interaction. Furthermore, we focus on the problem of hydrogen-like atoms to illustrate several methods that we will then systematically develop in the next chapter.

[1]A hydrogen-like atom is a bound state of two charged particles. The simplest example is hydrogen atom consisting of an electron and a proton. The other examples discussed extensively in this book are deuterium, where the proton is replaced by a deuteron, the bound state of a proton and a neutron, muonium, where the proton is replaced by an antimuon, an electron's heavier cousin with the opposite charge, positronium, where the proton is replaced by a positron, electron's antiparticle, see Chap. 7, and muonic hydrogen, where the electron is replaced by a muon. Under the term antiparticle we mean a particle with the same mass and spin as, but opposite charge to the "original" particle. Although this definition is too narrow, it suffices well in this book.

J. Zamastil, J. Benda, *Quantum Mechanics and Electrodynamics*,
DOI 10.1007/978-3-319-65780-6_3

3.1 A Particle in an Electromagnetic Field

In this section, we make use of the following relations. Within the Hamiltonian formalism, we substitute

$$E \to E - e\varphi\,, \qquad \mathbf{p} \to \mathbf{p} - e\mathbf{A} \tag{3.1}$$

to obtain the energy and momentum of a particle in an electromagnetic field. Assuming the Coulomb gauge $\nabla \cdot \mathbf{A} = 0$, the vector and scalar electromagnetic potentials, $\mathbf{A}$ and φ, respectively, solve the equations

$$\left(\frac{\partial^2}{\partial t^2} - \nabla^2\right)\mathbf{A}(\mathbf{r}, t) = \mathbf{j}(\mathbf{r}, t) - \nabla\frac{\partial\varphi}{\partial t}(\mathbf{r}, t)\,, \qquad -\nabla^2\varphi(\mathbf{r}, t) = \rho(\mathbf{r}, t)\,. \tag{6.11}$$

We derive these relations later in Chap. 6, Sect. 6.1. At this point, a detailed discussion would stray us too far from the topic, and a kind reader will surely believe the authors that these relations indeed hold.

Furthermore, we neglect the time retardation at this point and keep only the time-dependent source on the rhs, that is, we consider only

$$-\nabla^2\mathbf{A}(\mathbf{r}, t) = \mathbf{j}(\mathbf{r}, t) - \nabla\frac{\partial\varphi}{\partial t}(\mathbf{r}, t)\,, \qquad -\nabla^2\varphi(\mathbf{r}, t) = \rho(\mathbf{r}, t)\,. \tag{3.2}$$

instead of the entire Eq. (6.11). The operator on the lhs of Eq. (6.11) is responsible for propagation of the EM waves, while the rhs represents local changes in the charge distribution. We will show later, see Sect. 3.6.1, that the time derivative on the rhs can be eliminated. Having neglected the retardation, we can treat the electromagnetic field in terms of the classical theory. We may quantize only a physical object which we may ascribe a canonical coordinate and a canonical momentum. A static electromagnetic field does not carry any momentum, as we discuss in more detail in Chap. 6.

3.2 The Gross Structure

3.2.1 The Problem of Two Particles

In the nonrelativistic description, the total energy of hydrogen-like atoms equals the sum of the kinetic energy of the nucleus, the kinetic energy of the electron and the potential energy arising from the mutual interaction of the electron and the nucleus which depends only on the distance of the two mentioned particles. Having imposed the commutation relations, the Hamiltonian operator describing the system reads

$$\hat{\mathsf{H}} = \frac{\hat{\mathsf{p}}_1^2}{2m_1} + \frac{\hat{\mathsf{p}}_2^2}{2m_2} + V(|\hat{\boldsymbol{r}}_1 - \hat{\boldsymbol{r}}_2|)\,. \tag{3.3}$$

One can easily verify that the operator of total linear momentum commutes with the Hamiltonian

$$[\hat{\boldsymbol{p}}_1 + \hat{\boldsymbol{p}}_2, \hat{\mathsf{H}}] = 0\,.$$

As we will discuss in more detail later, the existence of operators that commute with the Hamiltonian substantially simplifies the process of solving the Schrödinger equation. In this case, for instance, it allows us to reduce the six-dimensional problem to two three-dimensional ones.

Within the framework of classical mechanics, one may transform such a problem of two particles to a one-particle problem by introducing the relative position of the two particles and the position of their center of mass, called the *Jacobi coordinates*

$$\mathbf{r}_R = \mathbf{r}_1 - \mathbf{r}_2, \qquad \mathbf{r}_T = \frac{m_1\mathbf{r}_1 + m_2\mathbf{r}_2}{m_1 + m_2}\,. \tag{3.4}$$

We find the operator of the squared momentum in terms of the new coordinates with the help of the chain rule for derivation of a composite function. We thus have for two particles moving in one direction

$$\hat{\mathsf{p}}_1^2 = -\frac{\partial^2}{\partial x_1^2} = -\frac{m_1^2}{(m_1+m_2)^2}\frac{\partial^2}{\partial x_T^2} - \frac{\partial^2}{\partial x_R^2} = \frac{m_1^2}{(m_1+m_2)^2}\hat{\mathsf{p}}_T^2 + \hat{\mathsf{p}}_R^2\,,$$

and similarly for $\hat{\mathsf{p}}_2^2$

$$\hat{\mathsf{p}}_2^2 = \frac{m_2^2}{(m_1+m_2)^2}\hat{\mathsf{p}}_T^2 + \hat{\mathsf{p}}_R^2\,.$$

Here, $\hat{\mathsf{p}}_T$ denotes the momentum of the center of mass and $\hat{\mathsf{p}}_R$ the relative momentum of the two particles. One can easily find that these relations between $\hat{\mathsf{p}}_1^2$, $\hat{\mathsf{p}}_2^2$ and $\hat{\mathsf{p}}_T^2$, $\hat{\mathsf{p}}_R^2$ hold also in three dimensions. Thus in terms of the new coordinates, the Hamiltonian operator (3.3) acquires the form

$$\hat{\mathsf{H}} = \frac{\hat{\mathsf{p}}_T^2}{2}\left(\frac{1}{m_1}\frac{m_1^2}{M^2} + \frac{1}{m_2}\frac{m_2^2}{M^2}\right) + \frac{\hat{\mathsf{p}}_R^2}{2}\left(\frac{1}{m_1} + \frac{1}{m_2}\right) + V(|\hat{\boldsymbol{r}}_R|) = \frac{\hat{\mathsf{p}}_T^2}{2M} + \frac{\hat{\mathsf{p}}_R^2}{2m_r} + V(|\hat{\boldsymbol{r}}_R|)\,.$$

Here, we set $M = m_1 + m_2$ and call the quantity

$$m_r = \frac{m_1 m_2}{m_1 + m_2} = \frac{m_1}{1 + \frac{m_1}{m_2}}$$

the *reduced mass of the system*. When one of the masses significantly exceeds the other (such as in the case of an electron and a nucleus), m_r nearly equals the mass of the lighter constituent. Since the Hamiltonian is a sum of two mutually commuting sets of terms, namely $\frac{\hat{\mathsf{p}}_T^2}{2M}$ and $\frac{\hat{\mathsf{p}}_R^2}{2m_r} + V(|\hat{\boldsymbol{r}}_R|)$, in the Schrödinger equation

$$\hat{\mathsf{H}}\psi(\mathbf{r}_T, \mathbf{r}_R) = E\psi(\mathbf{r}_T, \mathbf{r}_R)$$

we can separate the coordinate of the center of mass and the coordinate of the relative motion

$$\psi(\mathbf{r}_T, \mathbf{r}_R) = \psi(\mathbf{r}_T)\psi(\mathbf{r}_R)\,.$$

Denoting the energy corresponding to the free motion of the center of mass as E_T,

$$\frac{\hat{\mathsf{p}}_T^2}{2M}\psi(\mathbf{r}_T) = E_T\psi(\mathbf{r}_T)\,,$$

we obtain

$$\left(\frac{\hat{\mathsf{p}}_R^2}{2m_r} + V(|\mathbf{r}_R|)\right)\psi(\mathbf{r}_R) = (E - E_T)\psi(\mathbf{r}_R)\,. \tag{3.5}$$

We see that—as in classical mechanics—we can transform the problem of two particles, interaction of which depends only on their mutual distance, to a problem of one particle of the reduced mass m_r moving in an external potential field. The obtained energies are then diminished by the kinetic energy of the center of mass. In the next sections we will follow the common habit of expressing the energy of an atom in the center-of-mass frame where $E_T = 0$ as experimenters report their results in this frame.

3.2.2 *Electrostatic Potential*

The electrostatic potential caused by the nucleus is a solution to the Poisson equation, see Eq. (3.2),

$$-\nabla^2\varphi(\mathbf{r}) = \rho(\mathbf{r})\,. \tag{3.6}$$

In a very good approximation, one may consider the nucleus as a point—its size is typically of the order of magnitude 10^{-15} m, while the size of the whole atom reaches usually that of 10^{-10} m. With no loss of generality, we may place the origin, from which we determine the position vector $\mathbf{r}$ of a point where we find the potential $\varphi(r)$, so that it matches the position of the nucleus

$$\rho(\mathbf{r}) = Ze\delta(\mathbf{r})\,, \tag{3.7}$$

where e represents the elementary charge and Z the nuclear charge in the elementary charge units. A field created by a point charge is spherically symmetric, $\varphi(\mathbf{r}) = \varphi(r)$. If the function, upon which the Laplacian operator acts, depends only on the distance from the nucleus, the Laplacian operator acquires the form (see Eq. (3.21) below)

$$\nabla^2 \to \frac{\mathrm{d}^2}{\mathrm{d}r^2} + \frac{2}{r}\frac{\mathrm{d}}{\mathrm{d}r}.$$

We solve the equation

$$-\left(\frac{\mathrm{d}^2}{\mathrm{d}r^2} + \frac{2}{r}\frac{\mathrm{d}}{\mathrm{d}r}\right)\varphi(r) = 0,$$

for $r \neq 0$ with the initial guess $\varphi(r) = Ar^{\alpha}$. We obtain $\alpha \in \{0, -1\}$. However, only the latter satisfies Eq. (3.6) with the charge density given by Eq. (3.7). We find the constant A from the original equation by integrating it over volume including the central singularity,

$$\int -\nabla^2\varphi \,\mathrm{d}V = \int Ze\delta(\mathbf{r})\,\mathrm{d}V = Ze.$$

We rewrite the volume integral on the lhs using the Gauss's theorem (see, e.g., [1]) to a surface integral over the boundary of the considered volume. We further express this boundary in terms of a solid angle $\mathrm{d}\mathbf{S} = \mathbf{n}r^2\mathrm{d}\Omega$, and successively obtain

$$-\int \nabla^2\varphi\,\mathrm{d}^3\mathbf{r} = -\oint(\nabla\varphi)\cdot\mathrm{d}\mathbf{S} = -\int \mathbf{n}\cdot\nabla\varphi(r)r^2\,\mathrm{d}\Omega = \int \frac{A}{r^2}r^2\,\mathrm{d}\Omega = 4\pi A,$$

where we substituted $\varphi(r) = Ar^{-1}$ and $\nabla\varphi(r) = \mathbf{n}\frac{\partial\varphi}{\partial r}$ in the third step, see Eq. (3.15) below. We thus have for the potential energy of the electrostatic interaction in the Schrödinger equation (3.5)

$$V(r) = e\varphi(r) = -\frac{Ze^2}{4\pi r} = -\frac{Z\alpha}{r}. \tag{3.8}$$

We call the corresponding potential φ the *Coulomb potential*[2] and we will extensively focus on it in this book.

3.2.3 Units

The constant α introduced in the last relation is termed the *fine-structure constant*. It is a dimensionless number (thus independent of the choice of units) and its reciprocal value equals (see [3])

$$\alpha^{-1} = 137.035999074(44),$$

[2]The reason might probably be to stress that it is an electrostatic potential. Newton potential would be a more convenient name, though.

where the numbers in the parentheses indicate the uncertainty in determination of the last two figures. In the center-of-mass frame with the potential (3.8), the Schrödinger equation (3.5) acquires the form

$$\left(\frac{\hat{\mathrm{p}}^2}{2m_r} - \frac{Z\alpha}{r}\right)\psi = E\psi\,.$$

As in the case of the harmonic oscillator, we once again transform the equation to dimensionless (*atomic*) units by substituting $r = \kappa r_A$

$$\left(\frac{\hat{\mathrm{p}}_A^2}{2m_r\kappa^2} - \frac{Z\alpha}{r_A\kappa}\right)\psi = E\psi \quad \Rightarrow \quad \left(\frac{\hat{\mathrm{p}}_A^2}{2} - \frac{Z\alpha m_r\kappa}{r_A}\right)\psi = m_r\kappa^2 E\psi\,.$$

Now we exploit the freedom in the determination of κ and set $Z\alpha m_r\kappa = 1$. We thus obtain

$$\left(\frac{\hat{\mathrm{p}}_A^2}{2} - \frac{1}{r_A}\right)\psi = E_A\psi\,. \tag{3.9}$$

In the last equation, we have finally eliminated all of the constants, and so we have

$$E = m_e(Z\alpha)^2\frac{m_r}{m_e}E_A\,. \tag{3.10}$$

In natural units, we obtain energy in electronvolts. Experimenters, though, prefer to publish their data in terms of frequencies of transitions rather than of energies. The frequencies are usually listed in hertz, and we will now show how to change from electronvolts to hertz. Within the framework of SI units, we need to multiply the rhs of Eq. (3.10) with c^2 where c stands for the speed of light in vacuum. We thus successively obtain

$$\nu = \frac{E_{SI}}{h} = \frac{m_e c^2}{2\pi\hbar}\frac{m_r}{m_e}(Z\alpha)^2 E_A = \frac{m_e c\alpha^2}{4\pi\hbar}c\frac{m_r}{m_e}2Z^2E_A = R_\infty c\frac{m_r}{m_e}2Z^2E_A\,. \tag{3.11}$$

We call the constant R_∞ the *Rydberg constant* and its product with the speed of light in vacuum equals [3]

$$R_\infty c = 3.289841960364(17) \times 10^{15}\,\mathrm{Hz}\,.$$

Comparing Eqs. (3.10) and (3.11), we readily see that the transition from electronvolts to hertz

$$E(\mathrm{eV}) \rightarrow \nu(\mathrm{Hz})$$

corresponds to the substitution

$$m_e \to \frac{2R_\infty c}{\alpha^2}\,. \tag{3.12}$$

3.2.4 Spherical Coordinates

In case of an electron moving in a potential that depends only on the distance between the particle and the source of the acting force, it is advantageous to solve the Schrödinger equation in spherical coordinates. We write the transition from Cartesian to spherical coordinates as

$$\mathbf{x} = r\mathbf{n}\,, \tag{3.13}$$

where r is the distance between the particle and the origin of the acting force (e.g., the distance of an electron from the nucleus), and $\mathbf{n}$ is a unit vector pointing to any direction in space,

$$\mathbf{n} = (\sin\vartheta\cos\varphi, \sin\vartheta\sin\varphi, \cos\vartheta)\,. \tag{3.14}$$

Using the chain rule for derivative of a composite function, we find the form of the operator ∇ in spherical coordinates

$$\hat{\boldsymbol{p}}_A \to -\mathrm{i}\nabla\,, \quad \nabla_k = \frac{\partial}{\partial x_k} = n_k\frac{\partial}{\partial r} + \frac{\nabla_k^n}{r}\,, \tag{3.15}$$

where we introduced the angular differential operator ∇^n

$$\nabla^n = \left(-\frac{\sin\varphi}{\sin\vartheta}\frac{\partial}{\partial\varphi} + \cos\varphi\cos\vartheta\frac{\partial}{\partial\vartheta}, \frac{\cos\varphi}{\sin\vartheta}\frac{\partial}{\partial\varphi} + \sin\varphi\cos\vartheta\frac{\partial}{\partial\vartheta}, -\sin\vartheta\frac{\partial}{\partial\vartheta}\right)\,. \tag{3.16}$$

From Eqs. (3.14) and (3.16), we find the scalar product of the vector operators $\mathbf{n}$ and ∇^n

$$\mathbf{n}\cdot\nabla^n = 0\,. \tag{3.17}$$

The order of the operators matters since the components of $\mathbf{n}$ and ∇^n do not mutually commute. To find their commutation relation, we start with the commutation relation (confront with Eq. (1.85))

$$[\nabla_j, x_k] = \delta_{jk}$$

and substitute from Eqs. (3.15) and (3.13) for ∇_j and x_k, respectively. After rearranging the terms, we get

$$\left[n_j\frac{\partial}{\partial r}+\frac{\nabla_j^n}{r},rn_k\right]=n_jn_k\left[\frac{\partial}{\partial r},r\right]+[\nabla_j^n,n_k]=\delta_{jk}$$

$$\Rightarrow\left[\nabla_j^n,n_k\right]=\delta_{jk}-n_j\,n_k\,. \tag{3.18}$$

In the first equation, we exploited the fact that if one of any two operators acts solely on the angular variable while the other only on the radial variable, these two operators commute mutually, for instance $[\frac{\partial}{\partial r},n_k]=0$, $[\nabla_j^n,r]=0$, etc. Setting $j=k$ in Eq. (3.18) (and keeping in mind we are using the Einstein summation convention), we obtain

$$\left[\nabla_k^n,n_k\right]=2\,. \tag{3.19}$$

It follows then from Eqs. (3.17) and (3.19) that

$$\nabla_k^n\,n_k=2\,. \tag{3.20}$$

Substituting for ∇_k from Eq. (3.15) and considering Eqs. (3.17) and (3.20), we write for the square of the momentum

$$\hat{\mathsf{p}}_A^2\rightarrow-\nabla^2,\quad\nabla^2=\left(\frac{\partial^2}{\partial r^2}+\frac{2}{r}\frac{\partial}{\partial r}+\frac{(\nabla^n)^2}{r^2}\right)\,. \tag{3.21}$$

3.2.5 *Solution for s-States*

We now solve Eq. (3.9) for the special case of the so-called s-states, i.e., states with no angular dependence. That is, the wave function depends only on the radial variable r. Considering (3.21), we may then write Eq. (3.9) as (we use atomic units in this section)

$$\hat{\mathsf{H}}|\psi\rangle=-\frac{1}{2n^2}|\psi\rangle\,,\quad\hat{\mathsf{H}}=\frac{\hat{\mathsf{p}}_r^2}{2}-\frac{1}{\hat{\mathsf{r}}}\,, \tag{3.22}$$

where we introduced the operator of radial momentum $\hat{\mathsf{p}}_r$, which acquires the following form in the coordinate representation[3]

[3]Note that one is prone to make a mistake when finding $\hat{\mathsf{p}}_r^2$:

$$\hat{\mathsf{p}}_r = -\mathrm{i}\left(\frac{\partial}{\partial r} + \frac{1}{r}\right), \qquad \hat{\mathsf{p}}_r^2 = -\left(\frac{\partial^2}{\partial r^2} + \frac{2}{r}\frac{\partial}{\partial r}\right) \tag{3.23}$$

and where, without loss of generality, we rewrote the energy of a bound state into the form

$$E_A = -\frac{1}{2n^2}, \tag{3.24}$$

where n represents a real number for now. We will later show that it is a natural number.

We use several tricks to solve Eq. (3.22). First, we substitute $r \to nr$ and multiply the whole equation by r so as to transform the original eigenvalue spectrum of Hamiltonian operator to a spectrum of a different operator:

$$\hat{\mathsf{T}}_3|\psi\rangle = n|\psi\rangle, \quad \hat{\mathsf{T}}_3 = \frac{\hat{\mathsf{r}}\hat{\mathsf{p}}_r^2}{2} + \frac{\hat{\mathsf{r}}}{2}. \tag{3.25}$$

We denoted this operator $\hat{\mathsf{T}}_3$ for reasons that will shortly follow.

The benefit of this trick lies in the following. Spectrum of the Hamiltonian comprises a discrete ($E_A < 0$) and a continuous ($E_A > 0$) part. From a physical point of view, it is absolutely acceptable for we are then able to describe processes such as the photoelectric effect where an electron passes from the discrete to the continuous part of the spectrum, see Sect. 6.3. However, from a mathematical point of view, this implies that the discrete part of the spectrum of the Hamiltonian operator does not constitute a complete set of functions. Yet we know from Chap. 2 that a *complete* set of functions is requisite if we wish to determine energy levels of atoms with more than one electron employing the variational method. As we will shortly see, the operator $\hat{\mathsf{T}}_3$, unlike the Hamilton operator, has a complete discrete spectrum. Eigenstates of this operator are thus more favorable than those of the Hamiltonian (3.22) when searching for energy levels of many-electron atoms.

The second trick lies in noticing that the operators

$$\hat{\mathsf{W}}_1 = \hat{\mathsf{r}}, \qquad \hat{\mathsf{W}}_2 = \hat{\mathsf{r}}\hat{\mathsf{p}}_r, \qquad \hat{\mathsf{W}}_3 = \hat{\mathsf{r}}\hat{\mathsf{p}}_r^2 \tag{3.26}$$

are closed with respect to commutation

$$\left(\frac{\partial}{\partial r} + \frac{1}{r}\right)\left(\frac{\partial}{\partial r} + \frac{1}{r}\right) = \frac{\partial^2}{\partial r^2} + \frac{1}{r}\frac{\partial}{\partial r} + \frac{\partial}{\partial r}\frac{1}{r} + \frac{1}{r^2};$$

now

$$\frac{\partial}{\partial r}\frac{1}{r} = \frac{1}{r}\frac{\partial}{\partial r} + \left[\frac{\partial}{\partial r}, \frac{1}{r}\right] = \frac{1}{r}\frac{\partial}{\partial r} - \frac{1}{r^2}.$$

$$[\hat{W}_1, \hat{W}_2] = i\hat{W}_1\,, \qquad [\hat{W}_2, \hat{W}_3] = i\hat{W}_3\,, \qquad [\hat{W}_1, \hat{W}_3] = 2i\hat{W}_2\,. \tag{3.27}$$

Considering a linear combination of these operators

$$\hat{T}_1 = \frac{1}{2}(\hat{W}_3 - \hat{W}_1)\,, \qquad \hat{T}_2 = \hat{W}_2\,, \qquad \hat{T}_3 = \frac{1}{2}(\hat{W}_3 + \hat{W}_1)\,, \tag{3.28}$$

leads to commutation relations with a highly symmetric form

$$[\hat{T}_1, \hat{T}_2] = -i\hat{T}_3\,, \qquad [\hat{T}_2, \hat{T}_3] = i\hat{T}_1\,, \qquad [\hat{T}_3, \hat{T}_1] = i\hat{T}_2\,. \tag{3.29}$$

If we now introduce the ladder operators

$$\hat{T}_\pm = \hat{T}_1 \pm i\hat{T}_2\,, \tag{3.30}$$

we obtain

$$[\hat{T}_3, \hat{T}_\pm] = \pm\hat{T}_\pm\,. \tag{3.31}$$

Comparing the last equation with Eqs. (1.60) and (1.61), that is $[\hat{H}, \hat{a}] = -\hat{a}$ and $[\hat{H}, \hat{a}^+] = +\hat{a}^+$, respectively, we see that we transformed the original problem into that of a harmonic oscillator. The operator $\hat{T}_3$ now takes on the role of the Hamiltonian operator and likewise the operators $\hat{T}_\pm$ replace the non-Hermitian operators $\hat{a}$ and $\hat{a}^+$. The only difference is that instead of Eq. (1.59), $\hat{H} = \hat{a}^+\hat{a} + \frac{1}{2}$, we have the equation (prove that it holds!)

$$\hat{T}_3^2 - \hat{T}_1^2 - \hat{T}_2^2 = 0\,. \tag{3.32}$$

Hence instead of Eq. (1.67) we have

$$\begin{aligned} \hat{T}_+\hat{T}_-|n\rangle &= (\hat{T}_1 + i\hat{T}_2)(\hat{T}_1 - i\hat{T}_2)|n\rangle(\hat{T}_1^2 + \hat{T}_2^2 - i[\hat{T}_1, \hat{T}_2])|n\rangle \\ &= (\hat{T}_3^2 - \hat{T}_3)|n\rangle = (n^2 - n)|n\rangle\,, \end{aligned} \tag{3.33}$$

where we used Eqs. (3.29) and (3.32) in the third step and Eq. (3.25) in the fourth. Now we may completely adopt the solution for the harmonic oscillator. Namely from equations $\hat{T}_3(\hat{T}_\pm|n\rangle) = (n \pm 1)(\hat{T}_\pm|n\rangle)$, corresponding to Eqs. (1.63) and (1.64), we find that the individual eigenvalues of the operator $\hat{T}_3$ differ by unity. Once again, we demand a ground state exists

$$\hat{T}_-|n_{\min}\rangle = 0$$

and thus we obtain from Eq. (3.33)

$$n_{\min}(n_{\min} - 1) = 0\,. \tag{3.34}$$

The solution $n_{\min} = 0$ corresponds to infinite energy, see Eq. (3.24), hence is nonphysical. Therefore we have $n_{\min} = 1$. Taking into account that the individual n differs by unity, we may conclude that the *eigenvalues n are natural numbers* $n = 1, 2, 3, \ldots$

By substituting into Eq. (3.22), one can easily verify that the normalized ground state wave function is of the form

$$\langle \mathbf{r}|\, 1s\rangle = \psi_{1s}(\mathbf{r}) = R_{1s}(r)Y_{00}(\mathbf{n})\,, \quad R_{1s}(r) = R_{1,0}(r) = 2\mathrm{e}^{-r}\,, \quad Y_{00}(\mathbf{n}) = \frac{1}{\sqrt{4\pi}}\,. \tag{3.35}$$

The wave function is normalized so that the probability of finding an electron in the entire space equals unity, i.e.,

$$\langle 1s|\, 1s\rangle = \int \mathrm{d}^3r \langle 1s|\, \mathbf{r}\rangle \langle \mathbf{r}|\, 1s\rangle = \int \mathrm{d}\Omega \frac{1}{4\pi} \int_0^\infty \mathrm{d}r r^2 (2\mathrm{e}^{-r})^2 = 1\,.$$

$Y_{00}(\mathbf{n})$ captures the angular dependence of the s-states. However, there is none (from the definition), and thus $Y_{00}(\mathbf{n})$ is merely a constant given by normalization condition $\int |Y_{00}|^2\, \mathrm{d}\Omega = 1$. The probability of finding an electron in an s-state is thus the same for all angles and the angular part of s-orbitals has a shape of a sphere.

3.2.6 *Comparison with Experiment*

For the energy of a transition $1s - ns$ we have

$$\Delta E_A = \frac{1}{2}\left(1 - \frac{1}{n^2}\right)\,,$$

hence, see Eq. (3.11),

$$\nu_{1n} = R_\infty c \times Z^2 \left(1 - \frac{1}{n^2}\right) \frac{1}{1 + \frac{m_e}{m_n}}\,. \tag{3.36}$$

Now we can compare the theoretical values of ν with the experimental ones, namely for the transition $1s - 2s$. The Table 3.1 summarizes the results for hydrogen,

Table 3.1 Theoretical (3.36) and experimental values for the transition $1s - 2s$

System	Ratio m_e/m_n	ν_{12}^{theo} [10^{15} Hz]	ν_{12}^{exp} [10^{15} Hz]
p^+e^-	$5.4461702178(22) \times 10^{-4}$	2.4660384...	2.466061413187035(10)
$p^+n^0e^-$	$2.7244371095(11) \times 10^{-4}$	2.466709...	2.466732407521641(35)
μ^+e^-	$4.83633166(12) \times 10^{-3}$	2.4555058...	2.455528941(10)

deuterium, and muonium. The values of the mass ratios of the individual particles are adopted from [3], the only exception being muonium coming from [2].

If we neglect the effect of the motion of the nucleus, namely set $m_r \simeq m_e \ll m_n$, we obtain $\nu_{12}^{\rm theo} = 2.467381 \cdots \times 10^{15}$ Hz. We thus see that using the reduced mass instead of only the electron mass significantly improves the final result. We further see from the table that the agreement of the experimental and theoretical values for deuterium is similar to that of the hydrogen atom, while it is slightly worse for muonium. To elucidate this deviation, we need to study also other interactions between the nucleus and the electron than only the Coulombic, and we should include relativistic corrections as well. A long journey lies ahead of us...

3.3 The Hyperfine Structure

Apart from the electrostatic, we now include also the magnetic interaction in the Hamiltonian of the system. The reason is, both the nucleus and the electron possess an intrinsic magnetic moment and these moments act on each other via the magnetic force. We begin by finding the magnetic field caused by the nucleus.

3.3.1 *Magnetic Field of a Dipole*

The magnetic field of a steady electric current is given by the vector Poisson equation, see Eq. (3.2), where we neglect the time change of the gradient of the potential that arises from the change in the charge density with respect to time

$$-\nabla^2 \mathbf{A} = \mathbf{j} \,. \tag{3.37}$$

Assuming we know the so-called Green function

$$-\nabla_{\mathbf{r}_1}^2 G(\mathbf{r}_1, \mathbf{r}_2) = \delta(\mathbf{r}_1 - \mathbf{r}_2) \,, \tag{3.38}$$

we can write the general solution to Eq. (3.37) as

$$\mathbf{A}(\mathbf{r}_1) = \int G(\mathbf{r}_1, \mathbf{r}_2)\mathbf{j}(\mathbf{r}_2)\, \mathrm{d}V_2 \,. \tag{3.39}$$

One can easily verify this by acting with the operator $-\nabla^2$ on both sides of the last equation and by considering the penultimate equation and Eq. (1.75). As we showed above, see Eqs. (3.6)–(3.8), there is a simple solution to the Poisson equation for G

$$G(\mathbf{r}_1, \mathbf{r}_2) = \frac{1}{4\pi}\frac{1}{r_{12}} = \frac{1}{4\pi}\frac{1}{[(\mathbf{r}_1 - \mathbf{r}_2)\cdot(\mathbf{r}_1 - \mathbf{r}_2)]^{1/2}} \,. \tag{3.40}$$

We now move the origin to approximately the center of the area where the current density is located. Supposing this area is much smaller in size than the distance of the point where we calculate the field generated by this area, namely $r_2 \ll r_1$, we may expand the Green function in Eq. (3.39) into a series of powers of r_2/r_1,

$$G(\mathbf{r}_1, \mathbf{r}_2) = \frac{1}{4\pi}\frac{1}{[r_1^2 - 2\mathbf{r}_1\cdot\mathbf{r}_2 + r_2^2]^{1/2}} \simeq \frac{1}{4\pi}\left[\frac{1}{r_1} + \frac{\mathbf{r}_1\cdot\mathbf{r}_2}{r_1^3} + \ldots\right]. \tag{3.41}$$

This is the so-called multipole expansion to be discussed in more detail in Sects. 5.2.1 and 5.2.2. We have then

$$\mathbf{A}(\mathbf{r}_1) \simeq \frac{1}{4\pi}\left[\frac{1}{r_1}\int \mathbf{j}(\mathbf{r}_2)\mathrm{d}V_2 + \frac{1}{r_1^3}\int (\mathbf{r}_1\cdot\mathbf{r}_2)\,\mathbf{j}(\mathbf{r}_2)\mathrm{d}V_2\right]. \tag{3.42}$$

Acting with the divergence operator on both sides of Eq. (3.37) and recalling we are using Coulomb gauge $\nabla\cdot\mathbf{A} = 0$, we obtain the continuity equation for steady flow:

$$\nabla\cdot\mathbf{j} = 0. \tag{3.43}$$

We further obtain from the last equation the two identities

$$\begin{aligned} 0 &= \int x_i(\nabla\cdot\mathbf{j})\,\mathrm{d}V = \int\left[\frac{\partial}{\partial x_k}(x_i j_k) - \frac{\partial x_i}{\partial x_k}j_k\right]\mathrm{d}V \\ &= \oint x_i j_k\,\mathrm{d}S_k - \int j_i\,\mathrm{d}V = -\int j_i\,\mathrm{d}V \end{aligned} \tag{3.44}$$

and

$$\begin{aligned} 0 &= \int x_l x_i \frac{\partial}{\partial x_k} j_k\,\mathrm{d}V = \int\left[\frac{\partial}{\partial x_k}(x_l x_i j_k) - j_k\frac{\partial}{\partial x_k}(x_l x_i)\right]\mathrm{d}V \\ &= \oint x_i x_l j_k\,\mathrm{d}S_k + \int [x_i j_l + x_l j_i]\,\mathrm{d}V = \int [x_i j_l + x_l j_i]\,\mathrm{d}V\,. \end{aligned} \tag{3.45}$$

In the second steps we integrated by parts, in the third we used the Gauss's theorem. Finally, we exploited the fact that the surface integrals vanish for currents concentrated in a finite space.

We can thus rewrite the expression (3.42) for vector potential of a magnetic dipole into the form

$$\begin{aligned}
A_k(\mathbf{r}) &= \frac{x_i}{4\pi r^3}\int x'_i j_k \, \mathrm{d}V' = \frac{x_i}{8\pi r^3}\int \left((x'_i j_k + x'_k j_i) + (x'_i j_k - x'_k j_i)\right) \mathrm{d}V' \\
&= \frac{x_i}{8\pi r^3}\int (x'_i j_k - x'_k j_i) \, \mathrm{d}V' = \frac{x_i}{8\pi r^3}\int \varepsilon_{ikp}\varepsilon_{pqr} x'_q j_r \, \mathrm{d}V' \\
&= -\frac{1}{8\pi}\left[\frac{\mathbf{r}}{r^3}\times\int (\mathbf{r}_2\times\mathbf{j}(\mathbf{r}_2)) \, \mathrm{d}V_2\right]_k = \frac{1}{4\pi}\frac{(\boldsymbol{\mu}\times\mathbf{r})_k}{r^3}, \qquad (3.46)
\end{aligned}$$

where we considered Eqs. (3.44) and (3.45) in the first and third equality. In the fourth, we used the identity $\varepsilon_{ijk}\varepsilon_{ipq} = \delta_{jp}\delta_{kq} - \delta_{jq}\delta_{kp}$. Finally, in the last step, we introduced magnetic dipole moment

$$\boldsymbol{\mu} = \frac{1}{2}\int \mathbf{r}\times\mathbf{j}(\mathbf{r}) \, \mathrm{d}V .$$

We calculate the magnetic induction $\mathbf{B} = \nabla\times\mathbf{A}$ as well:

$$\begin{aligned}
B_k &= \varepsilon_{kpq}\frac{\partial}{\partial x_p}A_q = \varepsilon_{kpq}\varepsilon_{qrs}\frac{\partial}{\partial x_p}\left[\frac{1}{4\pi}\frac{\mu_r x_s}{r^3}\right] \\
&= \left(\delta_{kr}\delta_{ps} - \delta_{ks}\delta_{pr}\right)\frac{1}{4\pi}\left[\left(n_p\frac{\partial}{\partial r} + \frac{\nabla^n_p}{r}\right), \frac{\mu_r n_s}{r^2}\right] \\
&= \left(\delta_{kr}\delta_{ps} - \delta_{ks}\delta_{pr}\right)\frac{\mu_r}{4\pi}\left(n_p n_s\left(-\frac{2}{r^3}\right) + \frac{1}{r^3}\left[\nabla^n_p, n_s\right]\right),
\end{aligned}$$

where we substituted from Eq. (3.46) in the second step and from (3.15) and (1.85) in the third. Using Eq. (3.18), $[\nabla^{(n)}_p, n_s] = \delta_{ps} - n_p n_s$, we further write

$$B_k = \left(\delta_{kr}\delta_{ps} - \delta_{ks}\delta_{pr}\right)\frac{\mu_r}{4\pi}\left(\frac{\delta_{ps} - 3n_p n_s}{r^3}\right) = \frac{\mu_r}{4\pi r^3}(3n_k n_r - \delta_{kr}), \quad r \neq 0 .$$

However, we could have performed this step only for $r \neq 0$.[4] To find the magnetic induction at the point $r = 0$, we make a guess $B_k = \mathcal{A}_k\delta(\mathbf{r})$, where $\mathcal{A}_k$ is a constant we wish to find. We thus have

[4]It might seem rather odd that we focus on the point $r = 0$ and yet we have kept only the leading term when expanding the Coulomb potential, see Eq. (3.41). The expression (3.46) should thus hold solely for large r, and we should take into consideration other terms of the expansion (3.41) to accurately capture the situation for small r. Experiments show, though, that particles we consider as elementary (such as electrons, muons, etc.) behave as electric monopoles and magnetic dipoles. That is, the remaining terms of the expansion (3.41) are zero. Although in the case of a proton the other terms are not null, the dipole term still prevails for $r \approx (m\alpha)^{-1} \approx 10^{-10}$ m and an electron practically does not approach the proton any closer. In other words, the probability of finding an electron in the space between 0 and 10^{-1} atomic units from the proton is $\int_0^{0.1} \mathrm{d}r r^2 |R_{1s}(r)|^2 = 0.11\times 10^{-2}$, see Eq. (3.35). We thus see that one needs to distinguish between a physical and mathematical zero!

$$\left(\delta_{kr}\delta_{ps} - \delta_{ks}\delta_{pr}\right) \frac{\partial}{\partial x_p} \frac{1}{4\pi} \frac{\mu_r x_s}{r^3} = \mathcal{A}_k \delta(\mathbf{r}) .$$

Now we integrate over the area comprising the point $r = 0$,

$$\frac{\mu_r}{4\pi} \left(\delta_{kr}\delta_{ps} - \delta_{ks}\delta_{pr}\right) \int \left(\frac{\partial}{\partial x_p} \frac{n_s}{r^2}\right) \mathrm{d}V = \mathcal{A}_k \int \delta(\mathbf{r}) \mathrm{d}V ,$$

and use the Gauss's theorem

$$\mathcal{A}_k = \frac{\mu_r}{4\pi} \left(\delta_{kr}\delta_{ps} - \delta_{ks}\delta_{pr}\right) \oint \frac{n_s}{r^2} \, \mathrm{d}S_p = \frac{\mu_r}{4\pi} \left(\delta_{kr}\delta_{ps} - \delta_{ks}\delta_{pr}\right) \int n_s n_p \, \mathrm{d}\Omega .$$

Since the following relation holds[5]

$$\int n_s n_p \mathrm{d}\Omega = \frac{4\pi}{3} \delta_{sp} , \tag{3.47}$$

we obtain at last

$$\mathcal{A}_k = \frac{1}{3} \mu_r \left(\delta_{kr}\delta_{ps} - \delta_{ks}\delta_{pr}\right) \delta_{sp} = \frac{2}{3} \mu_k \qquad \Rightarrow \qquad B_k = \frac{2}{3} \mu_k \delta(\mathbf{r}) .$$

The final magnetic field of a dipole then reads

$$\mathbf{B} = \frac{1}{4\pi} \frac{3\mathbf{n}\mathbf{n} \cdot \boldsymbol{\mu} - \boldsymbol{\mu}}{r^3} + \frac{2}{3} \boldsymbol{\mu} \delta(\mathbf{r}) . \tag{3.48}$$

3.3.2 Hamiltonian of a Particle with Spin in an External Electromagnetic Field

The other part of the problem lies in determining how the magnetic dipole moment reacts to the surrounding magnetic field. To do so, we need to know the Hamiltonian of a particle with spin in an external electromagnetic field. Firstly, we consider only a free particle. In nonrelativistic approximation, its energy equals the kinetic energy of the particle $E = p^2/(2m)$, irrespective of whether it possesses spin or not. These should be the eigenvalues of the correct Hamiltonian for a free particle. The operators

[5] The easiest way to obtain this equation goes as follows. An object with two indices appears on the lhs, and the only plausible object with two indices on the rhs is the Kronecker delta, i.e., $\int n_s n_p \mathrm{d}\Omega = K\delta_{sp}$. Setting now $s = p$ and keeping in mind that we are using the Einstein summation convention and $\mathbf{n}$ is a unity vector, we obtain $4\pi = 3K$.

$$\hat{\mathsf{H}} = \frac{\hat{\mathsf{p}}^2}{2m}$$

and

$$\hat{\mathsf{H}} = \frac{\boldsymbol{\sigma} \cdot \hat{\boldsymbol{p}} \boldsymbol{\sigma} \cdot \hat{\boldsymbol{p}}}{2m} \tag{3.49}$$

obviously comply with this demand, see Eq. (1.25):

$$\sigma_i \hat{\mathsf{p}}_i \sigma_j \hat{\mathsf{p}}_j = (\delta_{ij} + \mathrm{i}\varepsilon_{ijk}\sigma_k)\hat{\mathsf{p}}_i \hat{\mathsf{p}}_j = \delta_{ij}\hat{\mathsf{p}}_i \hat{\mathsf{p}}_j = \hat{\mathsf{p}}^2 \,.$$

The former is a Hamiltonian operator of a scalar particle, while the latter, Eq. (3.49), describes a particle with spin 1/2.

In case of a particle in an electromagnetic field, we substitute according to Eq. (3.1), and obtain the Schrödinger equation for a particle with spin moving in external electric and magnetic fields, termed the *Pauli equation*,

$$\hat{\mathsf{H}}\psi = E\psi \,, \quad \hat{\mathsf{H}} = \frac{\boldsymbol{\sigma} \cdot (\hat{\boldsymbol{p}} - e\mathbf{A})\, \boldsymbol{\sigma} \cdot (\hat{\boldsymbol{p}} - e\mathbf{A})}{2m} + e\varphi \,. \tag{3.50}$$

In case of a free particle, the energy values do not differ whether the particle does or does not have spin. However, as soon as we add the electromagnetic potentials into the Hamiltonian operator, one has to proceed with more care:

$$\begin{aligned}
\sigma_i \sigma_j (\hat{\mathsf{p}}_i - eA_i)(\hat{\mathsf{p}}_j - eA_j) &= (\delta_{ij} + \mathrm{i}\varepsilon_{ijk}\sigma_k)(\hat{\mathsf{p}}_i - eA_i)(\hat{\mathsf{p}}_j - eA_j) \\
&= (\hat{\mathsf{p}}_i - eA_i)(\hat{\mathsf{p}}_i - eA_i) + \mathrm{i}\varepsilon_{ijk}\sigma_k \frac{1}{2}[\hat{\mathsf{p}}_i - eA_i, \hat{\mathsf{p}}_j - eA_j] \\
&= |\hat{\boldsymbol{p}} - e\mathbf{A}|^2 - e\boldsymbol{\sigma} \cdot \mathbf{B} \,,
\end{aligned}$$

where in the second step we used the fact that the product of two tensor quantities, one of them being symmetric and the other antisymmetric with respect to the interchange of two indices, equals zero. In the third step, we performed the following rearrangement, see Eq. (1.84),

$$\begin{aligned}
-\frac{\mathrm{i}e}{2}\varepsilon_{ijk}\sigma_k([\hat{\mathsf{p}}_i, A_j] + [A_i, \hat{\mathsf{p}}_j]) &= -\frac{e}{2}\varepsilon_{ijk}\sigma_k \left(\frac{\partial A_j}{\partial x_i} - \frac{\partial A_i}{\partial x_j} \right) \\
&= -e\varepsilon_{ijk}\sigma_k \frac{\partial A_j}{\partial x_i} = -e\sigma_k \left(\nabla \times \mathbf{A}\right)_k \,.
\end{aligned}$$

The magnetic interaction is much weaker in comparison to the electrostatic, as we will shortly see. It suffices thus to consider only the linear term of the vector potential **A**,

$$|\hat{\boldsymbol{p}} - e\mathbf{A}|^2 = (\hat{\boldsymbol{p}} - e\mathbf{A}) \cdot (\hat{\boldsymbol{p}} - e\mathbf{A}) \simeq \hat{\mathsf{p}}^2 - e(\mathbf{A} \cdot \hat{\boldsymbol{p}} + \hat{\boldsymbol{p}} \cdot \mathbf{A}) = \hat{\mathsf{p}}^2 - 2e\mathbf{A} \cdot \hat{\boldsymbol{p}} .$$

In the last step we used

$$[\hat{\mathsf{p}}_i, A_i] = -\mathrm{i}\nabla \cdot \mathbf{A} = 0 .$$

The first equality follows from Eq. (1.55), the second one from using Coulomb gauge. We thus finally have the Pauli Hamiltonian in a simplified form

$$\hat{\mathsf{H}} \simeq \frac{\hat{\mathsf{p}}^2}{2m} - \frac{e}{m}\mathbf{A} \cdot \hat{\boldsymbol{p}} - \frac{e}{m}\hat{\boldsymbol{S}} \cdot \mathbf{B} + e\varphi . \tag{3.51}$$

We now substitute into this Hamiltonian for the vector potential $\mathbf{A}$, magnetic induction $\mathbf{B}$ and scalar potential φ from Eqs. (3.46), (3.48), and (3.8), respectively:

$$\hat{\mathsf{H}} = \frac{\hat{\mathsf{p}}^2}{2m} - \frac{Z\alpha}{r} - \frac{e}{m}\frac{\boldsymbol{\mu} \cdot \hat{\boldsymbol{L}}}{4\pi r^3} - \frac{e}{4\pi m}\left(\hat{\boldsymbol{S}} \cdot \boldsymbol{\mu}\frac{8\pi}{3}\delta(\mathbf{r}) - \frac{1}{r^3}\left(\hat{\boldsymbol{S}} \cdot \boldsymbol{\mu} - 3\,\mathbf{n} \cdot \hat{\boldsymbol{S}}\,\mathbf{n} \cdot \boldsymbol{\mu}\right)\right) . \tag{3.52}$$

In this equation, we introduced the so-called orbital angular momentum

$$\hat{\boldsymbol{L}} = \hat{\boldsymbol{r}} \times \hat{\boldsymbol{p}} , \tag{3.53}$$

to which we will pay much attention in the next sections.

We now take a closer look on the derived expressions. The first two terms in Eq. (3.52) describe motion of an electron in a central electrostatic field. From the previous derivations, it should be evident that r denotes the relative distance of the electron and the nucleus. One might also suggest we add the kinetic energy of the nucleus $\hat{\mathsf{p}}^2/(2m_n)$ to these two terms. However, within the center-of-mass frame, it suffices to substitute m_r for m_e to include the nuclear kinetic energy, as we saw in Sect. 3.2.1. Therefore, setting $m = m_r$ in the first term results precisely in a Hamiltonian of a system of two particles which are mutually bound by an electrostatic force. The third term constitutes a scalar product of a nuclear magnetic moment and an orbital angular momentum of an electron. That is, it characterizes the interaction of the nuclear spin and orbital angular momentum of an electron, commonly termed as the *spin-other-orbit interaction*. The last term—the whole large parenthesis—describes a relatively complicated *spin-spin interaction*, as it contains spins of both the electron and the nucleus.

For the sake of the following, we now introduce an operator of nuclear spin $\hat{\boldsymbol{S}}_p$ defined by the substitution

$$\boldsymbol{\mu} \rightarrow -g_p\frac{Ze}{m_p}\hat{\boldsymbol{S}}_p . \tag{3.54}$$

The numerical constant g is the gyromagnetic ratio, and for the instance of a proton, the hydrogen nucleus, it equals $g_p = 2.792$. In case of elementary particles (such as an electron), we obtain $g = 1$ from the Pauli equation (3.50).[6] As usual, we divide the Hamiltonian into two parts: the first one, $\hat{\mathsf{H}}_0$, comprises the first two terms in Eq. (3.52), while the other one, $\hat{\mathsf{H}}_1$, includes the remaining terms. After substituting from Eq. (3.54) and changing to atomic units, $r \to r_A/m_r Z\alpha$, we obtain

$$\hat{\mathsf{H}} = \hat{\mathsf{H}}_0 + \hat{\mathsf{H}}_1\,,$$

$$\hat{\mathsf{H}}_0 = m_r(Z\alpha)^2\left(\frac{\hat{\mathsf{p}}_A^2}{2} - \frac{1}{r_A}\right),$$

$$\hat{\mathsf{H}}_1 = \frac{Z\alpha g_p}{m_e m_p}(m_r Z\alpha)^3\left[\hat{\boldsymbol{S}}_p\cdot\hat{\boldsymbol{L}}\frac{1}{r_A^3} + \hat{\boldsymbol{S}}_e\cdot\hat{\boldsymbol{S}}_p\frac{8\pi}{3}\delta(\mathbf{r}_A) - \frac{1}{r_A^3}\left(\hat{\boldsymbol{S}}_e\cdot\hat{\boldsymbol{S}}_p - 3\,\mathbf{n}\cdot\hat{\boldsymbol{S}}_e\,\mathbf{n}\cdot\hat{\boldsymbol{S}}_p\right)\right].$$

One can readily see that $\hat{\mathsf{H}}_0$ is of the order of $(Z\alpha)^2$ while $\hat{\mathsf{H}}_1$ of $(Z\alpha)^4$, hence the use of the perturbation method is justifiable. In addition, the ratio $\frac{m_r^3}{m_e m_p} = m_e\frac{m_e/m_p}{(1+m_e/m_p)^3}$ further reduces the relative magnitude of $\hat{\mathsf{H}}_1$ in comparison to $\hat{\mathsf{H}}_0$ by a factor of 10^3, see Table 3.1.

3.3.3 Hyperfine Splitting of the Hydrogen Ground State

Having laid foundations in general terms, we now illustrate the above presented theory on an example. Namely, we will calculate the energy of splitting in a hydrogen atom, i.e., we will consider nucleus as a single proton. Prior to including the magnetic interactions, each state is at least quadruply degenerate since there are four different ways of the electron and proton spins orientation. For instance, the ground state can be any of the following:

$$|1\rangle = |1s\rangle|p+\rangle|e+\rangle, \quad |2\rangle = |1s\rangle|p-\rangle|e+\rangle, \quad |3\rangle = |1s\rangle|p+\rangle|e-\rangle, \quad |4\rangle = |1s\rangle|p-\rangle|e-\rangle$$

or any linear combination thereof. As usual, we calculate the first-order perturbation to energy as the eigenvalue of the perturbation Hamiltonian which we project into the subspace of eigenvectors corresponding to the same eigenvalue of the unperturbed Hamiltonian. Multiplying Eq. (2.11) by the vector $\langle 1s|$, Eq. (3.35), leads to

$$\hat{\mathsf{h}}_1\left|j, m, \left(\frac{1}{2}, \frac{1}{2}\right)\right\rangle = E_1\left|j, m, \left(\frac{1}{2}, \frac{1}{2}\right)\right\rangle, \quad \hat{\mathsf{h}}_1 = \langle 1s|\hat{\mathsf{H}}_1|1s\rangle\,. \tag{3.55}$$

We will elucidate the way we denote the eigenvector of the operator $\hat{\mathsf{h}}_1$ later in Sect. 4.2. We find the operator $\hat{\mathsf{h}}_1$ in the coordinate representation,

[6]The fact that the gyromagnetic ratios of a proton and a neutron do not equal one (or zero, respectively) was one of the first clues that these two particles are not elementary.

$$\hat{\mathsf{h}}_1 = \frac{m_r^3}{m_e m_p}(Z\alpha)^4 g_p \left(\hat{\boldsymbol{S}}_p \cdot \int \psi_{1s} \frac{\hat{\boldsymbol{L}}}{r^3} \psi_{1s} \mathrm{d}V + \hat{\boldsymbol{S}}_e \cdot \hat{\boldsymbol{S}}_p \frac{8\pi}{3} \int \psi_{1s} \delta(\mathbf{r}) \psi_{1s} \mathrm{d}V \right.$$
$$\left. - \int \psi_{1s} \frac{1}{r^3} \left(\hat{\boldsymbol{S}}_e \cdot \hat{\boldsymbol{S}}_p - 3\, \mathbf{n} \cdot \hat{\boldsymbol{S}}_e\, \mathbf{n} \cdot \hat{\boldsymbol{S}}_p \right) \psi_{1s} \mathrm{d}V \right). \tag{3.56}$$

The first integral equals zero since the ground state of hydrogen-like atoms possesses no angular momentum. In other words, the wave function ψ_{1s} is independent of angles (it depends merely on r), while the operator $\hat{\boldsymbol{L}}$, as we will see later in Eq. (3.74), comprises only differentiation by angles, hence $\hat{\boldsymbol{L}}\psi_{1s} = 0$. It follows from the properties of the δ-function that the second integral equals $|\psi_{1s}(0)|^2 = 1/\pi$, see Eq. (3.35). The third integral, sometimes called the mean value of the *tensor interaction*, in the state $1s$, is also zero due to the angular integration

$$\int \psi_{1s} \frac{1}{r^3} \left(\hat{\boldsymbol{S}}_e \cdot \hat{\boldsymbol{S}}_p - 3\, \mathbf{n} \cdot \hat{\boldsymbol{S}}_e\, \mathbf{n} \cdot \hat{\boldsymbol{S}}_p \right) \psi_{1s}\, \mathrm{d}V$$
$$= (\hat{\boldsymbol{S}}_e)_i (\hat{\boldsymbol{S}}_p)_j \int \psi_{1s} (\delta_{ij} - 3 n_i n_j) \frac{1}{r^3} \psi_{1s} r^2\, \mathrm{d}r\, \mathrm{d}\Omega$$
$$= (\hat{\boldsymbol{S}}_e)_i (\hat{\boldsymbol{S}}_p)_j \int (\delta_{ij} - 3 n_i n_j)\, \mathrm{d}\Omega \int |\psi_{1s}|^2 \frac{\mathrm{d}r}{r} = 0,$$

see Eq. (3.47). Substituting into Eq. (3.55) then results in a more appealing equation

$$A \hat{\boldsymbol{S}}_e \cdot \hat{\boldsymbol{S}}_p \left| j, m, \left(\frac{1}{2}, \frac{1}{2} \right) \right\rangle = E_1 \left| j, m, \left(\frac{1}{2}, \frac{1}{2} \right) \right\rangle, \tag{3.57}$$

where the constant A equals

$$A = m_e \frac{\frac{m_e}{m_p}}{\left(1 + \frac{m_e}{m_p}\right)^3} \frac{8}{3} (Z\alpha)^4 g_p, \quad A_{\mathrm{SI}} = R_\infty c \frac{2}{\alpha^2} \frac{\frac{m_e}{m_p}}{\left(1 + \frac{m_e}{m_p}\right)^3} \frac{8}{3} (Z\alpha)^4 g_p. \tag{3.58}$$

The former follows from Eq. (3.56), the latter from Eq. (3.12).

We now commence solving Eq. (3.57). First of all, we find the action of the operator $\hat{\mathsf{h}}_1$ on the four basis states $|pX\rangle|eY\rangle$:

$$\hat{\mathsf{h}}_1|1\rangle = A \hat{\boldsymbol{S}}_e \cdot \hat{\boldsymbol{S}}_p |e+\rangle|p+\rangle = A \left(\frac{1}{2} \left(\hat{\mathsf{S}}^e_+ \hat{\mathsf{S}}^p_- + \hat{\mathsf{S}}^e_- \hat{\mathsf{S}}^p_+ \right) + \hat{\mathsf{S}}^e_z \hat{\mathsf{S}}^p_z \right) |e+\rangle|p+\rangle$$
$$= A \frac{1}{4} |e+\rangle|p+\rangle = A \frac{1}{4} |1\rangle,$$

$$\hat{\mathsf{h}}_1|2\rangle = A \hat{\boldsymbol{S}}_e \cdot \hat{\boldsymbol{S}}_p |e+\rangle|p-\rangle = A \left(\frac{1}{2} |e-\rangle|p+\rangle - \frac{1}{4} |e+\rangle|p-\rangle \right) = A \left(\frac{1}{2} |3\rangle - \frac{1}{4} |2\rangle \right),$$

$$\hat{\mathsf{h}}_1|3\rangle = A \hat{\boldsymbol{S}}_e \cdot \hat{\boldsymbol{S}}_p |e-\rangle|p+\rangle = A \left(\frac{1}{2} |e+\rangle|p-\rangle - \frac{1}{4} |e-\rangle|p+\rangle \right) = A \left(\frac{1}{2} |2\rangle - \frac{1}{4} |3\rangle \right),$$

$$\hat{\mathsf{h}}_1|4\rangle = \Lambda\hat{\boldsymbol{S}}_e \cdot \hat{\boldsymbol{S}}_p|e-\rangle|p-\rangle = \Lambda\frac{1}{4}|e-\rangle|p-\rangle = \Lambda\frac{1}{4}|4\rangle\,,$$

where we used the identity (1.35) and the action of electron and proton spin operators is given by Eqs. (1.32), (1.33), and (1.34).

It follows from these equations that only the two states $|1\rangle$ and $|4\rangle$ are eigenstates of the operator $\hat{\mathsf{h}}_1$. The energy correction E_1 to the states $|2\rangle$ and $|3\rangle$, on the other hand, is not well defined. However, we do not need to lose our heads. The operator $\hat{\mathsf{h}}_1$ mixes the two states $|2\rangle$ and $|3\rangle$ with each other, therefore their linear combination should be an eigenstate to this operator. Thus we now face the equation

$$\hat{\mathsf{h}}_1(c_2|2\rangle + c_3|3\rangle) = E_1(c_2|2\rangle + c_3|3\rangle)\,.$$

Acting with the operator on the states on the lhs and comparing the coefficients for the individual vectors $|2\rangle$, $|3\rangle$, which are mutually orthogonal,[7] leads to a system of linear equations:

$$\Lambda\left(-c_2\frac{1}{4} + c_3\frac{1}{2}\right) = c_2E_1\,,$$

$$\Lambda\left(c_2\frac{1}{2} - c_3\frac{1}{4}\right) = c_3E_1\,.$$

If we further require $|c_2|^2 + |c_3|^2 = 1$, the solutions read

$$c_2 = c_3 = \frac{1}{\sqrt{2}}\,, \qquad E_1^{(+)} = \frac{1}{4}\Lambda \quad \text{and}$$

$$c_2 = -c_3 = \frac{1}{\sqrt{2}}\,, \qquad E_1^{(-)} = -\frac{3}{4}\Lambda\,.$$

We thus conclude that the interactions of nuclear and electron spin splits the groundstate energy into two sublevels separated by Λ. The experimental value of this splitting was found to be [3]

$$\Lambda_{\text{exp}}(p^+e^-) = 1420.405751767(1)\ \text{MHz}\,, \tag{3.59}$$

which is about a million times smaller than the distance between the first and second s-state of the hydrogen atom. Therefore, we refer to this effect as the *hyperfine* splitting. The theoretical value given by Eq. (3.58) equals $\Lambda_{\text{theo}} = 1418.409\ldots$ MHz. For muonium, the experimental value reads

$$\Lambda_{\text{exp}}(\mu^+e^-) = 4463.30288(16)\ \text{MHz}\,, \tag{3.60}$$

[7] $\langle 2|3\rangle = \langle e+|e-\rangle\langle p-|p+\rangle = 0$, see Eq. (1.9).

while our theory predicts $\Lambda_{\text{theo}} = 4453.838\ldots$ MHz. Although these two values do not match each other perfectly (we neglected relativistic corrections and other details stemming from the nontrivial structure of the proton), their correspondence is more than satisfactory given the amount of work done. The spectral line at this frequency, which corresponds to the wavelength 21 cm, is generally known simply as the *hydrogen line*, or HI-line. This radiation is highly characteristic of hydrogen, and therefore many radiotelescopes focus on its presence.[8]

3.3.4 Classification of States Using the Integrals of Motion

The higher energy level, its energy being $\Lambda/4$, is triply degenerate (a so-called *triplet*), while the lower one stands alone (a so-called *singlet*). Within the spectroscopic notation, we denote these states as 1^3s and 1^1s, respectively.[9]

Why these very numbers and not any others? We will now discover the reason and smoothly thus proceed to the next section devoted to the angular momentum.

We know from classical mechanics that certain quantities are the so-called *integrals of motion*, that is, they are independent of time. In quantum mechanics, the Hamiltonian operator determines the time evolution of a system. If we demand the results of measuring an observable be independent of time, this observable must necessarily commute with the Hamiltonian. To see this, we let $[\hat{\mathsf{A}}, \hat{\mathsf{H}}] = 0$ and $\hat{\mathsf{A}}|A\rangle = A|A\rangle$, and we demand the system be in the state $|A\rangle$ at time $t = 0$. Any later in time, the system will be in the state $\mathrm{e}^{-\mathrm{i}\hat{\mathsf{H}}t}|A\rangle$. Now we act with the operator $\hat{\mathsf{A}}$ and successively obtain $\hat{\mathsf{A}}\mathrm{e}^{-\mathrm{i}\hat{\mathsf{H}}t}|A\rangle = \mathrm{e}^{-\mathrm{i}\hat{\mathsf{H}}t}\hat{\mathsf{A}}|A\rangle = A\mathrm{e}^{-\mathrm{i}\hat{\mathsf{H}}t}|A\rangle$. We thus see that the observable A does not change in time, i.e., it is an integral of motion.

Integrals of motion are of great importance in both classical and quantum mechanics, as they allow for solving problems that one can solve exactly and facilitate to great extent those that one can solve only approximately. We have already encountered both cases. For example, the fact that the operator of the total linear momentum $\hat{\boldsymbol{p}}_1 + \hat{\boldsymbol{p}}_2$ commutes with the Hamiltonian (3.3) enables us to reduce the Schrödinger equation with this Hamiltonian (3.3) from a six-variable to only a three-variable problem. In Chap. 2, we exploited the fact that the parity operator is an integral of motion for the Hamiltonian $\hat{\mathsf{H}} = \hat{\mathsf{p}}^2/2 + \hat{\mathsf{x}}^2/2 + \delta\hat{\mathsf{x}}^4$. As you recall, this markedly simplified the problem as one could restrict oneself to the class of functions with the same parity when looking for eigenstates and eigenvalues. This

[8] For the very same reason, it was used as a unit of length and time on information plates for extra-terrestrial civilizations on the probes Pioneer 10, 11 and Voyager 1, 2, and also the program SETI (*Search for Extra-Terrestrial Intelligence*) operated on this frequency.

[9] In light of this finding, one may ask to which of the four $2s - 1s$ transitions the values in Table 3.1 correspond. The numbers are spin-averaged values, meaning the spin-spin interaction was subtracted. Due to the substantially smaller magnitude of the hyperfine splitting in comparison to the $2s - 1s$ transition, one can perform that with sufficient precision.

simplification, as we will see, bears substantially more significance in cases of more complex problems, such as determining spectra of many-electron atoms.

From classical mechanics we know that the total momentum is an integral of motion, i.e., is independent of time. It appears therefore reasonable to consider the operator[10]

$$\hat{\boldsymbol{S}} = \hat{\boldsymbol{S}}_e + \hat{\boldsymbol{S}}_p ,$$

and find the commutator $[\hat{\boldsymbol{S}}, \hat{\mathsf{h}}_1]$:

$$\begin{aligned}[\hat{\boldsymbol{S}}, \hat{\mathsf{h}}_1] &= [\hat{\boldsymbol{S}}_e + \hat{\boldsymbol{S}}_p, \Lambda \hat{\boldsymbol{S}}_e \cdot \hat{\boldsymbol{S}}_p] = [\hat{\mathsf{S}}_i^e + \hat{\mathsf{S}}_i^p, \Lambda \hat{\mathsf{S}}_j^e \hat{\mathsf{S}}_j^p] = \Lambda \left([\hat{\mathsf{S}}_i^e, \hat{\mathsf{S}}_j^e]\hat{\mathsf{S}}_j^p + \hat{\mathsf{S}}_j^e [\hat{\mathsf{S}}_i^p, \hat{\mathsf{S}}_j^p] \right) \\ &= \Lambda \mathrm{i} \varepsilon_{ijk} \left(\hat{\mathsf{S}}_k^e \hat{\mathsf{S}}_j^p + \hat{\mathsf{S}}_j^e \hat{\mathsf{S}}_k^p \right) = \Lambda \mathrm{i} \varepsilon_{ijk} \left(\hat{\mathsf{S}}_k^e \hat{\mathsf{S}}_j^p - \hat{\mathsf{S}}_k^e \hat{\mathsf{S}}_j^p \right) = 0 .\end{aligned}$$

We thus see that the operator of total spin for s-states commutes with the Hamiltonian (in case of s-states, the spin-other-orbit interaction does not contribute). Obviously, the operator of total spin commutes also with $\hat{\mathsf{H}}_0$ because the Hamiltonian contains no spin-dependent terms. From the relations

$$\begin{aligned}[\hat{\mathsf{S}}_i^e, \hat{\mathsf{S}}_j^e] &= \mathrm{i}\varepsilon_{ijk} \hat{\mathsf{S}}_k^e , \\ [\hat{\mathsf{S}}_i^p, \hat{\mathsf{S}}_j^p] &= \mathrm{i}\varepsilon_{ijk} \hat{\mathsf{S}}_k^p , \\ [\hat{\mathsf{S}}_i^e, \hat{\mathsf{S}}_j^p] &= 0\end{aligned}$$

we easily find

$$[\hat{\mathsf{S}}_i, \hat{\mathsf{S}}_j] = \mathrm{i}\varepsilon_{ijk} \hat{\mathsf{S}}_k . \tag{3.61}$$

It appears the relation (3.61) be characteristic for any spin. Moreover, as we will see later, it does hold for any angular momentum. Furthermore, each of any three operators $\hat{\mathsf{S}}_i$ complying with the relation (3.61) commutes with the operator $\hat{\mathsf{S}}^2 = \hat{\boldsymbol{S}} \cdot \hat{\boldsymbol{S}}$:

$$\begin{aligned}[\hat{\mathsf{S}}^2, \hat{\mathsf{S}}_i] &= [\hat{\mathsf{S}}_j \hat{\mathsf{S}}_j, \hat{\mathsf{S}}_i] = \hat{\mathsf{S}}_j [\hat{\mathsf{S}}_j, \hat{\mathsf{S}}_i] + [\hat{\mathsf{S}}_j, \hat{\mathsf{S}}_i] \hat{\mathsf{S}}_j = \hat{\mathsf{S}}_j \mathrm{i}\varepsilon_{jik} \hat{\mathsf{S}}_k + \mathrm{i}\varepsilon_{jik} \hat{\mathsf{S}}_k \hat{\mathsf{S}}_j \\ &= \mathrm{i}\varepsilon_{jik} (\hat{\mathsf{S}}_j \hat{\mathsf{S}}_k + \hat{\mathsf{S}}_k \hat{\mathsf{S}}_j) = \mathrm{i}\varepsilon_{jik} (\hat{\mathsf{S}}_j \hat{\mathsf{S}}_k - \hat{\mathsf{S}}_j \hat{\mathsf{S}}_k) = 0 ,\end{aligned} \tag{3.62}$$

where we used nothing but the commutation relation (3.61).

We thus see that we are able to independently observe three quantities in the case of our system with the Hamiltonian $\hat{\mathsf{h}}_1 = \Lambda \hat{\boldsymbol{S}}_e \cdot \hat{\boldsymbol{S}}_p$ (and that they describe our system

[10]Precisely speaking, we consider $\hat{\boldsymbol{S}} = \hat{\boldsymbol{S}}_e \otimes \mathbf{1}_p + \mathbf{1}_e \otimes \hat{\boldsymbol{S}}_p$

to the utmost extent). The three observables are the energy, one of the components of the angular momentum, and the magnitude of the angular momentum (or its square). The set $\{\hat{\mathsf{h}}_1, \hat{\mathsf{S}}^2, \hat{\mathsf{S}}_z\}$ is commonly known as a *complete set of commuting operators.*

It follows from the discussion of the Stern-Gerlach experiments that if we can simultaneously measure any two observables, measuring one of them after the other cannot change the state of the system it is in after measuring the first one. This is possible only if the corresponding operators possess common eigenvectors. The mathematical proof reads as follows.

We assume that

$$[\hat{\mathsf{A}}, \hat{\mathsf{H}}] = 0 \tag{3.63}$$

and that there is only one solution $|\psi\rangle$ to the Schrödinger equation

$$\hat{\mathsf{H}}|\psi\rangle = E|\psi\rangle \tag{3.64}$$

for a given E. Then acting of the operator equality (3.63) on the state $|\psi\rangle$ and considering Eq. (3.64) lead to

$$\hat{\mathsf{H}}(\hat{\mathsf{A}}|\psi\rangle) = E(\hat{\mathsf{A}}|\psi\rangle)\,. \tag{3.65}$$

It then follows from the comparison of Eqs. (3.64) and (3.65)

$$\hat{\mathsf{A}}|\psi\rangle = A|\psi\rangle\,,$$

that is, the eigenstates of $\hat{\mathsf{H}}$ and $\hat{\mathsf{A}}$ *must* be the same.

In case of degenerate energy E, for instance doubly degenerate,

$$\hat{\mathsf{H}}|\psi_1\rangle = E|\psi_1\rangle\,, \qquad \hat{\mathsf{H}}|\psi_2\rangle = E|\psi_2\rangle\,, \tag{3.66}$$

we obtain the following relations using the very same arguments as above

$$\hat{\mathsf{H}}(\hat{\mathsf{A}}|\psi_1\rangle) = E(\hat{\mathsf{A}}|\psi_1\rangle)\,, \qquad \hat{\mathsf{H}}(\hat{\mathsf{A}}|\psi_2\rangle) = E(\hat{\mathsf{A}}|\psi_2\rangle)\,. \tag{3.67}$$

Comparison of Eqs. (3.66) and (3.67) then shows

$$\hat{\mathsf{A}}|\psi_1\rangle = c_{11}|\psi_1\rangle + c_{12}|\psi_2\rangle\,, \qquad \hat{\mathsf{A}}|\psi_2\rangle = c_{21}|\psi_1\rangle + c_{22}|\psi_2\rangle\,,$$

that is, eigenstates of $\hat{\mathsf{H}}$ and $\hat{\mathsf{A}}$ *can* but need not to be necessarily the same. It holds still, though, that at least their eigensubspaces match (i.e., sets of all linear combinations of eigenstates corresponding to the same eigenvalue).

What is the eigenvalue of $\hat{\mathsf{S}}^2$? The following relation must hold

$$\hat{\mathsf{S}}^2 = |\hat{\boldsymbol{\mathsf{S}}}_e + \hat{\boldsymbol{\mathsf{S}}}_p|^2 = \hat{\mathsf{S}}_e^2 + 2\hat{\boldsymbol{\mathsf{S}}}_e \cdot \hat{\boldsymbol{\mathsf{S}}}_p + \hat{\mathsf{S}}_p^2 ,$$

thus we have

$$\hat{\mathsf{h}}_1 = \Lambda \hat{\boldsymbol{\mathsf{S}}}_e \cdot \hat{\boldsymbol{\mathsf{S}}}_p = \Lambda \frac{1}{2}\left(\hat{\mathsf{S}}^2 - \hat{\mathsf{S}}_e^2 - \hat{\mathsf{S}}_p^2\right) = \Lambda \frac{1}{2}\left(\hat{\mathsf{S}}^2 - \frac{3}{2}\right), \tag{3.68}$$

where the last equality follows from the form of the spin operators (1.21), (1.22), and (1.23). For reasons we will discuss later, we usually write the eigenvalues of the operator $\hat{\mathsf{S}}^2$ in the form $S(S+1)$. After substituting them into Eq. (3.68), these eigenvalues must yield the eigenvalues $\Lambda/4$ and $-3\Lambda/4$, which is satisfied if S equals 1 or 0, respectively. It further holds that the operator of the spin projection along the z axis has the same eigenstates. One can verify this easily:

$$\hat{\mathsf{S}}_z|e+\rangle|p+\rangle = (\hat{\mathsf{S}}_z^e + \hat{\mathsf{S}}_z^p)|e+\rangle|p+\rangle = \frac{1}{2}|e+\rangle|p+\rangle + \frac{1}{2}|e+\rangle|p+\rangle = 1 \cdot |e+\rangle|p+\rangle,$$

$$\hat{\mathsf{S}}_z \frac{1}{\sqrt{2}}\left(|e+\rangle|p-\rangle \pm |e-\rangle|p+\rangle\right) = 0 \cdot \frac{1}{\sqrt{2}}\left(|e+\rangle|p-\rangle \pm |e-\rangle|p+\rangle\right)$$

and

$$\hat{\mathsf{S}}_z|e-\rangle|p-\rangle = (-1) \cdot |e-\rangle|p-\rangle .$$

We can thus construct a simple table with eigenvalues, see Table 3.2.

One can see that the triple degeneracy of the energy level with the energy deviation $\Lambda/4$ from the unperturbed ground state is related to the three different values of the total spin projection along the z axis. If we place the atom into a magnetic field, this triplet further splits into three different levels (Fig. 3.1). This effect, i.e., splitting of atom lines in an external magnetic field, is generally known as the *Zeeman effect* and we will focus on it later in Sect. 4.4.7.

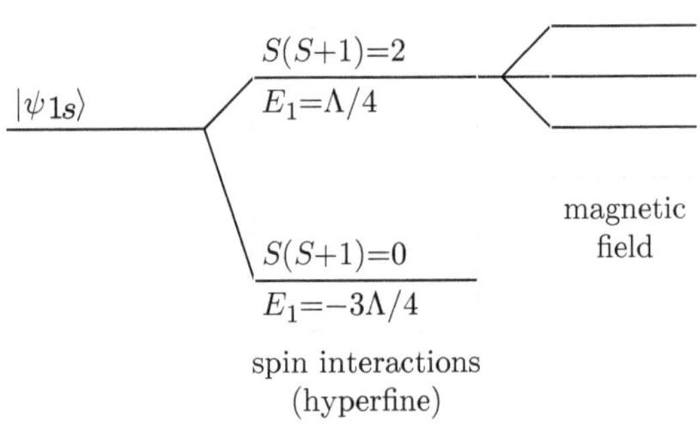

Fig. 3.1 Hyperfine splitting and the Zeeman effect

Table 3.2 Eigenvectors of the square and the projection onto one of the coordinate axes of the total spin operator

Spin state	Noted as	S	S_z
$\|e+\rangle\|p+\rangle$	$\|1,1\rangle$	1	1
$\frac{1}{\sqrt{2}}\left(\|e+\rangle\|p-\rangle + \|e-\rangle\|p+\rangle\right)$	$\|1,0\rangle$	1	0
$\|e-\rangle\|p-\rangle$	$\|1,-1\rangle$	1	-1
$\frac{1}{\sqrt{2}}\left(\|e+\rangle\|p-\rangle - \|e-\rangle\|p+\rangle\right)$	$\|0,0\rangle$	0	0

One can readily see from the table above that we denote the spin states according to the corresponding eigenvalues of the operators $\hat{\mathsf{S}}^2$ and $\hat{\mathsf{S}}_z$.[11] That is for example

$$\hat{\mathsf{S}}_z|1,1\rangle = |1,1\rangle\,, \quad \hat{\mathsf{S}}_z|1,0\rangle = 0\,, \quad \hat{\mathsf{S}}_z|1,-1\rangle = -|1,-1\rangle\,, \tag{3.69}$$

$$\hat{\mathsf{S}}^2|1,m\rangle = 1(1+1)|1,m\rangle\,, \quad \text{etc.}$$

We thus showed that two particles with half-integer spins (the so-called *fermions*) behave as particles with integer spin (the so-called *bosons*). This finding is of tremendous importance and, as we will see later, it is a special case of a more general rule which states that an even number of fermions behaves like a boson, while an odd number of fermions like a fermion.

Knowing whether a particle is a fermion or a boson plays an important role in statistical behavior of a great set of identical particles, as we will discuss later in this book. Generally speaking, fermions obey the Pauli exclusion principle, hence no two fermions may exist in the same state. This fact, for example, stands behind the chemical properties of atoms. On the other hand, the Pauli exclusion principle does not limit bosons in any way. At very low temperature, bosons exists in the same quantum state, which in turn allows for phenomena such as superconductivity and superfluidity.

We now introduce the ladder operators $\hat{\mathsf{S}}_\pm = \hat{\mathsf{S}}^e_\pm + \hat{\mathsf{S}}^p_\pm$, and easily find from Table 3.2 and Eqs. (1.33) and (1.34) that

$$\hat{\mathsf{S}}_+|1,1\rangle = 0\,, \quad \hat{\mathsf{S}}_+|1,0\rangle = \sqrt{2}|1,1\rangle\,, \quad \hat{\mathsf{S}}_+|1,-1\rangle = \sqrt{2}|1,0\rangle\,, \tag{3.70}$$

$$\hat{\mathsf{S}}_-|1,-1\rangle = 0\,, \quad \hat{\mathsf{S}}_-|1,0\rangle = \sqrt{2}|1,-1\rangle\,, \quad \hat{\mathsf{S}}_-|1,1\rangle = \sqrt{2}|1,0\rangle\,. \tag{3.71}$$

These relations prove indeed useful in the next chapter.

[11] A more accurate notation of these states would be $\left|j,m,\left(\frac{1}{2},\frac{1}{2}\right)\right\rangle$ where the halves stress that these states stem from two particles with spin 1/2 each, see Sect. 4.2.

Exercise 5: Projection of Spin $S = 1$ I

Consider the SG experiments for a beam of particles with spin 1. One can easily see that magnetic field splits the particle beam into three depending on their spin projection along the direction of the magnetic field. We now ask the very same question as in Exercise 1: if we assume only particles with spin projection +1 leave the first S-G apparatus, what is the probability of finding projection +1 at the second apparatus which is rotated by an arbitrary solid angle with respect to the first apparatus?

To the surprise of the kind reader, we will encounter the solution to this problem again later in Sect. 6.2.5.

Exercise 6: Projection of Spin $S = 1$ II

We now assume that only particles with spin projection 0 exit the first apparatus. The second apparatus, rotated by 90° with respect to the first one, catches particles with spin projection 0 along the new direction. What is the probability of detecting a particle with spin projection 0 at the third apparatus, which is placed in the same direction as the first one? Note: The probabilities of transitions of a system from a state with spin projection 0 along the z-axis to a state with spin projection 1 along the x-axis and from the same initial state to a state with spin projection 0 along the x-axis do not equal. Hence neither do the corresponding amplitudes.

3.4 Orbital Angular Momentum

3.4.1 Significance of Angular Momentum

We commence this chapter by revising our notation: $\hat{\boldsymbol{S}}$ represents the spin operator and $\hat{\boldsymbol{L}}$ the orbital angular momentum operator (for example of an electron in an atom), and we will use $\hat{\boldsymbol{J}}$ for a general angular momentum operator.

Angular momentum is of great importance already in classical mechanics. Finding the vector product of the Newton equation (we use atomic units in this section)

$$\frac{\mathrm{d}\mathbf{p}}{\mathrm{d}t} = -\nabla V(r) = -\mathbf{n}\frac{\mathrm{d}V}{\mathrm{d}r}, \quad \mathbf{p} = \frac{\mathrm{d}\mathbf{r}}{\mathrm{d}t}, \tag{3.72}$$

with the position vectors **r** leads to

$$\frac{\mathrm{d}}{\mathrm{d}t}\mathbf{L} = 0\,, \quad \mathbf{L} = \mathbf{r} \times \mathbf{p}\,, \tag{3.73}$$

where we use

$$\mathbf{r} \times \frac{\mathrm{d}\mathbf{p}}{\mathrm{d}t} = \frac{\mathrm{d}}{\mathrm{d}t}(\mathbf{r} \times \mathbf{p}) \quad \text{and} \quad -r\frac{\mathrm{d}V}{\mathrm{d}r}\mathbf{n} \times \mathbf{n} = 0\,.$$

This conservation law allows us to transform the three-dimensional equation (3.72) into a two-dimensional one. We assume that at the initial time $t = 0$ we have $z = 0$ and $\dot{z} = 0$. It follows thereof $L_x = (y\dot{z} - z\dot{y}) = 0$ and $L_y = (z\dot{x} - x\dot{z}) = 0$. However, both coordinates x and y and velocities $\dot{x}$ and $\dot{y}$ may acquire any value at arbitrary times. Therefore, the components L_x and L_y are conserved only if we have $z = \dot{z} = 0$ for all times.

In quantum mechanics one may thus expect the conserved components of angular momentum (3.53)

$$\hat{\mathsf{L}}_i = \varepsilon_{ijk}\hat{\mathsf{x}}_j\hat{\mathsf{p}}_k = -\mathrm{i}\varepsilon_{ijk}x_j\nabla_k = -\mathrm{i}\varepsilon_{ijk}rn_j\left(n_k\frac{\mathrm{d}}{\mathrm{d}r} + \frac{\nabla_k^n}{r}\right) = -\mathrm{i}\varepsilon_{ijk}n_j\nabla_k^n \tag{3.74}$$

to commute with any Hamiltonian operator with potential that depends only on the distance from the source,

$$[\hat{\boldsymbol{L}}, \hat{\mathsf{H}}] = 0, \qquad \hat{\mathsf{H}} = \frac{\hat{\mathsf{p}}^2}{2} + V(\hat{\mathsf{r}})\,, \tag{3.75}$$

and that it will be an indeed precious piece of information.

We see from Eq. (3.74) that the components of the angular momentum operator depend only on angular variables. Substituting for $\hat{\mathsf{p}}^2$ from Eq. (3.21) into Eq. (3.75) shows that the only angle-dependent operator in the Hamiltonian is the operator

$$-(\nabla^n)^2 = \hat{\boldsymbol{L}} \cdot \hat{\boldsymbol{L}} = \hat{\mathsf{L}}^2\,, \tag{3.76}$$

as follows from Eqs. (3.18) and (3.74). However, the components of the angular momentum operator commute with this operator. Why? In Eq. (3.62) we showed that any two operators complying with the commutation relation (3.61)

$$[\hat{\mathsf{L}}_i, \hat{\mathsf{L}}_j] = \mathrm{i}\varepsilon_{ijk}\hat{\mathsf{L}}_k\,, \tag{3.77}$$

also satisfy

$$[\hat{\boldsymbol{L}}, \hat{\mathsf{L}}^2] = 0\,. \tag{3.78}$$

One can easily check the commutation relations (3.77) by substituting from the definition (3.74) and using the canonical commutation relations (1.54).

We see from Eqs. (3.75) and (3.78) that components of angular momentum commute with both the operator of the square of the angular momentum and the Hamiltonian for a spherically symmetric potential. On the other hand, Eq. (3.77) shows that the components do not commute mutually. This situation is analogous to that of the hyperfine splitting discussed in the previous section. We can measure the energy, the magnitude of angular momentum, and only one of the angular momentum components. The operators $\{\hat{\mathsf{H}}, \hat{\mathsf{L}}^2, \hat{\mathsf{L}}_z\}$ form a complete set of commuting operators for a spinless particle moving in a spherically symmetric potential. Instead of a general spherically symmetric potential $V(r)$, we henceforth focus on the Coulomb potential which is of a particular interest to us

$$V(r) = -\frac{1}{r}\,.$$

We have already discussed in the previous section that one can find common eigenvectors for a complete set of commuting operators:

$$\hat{\mathsf{H}}|n,l,m\rangle = -\frac{1}{2n^2}|n,l,m\rangle\,, \tag{3.79}$$

$$\hat{\mathsf{L}}^2|n,l,m\rangle = l(l+1)|n,l,m\rangle \tag{3.80}$$

and

$$\hat{\mathsf{L}}_z|n,l,m\rangle = m|n,l,m\rangle\,, \tag{3.81}$$

where n, l, and m are called the principal, orbital, and magnetic quantum numbers, respectively. We write eigenvalues of the energy operators and of the operator of angular momentum magnitude in a form that will prove indeed useful, see the discussion in Sect. 3.2.5. We may certainly do so for one can write any negative real number as $-(2n^2)^{-1}$ where n is a real number, and any positive real number as $l(l+1)$, where l is a real number. We now rewrite these equations in the coordinate representation. After substituting Eqs. (3.21) and (3.76) into the Hamiltonian (3.75), we obtain from the first equation (3.79) the equation

$$\left[-\frac{1}{2}\left(\frac{\partial^2}{\partial r^2} + \frac{2}{r}\frac{\partial}{\partial r} - \frac{\hat{\mathsf{L}}^2}{r^2}\right) - \frac{1}{r}\right]\psi_{nlm} = -\frac{1}{2n^2}\psi_{nlm}\,. \tag{3.82}$$

In spherical coordinates, the operators $\hat{\mathsf{L}}_z$ and $\hat{\mathsf{L}}^2$ are independent of the radial variable r, see Eq. (3.74). Separating the radial and angular parts of the wave function

$$\langle \mathbf{r}|\,n,l,m\rangle = \psi_{nlm}(r,\vartheta,\varphi) = R_{nl}(r)Y_{lm}(\vartheta,\varphi)\,, \tag{3.83}$$

and using the fact that the operators $\hat{\mathsf{L}}^2$ a $\hat{\mathsf{L}}_z$ act only on the angular part of the wave function, the radial part drops out from Eqs. (3.80) and (3.81):

$$\hat{\mathsf{L}}^2 Y_{lm} = l(l+1)Y_{lm} \tag{3.84}$$

and

$$\hat{\mathsf{L}}_z Y_{lm} = mY_{lm} \,. \tag{3.85}$$

The functions Y_{lm} are commonly known as *spherical harmonics*. They are orthogonal to each other as they are eigenfunctions of Hermitian operators:

$$\langle l, m | l', m' \rangle = \int \langle l, m | \mathbf{n} \rangle \langle \mathbf{n} | l', m' \rangle \, \mathrm{d}\Omega = \int Y_{l,m}(\mathbf{n})^* Y_{l',m'}(\mathbf{n}) \, \mathrm{d}\Omega = \delta_{l,l'}\delta_{m,m'} \,. \tag{3.86}$$

For historical reasons, we denote the states corresponding to the orbital quantum number of values $l = 0, 1, 2, 3, 4$ as s, p, d, f, g.

After substitution of Eq. (3.84) into Eq. (3.82), integration over the angles and use of the orthonormality of the spherical harmonics (3.86), we finally obtain

$$\left[-\frac{1}{2} \left(\frac{\mathrm{d}^2}{\mathrm{d}r^2} + \frac{2}{r}\frac{\mathrm{d}}{\mathrm{d}r} - \frac{l(l+1)}{r^2} \right) - \frac{1}{r} \right] R_{nl} = -\frac{1}{2n^2} R_{nl} \,. \tag{3.87}$$

We thus see that integrals of motion, namely components of angular momentum, allow us to reformulate the three-dimensional problem (3.9) as three one-dimensional ones: (3.84), (3.85) and (3.87).

3.4.2 Angular Dependence of *p*-States

We now find the components of angular momentum in the coordinate representation and then derive thereof the shape of p-states using the relations (3.69), (3.70), and (3.71). As we will see later in this book, see Sect. 4.1, the relations (3.69), (3.70), and (3.71) hold for any type of angular momentum, irrespective of its origin. However, for the purpose of this section, we will only assume it without proof.

By substituting Eqs. (3.14) and (3.16) in Eq. (3.74), we obtain

$$\hat{\mathsf{L}}_z = -\mathrm{i}\frac{\partial}{\partial \varphi}$$

and

$$\hat{\mathsf{L}}_\pm = \hat{\mathsf{L}}_x \pm \mathrm{i}\hat{\mathsf{L}}_y = \mathrm{e}^{\pm \mathrm{i}\varphi} \left(\pm \frac{\partial}{\partial \vartheta} + \mathrm{i} \operatorname{cotg} \vartheta \frac{\partial}{\partial \varphi} \right) . \tag{3.88}$$

We have thus transformed the eigenvalue problem (3.85) into a differential equation for Y_{lm},

$$\hat{\mathsf{L}}_z Y_{lm} = -\mathrm{i}\frac{\partial Y_{lm}}{\partial\varphi} = mY_{lm} \Rightarrow Y_{lm}(\vartheta,\varphi) = P_{lm}(\vartheta)\mathrm{e}^{\mathrm{i}m\varphi}\,. \tag{3.89}$$

The wave function, directly proportional to Y_{lm}, must be an unambiguous function of space coordinates. Therefore, Y_{lm}, as a function of angles ϑ and φ, must be invariant with respect to the interchange $\varphi \to \varphi + 2\pi$, that is, to a rotation by a full circle:

$$Y_{lm}(\vartheta,\varphi) = Y_{lm}(\vartheta,\varphi+2\pi) = Y_{lm}(\vartheta,\varphi)\mathrm{e}^{2\pi\mathrm{i}m}\,. \tag{3.90}$$

We thus see that m must be an integer so that the exponential equals one. Therefore, in the case of orbital angular momentum, the magnetic quantum numbers m may be only an integer. Later, see Sect. 4.1, we show that the same applies also to the quantum numbers l.

We now determine the functions $P_{lm}(\vartheta)$, and hence also the whole functions $Y_{lm}(\vartheta,\varphi)$ for $l=1$ and successively for $m=-1,0,1$. To start with, according to Eq. (3.71) we have

$$\hat{\mathsf{L}}_-|1,-1\rangle = 0 \Rightarrow \hat{\mathsf{L}}_- Y_{1,-1}(\vartheta,\varphi) = 0\,.$$

Substitution for $\hat{\mathsf{L}}_-$ from Eq. (3.88) and $Y_{1,-1}(\vartheta,\varphi) = P_{1,-1}(\vartheta)\mathrm{e}^{\mathrm{i}(-1)\varphi}$ results in a differential equation for $P_{1,-1}$

$$\frac{\mathrm{d}P_{1,-1}}{\mathrm{d}\vartheta} = \operatorname{cotg}\vartheta\, P_{1,-1}\,,$$

with the solution

$$P_{1,-1} = K\sin\vartheta\,.$$

We find the normalization constant K from the normalization requirement

$$1 = \langle 1,-1|\,1,-1\rangle = \int Y^*_{1,-1}Y_{1,-1}\,\mathrm{d}\Omega = K^2\int\sin^2\vartheta\,\mathrm{d}\Omega = \frac{8\pi}{3}K^2\,,$$

hence

$$Y_{1,-1}(\vartheta,\varphi) = \sqrt{\frac{3}{8\pi}}\sin\vartheta\,\mathrm{e}^{-\mathrm{i}\varphi}\,. \tag{3.91}$$

We easily obtain other functions using Eq. (3.70):

$$Y_{1,0} = \frac{1}{\sqrt{2}}\hat{\mathsf{L}}_+ Y_{1,-1} = \frac{1}{\sqrt{2}} e^{i\varphi}\left(\frac{\partial}{\partial\vartheta} + i \operatorname{cotg}\vartheta \frac{\partial}{\partial\varphi}\right)\sqrt{\frac{3}{8\pi}} \sin\vartheta\, e^{-i\varphi} = \sqrt{\frac{3}{4\pi}} \cos\vartheta\,, \tag{3.92}$$

$$Y_{1,1} = \frac{1}{\sqrt{2}}\hat{\mathsf{L}}_+ Y_{1,0} = \frac{1}{\sqrt{2}} e^{i\varphi}\left(\frac{\partial}{\partial\vartheta} + i \operatorname{cotg}\vartheta \frac{\partial}{\partial\varphi}\right)\sqrt{\frac{3}{4\pi}} \cos\vartheta = -\sqrt{\frac{3}{8\pi}} \sin\vartheta\, e^{i\varphi}\,. \tag{3.93}$$

As physicists, we are satisfied. We have found the common eigenfunctions of the operator of angular momentum magnitude and of its third components for spin 1 in the coordinate representation. Absolute values of these functions then determine the angular probability distribution of the electron occurrence in a p-state.

Chemists, on the other hand, seek a trial function that describes a molecule as a linear combination of atomic functions. They give up the requirement these functions be eigenfunctions of $\hat{\mathsf{L}}_z$ in favor of a real basis $\{p_x, p_y, p_z\}$:

$$p_x = -\frac{1}{\sqrt{2}}\left(Y_{1,1} - Y_{1,-1}\right) = \sqrt{\frac{3}{4\pi}} \sin\vartheta \cos\varphi\,,$$

$$p_y = \frac{1}{i\sqrt{2}}\left(Y_{1,1} + Y_{1,-1}\right) = \sqrt{\frac{3}{4\pi}} \sin\vartheta \sin\varphi\,,$$

$$p_z = Y_{1,0} = \sqrt{\frac{3}{4\pi}} \cos\vartheta\,.$$

There is a general symmetry, see, e.g., Eqs. (3.91) and (3.93),

$$Y_{l,-m} = (-1)^m Y^*_{l,m}\,, \tag{3.94}$$

therefore, one can always find a real basis for d, f and other orbitals. One can also graph the angular distribution of the squares of the states p_x, p_y, p_z and obtain polar diagrams as in Fig. 3.2.

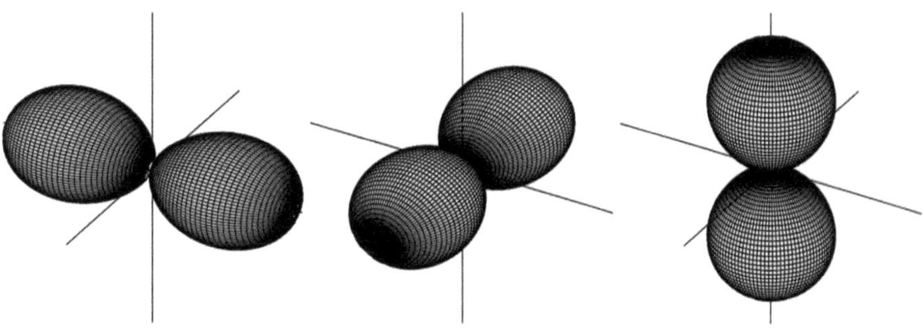

Fig. 3.2 *Left to right*: Graphs of p_x^2, p_y^2 and p_z^2

3.4.3 Accidental Degeneracy

The energy of stationary states of any spherically symmetric potential is independent of the magnetic number m, as one can see in Eq. (3.87). In case of the Coulomb potential, we encounter another degeneracy, the so-called *accidental degeneracy*: energy does not depend on the orbital quantum number l either. We will show the reasons later in Chap. 4.

After direct substitution in Eq. (3.87) one can easily verify that for the first excited state $n = 2$, there is a solution for the s-state ($l = 0$)

$$R_{20}(r) = R_{2s}(r) = \frac{1}{\sqrt{2}}\left(1 - \frac{r}{2}\right) e^{-r/2} \tag{3.95}$$

and also for the p-state ($l = 1$)

$$R_{21}(r) = R_{2p}(r) = \frac{1}{2\sqrt{6}} r e^{-r/2} . \tag{3.96}$$

However, we will leave a general solution of Eq. (3.87) for the next chapter.

3.5 Fine Structure

The sodium spectrum constitutes two very close yellow spectral lines at around the wavelength 589 nm arising from the splitting of a single line of the transition $3p \to 3s$. This splitting stems from the splitting of the $3p$ energy level due to the so-called spin-orbit interaction, that is, an interaction of the outer (orbital) and intrinsic (spin) angular momentum of an electron. As we have already mentioned, an electron behaves as an electric monopole and a magnetic dipole. Since an electron carries a charge, its orbital motion around the nucleus gives rise to a magnetic field, which in turn affects the magnetic dipole. This spin-orbital interaction is a relativistic effect, hence we will derive it from the Dirac equation, a relativistic equation for a particle with spin 1/2. However, instead of the splitting of sodium $3p$ level, we consider rather splitting of the hydrogen $2p$ level.[12]

3.5.1 Relativistic Corrections

It has been already Schrödinger who attempted to incorporate relativistic effects into quantum mechanics. Moreover, he obtained the simplest relativistic

[12] One will be surely able to solve the sodium case after reading Chap. 5.

quantum-mechanical equation, known as the Klein-Gordon equation, prior to the "Schrödinger" equation. He started by writing the relation between energy and momentum as $E^2 = m^2 + p^2$, substituted $\hat{\mathsf{p}}$ for p and thus obtained

$$(\hat{\mathsf{p}}^2 + m^2)|\psi\rangle = E^2|\psi\rangle\,. \tag{3.97}$$

However, this equation is inadequate for the description of a particle with spin one half. The reason is, Eq. (3.97) does not include any spin-dependent term at all.

In order to introduce spin into Eq. (3.97), we use a similar trick as in Eq. (3.49). We write

$$E - m = \frac{E^2 - m^2}{E + m} = \frac{p^2}{E + m} = \boldsymbol{\sigma}\cdot\mathbf{p}\frac{1}{E+m}\boldsymbol{\sigma}\cdot\mathbf{p}\,,$$

exactly as we did in Eq. (3.50). The relativistic equation for a free particle with spin reads then

$$(E - m)\psi = \left(\boldsymbol{\sigma}\cdot\hat{\boldsymbol{p}}\frac{1}{E+m}\boldsymbol{\sigma}\cdot\hat{\boldsymbol{p}}\right)\psi\,.$$

We perform the substitution (3.1) for a particle in an external electromagnetic field and thus obtain the *Dirac equation* for a two-component wave function of a particle moving in and external electromagnetic field

$$(E - e\varphi - m)\psi = \left[\boldsymbol{\sigma}\cdot(\hat{\boldsymbol{p}} - e\mathbf{A})\frac{1}{E - e\varphi + m}\boldsymbol{\sigma}\cdot(\hat{\boldsymbol{p}} - e\mathbf{A})\right]\psi\,. \tag{3.98}$$

The first and second components of the function $\psi(\mathbf{r})$ determine the amplitude of the probability density that a particle is located the point $\mathbf{r}$ and the spin projection along the z equals $+\frac{1}{2}$ and $-\frac{1}{2}$, respectively. We call such wave functions *spinors*.

One usually encounters the Dirac equation in a somewhat different form, namely for a four-component wave function. However, the above presented form serves the purposes of this section much better. We will revisit the Dirac equation again in Chap. 7 where we will use the more common form, see Eq. (7.19), and show its connection to the form (3.98).

We denote $\Delta E = E - m$, which represents the total energy without the rest mass, that is, ΔE stands for the bond energy. We then obtain

$$(\Delta E - e\varphi)\psi = \left[\boldsymbol{\sigma}\cdot(\hat{\boldsymbol{p}} - e\mathbf{A})\frac{1}{\Delta E - e\varphi + 2m}\boldsymbol{\sigma}\cdot(\hat{\boldsymbol{p}} - e\mathbf{A})\right]\psi\,.$$

We use the identity

$$\hat{\mathsf{A}}\hat{\mathsf{C}}\hat{\mathsf{A}} = \frac{1}{2}(\{\hat{\mathsf{A}}^2, \hat{\mathsf{C}}\} - [\hat{\mathsf{A}}, [\hat{\mathsf{A}}, \hat{\mathsf{C}}]]) \tag{3.99}$$

to rearrange the bracket. We thus have

$$2(\Delta E - e\varphi)\psi = \left(\left\{[\sigma \cdot (\hat{\boldsymbol{p}} - e\mathbf{A})]^2, \frac{1}{\Delta E - e\varphi + 2m}\right\}\right.$$
$$\left. - \left[\sigma \cdot (\hat{\boldsymbol{p}} - e\mathbf{A}), \left[\sigma \cdot (\hat{\boldsymbol{p}} - e\mathbf{A}), \frac{1}{\Delta E - e\varphi + 2m}\right]\right]\right)\psi .$$

The curly brackets denote the so-called *anticommutator* of two operators $\hat{\mathsf{A}}$ and $\hat{\mathsf{B}}$

$$\{\hat{\mathsf{A}}, \hat{\mathsf{B}}\} = \hat{\mathsf{A}}\hat{\mathsf{B}} + \hat{\mathsf{B}}\hat{\mathsf{A}} .$$

The rest energy of an electron m substantially exceeds the difference of its bond and potential energy $\Delta E - e\varphi$. The ionization energy of the hydrogen atom in a ground state amounts to 13.6 eV, while the rest mass of an electron equals 0.511 MeV, and the mean potential energy is of the same order of magnitude as the bond energy. This thus justifies the approximation

$$\frac{1}{2m + \Delta E - e\varphi} \simeq \frac{1}{2m}\left(1 - \frac{\Delta E - e\varphi}{2m}\right) , \tag{3.100}$$

where we considered the first two terms of the expansion of this fraction. We now substitute this expansion into our equation. The first of the two terms disappears within the commutator for it is a constant that commutes with both the momentum and the Pauli matrices. Thus we have

$$\Delta E\psi = \left(\frac{[\sigma \cdot (\hat{\boldsymbol{p}} - e\mathbf{A})]^2}{2m} + e\varphi - \frac{1}{8m^2}\left(\{(\hat{\boldsymbol{p}} - e\mathbf{A})^2, \Delta E - e\varphi\}\right.\right.$$
$$\left.\left. + \left[\sigma \cdot (\hat{\boldsymbol{p}} - e\mathbf{A}), \left[\sigma \cdot (\hat{\boldsymbol{p}} - e\mathbf{A}), e\varphi\right]\right]\right)\right)\psi . \tag{3.101}$$

Keeping only the first two terms on the rhs of this equation, that is, we neglect the terms proportional to m^{-2} and higher powers of m, we obtain the already discussed Pauli equation (3.50)—the nonrelativistic equation for a particle with spin $1/2$.

Since Eq. (3.101) is only approximate, see Eq. (3.100), there is no point in finding its exact solution, hence we will seek its solution employing the perturbation method. As has been already pointed out, the magnetic interactions are much smaller than the electrostatic one. Thus, we consider further only the effect of the latter, i.e., we set

$$\mathbf{A} = 0 .$$

Furthermore, we write the last equation in a symbolic form

$$\Delta E\psi = (\hat{\mathsf{H}}_0 + \lambda\hat{\mathsf{H}}_1)\psi ,$$

where $\hat{\mathsf{H}}_0$ denotes the nonrelativistic Hamiltonian and $\hat{\mathsf{H}}_1$ the relativistic correction

$$\hat{\mathsf{H}}_0 = \frac{\hat{\mathsf{p}}^2}{2m} + e\varphi\,, \quad \hat{\mathsf{H}}_1 = -\frac{1}{8m^2}\left(\{\hat{\mathsf{p}}^2, \Delta E - e\varphi\} + \left[\boldsymbol{\sigma}\cdot\hat{\boldsymbol{p}}, \left[\boldsymbol{\sigma}\cdot\hat{\boldsymbol{p}}, e\varphi\right]\right]\right)$$

and λ is a formal expansion parameter. We will see that for hydrogen-like atoms $\lambda = (Z\alpha)^2$. Even for a hydrogen-like 91 times ionized uranium, $Z = 92$, the factor equals $(Z\alpha)^2 \simeq 0.45$. We now find the first-order corrections to the nonrelativistic energies

$$\Delta E \simeq \Delta E_0 + \lambda \Delta E_1\,, \qquad \psi \simeq \psi_0 + \lambda\psi_1\,,$$

where ΔE_0 and ψ_0 stand for the nonrelativistic energies and wave functions

$$(\Delta E_0 - e\varphi)\,|\psi_0\rangle = \frac{\hat{\mathsf{p}}^2}{2}|\psi_0\rangle\,. \tag{3.102}$$

The expression for first-order corrections, Eq. (2.11), reads

$$(\hat{\mathsf{H}}_0 - \Delta E_0)|\psi_1\rangle = (\Delta E_1 - \hat{\mathsf{H}}_1)|\psi_0\rangle\,.$$

Multiplying this equation with $\langle\psi_0|$ leads to

$$\Delta E_1 = \langle\psi_0|\hat{\mathsf{H}}_1|\psi_0\rangle = \langle\psi_0|\hat{\mathsf{H}}_{\text{rel}}|\psi_0\rangle\,, \tag{3.103}$$

$$\hat{\mathsf{H}}_{\text{rel}} = -\frac{\hat{\mathsf{p}}^4}{8m^3} - \frac{1}{8m^2}\left[\boldsymbol{\sigma}\cdot\hat{\boldsymbol{p}}, \left[\boldsymbol{\sigma}\cdot\hat{\boldsymbol{p}}, e\varphi(\mathbf{r})\right]\right]\,, \tag{3.104}$$

where we substituted from Eq. (3.102) for $\Delta E_0 - e\varphi$.

We would have obtained the first term also had we treated the Klein-Gordon equation for a scalar particle in an external electromagnetic field with perturbation method. This term expresses the dependence of the inertial mass of an electron on its velocity. In the framework of classical physics, one can obtain this term from the expansion of the expression $\sqrt{p^2 + m^2} \simeq m + p^2/(2m) - p^4/(8m^3) + \ldots$. The second term captures the effect of spin. It is advantageous to further rearrange it

$$\begin{aligned}\left[\boldsymbol{\sigma}\cdot\hat{\boldsymbol{p}}, \left[\boldsymbol{\sigma}\cdot\hat{\boldsymbol{p}}, V(\mathbf{r})\right]\right] &= \sigma_i\sigma_j\left[\hat{\mathsf{p}}_i, \left[\hat{\mathsf{p}}_j, V\right]\right] + \left[\sigma_i, \sigma_j\right]\left[\hat{\mathsf{p}}_j, V\right]\hat{\mathsf{p}}_i \\ &= \left[\hat{\mathsf{p}}_i, \left[\hat{\mathsf{p}}_i, V\right]\right] + 2\mathrm{i}\varepsilon_{ijk}\sigma_k\left[\hat{\mathsf{p}}_j, V\right]\hat{\mathsf{p}}_i \\ &= -\nabla^2(V) - 4\mathrm{i}\hat{\boldsymbol{S}}\cdot\left([\hat{\boldsymbol{p}}, V]\times\hat{\boldsymbol{p}}\right)\,.\end{aligned} \tag{3.105}$$

In the first equality, we used the identity

$$[\hat{\mathsf{A}}\hat{\mathsf{B}}, \hat{\mathsf{C}}] = \hat{\mathsf{A}}[\hat{\mathsf{B}}, \hat{\mathsf{C}}] + [\hat{\mathsf{A}}, \hat{\mathsf{C}}]\hat{\mathsf{B}}\,,$$

where $\hat{\mathsf{A}} = \boldsymbol{\sigma}$, $\hat{\mathsf{B}} = \hat{\boldsymbol{p}}$ and $\hat{\mathsf{C}}$ contains the remaining terms. In the second step we exploited the identity for the Pauli matrices (1.25) and in the third we used Eqs. (1.55) and (1.24).

We now apply the foregoing considerations to the specific case of Coulomb potential

$$e\varphi = -\frac{Z\alpha}{r}\,.$$

We make transition to atomic (dimensionless) units, namely set $r = r_A/Z\alpha m$, $\hat{\mathsf{p}} = \hat{\mathsf{p}}_A Z\alpha m$. The expansion of the bond energy now takes the form

$$\Delta E = m(Z\alpha)^2\varepsilon\,, \quad \varepsilon \simeq \varepsilon_0 + (Z\alpha)^2\varepsilon_1\,,$$

where ε_0 is the nonrelativistic energy in the atomic units, Eq. (3.24). It follows for the Coulomb potential from Eq. (3.105)

$$\left[\boldsymbol{\sigma}\cdot\hat{\boldsymbol{p}}, \left[\boldsymbol{\sigma}\cdot\hat{\boldsymbol{p}}, -\frac{1}{r}\right]\right] = -4\pi\delta(\mathbf{r}) - 4\frac{\hat{\boldsymbol{S}}\cdot\hat{\boldsymbol{L}}}{r^3}\,, \tag{3.106}$$

where we used Eqs. (3.6)–(3.8) in the first term and Eqs. (3.15) and (3.74) in the second term. We substitute the last equation into Eq. (3.103) and from Eq. (3.102) preferably use the term with energy rather than momentum. We thus have an expression for the first-order corrections to hydrogen-like atoms given by relativistic effects

$$\varepsilon_1 = \frac{1}{2}\langle\psi_0|\left(-\left(\varepsilon_0 + \frac{1}{r_A}\right)^2 + \pi\delta(\mathbf{r}_A) + \frac{\hat{\boldsymbol{S}}\cdot\hat{\boldsymbol{L}}}{r_A^3}\right)|\psi_0\rangle\,. \tag{3.107}$$

3.5.2 *Fine Splitting of the Energy Level n* = 2

When considering the hydrogen atom in the nonrelativistic approximation, the state $2p\ |n = 2, l = 1, m\rangle|\pm\rangle$ is hextuply degenerate: three times for the orbital angular momentum projections ($m = 1, 0, 1$) and twice for the electron spin orientation. For the purpose of this section, we neglect the effects of the nuclear spin, as they are much smaller than the spin-orbit interaction, see Sect. 3.6. The first two terms in $\hat{\mathsf{H}}_1$, see Eq. (3.107), act on all these degenerate levels equally. The only term distinguishing between them is the last one. Since we are concerned only with the splitting of the $2p$ state at this moment, we will calculate only the expression

$$\langle\psi_0|\frac{\hat{\boldsymbol{S}}\cdot\hat{\boldsymbol{L}}}{2r^3}|\psi_0\rangle = \frac{1}{2}\int_0^\infty r^2 R_{2p}^2\frac{1}{r^3}\mathrm{d}r\left\langle j, m, \left(\frac{1}{2}, 1\right)\middle|\hat{\boldsymbol{S}}\cdot\hat{\boldsymbol{L}}\middle|j, m, \left(\frac{1}{2}, 1\right)\right\rangle$$

$$= \frac{1}{48}\left\langle j, m, \left(\frac{1}{2}, 1\right)\middle|\hat{\boldsymbol{S}}\cdot\hat{\boldsymbol{L}}\middle|j, m, \left(\frac{1}{2}, 1\right)\right\rangle, \qquad (3.108)$$

where we separated the radial and spin-angular degrees of freedom

$$\langle r|\,\psi_0\rangle = R_{2p}(r)\left|j, m, \left(\frac{1}{2}, 1\right)\right\rangle.$$

The vector $\left|j, m, \left(\frac{1}{2}, 1\right)\right\rangle$ is an eigenvector of the operator $\hat{\boldsymbol{S}}\cdot\hat{\boldsymbol{L}}$ which we are now about to find. We will explain the notation of this vector later in Sect. 4.2.

We use Eq. (1.35) to find how the operator $\hat{\boldsymbol{S}}\cdot\hat{\boldsymbol{L}}$ acts on the direct product of the orbital $|1, m\rangle$ and spin $|\pm\rangle$ states. To determine the action of spin operators, we employ Eqs. (1.32), (1.33) and (1.34), and for orbital operators Eqs. (3.69), (3.70), and (3.71), where we substitute $\hat{\mathsf{L}}$ for $\hat{\mathsf{S}}$. We thus obtain the following relations:

$$\begin{aligned}
|I\rangle &= |1, 1\rangle|+\rangle & \hat{\boldsymbol{S}}\cdot\hat{\boldsymbol{L}}|I\rangle &= \frac{1}{2}|I\rangle\,,\\
|II\rangle &= |1, 1\rangle|-\rangle & \hat{\boldsymbol{S}}\cdot\hat{\boldsymbol{L}}|II\rangle &= -\frac{1}{2}|II\rangle + \frac{1}{\sqrt{2}}|III\rangle\,,\\
|III\rangle &= |1, 0\rangle|+\rangle & \hat{\boldsymbol{S}}\cdot\hat{\boldsymbol{L}}|III\rangle &= \frac{1}{\sqrt{2}}|II\rangle\,,\\
|IV\rangle &= |1, 0\rangle|-\rangle & \hat{\boldsymbol{S}}\cdot\hat{\boldsymbol{L}}|IV\rangle &= \frac{1}{\sqrt{2}}|V\rangle\,,\\
|V\rangle &= |1, -1\rangle|+\rangle & \hat{\boldsymbol{S}}\cdot\hat{\boldsymbol{L}}|V\rangle &= -\frac{1}{2}|V\rangle + \frac{1}{\sqrt{2}}|IV\rangle\,,\\
|VI\rangle &= |1, -1\rangle|-\rangle & \hat{\boldsymbol{S}}\cdot\hat{\boldsymbol{L}}|VI\rangle &= \frac{1}{2}|VI\rangle\,.
\end{aligned}$$

The states $|I\rangle$ and $|VI\rangle$ are eigenstates of this operator with the eigenvalue of $\frac{1}{2}$. The other states mix only in pairs ($|II\rangle$ with $|III\rangle$ and $|IV\rangle$ with $|V\rangle$) so one can expect the eigenvectors to be linear combinations of each of the pairs. We find the coefficients:

$$\hat{\boldsymbol{S}}\cdot\hat{\boldsymbol{L}}\,(c_2|II\rangle + c_3|III\rangle) = \lambda\,(c_2|II\rangle + c_3|III\rangle)\,,$$

where on the lhs we insert from the above found expressions

$$c_2\left(\frac{1}{\sqrt{2}}|III\rangle - \frac{1}{2}|II\rangle\right) + c_3\frac{1}{\sqrt{2}}|II\rangle = c_2\lambda|II\rangle + c_3\lambda|III\rangle\,.$$

We compare the coefficients for each of the orthogonal vectors and obtain the eigenvalue problem:

$$\begin{pmatrix} -\frac{1}{2} & \frac{1}{\sqrt{2}} \\ \frac{1}{\sqrt{2}} & 0 \end{pmatrix}\begin{pmatrix} c_2 \\ c_3 \end{pmatrix} = \lambda\begin{pmatrix} c_2 \\ c_3 \end{pmatrix}.$$

Imposing the normalization requirement $|c_2|^2 + |c_3|^2 = 1$, the solutions read

$$\lambda = \frac{1}{2}, \qquad c_2 = \frac{1}{\sqrt{3}}, \qquad c_3 = \sqrt{\frac{2}{3}}$$

and

$$\lambda = -1, \qquad c_2 = \sqrt{\frac{2}{3}}, \qquad c_3 = -\frac{1}{\sqrt{3}}.$$

We thus have the eigenstates

$$|\overline{II}\rangle = \frac{1}{\sqrt{3}}|II\rangle + \sqrt{\frac{2}{3}}|III\rangle \quad \text{and} \quad |\overline{III}\rangle = \sqrt{\frac{2}{3}}|II\rangle - \frac{1}{\sqrt{3}}|III\rangle$$

with the eigenvalues $\frac{1}{2}$ and -1, respectively. In a similar manner, we obtain symmetric relations for the other pair

$$|\overline{V}\rangle = \frac{1}{\sqrt{3}}|V\rangle + \sqrt{\frac{2}{3}}|IV\rangle \quad \text{and} \quad |\overline{IV}\rangle = \sqrt{\frac{2}{3}}|V\rangle - \frac{1}{\sqrt{3}}|IV\rangle,$$

again with the eigenvalues $\frac{1}{2}$ and -1, respectively.

We now revise our findings. The spin-orbital interaction splits the initially hextuply degenerate state into a quadruply and a doubly degenerate states, commonly referred to as a quadruplet and a doublet, respectively. The splitting of the energy level amounts to $\varepsilon_1 = \frac{1}{48}\left(\frac{1}{2} - (-1)\right) = \frac{1}{32}$ in atomic units. We can easily find the energy in electronvolts by returning to E, namely

$$\Delta E = \varepsilon_1 m (Z\alpha)^4 .$$

The difference in terms of frequency equals, according to Eq. (3.12),

$$\Delta\nu = R_\infty c \frac{\Delta E}{m\frac{\alpha^2}{2}} = R_\infty c \frac{Z^4\alpha^2}{16} .$$

For the hydrogen atom ($Z = 1$) we find that $\Delta\nu_{\text{theo}} = 10{,}949\,\text{MHz}$, while the experimental value is [3]

$$\Delta\nu_{\text{exp}}(2p_{3/2} - 2p_{1/2}) = 10{,}969.041475\,(99)\,\text{MHz} \tag{3.109}$$

Once again, we have arrived at a reasonable match. The wavelength of the radiation that arises from the transition between these two states then equals approximately 2.7 cm.

3.5.3 Classification of States Using the Integrals of Motion

As in the case of the spin-spin interaction, we can ask why one of the levels is quadruply degenerate while the other only doubly. We introduce an operator of the total angular momentum and its square

$$\hat{\boldsymbol{J}} = \hat{\boldsymbol{S}} + \hat{\boldsymbol{L}}, \qquad \hat{\mathrm{J}}^2 = \hat{\mathrm{L}}^2 + \hat{\mathrm{S}}^2 + 2\hat{\boldsymbol{L}} \cdot \hat{\boldsymbol{S}}.$$

We then find that

$$\hat{\boldsymbol{L}} \cdot \hat{\boldsymbol{S}} = \frac{1}{2}\left(\hat{\mathrm{J}}^2 - \hat{\mathrm{L}}^2 - \hat{\mathrm{S}}^2\right). \tag{3.110}$$

Since the orbital angular momentum of our system is $l = 1$, we have $l(l+1) = 1(1+1) = 2$ and similarly for the square of the electron spin $S(S+1) = \frac{1}{2}\left(\frac{1}{2}+1\right) = \frac{3}{4}$. We also know that $\hat{\boldsymbol{L}} \cdot \hat{\boldsymbol{S}}$ should equal one half in four cases and minus one in the other two cases

$$\hat{\boldsymbol{L}} \cdot \hat{\boldsymbol{S}} \left| j, m, \left(\frac{1}{2}, 1\right) \right\rangle = \frac{1}{2}\left(j(j+1) - \frac{11}{4}\right) \left| j, m, \left(\frac{1}{2}, 1\right) \right\rangle = \begin{cases} 1/2 \\ -1 \end{cases} \left| j, m, \left(\frac{1}{2}, 1\right) \right\rangle. \tag{3.111}$$

Therefore, the eigenvalue of the operator $\hat{\mathrm{J}}^2$ must necessarily equal either

$$j(j+1) = \frac{15}{4} = \frac{3}{2}\left(\frac{3}{2}+1\right),$$

or

$$j(j+1) = \frac{3}{4} = \frac{1}{2}\left(\frac{1}{2}+1\right).$$

In other words, from an outside point of view, the system behaves either as a particle with spin $\frac{3}{2}$ or $\frac{1}{2}$. From the action of the operator $\hat{\mathrm{J}}_z = \hat{\mathrm{L}}_z + \hat{\mathrm{S}}_z$ on the eigenvectors of the operator $\hat{\boldsymbol{L}} \cdot \hat{\boldsymbol{S}}$ we easily find that the quadruple degeneracy of the higher energy level arises from the existence of four states with four possible projections $-\frac{3}{2}, -\frac{1}{2}, \frac{1}{2}, \frac{3}{2}$ of the total angular moment along the z-axis. The doubly degenerate lower level, on the other hand, results from the existence of two states with two possible projections $-\frac{1}{2}, \frac{1}{2}$ along the z-axis. The spectroscopic notation of these spectral lines is derived from it, namely $2p_{3/2}$ for the former and $2p_{1/2}$ for the latter.

Exercise 7: Fine Splitting

Calculate the expression (3.107) for the states $2s_{1/2}$ and $2p_{1/2}$. Consider

$$\langle \mathbf{r}|2s_{1/2}\rangle = R_{20}(r)Y_{00}(\mathbf{n})|+\rangle$$

for the former and

$$\langle \mathbf{r}|2p_{1/2}\rangle = R_{21}(r)\left[\sqrt{\frac{2}{3}}Y_{11}(\mathbf{n})|-\rangle - \frac{1}{\sqrt{3}}Y_{10}(\mathbf{n})|+\rangle\right]$$

for the latter.

3.6 Hamiltonian of Two Particles with Precision to α^4

In this section, we summarize and complete individual corrections to the Hamiltonian (3.3). We focus on those which we discussed in this chapter and which contribute to the spacing of the energy levels of atoms in the order of α^4.

We used hydrogen-like atoms as an example to illustrate how the nonrelativistic Hamiltonian constituted by the sum of kinetic energy of particles and electrostatic potential energy defines the gross structure of atomic spectra. The spacing of the energy levels caused by this Hamiltonian is proportional to α^2, see, e.g., Eq. (3.10). To comprehend finer details of atomic spectra, one needs to take into consideration magnetic interactions and effects of relativistic kinematics as well. The contribution of these effects to the energy splitting of atomic energy levels is suppressed by the factor of at least α^2 in comparison to the nonrelativistic Hamiltonian. The magnetic effects arising from the nuclear spin are further diminished by the ratio of the electron and nuclear masses m_e/m_p when compared to those of relativistic kinematics. In the case of more complicated atoms, such as helium, the interaction of an electron with spins of the other electrons reaches the same order as the effects of relativistic kinematics. Therefore, both the interaction of electron spin with orbital angular momentum and the spin-spin interaction among electrons determine the fine structure of such atoms. For clarity, it is definitely worth listing all terms which contribute to the energy spacing of atomic levels to the order of α^4 in two-electron atoms. One surely then manages oneself to generalize them for many-electron atoms.

Although we have already considered magnetic interactions, one needs to include also quantum analogy of the interaction of two currents to achieve precision of the order of α^4. To do this properly, we need to consider a more accurate expression for vector potential than Eq. (3.37) which we obtained in the steady flow approximation.

We thus begin by finding the magnetic field of a moving charge and neglect the retardation effect. Subsequently, we list all terms that contribute to the α^4-order-precise Hamiltonian for two electrons in an external electrostatic field.[13] Finally, we conclude this chapter by discussing the two most interesting cases, namely helium- and hydrogen-like atoms.

3.6.1 *Magnetic Field of a Moving Charge*

To start with, we rewrite Eq. (3.2) for vector potential into a more convenient form. Formally, one can solve the Poisson equation for electrostatic potential

$$\varphi = -\nabla^{-2}\rho\,.$$

Using the continuity equation for charge density

$$\frac{\partial \rho}{\partial t} + \nabla \cdot \mathbf{j} = 0\,, \tag{3.112}$$

we can write the time derivation of the gradient of the potential in the form

$$\nabla\frac{\partial \varphi}{\partial t} = -\nabla\frac{1}{\nabla^2}\frac{\partial \rho}{\partial t} = \nabla\frac{1}{\nabla^2}\nabla\cdot\mathbf{j}\,.$$

We then substitute this mysteriously looking set of triangles on the rhs of Eq. (3.2) for vector potential

$$-\nabla^2\mathbf{A}(x) = \mathbf{j}_\perp(x), \qquad \mathbf{j}_\perp = \left(1 - \frac{\nabla\nabla\cdot}{\nabla^2}\right)\mathbf{j}\,.$$

Note that acting with the divergence operator on both sides of the last equation yields automatically zeros on both sides. This is unlike Eq. (3.37) where the action of the divergence operator resulted in a condition for the current (3.43).

We can express a general solution to the last equation in terms of a time-independent Green function (3.40) as, see Eq. (3.39),

$$\mathbf{A}(\mathbf{r}_1) = \int \mathrm{d}^3\mathbf{r}_2 G(\mathbf{r}_1,\mathbf{r}_2)\mathbf{j}_\perp(\mathbf{r}_2) = \int \mathrm{d}^3\mathbf{r}_2 G(\mathbf{r}_1,\mathbf{r}_2)\left(1 - \frac{\nabla_2\nabla_2\cdot}{\nabla_2^2}\right)\mathbf{j}(\mathbf{r}_2)\,. \tag{3.113}$$

We now write the time-independent Green function using its Fourier transform

[13]A kind reader surely manages to generalize these findings for the case when an external magnetic field is present.

$$G(\mathbf{r}_1, \mathbf{r}_2) = \frac{1}{4\pi|\mathbf{r}_1 - \mathbf{r}_2|} = \frac{1}{(2\pi)^3} \int \frac{\mathrm{d}^3\mathbf{k}}{\omega^2} \mathrm{e}^{\mathrm{i}\mathbf{k}\cdot(\mathbf{r}_1-\mathbf{r}_2)} . \tag{3.114}$$

One can easily verify that this equation holds by substituting the last equation and the three-dimensional generalization of (1.74)

$$\delta(\mathbf{r}_1 - \mathbf{r}_2) = \int \mathrm{d}^3\mathbf{k} \frac{\mathrm{e}^{\mathrm{i}\mathbf{k}\cdot(\mathbf{r}_1-\mathbf{r}_2)}}{(2\pi)^3} \tag{3.115}$$

into Eq. (3.38).

We now substitute for the current density $\mathbf{j}$ its Fourier transform

$$\mathbf{j}(\mathbf{r}) = \frac{1}{(2\pi)^3} \int \mathrm{d}^3\mathbf{k}\mathrm{e}^{\mathrm{i}\mathbf{k}\cdot\mathbf{r}}\mathbf{j}(\mathbf{k}) \tag{3.116}$$

and the second form of the Green function into Eq. (3.113):

$$\mathbf{A}(\mathbf{r}_1) = \int \mathrm{d}^3\mathbf{r}_2 \int \frac{\mathrm{d}^3\mathbf{k}_1}{(2\pi)^3} \int \frac{\mathrm{d}^3\mathbf{k}_2}{(2\pi)^3} \frac{\mathrm{e}^{\mathrm{i}\mathbf{k}_1\cdot(\mathbf{r}_1-\mathbf{r}_2)}}{\omega_1^2} \left(1 - \frac{\mathbf{k}_2\mathbf{k}_2\cdot}{\omega_2^2}\right) \mathbf{j}(\mathbf{k}_2)\mathrm{e}^{\mathrm{i}\mathbf{k}_2\cdot\mathbf{r}_2} .$$

One readily sees that the Fourier transform of the mysteriously looking set of triangles is a *transverse projector*. It isolates only the part of the wave function which changes perpendicularly to the line of sight $\left(1 - \frac{\mathbf{k}\mathbf{k}\cdot}{\omega^2}\right)\mathbf{k} = 0$. Therefore, one can detect only the part of electromagnetic field which results from a charge accelerating in the plane that is perpendicular to the direction connecting the positions of the charge and the observer.

Integration over $\mathbf{r}_2$ yields a δ-function in momenta, see Eq. (3.115), therefore

$$\mathbf{A}(\mathbf{r}_1) = \int \frac{\mathrm{d}^3\mathbf{k}}{(2\pi)^3} \frac{\mathrm{e}^{\mathrm{i}\mathbf{k}\cdot\mathbf{r}_1}}{\omega^2} \left(1 - \frac{\mathbf{k}\mathbf{k}\cdot}{\omega^2}\right) \mathbf{j}(\mathbf{k}) .$$

We assume the current is constituted merely by a point particle

$$\mathbf{j}(\mathbf{r}_2) = e_2\mathbf{v}_2\delta(\mathbf{r}_2 - \mathbf{q}_2) .$$

Fourier transformation of the current caused by a point particle reads, see Eq. (3.115),

$$\mathbf{j}(\mathbf{k}) = e\mathbf{v}_2\mathrm{e}^{-\mathrm{i}\mathbf{k}\cdot\mathbf{q}_2} .$$

After its substitution we find for the vector potential

$$\mathbf{A}(\mathbf{r}_1) = \int \frac{\mathrm{d}^3\mathbf{k}}{(2\pi)^3} \frac{\mathrm{e}^{\mathrm{i}\mathbf{k}\cdot\mathbf{r}_{12}}}{\omega^2} \left(1 - \frac{\mathbf{k}\mathbf{k}\cdot}{\omega^2}\right) e_2\mathbf{v}_2 . \tag{3.117}$$

We have thus obtained a classical result for the vector potential caused by a moving point particle where we neglected the time retardation.

We further rearrange this result. As one may notice, we need to find the integral over wave vectors $\mathbf{k}$:

$$\int \frac{\mathrm{d}^3\mathbf{k}}{(2\pi)^3\omega^2} \mathrm{e}^{\mathrm{i}\omega r\mathbf{n}\cdot\boldsymbol{\eta}} \eta_i\eta_j = \frac{1}{4\pi r}(A\delta_{ij} + Bn_in_j)\,. \tag{3.118}$$

where we changed to spherical coordinates $\mathbf{r} = r\mathbf{n}$ and $\mathbf{k} = \omega\boldsymbol{\eta}$, and A and B are numbers we are about to find. We arrived at this equation employing the following reasoning. There is an object with two Cartesian indices on the lhs which is symmetrical therein. On the rhs, the simplest such object constituted by available indexed objects is the Kronecker δ_{ij} and a direction vector n_i. We find the factor $\frac{1}{r}$ from the substitution $\omega \to \omega/r$ and for esthetic reasons we factor out $\frac{1}{4\pi}$. We now find the numbers A and B by calculating the integrals for two specific cases.

Firstly, if $i = j$ (note we are using the Einstein summation convention!) then

$$3A + B = 1\,.$$

Secondly, if $i = j = 3$ and $\mathbf{n} = (0, 0, 1)$ then[14]

$$A + B = 0\,.$$

The desired integral then equals

$$(3.118) = \frac{1}{4\pi r}\frac{\delta_{ij} - n_in_j}{2}\,.$$

After substitution of this integral into the expression (3.117) we eventually obtain the sought vector potential in the form

---

[14]In this case

$$\begin{aligned}\int \frac{\mathrm{d}^3\mathbf{k}}{(2\pi)^3\omega^2} \mathrm{e}^{\mathrm{i}\omega r\cos\vartheta}\cos^2\vartheta &= \frac{1}{(2\pi)^3 r}\int_0^{2\pi}\mathrm{d}\varphi \int_0^\infty \mathrm{d}\omega \int_0^\pi \mathrm{d}\vartheta \sin\vartheta\, \mathrm{e}^{\mathrm{i}\omega\cos\vartheta}\cos^2\vartheta \\ &= \frac{1}{(2\pi)^2 r}\int_0^\infty \mathrm{d}\omega \left(-\frac{\mathrm{d}^2}{\mathrm{d}\omega^2}\right)\int_0^\pi \mathrm{d}\vartheta \sin\vartheta\, \mathrm{e}^{\mathrm{i}\omega\cos\vartheta} \\ &= \frac{-2}{(2\pi)^2 r}\int_0^\infty \mathrm{d}\omega \frac{\mathrm{d}^2}{\mathrm{d}\omega^2}\frac{\sin\omega}{\omega} = 0\,.\end{aligned}$$

The last equality holds since

$$\left[\frac{\mathrm{d}}{\mathrm{d}\omega}\frac{\sin\omega}{\omega}\right]_0^\infty = 0\,.$$

$$\mathbf{A}(\mathbf{r}_1) = \frac{e_2}{4\pi} \frac{1}{2r_{12}} \left[\mathbf{v}_2 + \mathbf{n}_{12} \, \mathbf{n}_{12} \cdot \mathbf{v}_2 \right] . \tag{3.119}$$

Here, $\mathbf{n}_{12}$ represents a unit vector in the direction $\mathbf{r}_{12} = \mathbf{r}_1 - \mathbf{r}_2$.

3.6.2 Hamiltonian of Two Particles in an External Electromagnetic Field

If we consider both magnetic and relativistic effects, we obtain the Hamiltonian for two particles moving in an external field in the form

$$\hat{\mathsf{H}} = \hat{\mathsf{H}}_0 + \hat{\mathsf{H}}_1, \qquad \hat{\mathsf{H}}_1 = (\hat{\mathsf{H}}_1)_{\text{mag}} + (\hat{\mathsf{H}}_1)_{\text{rel}} .$$

Here, $\hat{\mathsf{H}}_0$ stands for the sum of kinetic energies of both particles and of the electrostatic interaction which comprises both the mutual interaction of the two particles and their interaction with an external field

$$\hat{\mathsf{H}}_0 = \frac{\hat{\mathsf{p}}_1^2}{2m_1} + \frac{\hat{\mathsf{p}}_2^2}{2m_2} + e\varphi(r_1, r_2, r_{12}) , \tag{3.120}$$

where $\mathbf{r}_{12} = \mathbf{r}_1 - \mathbf{r}_2$. The term $(\hat{\mathsf{H}}_1)_{\text{mag}}$ captures the most important magnetic interactions. We find them by substitution into the Pauli Hamiltonian (3.51)

$$(\hat{\mathsf{H}}_1)_{\text{mag}} = -\frac{e_1}{m_1} \left[\mathbf{A}(\mathbf{r}_1) \cdot \hat{\boldsymbol{p}}_1 + g_1 \hat{\boldsymbol{S}}_1 \cdot \mathbf{B}(\mathbf{r}_1) \right] ,$$

for the vector potential from Eqs. (3.46) and (3.119) and by interchange (3.54) and $\mathbf{v}_2 \to \hat{\boldsymbol{p}}_2 / m_2$ [15]

$$\mathbf{A}(\mathbf{r}_1) \to \frac{e_2 g_2}{4\pi m_2} \frac{\hat{\boldsymbol{S}}_2 \times \mathbf{r}_{12}}{r_{12}^3} + \frac{e_2}{4\pi m_2} \frac{1}{2r_{12}} \left[\hat{\boldsymbol{p}}_2 + \mathbf{n}_{12} \, \mathbf{n}_{12} \cdot \hat{\boldsymbol{p}}_2 \right] .$$

and similarly by substitution for the magnetic induction from Eq. (3.48). A kind reader surely noticed that we replace all of the ratios e/m, which appear in front of spin operators, by eg/m so that the expression holds for nonelementary particles as well.[16] The term $(\hat{\mathsf{H}}_1)_{\text{rel}}$ comprises the most important effects of relativistic

[15] The placing of the operators $\hat{\boldsymbol{p}}_1$ and $\hat{\boldsymbol{p}}_2$ is arbitrary as both $(\hat{\mathsf{p}}_1)_i$ and $(\hat{\mathsf{p}}_2)_j$ commute with the expression $(\delta_{ij} + (n_{12})_i (n_{12})_j)/2r_{12}$. The reason thereof is we use the gauge $\nabla \cdot \mathbf{A} = 0$.

[16] We will show later in Chap. 7 that even when the particles indeed are elementary, the gyromagnetic ratio g differs from 1 as a result of effects of quantum electrodynamics. The deviation is very small,

kinematics which we can find by summing the Hamiltonians (3.104) for each particle.

The total Hamiltonian $\hat{\mathsf{H}}_1$ constitutes a sum of several terms

$$\hat{\mathsf{H}}_1 = (\hat{\mathsf{H}}_1)_{\mathrm{imv}} + (\hat{\mathsf{H}}_1)_{\mathrm{con}} + (\hat{\mathsf{H}}_1)_{\mathrm{so}} + (\hat{\mathsf{H}}_1)_{\mathrm{ssc}} + (\hat{\mathsf{H}}_1)_{\mathrm{sst}} + (\hat{\mathsf{H}}_1)_{\mathrm{soo}} + (\hat{\mathsf{H}}_1)_{\mathrm{cc}}\,, \qquad (3.121)$$

where

$$(\hat{\mathsf{H}}_1)_{\mathrm{imv}} = -\frac{\hat{\mathsf{p}}_1^4}{8m_1^3} - \frac{\hat{\mathsf{p}}_2^4}{8m_2^3} \qquad (3.122)$$

results from the dependence of the inertial mass of the particle on its velocity,

$$(\hat{\mathsf{H}}_1)_{\mathrm{con}} = \frac{e_1}{8m_1^2}\nabla_1^2\varphi + \frac{e_2}{8m_2^2}\nabla_2^2\varphi$$

is the spin-independent "contact" term, although stemming from the spin of a particle,

$$(\hat{\mathsf{H}}_1)_{\mathrm{so}} = \frac{\mathrm{i}}{2}\left\{\frac{e_1}{m_1^2}\hat{\boldsymbol{S}}_1\cdot([\hat{\boldsymbol{p}}_1,\varphi]\times\hat{\boldsymbol{p}}_1) + \frac{e_2}{m_2^2}\hat{\boldsymbol{S}}_2\cdot([\hat{\boldsymbol{p}}_2,\varphi]\times\hat{\boldsymbol{p}}_2)\right\} \qquad (3.123)$$

is a term describing the spin-orbit interaction,

$$(\hat{\mathsf{H}}_1)_{\mathrm{ssc}} = -\frac{e_1e_2g_1g_2}{m_1m_2 4\pi}\hat{\boldsymbol{S}}_1\cdot\hat{\boldsymbol{S}}_2\frac{8\pi}{3}\delta(\mathbf{r}_{12}) \qquad (3.124)$$

represents the "contact" part of the spin-spin interaction,

$$(\hat{\mathsf{H}}_1)_{\mathrm{sst}} = \frac{e_1e_2g_1g_2}{m_1m_2 4\pi}\frac{1}{r_{12}^3}\left(\hat{\boldsymbol{S}}_1\cdot\hat{\boldsymbol{S}}_2 - 3\mathbf{n}_{12}\cdot\hat{\boldsymbol{S}}_1\mathbf{n}_{12}\cdot\hat{\boldsymbol{S}}_2\right) \qquad (3.125)$$

stands for the tensor part of the spin-spin interaction,

$$(\hat{\mathsf{H}}_1)_{\mathrm{soo}} = -\frac{e_1e_2}{m_1m_2 4\pi}\frac{1}{r_{12}^3}\left[g_2\hat{\boldsymbol{S}}_2\cdot(\mathbf{r}_{12}\times\hat{\boldsymbol{p}}_1) + g_1\hat{\boldsymbol{S}}_1\cdot(\mathbf{r}_{21}\times\hat{\boldsymbol{p}}_2)\right] \qquad (3.126)$$

describes the spin-other-orbit interaction, and finally

$$(\hat{\mathsf{H}}_1)_{\mathrm{cc}} = -\frac{e_1e_2}{m_1m_2 4\pi}\frac{1}{2r_{12}}\left[\hat{\boldsymbol{p}}_1\cdot\hat{\boldsymbol{p}}_2 + \mathbf{n}_{12}\cdot\hat{\boldsymbol{p}}_1\,\mathbf{n}_{12}\cdot\hat{\boldsymbol{p}}_2\right] \qquad (3.127)$$

captures the current-current interaction.

---

though, as it is proportional to α, see Eq. (7.204). Within our current precision, we can surely neglect this minor deviation.

We obtained the second term in Eq. (3.126) by swapping the roles of the first and second particles: Hamiltonian $(\hat{\mathsf{H}}_1)_{\rm mag}$ describes the reaction of the first particle on the magnetic field caused by the other particle. Similarly, we could consider the reaction of the second particle on the magnetic field stemming from the first particle. In case of spin–spin and current–current interaction, this reasoning would result in including the same interaction twice. In case of spin–other–orbit interaction, on the other hand, one must consider both situations.

The total Hamiltonian $\hat{\mathsf{H}}_1$ discussed in this section is commonly termed as the *Breit Hamiltonian.*

3.6.3 Helium-Like Atoms

We start with Hamiltonian of a helium-like atom in the nonrelativistic approximation

$$\hat{\mathsf{H}} = \frac{\hat{\mathsf{p}}_n^2}{2m_n} + \frac{\hat{\mathsf{p}}_1^2}{2m_e} + \frac{\hat{\mathsf{p}}_2^2}{2m_e} - \frac{Z\alpha}{|\mathbf{r}_1 - \mathbf{r}_n|} - \frac{Z\alpha}{|\mathbf{r}_2 - \mathbf{r}_n|} + \frac{\alpha}{|\mathbf{r}_1 - \mathbf{r}_2|}\,. \tag{3.128}$$

The quantities with the index n are related to the nucleus, the other numbered quantities to the individual electrons. The last term in the potential energy captures the electrostatic repulsion of electrons and the other two terms the electrostatic attraction of the electrons to the nucleus.

The reader can check that the operator of the total linear momentum $\hat{\boldsymbol{p}}_n + \hat{\boldsymbol{p}}_1 + \hat{\boldsymbol{p}}_2$ commutes with the Hamiltonian (3.128). Owing to existence of three operators commuting with the Hamiltonian and with each other we can—as in the case of the hydrogen-like atoms—decrease the dimension of the problem by three. When we introduce the *Jacobi coordinates*

$$\mathbf{r}_{R1} = \mathbf{r}_1 - \mathbf{r}_n, \quad \mathbf{r}_{R2} = \mathbf{r}_2 - \mathbf{r}_n, \quad \mathbf{r}_T = \frac{1}{M}(m_n \mathbf{r}_n + m_e(\mathbf{r}_1 + \mathbf{r}_2)), \quad M = m_n + 2m_e\,,$$

we obtain the Hamiltonian in the form

$$\hat{\mathsf{H}} = \frac{\hat{\mathsf{p}}_T^2}{2M} + \frac{1}{2m_r}(\hat{\mathsf{p}}_{R1}^2 + \hat{\mathsf{p}}_{R2}^2) + \frac{\hat{\boldsymbol{p}}_{R1} \cdot \hat{\boldsymbol{p}}_{R2}}{m_n} - \frac{Z\alpha}{r_{R1}} - \frac{Z\alpha}{r_{R2}} + \frac{\alpha}{|\mathbf{r}_{R1} - \mathbf{r}_{R2}|}\,,$$

where $\hat{\mathsf{p}}_T^2 = -\nabla_T^2$, $\hat{\mathsf{p}}_{R1}^2 = -\nabla_{R1}^2$, $\hat{\mathsf{p}}_{R2}^2 = -\nabla_{R2}^2$, and $m_r = m_e(1 + m_e/m_n)^{-1}$ is the reduced mass of the nucleus-electron system. Further, we change to the center-of-mass system, $\mathbf{p}_T = 0$. Equation (3.120) now takes the form

$$\hat{\mathsf{H}}_0 = \frac{1}{2m_r}(\hat{\mathsf{p}}_1^2 + \hat{\mathsf{p}}_2^2) + e\varphi(r_1, r_2, r_{12})\,, \quad e\varphi(r_1, r_2, r_{12}) = \frac{\alpha}{r_{12}} - \frac{Z\alpha}{r_1} - \frac{Z\alpha}{r_2}\,, \tag{3.129}$$

in which we simplified the indices $R1$, $R2$ to 1, 2 and introduced $r_{12} = |\mathbf{r}_1 - \mathbf{r}_2|$.

The fine structure and corrections to the gross structure are determined by all terms in the Hamiltonian (3.121), where we set $g_1 = g_2 = 1$, plus the term

$$(\hat{\mathsf{H}}_1)_{\mathrm{nm}} = \frac{\hat{\boldsymbol{p}}_1 \cdot \hat{\boldsymbol{p}}_2}{m_n}, \tag{3.130}$$

commonly known as the *mass polarization operator*. After transition to atomic units, $r \to r_A/(m_r Z\alpha)$, we find that the contribution of this term is suppressed by the factor of $\frac{m_r}{m_n}$ in comparison to the terms in Eq. (3.120). Calculation of the factors for helium[17] $\frac{m_r}{m_n} \simeq 1.38 \times 10^{-4}$ and $(2\alpha)^2 \simeq 2.13 \times 10^{-4}$ shows that corrections arising from the nuclear motion are of about the same importance as relativistic and magnetic effects.

With potential given by Eq. (3.129) we can write the spin-orbit interaction (3.123) as

$$(\hat{\mathsf{H}}_1)_{\mathrm{so}} = (\hat{\mathsf{H}}_1)_{\mathrm{son}} + (\hat{\mathsf{H}}_1)_{\mathrm{soe}},$$

where

$$(\hat{\mathsf{H}}_1)_{\mathrm{son}} = \frac{Z\alpha}{2m_e^2}\left[\frac{\hat{\boldsymbol{S}}_1 \cdot (\mathbf{r}_1 \times \hat{\boldsymbol{p}}_1)}{r_1^3} + \frac{\hat{\boldsymbol{S}}_2 \cdot (\mathbf{r}_2 \times \hat{\boldsymbol{p}}_2)}{r_2^3}\right]$$

describes the interaction of electron spins with their orbital angular momenta with respect to the nucleus, and

$$(\hat{\mathsf{H}}_1)_{\mathrm{soe}} = -\frac{\alpha}{2m_e^2}\left[\frac{\hat{\boldsymbol{S}}_1 \cdot (\mathbf{r}_{12} \times \hat{\boldsymbol{p}}_1)}{r_{12}^3} + \frac{\hat{\boldsymbol{S}}_2 \cdot (\mathbf{r}_{21} \times \hat{\boldsymbol{p}}_2)}{r_{12}^3}\right]$$

expresses the interaction of electron spins with their orbital angular momentum with respect to each other.

3.6.4 Hydrogen-Like Atoms

In case of a hydrogen-like atom, we have $r_{12} = r$ and $\hat{\boldsymbol{p}}_2 = -\hat{\boldsymbol{p}}_1 = -\hat{\boldsymbol{p}}$ in the center-of-mass frame, and we further set $e_2 = -Ze_1 = -Ze$. The potential energy of the electrostatic interaction equals $e\varphi = -\frac{Z\alpha}{r}$. The spin-orbit interaction (3.123) and the current-current interaction (3.127) then acquire the form

[17]The nucleus of helium is an α-particle, a bound-state of two protons and two neutrons.

$$(\hat{\mathsf{H}}_1)_{so} = \frac{Z\alpha}{2r^3}\hat{\boldsymbol{L}} \cdot \left[\frac{1}{m_1^2}\hat{\boldsymbol{S}}_1 + \frac{1}{m_2^2}\hat{\boldsymbol{S}}_2\right]$$

and

$$(\hat{\mathsf{H}}_1)_{cc} = -\frac{Z\alpha}{m_1 m_2}\frac{1}{2r}\left[\hat{\mathsf{p}}^2 + (\mathbf{n}\cdot\hat{\boldsymbol{p}})^2\right].$$

Exercise 8: Correction to the Gross Structure I

Calculate the total shift of the energy levels 1s, 2s, and $2p_{1/2}$ given by the Breit Hamiltonian $\hat{\mathsf{H}}_1$ with precision to α^4. Neglect the interaction of an electron with the nuclear spin. Compare the calculated values with the experimental ones in Table 3.1.
Hint: First of all, change to atomic units $r = r_A/(m_r Z\alpha)$.

A kind reader may easily verify that in case of "normal" hydrogen-like atoms, where the mass of the nucleus markedly exceeds that of an electron, the fine structure and corrections to the gross structure are predominantly given by relativistic effects considered in Sect. 3.5.2, see, e.g., Eq. (3.107).

In the case of a positronium, where the nucleus is constituted by a positron, i.e., a particle with the same mass and spin as and opposite charge to an electron, hence $g_1 = g_2 = g$, $m_1 = m_2 = m$, $Z = 1$, and $e_2 = -e_1 = -e$. Equation (3.120) now takes the form

$$\hat{\mathsf{H}}_0 = \frac{m\alpha^2}{2}\left(\frac{\hat{\mathsf{p}}_A^2}{2} - \frac{1}{r_A}\right) \tag{3.131}$$

and all terms in Eq. (3.121) contribute equally. After the transition to atomic units $r \to r_A/(m_r\alpha)$, $m_r = m/2$, the Breit Hamiltonian (3.121) takes the form

$$\begin{aligned}\hat{\mathsf{H}}_1 = \frac{m\alpha^4}{2^3}\Bigg\{&-\frac{\hat{\mathsf{p}}_A^4}{8} - \frac{1}{2r_A}\left[\hat{\mathsf{p}}_A^2 + (\mathbf{n}\cdot\hat{\boldsymbol{p}}_A)^2\right]\\ &+ \delta(\mathbf{r}_A)\pi\left[1 + \frac{8}{3}g^2\hat{\boldsymbol{S}}_e\cdot\hat{\boldsymbol{S}}_p + \hat{\mathsf{S}}^2\right]\\ &+ \frac{1}{r_A^3}\left[\hat{\boldsymbol{L}}\cdot\hat{\boldsymbol{S}}\frac{1+2g}{2} - g^2\left(\hat{\boldsymbol{S}}_e\cdot\hat{\boldsymbol{S}}_p - 3\mathbf{n}\cdot\hat{\boldsymbol{S}}_e\mathbf{n}\cdot\hat{\boldsymbol{S}}_p\right)\right]\Bigg\}.\end{aligned} \tag{3.132}$$

Here we introduced the total spin operator

$$\hat{\boldsymbol{S}} = \hat{\boldsymbol{S}}_e + \hat{\boldsymbol{S}}_p$$

and added the effective interaction $\frac{\pi\alpha}{m^2}\delta(\mathbf{r})\hat{\mathsf{S}}^2$, caused by virtual electron-positron annihilation. We will derive this interaction later on, see Sect. 7.3.2, Eq. (7.105). A dear reader may take a look on the last diagram of Fig. 7.2 to get a feeling what the term "virtual electron-positron annihilation" corresponds to.

3.6.5 *Final Notes*

As we have seen, as soon as we are forced to consider both the spin-spin and spin-orbital interactions, calculation becomes substantially more complex. Therefore, to determine the fine structure of the positronium spectral lines, not to mention more complex atoms, one needs to opt for a systematic approach developed in the next chapter.

There is no point in attempting to include the effect of $\hat{\mathsf{H}}_1$ beyond the first order of the perturbation method. The reason is, new effects of quantum electrodynamics come into play at the order of α^5. In fact for positronium they enter already at the order α^4. We will focus in depth on the leading quantum-electrodynamical effects in Chaps. 6 and 7.

References

1. R.P. Feynman, R.B. Leighton, M. Sands, *Feynman Lectures on Physics 1, 2* (Addison-Wesley, Reading, 1977)
2. V. Meyer et al., Phys. Rev. Lett. **84**, 1136 (2000)
3. P.J. Mohr, B.N. Taylor, D.B. Newell, Rev. Mod. Phys. **84**, 1527 (2012)

Chapter 4
Treasures Hidden in Commutators

In this chapter, we focus on a topic usually called the application of algebraic methods or Lie algebras within quantum mechanics. We have already demonstrated that one can determine very efficiently the spectrum of the harmonic oscillator owing to the closure of the set of three operators, namely the Hamiltonian $\hat{\mathsf{H}}$ and the ladder operators $\hat{\mathsf{a}}$ and $\hat{\mathsf{a}}^+$, under the operation of commutation. We now show that this method can be extended to the problem of angular momentum, the addition of angular momenta, the hydrogen atom, and a free particle. We will further see that the application of the algebraic method does not only represent a very elegant way of solving the listed problems, it also directly leads to a variety of relations for radial and angular parts of the hydrogen wave functions, which would be very laborious to obtain otherwise. In the next chapter, we will subsequently demonstrate that these relations prove to be indeed useful when determining the spectra of many-electron atoms.

4.1 A General Solution To Angular Momentum

We now solve the problem of angular momentum in quantum mechanics "once and for all."[1] The angular momentum is any observable complying with the commutation relations

$$[\hat{\mathsf{J}}_i, \hat{\mathsf{J}}_j] = \mathrm{i}\varepsilon_{ijk}\hat{\mathsf{J}}_k\,, \tag{4.1}$$

irrespective of its origin. We have already shown, see Eq. (1.26), that these relations hold for both the intrinsic angular momentum of a particle with spin 1/2 and the angular momentum of a particle orbiting around a center of attraction, see Eq. (3.77).

[1] This section and the next one present the very fundamentals of quantum mechanics. One may find a different exposition of this topic in any other textbook, see for example [8, 10, 11].

J. Zamastil, J. Benda, *Quantum Mechanics and Electrodynamics*,
DOI 10.1007/978-3-319-65780-6_4

Moreover, as we also noted before, see Eq. (3.61), the total angular momentum created as a sum of two independent angular momenta obeys Eq. (4.1).

One conclusion that can be drawn from the relation (4.1) is that one cannot find common eigenvectors of the individual components of the angular momentum. Therefore, we cannot measure them simultaneously. On the other hand, the square of the magnitude of the angular momentum commutes with each of the components, see Eq. (3.62),

$$[\hat{J}^2, \hat{J}_k] = 0\,, \qquad \hat{J}^2 = \hat{J}_k\hat{J}_k\,, \tag{4.2}$$

and therefore one can always measure the magnitude of the angular momentum and a projection along one of the coordinate axes. A common convention chooses the third axis;

$$\hat{J}^2|j,m\rangle = j(j+1)|j,m\rangle\,, \qquad \hat{J}_z|j,m\rangle = m|j,m\rangle\,. \tag{4.3}$$

These equations define the eigenvectors $|j,m\rangle$ and the eigenvalues $j(j+1)$ and m. We have already encountered a few special cases of these eigenvectors and eigenvalues: for $j = \frac{1}{2}$ there are two projections $m \in \{-\frac{1}{2}, +\frac{1}{2}\}$, for $j = 1$ there are three projections $m \in \{-1, 0, +1\}$, and for $j = \frac{3}{2}$ there are four projections $m \in \{-\frac{3}{2}, -\frac{1}{2}, +\frac{1}{2}, +\frac{3}{2}\}$. The states we previously labeled $|\pm\rangle$, e.g., in Eqs. (1.32), (1.33), and (1.34), correspond to the states $|+\rangle = \left|\frac{1}{2}, +\frac{1}{2}\right\rangle$ and $|-\rangle = \left|\frac{1}{2}, -\frac{1}{2}\right\rangle$ in the new notation.

We will now show that the information contained in the commutation relation (4.1) and the definition (4.3) already suffices for the deduction that the eigenvalues m increase by one from $m = -j$ to $m = +j$. This finding then immediately leads to the conclusion that j can acquire only integer or half-integer values; the proof thereof is left to the reader. Subsequently, we will find an expression for a general action of the remaining angular momentum components $\hat{J}_x$ and $\hat{J}_y$ on the states $|j,m\rangle$, namely by generalizing the relations (1.33), (1.34), (3.70), and (3.71) to an arbitrary j.

As already discussed in the cases of $j = \frac{1}{2}$ and $j = 1$, it is advantageous to define the ladder operators

$$\hat{J}_\pm = \hat{J}_x \pm \mathrm{i}\hat{J}_y\,. \tag{4.4}$$

Their commutator with the third component of the angular momentum is

$$[\hat{J}_z, \hat{J}_\pm] = [\hat{J}_z, \hat{J}_x] \pm \mathrm{i}[\hat{J}_z, \hat{J}_y] = \mathrm{i}\hat{J}_y \pm \hat{J}_x = \pm(\hat{J}_x \pm \mathrm{i}\hat{J}_y) = \pm\hat{J}_\pm\,,$$

and with the squared angular momentum

$$[\hat{J}^2, \hat{J}_\pm] = [\hat{J}^2, \hat{J}_x] \pm \mathrm{i}[\hat{J}^2, \hat{J}_y] = 0\,.$$

It follows from the definitions (4.3) that the operators act on the common eigenstate according to

$$[\hat{J}_z, \hat{J}_\pm]|j,m\rangle = \pm\hat{J}_\pm|j,m\rangle \Rightarrow \hat{J}_z\left(\hat{J}_\pm|j,m\rangle\right) = (m\pm 1)\left(\hat{J}_\pm|j,m\rangle\right),$$

and

$$[\hat{J}^2, \hat{J}_\pm]|j,m\rangle = 0 \Rightarrow \hat{J}^2\left(\hat{J}_\pm|j,m\rangle\right) = j(j+1)\left(\hat{J}_\pm|j,m\rangle\right).$$

The logic is the same as in Eqs. (1.63) and (1.64). The role of $\hat{H}$ is replaced by $\hat{J}_z$, and the role of $\hat{a}$ and $\hat{a}^+$ by $\hat{J}_-$ and $\hat{J}_+$, respectively. By comparison of the last two equations with Eq. (4.3), we readily see that the state $\hat{J}_\pm|j,m\rangle$ is again an eigenstate of both the operators $\hat{J}_z$ and $\hat{J}^2$ with eigenvalues $m\pm 1$ and $j(j+1)$, respectively; thus

$$\hat{J}_\pm|j,m\rangle = \alpha^\pm(j,m)|j,m\pm 1\rangle. \tag{4.5}$$

Finally, we determine the coefficients α by the same procedure as in case of the harmonic oscillator. Confront the following steps with those following Eq. (1.66).

Owing to the freedom in the phase of eigenvectors, we may choose α such as it is real. Next, we use the definition (4.4) of the ladder operators to find the action of $\hat{J}_+\hat{J}_-$

$$\begin{aligned}\hat{J}_+\hat{J}_-|j,m\rangle &= (\hat{J}_x + i\hat{J}_y)(\hat{J}_x - i\hat{J}_y)|j,m\rangle = \left(\hat{J}_x^2 + \hat{J}_y^2 - i[\hat{J}_x, \hat{J}_y]\right)|j,m\rangle \\ &= \left(\hat{J}^2 - \hat{J}_z^2 + \hat{J}_z\right)|j,m\rangle = (j(j+1) - m(m-1))|j,m\rangle,\end{aligned}$$

where we used Eq. (4.1) and the definition of the square of the angular momentum, Eq. (4.2), in the third equality. In the fourth step we used Eq. (4.3). It holds as well that, see Eq. (4.5),

$$\hat{J}_+\hat{J}_-|j,m\rangle = \hat{J}_+\alpha^-(j,m)|j,m-1\rangle = \alpha^+(j,m-1)\alpha^-(j,m)|j,m\rangle.$$

By comparison of the last two equations, we obtain

$$\alpha^+(j,m-1)\alpha^-(j,m) = j(j+1) - m(m-1).$$

One can readily see from the definition of the ladder operators that the operators $\hat{J}_+$ and $\hat{J}_-$ are Hermitian conjugates of each other; hence

$$\begin{aligned}\langle j,m|\hat{J}_+|j,m-1\rangle &= \langle j,m|\left(\hat{J}_+|j,m-1\rangle\right) = \alpha^+(j,m-1)\langle j,m|j,m\rangle \\ &= \left(\langle j,m|\hat{J}_+\right)|j,m-1\rangle = \left(\hat{J}_-|j,m\rangle\right)^+|j,m-1\rangle \\ &= \alpha^-(j,m)\langle j,m-1|j,m-1\rangle\,.\end{aligned}$$

Requiring further the vectors $|j,m\rangle$ be also orthonormal (and not only orthogonal) leads to

$$\langle j,m|j',m'\rangle = \delta_{j,j'}\delta_{m,m'}\,, \tag{4.6}$$

then necessarily

$$\alpha^-(j,m) = \alpha^+(j,m-1)\,.$$

Hence, the coefficients α acquire values

$$\alpha^-(j,m) = \alpha^+(j,m-1) = \sqrt{j(j+1)-m(m-1)}\,, \tag{4.7}$$

and the action of the ladder operators can be written in a final, slightly reorganized form as

$$\hat{J}_\pm|j,m\rangle = \sqrt{(j+1\pm m)(j\mp m)}|j,m\pm 1\rangle\,. \tag{4.8}$$

For $j=\frac{1}{2}$ and $m\in\{-\frac{1}{2},+\frac{1}{2}\}$, we obtain from the last equation the relations (1.33) and (1.34), and for $j=1$ and $m\in\{-1,0,+1\}$ we get the relations (3.70) and (3.71).

What values may m acquire? Since it is an eigenvalue which is to represent a projection of the angular momentum of a finite magnitude $j(j+1)$ along the z axis, there must be a smallest (m_{min}) and largest (m_{max}) possible value. Furthermore, the action of the ladder operators on the states with the extremal projections must not lead to a state that would lie outside the interval bounded by m_{min} and m_{max}. Hence we require

$$\hat{J}_+|j,m_{\text{max}}\rangle = 0 \quad \text{and} \quad \hat{J}_-|j,m_{\text{min}}\rangle = 0\,.$$

Comparison of the last equations and Eq. (4.8) yields two possible results, $m_{\text{max}}\in\{j,-j-1\}$ and $m_{\text{min}}\in\{-j,j+1\}$, for the limiting values of the projections. However, only the former is meaningful, as m_{min} has to be less than m_{max}, hence

$$m_{\text{min}} = -j, \quad m_{\text{max}} = j\,.$$

Starting merely with the innocently looking commutation relation (4.1), we reached very nontrivial conclusions: we were able to determine the allowed projections of the angular momentum along a given axis, the values of the magnitude of the angular momentum, and the action of all of the components of angular momentum operator on the pertinent eigenstates. The commutators are indeed a powerful tool, as we are to see many more times in the next sections.

4.2 Addition of Angular Momenta

We now consider a system with angular momentum $\hat{\boldsymbol{J}}$ arising from two distinct contributions (such as two orbiting electrons in atom, two particles with spin, one particle with spin and an orbital momentum, etc.) $\hat{\boldsymbol{J}}_1$ and $\hat{\boldsymbol{J}}_2$[2]:

$$\hat{\boldsymbol{J}} = \hat{\boldsymbol{J}}_1 + \hat{\boldsymbol{J}}_2 . \tag{4.9}$$

We can find common eigenstates for the complete set of commuting operators $\{\hat{J}_1^2, \hat{J}_2^2, \hat{J}_z, \hat{J}^2\}$,[3] namely

$$\hat{J}^2 |j, m, (j_1, j_2)\rangle = j(j+1) |j, m, (j_1, j_2)\rangle , \tag{4.10}$$

$$\hat{J}_z |j, m, (j_1, j_2)\rangle = m |j, m, (j_1, j_2)\rangle , \tag{4.11}$$

$$\hat{J}_1^2 |j, m, (j_1, j_2)\rangle = j_1(j_1+1) |j, m, (j_1, j_2)\rangle , \tag{4.12}$$

$$\hat{J}_2^2 |j, m, (j_1, j_2)\rangle = j_2(j_2+1) |j, m, (j_1, j_2)\rangle . \tag{4.13}$$

These states form a complete and orthonormal basis in the space spanned by possible states of two "particles" with the angular momenta j_1 and j_2.[4] We expect the set of operators $\{\hat{J}_1^2, \hat{J}_2^2, \hat{J}_z, \hat{J}^2\}$ to act on the same abstract space as the set $\{\hat{J}_1^2, \hat{J}_2^2, \hat{J}_{1z}, \hat{J}_{2z}\}$, with the same number of degrees of freedom. Therefore, it should be possible to compose the new "two-particle" basis states $|j, m, (j_1, j_2)\rangle$ from products of the basis states of the individual constituents of the system[5]:

$$|j_1, m_1, j_2, m_2\rangle = |j_1, m_1\rangle |j_2, m_2\rangle , \tag{4.14}$$

$$\hat{J}_1^2 |j_1, m_1\rangle = j_1(j_1+1) |j_1, m_1\rangle , \qquad \hat{J}_{1z} |j_1, m_1\rangle = m_1 |j_1, m_1\rangle \tag{4.15}$$

and

$$\hat{J}_2^2 |j_2, m_2\rangle = j_2(j_2+1) |j_2, m_2\rangle , \qquad \hat{J}_{2z} |j_2, m_2\rangle = m_2 |j_2, m_2\rangle , \tag{4.16}$$

---

[2]To be more precise, we should express the compound angular momentum $\hat{\boldsymbol{J}}$ as a sum of tensor products $\hat{\boldsymbol{J}} = \hat{\boldsymbol{J}}_1 \otimes \mathbf{1}_2 + \mathbf{1}_1 \otimes \hat{\boldsymbol{J}}_2$ to avoid ambiguity when applying $\hat{\boldsymbol{J}}$ on the state $|j_1, m_1\rangle |j_2, m_2\rangle$.

[3]The reader can easily prove by himself that they all commute with each other.

[4]We inserted the word particles into quotation marks since one can have in mind the spin and orbital states of the same particle as well.

[5]There are $(2j_1 + 1)(2j_2 + 1)$ such products due to the restriction on the possible values of the projections m_1 and m_2.

see Eq. (4.3). Our task now consists of finding the connection between j, m and j_1, j_2, m_1, m_2, and also of finding the relation between the basis states (4.10) and (4.14). For the sake of definiteness, we consider $j_1 \leq j_2$ henceforth.

One of the tasks is trivial: when we act with the third component of the total angular momentum on an arbitrary product $|j_1, m_1\rangle |j_2, m_2\rangle$, we obtain

$$\begin{aligned}\hat{J}_z|j_1, m_1\rangle|j_2, m_2\rangle &= (\hat{J}_{1z}|j_1, m_1\rangle)|j_2, m_2\rangle + |j_1, m_1\rangle(\hat{J}_{2z}|j_2, m_2\rangle) \\ &= (m_1 + m_2)|j_1, m_1\rangle|j_2, m_2\rangle\,.\end{aligned}$$

Therefore, considering the definition of the magnetic quantum number, we see that $m = m_1 + m_2$, which also imposes a restriction on the values of m. As the values of $m_{1,2}$ lie between $-j_{1,2}$ and $+j_{1,2}$, their sum must lie within the interval spanned by $-(j_1 + j_2)$ and $+(j_1 + j_2)$.

It follows from the fact that the operators $\hat{J}^2$ and $\hat{J}_z$ commute with each other that the operator $\hat{J}^2$ does not mix states with different eigenvalues of the operator $\hat{J}_z$. This means, if we express the states in (4.10) as linear combinations of the states (4.14), we only need to sum over the states with $m_1 + m_2 = m$ for a given m. That is,

$$|j, m, (j_1, j_2)\rangle = \sum_{i=-j_1}^{j_1} c_i |j_1, i, j_2, m - i\rangle\,, \tag{4.17}$$

where we have from the orthonormality of the states (4.14) that

$$c_i = \langle j_1, i, j_2, m - i | j, m, (j_1, j_2)\rangle.$$

The coefficients c_i are most often written in the form

$$c_i = (j_1, i, j_2, m - i | j, m)$$

and called the *Clebsch-Gordan (CG) coefficients*. As we know the allowed range of values of m, we are able to construct a table of states, Table 4.1, with rows containing the states that can be used in the summation (4.17) for a given m.

Table 4.1 Addition of angular momenta I

m	States that can be used in composition	For $j_1 = j_2 = \frac{1}{2}$
$j_1 + j_2$	$\|j_1, j_1\rangle\|j_2, j_2\rangle$	$\|+\rangle\|+\rangle$
$j_1 + j_2 - 1$	$\|j_1, j_1 - 1\rangle\|j_2, j_2\rangle \quad \|j_1, j_1\rangle\|j_2, j_2 - 1\rangle$	$\|-\rangle\|+\rangle \quad \|+\rangle\|-\rangle$
$j_1 + j_2 - 2$	$\|j_1, j_1 - 2\rangle\|j_2, j_2\rangle \quad \|j_1, j_1 - 1\rangle\|j_2, j_2 - 1\rangle \quad \|j_1, j_1\rangle\|j_2, j_2 - 2\rangle$	$\|-\rangle\|-\rangle$
$\vdots$	$\vdots$	
$-(j_1 + j_2)$	$\|j_1, -j_1\rangle\|j_2, -j_2\rangle$	$\|-\rangle\|-\rangle$

Since m may acquire only values within the interval spanned by $-j$ and $+j$, there is a limit on the *maximum* value of the total j, namely $j_{\max} = j_1 + j_2$. It follows from the general solution of the angular momentum that for a given j the projection of the angular momentum along any axis may equal $2j + 1$ different values $-j, -j + 1, -j + 2, \ldots, +j - 1, +j$. Note that there are no other states but those listed in Table 4.1, therefore we must be able to construct each of the eigenstates of $\hat{\mathbf{J}}^2$ corresponding to the eigenvalue $j_{\max}(j_{\max} + 1)$ with the above listed $2j_{\max} + 1$ projections using the states from the table. One can do so easily in case of the states with the maximal and minimal projections; there is merely one state to use for each of them, namely $|j_1, j_1\rangle|j_2, j_2\rangle$ and $|j_1, -j_1\rangle|j_2, -j_2\rangle$, respectively. In all of the other cases, we seek (and always find) linear combinations of the states listed in Table 4.1 which form eigenvectors of $\hat{\mathbf{J}}^2$ pertaining to the eigenvalue $j_{\max}(j_{\max} + 1)$ for a given projection m. That is, we search for a suitable linear combination of the two vectors for $m = j_1 + j_2 - 1$, a combination of three for $m = j_1 + j_2 - 2$, and so on.

A linear combination of the two vectors displayed in Table 4.1 for $m = j_1 + j_2 - 1$, which is orthogonal to the linear combination constituting an eigenvector of $\hat{\mathbf{J}}^2$ with the eigenvalue $j_{\max}(j_{\max} + 1)$, must yield an eigenvector of $\hat{\mathbf{J}}^2$ with the eigenvalue $(j_{\max} - 1)j_{\max}$. Why? It follows from the general solution of angular momentum that the projection m cannot be greater than j. Altogether, we have to find $2(j_{\max} - 1) + 1$ eigenvectors of $\hat{\mathbf{J}}^2$ with the eigenvalue $(j_{\max} - 1)j_{\max}$. For the instance of two particles with spin 1/2, we have found $j_{\max} = 1$. The three states corresponding to the projections $m = 1, 0, -1$ have been found in the form $|1, 1\rangle = |+\rangle|+\rangle$, $|1, 0\rangle = (|+\rangle|-\rangle + |-\rangle|+\rangle)/\sqrt{2}$ and $|1, -1\rangle = |-\rangle|-\rangle$. Orthogonal linear combination to the state with $j_{\max} = 1$, $m = 0$, the combination $|0, 0\rangle = (|+\rangle|-\rangle - |-\rangle|+\rangle)/\sqrt{2}$, is eigenstate of $\hat{\mathbf{J}}^2$ with eigenvalues $j_{max} - 1 = 0$.

Linear combination of three vectors displayed in Table 4.1 for $m = j_1 + j_2 - 2$, orthogonal to two linear combinations providing eigenvectors of $\hat{\mathbf{J}}^2$ with eigenvalues of $j_{\max}(j_{\max} + 1)$ and $(j_{\max} - 1)j_{\max}$, must yield eigenvector of $\hat{\mathbf{J}}^2$ with eigenvalue $(j_{\max} - 2)(j_{\max} - 1)$, and so on. For clarity, Table 4.2 lists all possible combined angular states that can be constructed using the states in Table 4.1. In the second table, we used a shorter notation $|j, m\rangle \equiv |j, m, (j_1, j_2)\rangle$.

Table 4.2 Addition of angular momenta II

m	States that can be composed			For $j_1 = j_2 = \frac{1}{2}$	
$j_1 + j_2$	$\|j_1 + j_2, m\rangle$			$\|1, 1\rangle$	
$j_1 + j_2 - 1$	$\|j_1 + j_2, m\rangle$	$\|j_1 + j_2 - 1, m\rangle$		$\|1, 0\rangle$	$\|0, 0\rangle$
$j_1 + j_2 - 2$	$\|j_1 + j_2, m\rangle$	$\|j_1 + j_2 - 1, m\rangle$	$\|j_1 + j_2 - 2, m\rangle$	$\|1, -1\rangle$	
$\vdots$	$\vdots$				
$-(j_1 + j_2)$	$\|j_1 + j_2, m\rangle$			$\|1, -1\rangle$	

What is the minimal j where the procedure terminates? Irrespective of whether we describe the state space using the basis $\{|j_1, m_1\rangle|j_2, m_2\rangle\}$ or $\{|j, m, (j_1, j_2)\rangle\}$, it must always maintain the same dimension. This means, for a given j_1 and j_2, there is an equality between the dimensions of the first and the second basis

$$(2j_1 + 1)(2j_2 + 1) = \sum_{j=j_{\text{min}}}^{j_{\text{max}}} (2j + 1)\,, \tag{4.18}$$

where $j_{\text{max}} = j_1 + j_2$. The sum on the rhs may be written as $(j_{\text{max}} + 1)^2 - j_{\text{min}}^2$, as one can easily verify. We obtain thereof $j_{\text{min}}^2 = (j_1 - j_2)^2$, and after imposing the restriction on $j_{\text{min}} \geq 0$ we finally find $j_{\text{min}} = |j_1 - j_2|$.

We have thus arrived at a very important conclusion: When adding two angular momenta with quantum numbers j_1, m_1 and j_2, m_2, the resulting j acquires values $|j_1-j_2|, |j_1-j_2|+1, \ldots, j_1+j_2-1, j_1+j_2$ and the resulting m is given by $m = m_1+m_2$. The restriction on j is widely known as the *triangle inequality*.

There is one last step left: to determine the value of the Clebsch-Gordan coefficients in Eq. (4.17). One can do so by expressing Eq. (4.10) using the vectors (4.14)

$$\hat{\mathrm{J}}^2|j, m, (j_1, j_2)\rangle = j(j+1)|j, m, (j_1, j_2)\rangle = j(j+1) \sum_{i=-j_1}^{j_1} c_i |j_1, i\rangle|j_2, m - i\rangle.$$

At the same time, we can express $\hat{\mathrm{J}}^2$ using the one-particle operators, see Eq. (4.9),

$$\hat{\mathrm{J}}^2 = \hat{\mathrm{J}}_1^2 + 2\hat{\boldsymbol{J}}_1 \cdot \hat{\boldsymbol{J}}_2 + \hat{\mathrm{J}}_2^2\,;$$

then

$$\begin{aligned}
&\hat{\mathrm{J}}^2|j, m, (j_1, j_2)\rangle \\
&= \left(\hat{\mathrm{J}}_1^2 + \hat{\mathrm{J}}_2^2 + \hat{\mathrm{J}}_{1+}\hat{\mathrm{J}}_{2-} + \hat{\mathrm{J}}_{1-}\hat{\mathrm{J}}_{2+} + 2\hat{\mathrm{J}}_{1z}\hat{\mathrm{J}}_{2z}\right) \sum_{i=-j_1}^{j_1} c_i |j_1, i\rangle|j_2, m - i\rangle \\
&= \sum_{i=-j_1}^{j_1} c_i \left(j_1(j_1 + 1) + j_2(j_2 + 1) + 2i(m - i)\right)|j_1, i\rangle|j_2, m - i\rangle \\
&\quad + \sum_{i=-j_1}^{j_1} c_i \alpha^+(j_1, i)\alpha^-(j_2, m - i)|j_1, i + 1\rangle|j_1, m - i - 1\rangle \\
&\quad + \sum_{i=-j_1}^{j_1} c_i \alpha^-(j_1, i)\alpha^+(j_2, m - i)|j_1, i - 1\rangle|j_1, m - i + 1\rangle\,.
\end{aligned} \tag{4.19}$$

We used the relations (4.15), (4.16), and (4.8) in the second equality. By shifting the summation index (while requiring $c_i = 0$ when the summation index is out of the allowed range), we obtain

$$0 = \sum_{i=-j_1}^{j_1} \big[(j_1(j_1+1) + j_2(j_2+1) - j(j+1) + 2i(m-i))c_i$$
$$+ \alpha^+(j_1, i-1)\alpha^-(j_2, m-i+1)c_{i-1} +$$
$$+ \alpha^-(j_1, i+1)\,\alpha^+(j_2, m-i-1)c_{i+1}\big]\, |j_1, i\rangle |j_2, m-i\rangle\,.$$

Owing to the orthonormality of the states $|j_1, i\rangle |j_2, m-i\rangle$, this produces a set of $2j_1 + 1$ equations for $2j_1 + 1$ coefficients c_i

$$\begin{aligned}(j_1(j_1+1) + j_2(j_2+1) - j(j+1) + 2i(m-i))c_i&\\ + \alpha^+(j_1, i-1)\alpha^-(j_2, m-i+1)c_{i-1}&\\ + \alpha^-(j_1, i+1)\alpha^+(j_2, m-i-1)c_{i+1} &= 0\,,\end{aligned} \tag{4.20}$$

where $\alpha^{\pm}$ are given by Eq. (4.7). However, these equations provide us with $2j_1$ coefficients c_i only; the last equation does not offer any new piece of information. Why is it so? The reason is we are searching for eigenvectors of an operator, $\hat{\mathrm{J}}^2$ in this case, see Eq. (4.10), and these are given but for the normalization. Therefore, we need to consider an additional equation to the set (4.20), namely the normalization condition

$$\begin{aligned}\langle j, m, (j_1,j_2) | j, m, (j_1,j_2)\rangle &= \sum_i \sum_k c_i^* c_k \langle j_1, i, j_2, m-i | j_1, k, j_2, m-k\rangle\\ &= \sum_i |c_i|^2 = 1\,,\end{aligned} \tag{4.21}$$

where the first equality follows from Eq. (4.17) and the second from Eq. (4.6). This way, we obtain the final system of equations whose solution is unique but for a complex phase which is commonly chosen so that the coefficients c_i are real.

Example: Consider a system with $j_1 = j_2 = 1$, for example two electrons in a p-state. For the sake of simplicity, we focus only on the states with $m = 0$. The set of equations (4.20) for the CG coefficients simplifies after the substitution of the particular values of j_1, j_2, and m to

$$\begin{aligned}i = -1 &: (2 - j(j+1))c_{-1} + 2c_0 = 0\,,\\ i = 0 &: (4 - j(j+1))c_0 + 2c_{-1} + 2c_{+1} = 0\,,\\ i = +1 &: (2 - j(j+1))c_{+1} + 2c_0 = 0\,.\end{aligned}$$

By substituting the three possible values of j, i.e., $j = 0, 1, 2$, and using the normalization condition

$$c_{-1}^2 + c_0^2 + c_1^2 = 1$$

we find the three composed states

$$\begin{aligned}
|0, 0, (1, 1)\rangle &= \frac{1}{\sqrt{3}} \left(|1, -1\rangle|1, 1\rangle - |1, 0\rangle|1, 0\rangle + |1, 1\rangle|1, -1\rangle\right), \\
|1, 0, (1, 1)\rangle &= \frac{1}{\sqrt{2}} \left(|1, -1\rangle|1, 1\rangle - |1, 1\rangle|1, -1\rangle\right), \\
|2, 0, (1, 1)\rangle &= \frac{1}{\sqrt{6}} \left(|1, -1\rangle|1, 1\rangle + 2|1, 0\rangle|1, 0\rangle + |1, 1\rangle|1, -1\rangle\right).
\end{aligned}$$

One can easily verify that these vectors are mutually orthogonal. They must be since they are the eigenvectors of the Hermitian operator $\hat{\mathrm{J}}^2$ corresponding to different eigenvalues $j(j + 1)$, where $j = 0, 1, 2$.

How does the inverse transformation

$$|j_1, i, j_2, m - i\rangle = \sum_{j=|j_1-j_2|}^{j_1+j_2} d_j |j, m, (j_1, j_2)\rangle \tag{4.22}$$

read? This is not a purely academical question, we will need this formula later. It follows from the orthonormality of the states $|j, m, (j_1, j_2)\rangle$ and the property $c_i \in \mathbb{R}$ that

$$d_j = \langle j, m, (j_1, j_2)|j_1, i, j_2, m - i\rangle = (\langle j_1, i, j_2, m - i|j, m, (j_1, j_2)\rangle)^+ = c_i^* = c_i\,.$$

Example: The transform (4.17) is unitary. Moreover, since it is also real, it is orthogonal. Therefore, when we rewrite the previous example into a matrix notation,

$$\begin{pmatrix} |0, 0, (1, 1)\rangle \\ |1, 0, (1, 1)\rangle \\ |2, 0, (1, 1)\rangle \end{pmatrix} = \begin{pmatrix} \frac{1}{\sqrt{3}} & \frac{-1}{\sqrt{3}} & \frac{1}{\sqrt{3}} \\ \frac{1}{\sqrt{2}} & 0 & \frac{-1}{\sqrt{2}} \\ \frac{1}{\sqrt{6}} & \frac{2}{\sqrt{6}} & \frac{1}{\sqrt{6}} \end{pmatrix} \begin{pmatrix} |1, -1\rangle|1, 1\rangle \\ |1, 0\rangle|1, 0\rangle \\ |1, 1\rangle|1, -1\rangle \end{pmatrix},$$

the matrix of the transform is orthogonal. The inverse matrix to an orthogonal matrix is its transpose:

$$\begin{pmatrix} |1,-1\rangle|1,1\rangle \\ |1,0\rangle|1,0\rangle \\ |1,1\rangle|1,-1\rangle \end{pmatrix} = \begin{pmatrix} \frac{1}{\sqrt{3}} & \frac{1}{\sqrt{2}} & \frac{1}{\sqrt{6}} \\ \frac{-1}{\sqrt{3}} & 0 & \frac{2}{\sqrt{6}} \\ \frac{1}{\sqrt{3}} & \frac{-1}{\sqrt{2}} & \frac{1}{\sqrt{6}} \end{pmatrix} \begin{pmatrix} |0,0,(1,1)\rangle \\ |1,0,(1,1)\rangle \\ |2,0,(1,1)\rangle \end{pmatrix} .$$

However, when we compare Eqs. (4.17) and (4.22) it seems that the matrices of the transforms should be the same. Is our result not a contradiction? No, it is indeed not. We write the transforms (4.17) and (4.22) in an abstract form

$$|\psi_i\rangle = \sum_j A_{ij}|\chi_j\rangle, \qquad A_{ij} = \langle\chi_j|\,\psi_i\rangle \tag{4.23}$$

and

$$|\chi_j\rangle = \sum_k B_{jk}|\psi_k\rangle \, ,$$

where

$$B_{jk} = \langle\psi_k|\,\chi_j\rangle = \langle\chi_j|\,\psi_k\rangle^* = A^*_{kj} = A_{kj} \, ,$$

hence

$$|\chi_j\rangle = \sum_k A_{kj}|\psi_k\rangle \, . \tag{4.24}$$

Recalling the definition of matrix multiplications (note that the order of the indices is important!), we see that the matrix of the transform (4.24), or (4.22), is the transposed matrix of the transform (4.23), or (4.17).

We have thus learned how to add two angular momenta. Needless to say, one can follow a similar procedure to combine a greater number of angular momenta. In brief, one starts by adding two of them and subsequently combines the obtained result with the third angular momentum, etc. The reader is now able to prove all by himself the statement under Fig. 3.1, which claims that an even number of fermions behaves as a boson and an odd number of fermions as a fermion. In other words, if we add an even number of particles with half-integer j we obtain a compound particle with an integer j, and if we add an odd number of particles with a half-integer j we obtain a compound particle with half-integer j. For instance, the helium atom ${}^4_2$He constitutes two neutrons, two protons, and two electrons. Hence as a whole, it is a boson. At a sufficiently low temperature, these atoms can transit to the same quantum states which then macroscopically manifests as a superfluidity. In contrast, the helium isotope ${}^3_2$He constitutes only one neutron, two protons, and two electrons. It is thus a fermion and two such atoms may never populate the same quantum state. This form of helium becomes superfluid, too, though only when the sufficiently cooled atoms make pairs. These pairs then behave as bosons and can accumulate in the same state. Similarly, two electrons in a superconductor form the so-called *Cooper pairs* at low temperatures and behave as bosons.

Let us add one more remark here. In the literature and symbolic computer programs, one can meet a slight variation of the Clebsch-Gordan coefficients, the so-called *Wigner 3j-symbols*, which are defined in the following way:

$$\begin{pmatrix} j_1 & j_2 & j_3 \\ m_1 & m_2 & m_3 \end{pmatrix} = (-1)^{j_1-j_2-m_3} \frac{(j_1, m_1, j_2, m_2 | j_3, -m_3)}{\sqrt{2j_3+1}} .$$

4.3 The Runge-Lenz Vector

4.3.1 The Runge-Lenz Vector in Classical Mechanics

It belongs to common knowledge that the angular momentum conserves, see Eq. (3.73), in every central field. However, in the special case of the Coulomb field,

$$V(r) = -Zr^{-1} ,$$

there exists another important conserved vector. If we take vector product of the angular momentum and Eq. (3.72) we obtain

$$\mathbf{L} \times \frac{\mathrm{d}\mathbf{p}}{\mathrm{d}t} = -\frac{Z\mathbf{L} \times \mathbf{r}}{r^3} . \tag{4.25}$$

The rhs features a double cross product which can be simplified:

$$\begin{aligned}(\mathbf{L} \times \mathbf{r})_i &= ((\mathbf{r} \times \mathbf{p}) \times \mathbf{r})_i = \varepsilon_{ijk}\varepsilon_{jrq}x_r p_q x_k = (-\delta_{ir}\delta_{kq} + \delta_{iq}\delta_{kr})x_r p_q x_k \\ &= -x_i p_k x_k + x_k p_i x_k = r^2(-n_i n_k p_k + p_i) .\end{aligned}$$

One can express the momentum also in the following way[6]:

$$p_i = \frac{\mathrm{d}x_i}{\mathrm{d}t} = \frac{\mathrm{d}(rn_i)}{\mathrm{d}t} = \frac{\mathrm{d}r}{\mathrm{d}t}n_i + r\frac{\mathrm{d}n_i}{\mathrm{d}t},$$

which in turn yields

$$(\mathbf{L}\times\mathbf{r})_i = r^2\left(-n_in_k\left(n_k\frac{\mathrm{d}r}{\mathrm{d}t} + r\frac{\mathrm{d}n_k}{\mathrm{d}t}\right) + \frac{\mathrm{d}r}{\mathrm{d}t}n_i + r\frac{\mathrm{d}n_i}{\mathrm{d}t}\right).$$

The first and the third terms cancel each other out. The second term is equal to zero due to $n_k\frac{\mathrm{d}n_k}{\mathrm{d}t} = \frac{1}{2}\frac{\mathrm{d}}{\mathrm{d}t}n_kn_k = 0$ since $n_kn_k = 1$. Therefore, we are left with

$$\mathbf{L}\times\mathbf{r} = r^3\frac{\mathrm{d}\mathbf{n}}{\mathrm{d}t}.$$

If we return to Eq. (4.25) and realize that owing to Eq. (3.73) it must hold that $\mathbf{L}\times\frac{\mathrm{d}\mathbf{p}}{\mathrm{d}t} = \frac{\mathrm{d}}{\mathrm{d}t}\mathbf{L}\times\mathbf{p}$, we arrive at

$$\frac{\mathrm{d}\mathbf{X}}{\mathrm{d}t} = 0, \qquad \mathbf{X} = \mathbf{L}\times\mathbf{p} + Z\mathbf{n}.$$

Namely, the vector **X** does not change with time, it is an integral of motion. This vector is commonly referred to as the *Runge-Lenz vector* and its physical significance is the following. First, its scalar product with the angular momentum equals zero:

$$\mathbf{L}\cdot\mathbf{X} = \mathbf{L}\cdot(\mathbf{L}\times\mathbf{p}) + Z\mathbf{L}\cdot\mathbf{n} = 0 + 0.$$

We can readily see that the vector **X** lies in the orbiting plane of the body since the angular momentum vector **L** is perpendicular to the orbiting plane of the body. Second, the scalar product of the Runge-Lenz vector with the position vector reads

$$\mathbf{r}\cdot\mathbf{X} = \mathbf{r}\cdot(\mathbf{L}\times\mathbf{p}) + Zr = -L^2 + Zr. \tag{4.26}$$

Exploiting the freedom in the choice of a coordinate system, we can pick it so that the angular momentum points in the direction of the axis z. The motion of the body is then restricted to the plane xy. If we further place the axis x in the direction of the vector **X** and introduce the polar coordinates

$$x = r\cos\varphi, \quad y = r\sin\varphi, \tag{4.27}$$

Eq. (4.26) acquires the form

$$rX\cos\varphi = Zr - L^2, \tag{4.28}$$

[6]Recall, see Eq. (3.72), that we are considering particle of unit mass.

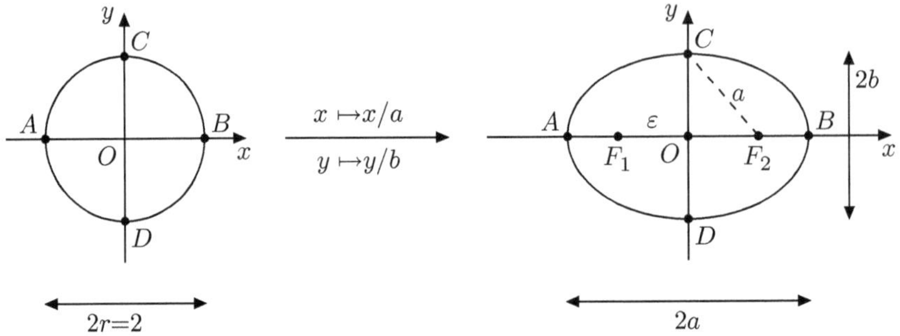

Fig. 4.1 The ellipse and its prominent parameters

which is an equation of an ellipse. To prove this statement, we start with the equation of an ellipse in the Cartesian coordinates, that is

$$\left(\frac{x}{a}\right)^2 + \left(\frac{y}{b}\right)^2 = 1\,, \tag{4.29}$$

where a and b are called the semi-axes of the ellipse. In terms of geometry, an ellipse is a set of points which have a constant sum of their respective distances from the two focal points $F_1 = (-\varepsilon, 0)$ and $F_2 = (\varepsilon, 0)$. The parameter ε is termed the excentricity of the ellipse and it assesses the difference between the ellipse and a perfect circle; that is, the value $\varepsilon = 0$ corresponds to a circle. The reader can prove on his or her own that the set of points (x, y) satisfying Eq. (4.29) does comply with this geometric interpretation. While doing so, the reader can also deduce that

$$\varepsilon^2 = a^2 - b^2, \tag{4.30}$$

see Fig. 4.1.

We now move the coordinate origin into one of the focal points, $x \to x - \varepsilon$, and introduce polar coordinates (4.27). Equation (4.29) can be then—with the aid of algebraic operations and the formula (4.30)—manipulated to

$$(r\varepsilon \cos\varphi + b^2)^2 = (ra)^2 \;\Rightarrow r\varepsilon \cos\varphi = ra - b^2\,.$$

By comparison of the last equation with Eq. (4.28), we see that the magnitude of the Runge-Lenz vector **X** is equal to the excentricity ε of the ellipse with the semi-axes of the magnitudes $a = Z$ and $b = L$, respectively. The Runge-Lenz vector points from the center of the ellipse to one of the focal points. The conservation of this vector in time means that the classical motion in the Coulomb field takes place along an ellipse and that this ellipse maintains its shape and orientation, i.e., it does not rotate or deform. We thus derived the first Kepler law without solving the Newton equations!

4.3.2 The Runge-Lenz Vector in Quantum Mechanics

It is natural to expect that since the Runge-Lenz vector is an integral of motion for the classical motion in the Coulomb field, the corresponding operator

$$\hat{\boldsymbol{X}} = \frac{1}{2}(\hat{\boldsymbol{L}} \times \hat{\boldsymbol{p}} - \hat{\boldsymbol{p}} \times \hat{\boldsymbol{L}}) + \hat{\boldsymbol{n}} \tag{4.31}$$

commutes with the Hamiltonian of the hydrogen atom,

$$[\hat{\boldsymbol{X}}, \hat{\mathsf{H}}] = 0\,, \qquad \hat{\mathsf{H}} = \frac{\hat{\mathsf{p}}^2}{2} - \frac{1}{r}\,. \tag{4.32}$$

This expectation is rightful and its proof is left to the reader. The cross product $\hat{\boldsymbol{L}} \times \hat{\boldsymbol{p}}$ in Eq. (4.31) is antisymmetrized in order to make the resulting operator Hermitian. The reason is that in general, a product of two Hermitian operators needs not necessarily yield a Hermitian operator: $(\hat{\mathsf{A}}\hat{\mathsf{B}})^+ = \hat{\mathsf{B}}^+\hat{\mathsf{A}}^+ = \hat{\mathsf{B}}\hat{\mathsf{A}}$, which is not the same as $\hat{\mathsf{A}}\hat{\mathsf{B}}$. On the other hand, the antisymmetrized form produces: $(\hat{\mathsf{A}}\hat{\mathsf{B}} + \hat{\mathsf{B}}\hat{\mathsf{A}})^+ = \hat{\mathsf{B}}\hat{\mathsf{A}} + \hat{\mathsf{A}}\hat{\mathsf{B}}$. The symmetrization of the components of $\hat{\boldsymbol{L}}$ and $\hat{\boldsymbol{p}}$ corresponds to the antisymmetrization of the cross product $\varepsilon_{ijk}(\hat{\mathsf{L}}_j\hat{\mathsf{p}}_k + \hat{\mathsf{p}}_k\hat{\mathsf{L}}_j) = \varepsilon_{ijk}\hat{\mathsf{L}}_j\hat{\mathsf{p}}_k - \varepsilon_{ikj}\hat{\mathsf{p}}_k\hat{\mathsf{L}}_j = (\hat{\boldsymbol{L}} \times \hat{\boldsymbol{p}} - \hat{\boldsymbol{p}} \times \hat{\boldsymbol{L}})_i$.

There is an important consequence arising from Eq. (4.32): the components of the vector operator $\hat{\boldsymbol{X}}$ do not mix states with different energy values. Namely,

$$0 = [\hat{\boldsymbol{X}}, \hat{\mathsf{H}}] = \langle n', l', m' | [\hat{\boldsymbol{X}}, \hat{\mathsf{H}}] | n, l, m \rangle = \left(\frac{1}{2n'^2} - \frac{1}{2n^2} \right) \langle n', l', m' | \hat{\boldsymbol{X}} | n, l, m \rangle\,. \tag{4.33}$$

Hence, we have $\langle n', l', m' | \hat{\boldsymbol{X}} | n, l, m \rangle = 0$ if $n' \neq n$.

In our further considerations it will prove advantageous to separate parts of the operator $\hat{\boldsymbol{X}}$ acting on the angular and radial degrees of freedom. Therefore, we substitute the expression for angular momentum (3.74) into Eq. (4.31),

$$\begin{aligned}
\hat{\mathsf{X}}_s &= \frac{1}{2}\left[\varepsilon_{siq}\varepsilon_{ijk}\hat{\mathsf{x}}_j\hat{\mathsf{p}}_k\hat{\mathsf{p}}_q - \varepsilon_{siq}\varepsilon_{qjk}\hat{\mathsf{p}}_i\hat{\mathsf{x}}_j\hat{\mathsf{p}}_k\right] + \hat{\mathsf{n}}_s \\
&= \frac{1}{2}\left[-(\delta_{sj}\delta_{qk} - \delta_{sk}\delta_{qj})\hat{\mathsf{x}}_j\hat{\mathsf{p}}_k\hat{\mathsf{p}}_q - (\delta_{sj}\delta_{ik} - \delta_{sk}\delta_{ij})\hat{\mathsf{p}}_i\hat{\mathsf{x}}_j\hat{\mathsf{p}}_k\right] + \hat{\mathsf{n}}_s \\
&= \frac{1}{2}(-\hat{\mathsf{x}}_s\hat{\mathsf{p}}^2 + \hat{\mathsf{x}}_q\hat{\mathsf{p}}_s\hat{\mathsf{p}}_q - \hat{\mathsf{p}}_k\hat{\mathsf{x}}_s\hat{\mathsf{p}}_k + \hat{\mathsf{p}}_j\hat{\mathsf{x}}_j\hat{\mathsf{p}}_s) + \hat{\mathsf{n}}_s\,.
\end{aligned}$$

The third and the fourth term can be rewritten by swapping the leading two factors and adding an appropriate commutator, so that we obtain

$$\hat{\mathsf{X}}_s = -\hat{\mathsf{x}}_s\hat{\mathsf{p}}^2 + (\hat{\mathsf{x}}_j\hat{\mathsf{p}}_j - \mathrm{i})\hat{\mathsf{p}}_s + \hat{\mathsf{n}}_s\,.$$

For the participating operators we have, in coordinate representation, Eqs. (3.13), $\hat{\mathsf{x}}_k \rightarrow rn_k$, (3.15), $\hat{\mathsf{p}}_k \rightarrow -\mathrm{i}\left(n_k \frac{\partial}{\partial r} + \frac{\nabla^n_k}{r}\right)$, (3.21) and (3.76), $\hat{\mathsf{p}}^2 \rightarrow -\left(\frac{\partial^2}{\partial r^2} + \frac{2}{r}\frac{\partial}{\partial r} - \frac{\hat{\mathsf{L}}^2}{r^2}\right)$, thus we can write

$$\hat{\boldsymbol{X}} = \mathbf{n}\left(\frac{\partial}{\partial r} - \frac{\hat{\mathsf{L}}^2}{r} + 1\right) - \nabla^n \frac{\partial}{\partial r} . \tag{4.34}$$

Note that this operator is a first-order differential operator in r. We now find the action of the components of the Runge-Lenz vector on the common eigenstates of the operators $\hat{\mathsf{H}}, \hat{\mathsf{L}}^2, \hat{\mathsf{L}}_z$, namely

$$\hat{\mathsf{X}}_j |n,l,m\rangle = \sum_{l',m'} d^{(j)}_{l',m'} \left|n,l',m'\right\rangle .$$

These equations will prove indeed valuable as they will provide first-order differential relations for the radial functions. Extracting the radial functions from these equations will be substantially easier than solving the Schrödinger equation (3.87) directly.

4.4 Matrix Elements of Vector Operators

4.4.1 Motivation

Let us be a little more general and find matrix elements of the vector operators of the type[7]

$$\hat{\boldsymbol{V}} = \mathbf{n}\,\hat{\mathsf{f}}_n(r, \hat{\mathsf{L}}^2) + \nabla^n \hat{\mathsf{f}}_\nabla(r) \tag{4.35}$$

between spherical harmonics. Here, $\hat{\mathsf{f}}_n$ and $\hat{\mathsf{f}}_\nabla$ can be arbitrary operators which depend on the coordinate r or on the squared magnitude of the angular momentum $\hat{\mathsf{L}}^2$, respectively. The Runge-Lenz vector is obviously a special case of such an operator, see Eq. (4.34). The operators of the momentum and of the coordinate, $\hat{\boldsymbol{p}}$ and $\hat{\boldsymbol{x}}$, are another special instances of this operator. As we will see later in Chap. 6, the probability that an atom changes its state from the initial state $|i\rangle$ to the final state $|f\rangle$ and radiates a photon during that transition is proportional, in a good approximation, to the squared magnitude of the matrix element

$$(\hat{\boldsymbol{x}})_{if} = \langle i|\hat{\boldsymbol{x}}|f\rangle = \int_0^\infty R_{n_i,l_i} r R_{n_f,l_f}\, r^2 \mathrm{d}r \int Y^*_{l_i,m_i} \mathbf{n} Y_{l_f,m_f}\, \mathrm{d}\Omega . \tag{4.36}$$

[7] The derivation in this section is inspired by that in [1].

The knowledge of the matrix elements of the coordinate operator between the spherical harmonics dictates which transitions are allowed and which forbidden. The herein presented derivation may seem rather lengthy, therefore an impatient reader can skip over to the final equations (4.50)–(4.52).

4.4.2 Commutation Relations

The sought matrix elements can be almost completely determined merely from the commutation relations

$$[\hat{\mathsf{L}}_i, n_j] = \mathrm{i}\varepsilon_{ijk} n_k\,, \qquad [\hat{\mathsf{L}}_i, \nabla^n_j] = \mathrm{i}\varepsilon_{ijk} \nabla^n_k\,, \tag{4.37}$$

$$[\hat{\mathsf{L}}^2, n_k] = 2(n_k - \nabla^n_k)\,, \qquad [\hat{\mathsf{L}}^2, \nabla^n_k] = -2n_k \hat{\mathsf{L}}^2\,. \tag{4.38}$$

The first one can be verified easily (try the rest yourself!)

$$\begin{aligned}[\hat{\mathsf{L}}_i, n_j] &= [-\mathrm{i}\varepsilon_{ipq} n_p \nabla^n_q, n_j] = -\mathrm{i}\varepsilon_{ipq} n_p [\nabla^n_q, n_j] \\ &= -\mathrm{i}\varepsilon_{ipq}(\delta_{qj} - n_q n_j) n_p = \mathrm{i}\varepsilon_{ijp} n_p + \mathrm{i}\varepsilon_{ipq} n_p n_q n_j\,.\end{aligned}$$

In the last term, we contracted an antisymmetric and a symmetric tensor, which yields zero, hence leaving only the first term. For verification of the other relations, it is useful to know the commutators of the components of the angular operator ∇^n. Employing Eqs. (3.15) and (3.18) leads to

$$\begin{aligned}0 = \left[\frac{\partial}{\partial x^i}, \frac{\partial}{\partial x^j}\right] &= \left[n_i \frac{\partial}{\partial r} + \frac{\nabla^n_i}{r}, n_j \frac{\partial}{\partial r} + \frac{\nabla^n_j}{r}\right] \\ &= \left[n_i \frac{\partial}{\partial r}, \frac{\nabla^n_j}{r}\right] + \left[\frac{\nabla^n_i}{r}, n_j \frac{\partial}{\partial r}\right] + \left[\frac{\nabla^n_i}{r}, \frac{\nabla^n_j}{r}\right] \\ &= -\left[\frac{\nabla^n_j}{r}, n_i \frac{\partial}{\partial r}\right] + \left[\frac{\nabla^n_i}{r}, n_j \frac{\partial}{\partial r}\right] + \left[\nabla^n_i, \nabla^n_j\right]\frac{1}{r^2} \\ &= -n_i \nabla^n_j \left[\frac{1}{r}, \frac{\partial}{\partial r}\right] - \frac{1}{r}\left[\nabla^n_j, n_i\right]\frac{\partial}{\partial r} + n_j \nabla^n_i \left[\frac{1}{r}, \frac{\partial}{\partial r}\right] \\ &\quad + \frac{1}{r}\left[\nabla^n_i, n_j\right]\frac{\partial}{\partial r} + \left[\nabla^n_i, \nabla^n_j\right]\frac{1}{r^2} \\ &= -\frac{1}{r^2}(n_i \nabla^n_j - n_j \nabla^n_i) - \frac{1}{r}\frac{\partial}{\partial r}(\delta_{ji} - n_j n_i - \delta_{ij} + n_i n_j) + \left[\nabla^n_i, \nabla^n_j\right]\frac{1}{r^2} \\ &= -\frac{1}{r^2}(n_i \nabla^n_j - n_j \nabla^n_i) + \left[\nabla^n_i, \nabla^n_j\right]\frac{1}{r^2}\,.\end{aligned}$$

It follows thereof

$$[\nabla_i^n, \nabla_j^n] = n_i \nabla_j^n - n_j \nabla_i^n \,.$$

We thus learn from the relation (4.37) that every vector operator $\hat{\boldsymbol{V}}$ of the type (4.35) satisfies

$$[\hat{\mathsf{L}}_i, \hat{\mathsf{V}}_j] = \mathrm{i}\varepsilon_{ijk}\hat{\mathsf{V}}_k \,. \tag{4.39}$$

Next, we will show how this commutation relation leads to selection rules for the magnetic quantum number m.

4.4.3 Selection Rules in m

Equation (4.39) yields for $i = j = 3$

$$[\hat{\mathsf{L}}_3, \hat{\mathsf{V}}_3] = 0 \,.$$

This operator equality holds true when we apply it on the eigenvectors of the operators $\hat{\mathsf{L}}_3$ and $\hat{\mathsf{L}}^2$,

$$\langle l', m' \big| [\hat{\mathsf{L}}_3, \hat{\mathsf{V}}_3] \big| l, m \rangle = 0 \,.$$

From the definition (4.3), we obtain

$$(m' - m)\langle l', m' \big| \hat{\mathsf{V}}_3 \big| l, m \rangle = 0 \,.$$

In other words, if $m' \neq m$ then necessarily $\langle l', m' | \hat{\mathsf{V}}_3 | l, m \rangle = 0$. Next, let us inspect the commutator

$$[\hat{\mathsf{L}}_3, \hat{\mathsf{V}}_\pm] = [\hat{\mathsf{L}}_3, \hat{\mathsf{V}}_1] \pm \mathrm{i}[\hat{\mathsf{L}}_3, \hat{\mathsf{V}}_2] = \mathrm{i}\hat{\mathsf{V}}_2 \pm \mathrm{i}(-\mathrm{i}\hat{\mathsf{V}}_1) = \pm\hat{\mathsf{V}}_1 + \mathrm{i}\hat{\mathsf{V}}_2 = \pm\hat{\mathsf{V}}_\pm \,,$$

where in the second equality Eq. (4.39) was used. Based on the same arguments as before, we obtain

$$\begin{aligned}
\langle l', m' \big| [\hat{\mathsf{L}}_3, \hat{\mathsf{V}}_\pm] \big| l, m \rangle &= \pm \langle l', m' \big| \hat{\mathsf{V}}_\pm \big| l, m \rangle \\
\Rightarrow (m' - m)\langle l', m' \big| \hat{\mathsf{V}}_\pm \big| l, m \rangle &= \pm \langle l', m' \big| \hat{\mathsf{V}}_\pm \big| l, m \rangle \\
\Rightarrow (m' - m \mp 1)\langle l', m' \big| \hat{\mathsf{V}}_\pm \big| l, m \rangle &= 0 \,.
\end{aligned}$$

Therefore, we have $\langle l', m' | \hat{\mathsf{V}}_\pm | l, m \rangle = 0$ whenever m' differs from $m \pm 1$.

4.4.4 Selection Rules in *l*

As our next step, we show how the selection rules for the orbital quantum number l follow from the commutation relations (4.38). After applying the first equation (4.38) on the eigenvectors of the operators $\hat{\mathsf{L}}_3$ and $\hat{\mathsf{L}}^2$, we obtain

$$\langle l', m' | [\hat{\mathsf{L}}^2, n_k] | l, m\rangle = 2\langle l', m' | (n_k - \nabla_k^n) | l, m\rangle ,$$

hence

$$\big(l'(l'+1) - l(l+1)\big)\langle l', m' | n_k | l, m\rangle = 2\langle l', m' | (n_k - \nabla_k^n) | l, m\rangle ,$$

and we express the term with the angular nabla,

$$\langle l', m' | \nabla_k^n | l, m\rangle = \frac{1}{2}\big[2 + l(l+1) - l'(l'+1)\big]\langle l', m' | n_k | l, m\rangle . \tag{4.40}$$

Similarly, we obtain from the second equation (4.38)

$$\langle l', m' | [\hat{\mathsf{L}}^2, \nabla_k^n] | l, m\rangle = -2\langle l', m' | n_k \hat{\mathsf{L}}^2 | l, m\rangle .$$

Thus, after expanding the commutator we arrive at

$$\big(l'(l'+1) - l(l+1)\big)\langle l', m' | \nabla_k^n | l, m\rangle = -2l(l+1)\langle l', m' | n_k | l, m\rangle . \tag{4.41}$$

Equations (4.40) and (4.41) can be further rearranged to yield a restriction on the matrix elements of the operator n_k in the form

$$\langle l', m' | n_k | l, m\rangle (l' + l + 2)(l' + l)(l' + 1 - l)(l' - 1 - l) = 0 .$$

Since l is always nonnegative, the first two brackets are always nonzero (except for the trivial case $l = l' = 0$). Thus, we reach the result: we have $\langle l', m' | n_k | l, m\rangle = 0$ for l' different from $l \pm 1$. In addition, it also follows from Eqs. (4.40) and (4.41) that in such a case the element $\langle l', m' | \nabla_k^n | l, m\rangle$ is equal to zero as well. In consequence, all electromagnetic transitions other than between an s- and p-state, or p- and d-state, etc., that is always by one angular degree, are forbidden.[8]

[8]This conclusion is true only for the so-called dipole radiation, see Sect. 6.2.4.

4.4.5 Nonzero Matrix Elements: Dependence on m

We will continue our analysis of the action of a general vector operator $\hat{\boldsymbol{V}}$ on spherical harmonics. We managed to discover the selection rules so far; however, we desire to find the complete expressions for the resulting vectors $\hat{\mathsf{V}}_3|l,m\rangle$ and $\hat{\mathsf{V}}_\pm|l,m\rangle$ as well. The reason is, their knowledge substantially simplifies the calculation of both the radial and angular wave functions of the hydrogen atom. We begin by determining the commutator $[\hat{\mathsf{L}}_+,\hat{\mathsf{V}}_+]$ from Eq. (4.39),

$$[\hat{\mathsf{L}}_+,\hat{\mathsf{V}}_+] = [\hat{\mathsf{L}}_1 + \mathrm{i}\hat{\mathsf{L}}_2, \hat{\mathsf{V}}_1 + \mathrm{i}\hat{\mathsf{V}}_2] = \mathrm{i}[\hat{\mathsf{L}}_1,\hat{\mathsf{V}}_2] + \mathrm{i}[\hat{\mathsf{L}}_2,\hat{\mathsf{V}}_1] = \mathrm{i}(\mathrm{i}\hat{\mathsf{V}}_3) + \mathrm{i}(-\mathrm{i}\hat{\mathsf{V}}_3) = 0\,.$$

Subsequently, we apply this equation once again on the eigenstates of the operators $\hat{\mathsf{L}}^2$ and $\hat{\mathsf{L}}_3$,

$$\langle l',m'|[\hat{\mathsf{L}}_+,\hat{\mathsf{V}}_+]|l,m\rangle = 0\,. \tag{4.42}$$

Using the Hermitian conjugate of Eq. (4.8) for $\hat{\mathsf{L}}_\pm$,

$$\langle l,m|\hat{\mathsf{L}}_\mp = \langle l,m\pm 1|\sqrt{(l\mp m)(l+1\pm m)}\,, \tag{4.43}$$

we obtain from Eq. (4.42)

$$\begin{aligned} 0 = &\sqrt{(l'+m')(l'+1-m')}\langle l',m'-1|\hat{\mathsf{V}}_+|l,m\rangle \\ &- \sqrt{(l-m)(l+1+m)}\langle l',m'|\hat{\mathsf{V}}_+|l,m+1\rangle\,. \end{aligned}$$

With the choice $m' = m+2$ and $l' = l-1$, the last equation takes on the form

$$\begin{aligned} 0 = &\sqrt{(l+m+1)(l-m-2)}\langle l-1,m+1|\hat{\mathsf{V}}_+|l,m\rangle \\ &- \sqrt{(l-m)(l+1+m)}\langle l-1,m+2|\hat{\mathsf{V}}_+|l,m+1\rangle\,; \end{aligned}$$

therefore

$$\frac{\langle l-1,m+2|\hat{\mathsf{V}}_+|l,m+1\rangle}{\langle l-1,m+1|\hat{\mathsf{V}}_+|l,m\rangle} = \sqrt{\frac{l-m-2}{l-m}} = \sqrt{\frac{(l-m-2)(l-m-1)}{(l-m-1)(l-m)}}\,.$$

On both sides of the last equation, there is a ratio of two expressions which differ solely by the interchange $m \leftarrow m+1$. Hence, the dependency of the matrix element on m then must read

$$\langle l-1,m+1|\hat{\mathsf{V}}_+|l,m\rangle = c_l\sqrt{(l-m-1)(l-m)}\,, \tag{4.44}$$

where c_l depends on l and possibly also on other quantum numbers except for m. In literature, the coefficients c_l are frequently termed as the *reduced matrix elements*.

We have thus successfully completed one sixth of our task. As our next step, we find from Eq. (4.39),

$$[\hat{L}_-, \hat{V}_+] = [\hat{L}_1, \hat{V}_1] + i[\hat{L}_1, \hat{V}_2] - i[\hat{L}_2, \hat{V}_1] + [\hat{L}_2, \hat{V}_2] = -2\hat{V}_3\,,$$

that $\hat{V}_3 = -[\hat{L}_-, \hat{V}_+]/2$. Using this operator equality on the eigenstates of the operators $\hat{L}^2$ and $\hat{L}_3$, and employing Eqs. (4.8), (4.43), and (4.44) we successively obtain

$$\begin{aligned}
\langle l-1, m|\hat{V}_3|l, m\rangle &= \langle l-1, m|\left(-\frac{1}{2}\right)[\hat{L}_-, \hat{V}_+]|l, m\rangle \\
&= -\frac{1}{2}\langle l-1, m|(\hat{L}_-\hat{V}_+ - \hat{V}_+\hat{L}_-)|l, m\rangle \\
&= -\frac{1}{2}\sqrt{(l-1-m)(l-1+1+m)}\langle l-1, m+1|\hat{V}_+|l, m\rangle \\
&\quad + \frac{1}{2}\sqrt{(l+m)(l+1-m)}\langle l-1, m|\hat{V}_+|l, m-1\rangle \\
&= -\frac{1}{2}\sqrt{(l-1-m)(l+m)}c_l\sqrt{(l-m-1)(l-m)}+ \\
&\quad + \frac{1}{2}\sqrt{(l+m)(l+1-m)}c_l\sqrt{(l-m)(l-m+1)} \\
&= c_l\sqrt{(l-m)(l+m)}\,. \qquad (4.45)
\end{aligned}$$

One can also easily derive from Eq. (4.39) that $[\hat{L}_-, \hat{V}_3] = \hat{V}_-$. It thus follows (the reader will be able to reproduce the missing steps on his or her own by now),

$$\langle l-1, m-1|\hat{V}_-|l, m\rangle = -c_l\sqrt{(l+m-1)(l+m)}\,. \qquad (4.46)$$

We assume that the coefficients c_l are real and will comment thereon in detail at the end of this section. For $\hat{V}_3$ is a Hermitian operator, the Hermitian conjugation of Eq. (4.45) together with the interchange $l \to l+1$ lead to

$$\langle l+1, m|\hat{V}_3|l, m\rangle = c_{l+1}\sqrt{(l+1-m)(l+1+m)}\,. \qquad (4.47)$$

We proceed in a similar manner in case of Eq. (4.44), with the exception of performing a simultaneous replacement $l \to l+1$ and $m \to m-1$, and Hermitian conjugation:

$$\langle l+1, m-1|\hat{V}_-|l, m\rangle = c_{l+1}\sqrt{(l-m+2)(l-m+1)}\,. \qquad (4.48)$$

Finally, the replacements $l \to l+1$ and $m \to m+1$ and Hermitian conjugation lead from Eq. (4.46) to

$$\langle l+1, m+1|\hat{\mathsf{V}}_+|l, m\rangle = -c_{l+1}\sqrt{(l+m+1)(l+m+2)}\,. \tag{4.49}$$

Recalling now the selection rules, these six elements from (4.44) to (4.49) are the only matrix elements of the operators $\hat{\mathsf{V}}_\pm$ and $\hat{\mathsf{V}}_3$ which do not necessarily vanish. Therefore, we can write

$$\begin{aligned}\hat{\mathsf{V}}_3|l, m\rangle &= c_l\sqrt{(l+m)(l-m)}|l-1, m\rangle \\ &\quad + c_{l+1}\sqrt{(l+1+m)(l+1-m)}|l+1, m\rangle\,, \end{aligned} \tag{4.50}$$

$$\begin{aligned}\hat{\mathsf{V}}_+|l, m\rangle &= c_l\sqrt{(l-m-1)(l-m)}|l-1, m+1\rangle \\ &\quad - c_{l+1}\sqrt{(l+m+2)(l+m+1)}|l+1, m+1\rangle\,, \end{aligned} \tag{4.51}$$

$$\begin{aligned}\hat{\mathsf{V}}_-|l, m\rangle &= -c_l\sqrt{(l+m-1)(l+m)}|l-1, m-1\rangle \\ &\quad + c_{l+1}\sqrt{(l-m+2)(l-m+1)}|l+1, m-1\rangle\,. \end{aligned} \tag{4.52}$$

In case of $[\hat{\boldsymbol{V}}, \hat{\mathsf{H}}] = 0$, the operator $\hat{\boldsymbol{V}}$ does not mix quantum states with different n, see for example Eq. (4.33), and we can write

$$\begin{aligned}\hat{\mathsf{V}}_3|n, l, m\rangle &= c_{l,n}\sqrt{(l+m)(l-m)}|n, l-1, m\rangle \\ &\quad + c_{l+1,n}\sqrt{(l+1+m)(l+1-m)}|n, l+1, m\rangle\,, \end{aligned} \tag{4.53}$$

where (in contrast to Eq. (4.50)) the coefficients c may depend on the principal quantum number n as well.

Let us return to the assumption that the coefficients c_l are real. We will see that it does not lead to a contradiction in the case of a purely angular operator $\hat{\boldsymbol{V}}$. The operator $\hat{\boldsymbol{V}}$, which comprises both radial and angular operators, such as the Runge-Lenz vector, Eq. (4.34), acts on the states which may be expressed as products of radial and angular parts. Choosing an appropriate complex phase of the radial part of the wave function, we may require the coefficients $c_{l,n}$ be real. For instance, it must be possible to choose the radial functions real since the radial Hamiltonian in Eq. (3.87) is real. Therefore, the coefficients $c_{l,n}$ must be real as well. As we will see later on in Sect. 4.5.2, this finding is of great importance.

4.4.6 Generalization

The above considerations may be generalized even further. An arbitrary vector operator $\hat{\boldsymbol{V}}$ complying with

$$[\hat{\mathsf{J}}_i, \hat{\mathsf{V}}_j] = \mathrm{i}\varepsilon_{ijk}\hat{\mathsf{V}}_k\,, \tag{4.54}$$

where $\hat{\mathsf{J}}_i$ are the components of an arbitrary angular momentum obeying the commutation relations (4.1), also satisfies the identity

$$[\hat{\mathsf{J}}^2, [\hat{\mathsf{J}}^2, \hat{\boldsymbol{V}}]] = 2(\{\hat{\mathsf{J}}^2, \hat{\boldsymbol{V}}\} - \{\hat{\boldsymbol{J}}, \hat{\boldsymbol{J}} \cdot \hat{\boldsymbol{V}}\}),$$

the proof being straightforward, though laborious. From the last expression, we obtain the following formula for the matrix elements of the operator $\hat{\boldsymbol{V}}$ between the eigenstates of $\hat{\mathsf{J}}^2$ and $\hat{\mathsf{J}}_z$

$$[(j'-j)^2 - 1][(j'+j+1)^2 - 1]\langle j', m' | \hat{\boldsymbol{V}} | j, m\rangle = -2\langle j', m' | \{\hat{\boldsymbol{J}}, \hat{\boldsymbol{J}} \cdot \hat{\boldsymbol{V}}\} | j, m\rangle, \qquad (4.55)$$

which can be found using similar consideration as in the previous section. From Eq. (4.54), one can also easily deduce that

$$[\hat{\boldsymbol{J}}, \hat{\boldsymbol{J}} \cdot \hat{\boldsymbol{V}}] = 0.$$

Considering Eq. (4.55) for $j' = j$ yields

$$\langle j, m' | \hat{\boldsymbol{V}} | j, m\rangle = \frac{1}{j(j+1)} \langle j, m' | \hat{\boldsymbol{J}} \hat{\boldsymbol{J}} \cdot \hat{\boldsymbol{V}} | j, m\rangle. \qquad (4.56)$$

This implies that when $\hat{\boldsymbol{J}} \cdot \hat{\boldsymbol{V}} = 0$ the matrix element $\langle j, m' | \hat{\boldsymbol{V}} | j, m\rangle$ equals zero.

If $j' \neq j$ then the rhs of Eq. (4.55) vanishes and then lhs has to vanish as well. This is possible only if $\langle j', m' | \hat{\boldsymbol{V}} | j, m\rangle = 0$ if $j' \neq j \pm 1$. To sum up, the matrix elements $\langle j', m' | \hat{\boldsymbol{V}} | j, m\rangle$ vanish whenever $j' \neq j \pm 1$ or $j' \neq j$. Equations (4.44)–(4.49) were derived merely from the commutation relation (4.39). Therefore, they hold true for a general operator $\hat{\boldsymbol{V}}$ obeying Eq. (4.54), with l replaced by j. The generalizations of Eqs. (4.50)–(4.52) for an arbitrary operator obeying Eq. (4.54) then reads:

$$\begin{aligned}
\hat{\mathsf{V}}_3 | j, m\rangle &= c_j \sqrt{(j+m)(j-m)} | j-1, m\rangle \\
&\quad + c_{j+1} \sqrt{(j+1+m)(j+1-m)} | j+1, m\rangle \\
&\quad + m a_j | j, m\rangle, \\
\hat{\mathsf{V}}_+ | j, m\rangle &= c_j \sqrt{(j-m-1)(j-m)} | j-1, m+1\rangle \\
&\quad - c_{j+1} \sqrt{(j+m+2)(j+m+1)} | j+1, m+1\rangle \\
&\quad + \sqrt{(j-m)(j+m+1)} a_j | j, m+1\rangle, \\
\hat{\mathsf{V}}_- | j, m\rangle &= -c_j \sqrt{(j+m-1)(j+m)} | j-1, m-1\rangle \\
&\quad + c_{j+1} \sqrt{(j-m+2)(j-m+1)} | j+1, m-1\rangle \\
&\quad + \sqrt{(j+m)(j-m+1)} a_j | j, m-1\rangle,
\end{aligned}$$

where

$$a_j = \frac{1}{j(j+1)}\langle j, m|\hat{\boldsymbol{J}} \cdot \hat{\boldsymbol{V}}|j, m\rangle \, .$$

4.4.7 The Zeeman Effect

The above presented generalization finds an indeed nice application in the so-called *anomalous Zeeman effect*. This effect is nothing else but splitting of atomic spectral lines arising from the combined influence of the spin-orbit interaction and the interaction of the spin and orbital angular momenta of an electron with the external magnetic field.[9] With all the knowledge acquired so far, the reader will be able to study this effect independently with the aid of the following instructions.

1. Apart from the kinetic and electrostatic potential energy, consider only the terms describing the spin-orbit interaction on the rhs of Eq. (3.101)

$$\hat{\mathsf{H}} = \frac{[\sigma \cdot (\hat{\boldsymbol{p}} - e\mathbf{A})]^2}{2m} + e\varphi - \frac{1}{8m^2}\left[\sigma_i, \sigma_j\right]\left[\hat{\mathsf{p}}_j, e\varphi\right]\hat{\mathsf{p}}_i \, .$$

   Further, consider a general electrostatic potential $e\varphi$, though spherically symmetrical, $\varphi = \varphi(\hat{\mathsf{r}})$.
2. As usually, neglect the term $e^2\mathbf{A} \cdot \mathbf{A}$. The Hamilton operator $\hat{\mathsf{H}}$ can then be cast into the form

$$\hat{\mathsf{H}} = \hat{\mathsf{H}}_0 + \hat{\mathsf{H}}_1, \quad \hat{\mathsf{H}}_0 = \frac{\hat{\mathsf{p}}^2}{2m} + e\varphi \, ,$$

$$\hat{\mathsf{H}}_1 = \frac{\hat{\boldsymbol{S}} \cdot \hat{\boldsymbol{L}}}{2m^2}\frac{1}{r}\frac{\mathrm{d}(e\varphi)}{\mathrm{d}r} - \frac{e}{m}\left(\mathbf{A} \cdot \hat{\boldsymbol{p}} + \mathbf{B} \cdot \hat{\boldsymbol{S}}\right) .$$

3. Choose the z-axis in the direction of the magnetic field. Then

$$\mathbf{B} = B(0, 0, 1) \, , \quad \mathbf{A} = \frac{B}{2}(-y, x, 0) \, ,$$

   where the vector potential $\mathbf{A}$ was chosen so that $\mathbf{B} = \nabla \times \mathbf{A}$ and $\nabla \cdot \mathbf{A} = 0$; hence

$$\hat{\mathsf{H}}_1 = \frac{\hat{\boldsymbol{S}} \cdot \hat{\boldsymbol{L}}}{2m^2}\frac{1}{r}\frac{\mathrm{d}(e\varphi)}{\mathrm{d}r} - \frac{eB}{2m}\left(\hat{\mathsf{L}}_z + 2\hat{\mathsf{S}}_z\right) .$$

[9] The *normal* Zeeman effect occurs under the very same conditions with the only exception that the spin-orbit interaction is negligible. The terminology, stemming from the historical development, was coined somewhat unfortunately, though.

4. Change to the atomic units $r \to r_A/(mZ\alpha)$ and introduce the notation

$$V(r_A) = \frac{e\varphi\left(\frac{r_A}{mZ\alpha}\right)}{m(Z\alpha)^2}$$

and

$$\gamma_{\mathrm{B}} = \frac{eB}{(m\alpha)^2} \,. \tag{4.57}$$

The Hamiltonian $\hat{\mathsf{H}}_1$, which comprises the spin-orbit interaction and the interaction with the external magnetic field, then takes on the form

$$\frac{\hat{\mathsf{H}}_1}{m(Z\alpha)^2} = \frac{(Z\alpha)^2}{2}\hat{\boldsymbol{S}}\cdot\hat{\boldsymbol{L}}\frac{1}{r_A}\frac{\mathrm{d}V(r_A)}{\mathrm{d}r_A} - \frac{\gamma_{\mathrm{B}}}{2Z^2}\left(\hat{\mathsf{L}}_z + 2\hat{\mathsf{S}}_z\right) .$$

As shown in Eq. (0.7), the dimensionless strength equals $\gamma_B \simeq 4.254 \times 10^{-6}$ for the magnetic field of the intensity 1 T. To realize how strong the magnetic field is, we can compare it to a few examples: the protons in the Large Hadron Collider (LHC) are kept on their orbit by superconducting magnets maintaining the field of about 8 T, and the highest recorded magnetic intensity ever reached on Earth is 34 T. Considering that $\alpha^2/2 \simeq 2.7 \times 10^{-5}$, a situation when the two interactions are of a comparable magnitude is certainly plausible. However, within the conditions we commonly encounter on our Earth, the first-order perturbation method suffices.

5. Choose the eigenvectors of $\hat{\mathsf{H}}_0$ in the form

$$\left\langle r \middle| n, j, m, \left(\frac{1}{2}, l\right)\right\rangle = R_{n,l}(r)\left| j, m, \left(\frac{1}{2}, l\right)\right\rangle, \tag{4.58}$$

where the states $\left|j, m, \left(\frac{1}{2}, l\right)\right\rangle$ are called the *spherical spinors*. We already encountered their simplest instance for $l = 1$ in Sect. 3.5.2. For the first-order perturbation method, we will need the matrix elements

$$\begin{aligned}
&\left\langle n, j, m, \left(\frac{1}{2}, l\right)\middle|\hat{\mathsf{H}}_1\middle| n, j, m, \left(\frac{1}{2}, l\right)\right\rangle \\
&= \frac{(Z\alpha)^2}{2}\int_0^\infty \mathrm{d}r r R_{n,l}^2(r)\frac{\mathrm{d}V}{\mathrm{d}r}\left\langle j, m, \left(\frac{1}{2}, l\right)\middle|\hat{\boldsymbol{S}}\cdot\hat{\boldsymbol{L}}\middle| j, m, \left(\frac{1}{2}, l\right)\right\rangle \\
&- \frac{\gamma_{\mathrm{B}}}{2Z^2}\left\langle j, m, \left(\frac{1}{2}, l\right)\middle|\left(\hat{\mathsf{L}}_z + 2\hat{\mathsf{S}}_z\right)\middle| j, m, \left(\frac{1}{2}, l\right)\right\rangle .
\end{aligned}$$

With the aid of Eq. (3.110) and the defining formulas for $\left|j,m,\left(\frac{1}{2},l\right)\right\rangle$, Eqs. (4.10), (4.12), and (4.13), the spin-orbit interaction matrix element equals

$$\left\langle j,m,\left(\frac{1}{2},l\right)\right|\hat{\boldsymbol{S}}\cdot\hat{\boldsymbol{L}}\left|j,m,\left(\frac{1}{2},l\right)\right\rangle = \frac{j(j+1)-l(l+1)}{2} - \frac{3}{8}$$

6. Now the key point comes. We can write for the matrix elements of the angular momentum projections along the direction of the magnetic field

$$\left\langle j,m,\left(\frac{1}{2},l\right)\right|\left(\hat{\mathsf{L}}_z+2\hat{\mathsf{S}}_z\right)\left|j,m,\left(\frac{1}{2},l\right)\right\rangle$$
$$= m + \left\langle j,m,\left(\frac{1}{2},l\right)\right|\hat{\mathsf{S}}_z\left|j,m,\left(\frac{1}{2},l\right)\right\rangle,$$

where we used Eqs. (4.9) and (4.11). However, how to calculate the last matrix element? We could use the expansion (4.17) in terms of orbital and spin states as we have already determined the action of $\hat{\mathsf{S}}_z$ on the spin state. Nevertheless, there is a more elegant way. One can easily find that

$$[\hat{\mathsf{J}}_i,\hat{\mathsf{S}}_j] = [\hat{\mathsf{L}}_i+\hat{\mathsf{S}}_i,\hat{\mathsf{S}}_j] = [\hat{\mathsf{S}}_i,\hat{\mathsf{S}}_j] = \mathrm{i}\varepsilon_{ijk}\hat{\mathsf{S}}_k\,,$$

thus $\hat{\boldsymbol{S}}$ behaves as a vector operator with respect to $\hat{\boldsymbol{J}}$! Owing to the relation

$$\hat{\boldsymbol{S}}\cdot\hat{\boldsymbol{J}} = \frac{1}{2}(\hat{\mathsf{J}}^2+\hat{\mathsf{S}}^2-\hat{\mathsf{L}}^2)\,,$$

we have from Eq. (4.56)

$$\left\langle j,m,\left(\frac{1}{2},l\right)\right|\hat{\mathsf{S}}_z\left|j,m,\left(\frac{1}{2},l\right)\right\rangle = \frac{j(j+1)-l(l+1)+3/4}{2j(j+1)}m\,.$$

Exercise 9: The Zeeman Effect

Determine (numerically in hertz) the splitting of the 2p level of the hydrogen atom when it is inserted into a magnetic field of intensity 0.5 T.

4.4.8 Nonzero Matrix Elements: Dependence on *l* and *n*

The only remaining task now is to determine the numbers $c_{l,(n)}$. One can do so in the following way. On one hand, we can calculate $(\hat{\mathsf{V}}_+\hat{\mathsf{V}}_- + \hat{\mathsf{V}}_3^2)|(n), l, m\rangle$ directly from Eqs. (4.50) to (4.52). After straightforward, though somewhat lengthy manipulations, we arrive at

$$(\hat{\mathsf{V}}_+\hat{\mathsf{V}}_- + \hat{\mathsf{V}}_3^2)|(n), l, m\rangle$$
$$= \left(c_{l,(n)}^2(2l-1)(l+m) + c_{l+1,(n)}^2(2l+3)(l+1-m)\right)|(n), l, m\rangle\,. \qquad (4.59)$$

On the other hand, we can use the defining equations $\hat{\mathsf{V}}_\pm = \hat{\mathsf{V}}_1 \pm \mathrm{i}\hat{\mathsf{V}}_2$ and rewrite the lhs of the last equation as

$$(\hat{\mathsf{V}}_+\hat{\mathsf{V}}_- + \hat{\mathsf{V}}_3^2)|(n), l, m\rangle = \left(\hat{\mathsf{V}}_1^2 + \hat{\mathsf{V}}_2^2 + \hat{\mathsf{V}}_3^2 - \mathrm{i}[\hat{\mathsf{V}}_1, \hat{\mathsf{V}}_2]\right)|(n), l, m\rangle$$
$$= \left(\hat{\mathsf{V}}^2 - \mathrm{i}[\hat{\mathsf{V}}_1, \hat{\mathsf{V}}_2]\right)|(n), l, m\rangle\,. \qquad (4.60)$$

Thus, in order to find the coefficients $c_{l,(n)}$, we need to know the squared magnitude of the vector operator and the commutator of the vector components with each other. These commutation relations, however, depend on the specific form of every individual operator $\hat{\mathsf{V}}$. In other words, it depends on the particular form of operators $\hat{\mathsf{f}}_n$ and $\hat{\mathsf{f}}_\nabla$ appearing in Eq. (4.35).

4.4.9 Spherical Harmonics

If the vector operator of choice is the direction vector $\hat{\boldsymbol{V}} = \mathbf{n}$ then obviously $n_i n_i = 1$, $[n_i, n_j] = 0$. It follows then from Eqs. (4.59) and (4.60)

$$1 = c_l^2(2l-1)(l+m) + c_{l+1}^2(l+1-m)(2l+3)\,. \qquad (4.61)$$

Since c_l is independent of m, the same constant must appear in front of the same powers of m on both sides of the last equation. Due to this requirement, we obtain two equations: one for c_l and one for c_{l+1}. These equations must be, and they are, compatible;

$$\left.\begin{aligned} 1 &= c_l^2(2l-1)l + c_{l+1}^2(l+1)(2l+3) \\ 0 &= c_l^2(2l-1) - c_{l+1}^2(2l+3) \end{aligned}\right\} \Rightarrow c_l = \frac{1}{\sqrt{(2l-1)(2l+1)}}\,.$$

We can now express the action of the components of an arbitrarily oriented unit vector on the spherical harmonics as

$$n_3 Y_{lm}(\mathbf{n}) = \sqrt{\frac{(l-m)(l+m)}{(2l-1)(2l+1)}}\, Y_{l-1,m}(\mathbf{n}) + \sqrt{\frac{(l+1+m)(l+1-m)}{(2l+1)(2l+3)}}\, Y_{l+1,m}(\mathbf{n})\,, \tag{4.62}$$

$$n_+ Y_{lm}(\mathbf{n}) = \sqrt{\frac{(l-m-1)(l-m)}{(2l-1)(2l+1)}}\, Y_{l-1,m+1}(\mathbf{n}) - \sqrt{\frac{(l+m+2)(l+m+1)}{(2l+1)(2l+3)}}\, Y_{l+1,m+1}(\mathbf{n}) \tag{4.63}$$

and

$$n_- Y_{lm}(\mathbf{n}) = -\sqrt{\frac{(l+m-1)(l+m)}{(2l-1)(2l+1)}}\, Y_{l-1,m-1}(\mathbf{n}) + \sqrt{\frac{(l-m+2)(l-m+1)}{(2l+1)(2l+3)}}\, Y_{l+1,m-1}(\mathbf{n})\,. \tag{4.64}$$

These formulas find a wide range of applications, such as in the determination of the form of spherical harmonics. From Eqs. (4.62) to (4.64), one can calculate all other Y_{lm} if given the starting Y_{00}, see Eq. (3.35), in an elegant way,

$$\begin{aligned} n_3 Y_{00}(\mathbf{n}) &= \frac{1}{\sqrt{3}} Y_{10}(\mathbf{n})\,, \\ n_3 Y_{10}(\mathbf{n}) &= \frac{1}{\sqrt{3}} Y_{00}(\mathbf{n}) + \frac{2}{\sqrt{3\cdot 5}} Y_{20}(\mathbf{n})\,, \\ n_+ Y_{00}(\mathbf{n}) &= -\sqrt{\frac{2}{3}} Y_{11}(\mathbf{n})\,, \\ n_- Y_{00}(\mathbf{n}) &= \sqrt{\frac{2}{3}} Y_{1,-1}(\mathbf{n}) \end{aligned} \tag{4.65}$$

and so on, using neither differentiation nor integration. Compare the results with Eqs. (3.91), (3.92), and (3.93).

One last comment: Once we know the matrix elements of the operator $\mathbf{n}$ (which are easy to obtain), we can use Eq. (4.40) to calculate the matrix elements of the differential operator ∇^n. To be more specific, we substitute $l' = l + 1$ and $l' = l - 1$ into Eq. (4.40) and obtain

$$\langle l+1, m' | \nabla_k^n | l, m \rangle = -l \langle l+1, m' | n_k | l, m \rangle \tag{4.66}$$

and

$$\langle l-1, m' | \nabla_k^n | l, m \rangle = (l+1) \langle l-1, m' | n_k | l, m \rangle \,. \tag{4.67}$$

Even if *that* were the only application of the presented algebraic construction (which is surely not), it did pay off. . .

Finally, we have all the requisite tools to tackle the fine-structure of positronium spectral lines.

Exercise 10: Structure of Positronium Spectral Lines

Consider the relativistic and magnetic corrections in positronium. Let us recall that the Hamiltonian with accuracy up to α^4 is given by the nonrelativistic Hamiltonian $\hat{H}_0$, Eq. (3.131), and the Breit Hamiltonian $\hat{H}_1$, Eq. (3.132). Verify that the total angular momentum

$$\hat{\boldsymbol{J}} = \hat{\boldsymbol{L}} + \hat{\boldsymbol{S}}, \quad \hat{\boldsymbol{S}} = \hat{\boldsymbol{S}}_e + \hat{\boldsymbol{S}}_p$$

commutes with the total Hamiltonian $\hat{H} = \hat{H}_0 + \hat{H}_1$. When considering the degenerate level $n = 2$, there are four classes of states, namely 2^1s, 2^3s, 2^1p and 2^3p. Reflect that the matrix elements of the Hamiltonian $\hat{H}_1$ between four classes of states 2^1s, 2^3s, 2^1p, and 2^3p vanish and we can thus restrict our further considerations to only one of the classes of states. For instance, for the projection of the total angular momentum $m = 0$, there are three possible states 2^3p,

$$R_{2p}(r) Y_{1,1}(\mathbf{n}) |1, -1\rangle \,, \quad |1,-1\rangle = |-\rangle_e |-\rangle_p \,,$$

$$R_{2p}(r) Y_{1,0}(\mathbf{n}) |1, 0\rangle \,, \quad |1,0\rangle = \frac{1}{\sqrt{2}} (|+\rangle_e |-\rangle_p + |-\rangle_e |+\rangle_p)$$

and

$$R_{2p}(r) Y_{1,-1}(\mathbf{n}) |1, 1\rangle \,, \quad |1,1\rangle = |+\rangle_e |+\rangle_p \,.$$

We can compose states with $j = 0, 1, 2$ from these three states, see Sect. 4.2. Determine the splitting between the p-states with definite values of j with accuracy up to α^4, i.e., using the first order of the perturbation *method. Set the gyromagnetic ratio g to 1. Compare the results with the experimental values [5]*

$$\nu_{\text{exp}}(2^3s - 2^3p_0) = 18{,}504.1\,(10.0)\text{MHz}$$
$$\nu_{\text{exp}}(2^3s - 2^3p_1) = 13{,}001.3\,(3.9)\text{MHz}$$

and

$$\nu_{\text{exp}}(2^3s - 2^3p_2) = 8619.6\,(2.7)\text{MHz}\,.$$

The kind reader will surely be able to derive from these three numbers the experimental values for the intervals $2^3p_0 - 2^3p_1$ *and* $2^3p_1 - 2^3p_2$.
Hint: the spin-dependent operators in $\hat{\mathrm{H}}_1$ *can be rewritten as*

$$\hat{\boldsymbol{S}}_e \cdot \hat{\boldsymbol{S}}_p = \frac{1}{2}(\hat{\mathrm{S}}^2 - \hat{\mathrm{S}}_e^2 - \hat{\mathrm{S}}_p^2) \to \frac{1}{2}\left(s(s+1) - \frac{3}{2}\right),$$
$$\hat{\boldsymbol{S}} \cdot \hat{\boldsymbol{L}} = \frac{1}{2}(\hat{\mathrm{J}}^2 - \hat{\mathrm{S}}^2 - \hat{\mathrm{L}}^2) \to \frac{1}{2}[j(j+1) - s(s+1) - l(l+1)]$$

and

$$\mathbf{n}\cdot\hat{\boldsymbol{S}}_e\mathbf{n}\cdot\hat{\boldsymbol{S}}_p = \frac{1}{2}\left[(\mathbf{n}\cdot\hat{\boldsymbol{S}})^2 - (\mathbf{n}\cdot\hat{\boldsymbol{S}}_e)^2 - (\mathbf{n}\cdot\hat{\boldsymbol{S}}_p)^2\right] = \frac{1}{2}\left[(\mathbf{n}\cdot\hat{\boldsymbol{S}})^2 - \frac{1}{2}\right].$$

To calculate the matrix elements of $n_i n_j$ *between spherical harmonics, use repeatedly Eqs. (4.62)–(4.64) and orthornormality relations for spherical harmonics, Eq. (3.86).*

Exercise 11: Spherical Spinors

For the spherical spinors introduced in Eq. (4.58), verify that

$$\boldsymbol{\sigma}\cdot\mathbf{n}|j,m\rangle^{\pm} = -|j,m\rangle^{\mp},$$

where

$$|j,m\rangle^{\pm} = \left|j, m, \left(\frac{1}{2}, j \mp \frac{1}{2}\right)\right\rangle.$$

To prove this, use Eqs. (4.62), (4.63), (4.64), and (4.20).

4.5 The Hydrogen Atom: A General Solution

Now it is finally the time to solve the problem of the hydrogen atom bound states in its utmost generality. We will do so by finding the eigenvalues of the Hamiltonian (4.32), $\hat{\mathsf{H}} = \frac{\hat{\mathsf{p}}^2}{2} - \frac{1}{r}$. As we discussed in Sect. 3.4, the Hamiltonian $\hat{\mathsf{H}}$, squared magnitude of the angular momentum $\hat{\boldsymbol{L}}^2$, and the projection of the angular momentum along one of the axes $\hat{\boldsymbol{L}}_z$ form a complete set of commuting operators. In Sect. 3.4, we also showed how to obtain the differential equation for the radial parts of the bound state wave functions, Eq. (3.87). Here, we will explain how to avoid the direct solving of this equation and how to find the energy eigenvalues and the corresponding wave functions merely with the knowledge of the matrix elements of the Runge-Lenz vector.

4.5.1 *Matrix Elements of the Runge-Lenz Vector*

When we substitute $\hat{\boldsymbol{V}} = \hat{\boldsymbol{X}}$ into Eq. (4.53), where $\hat{\boldsymbol{X}}$ is the Runge-Lenz vector operator (4.34), we obtain

$$\hat{\mathsf{X}}_3|n,l,m\rangle = c_{l,n}\sqrt{l^2-m^2}|n,l-1,m\rangle + c_{l+1,n}\sqrt{(l+1)^2-m^2}|n,l+1,m\rangle\,. \tag{4.68}$$

Comparison of Eqs. (4.59) and (4.60) leads to the equation for the coefficients $c_{l,n}$

$$[c_{l,n}^2(2l-1)(l+m) + c_{l+1,n}^2(l+1-m)(2l+3)]|n,l,m\rangle = (\hat{\mathsf{X}}^2 - \mathrm{i}[\hat{\mathsf{X}}_1,\hat{\mathsf{X}}_2])|n,l,m\rangle\,. \tag{4.69}$$

Since (prove by yourself!)

$$\hat{\mathsf{X}}^2 = 1 + 2\hat{\mathsf{H}}(\hat{\mathsf{L}}^2+1)$$

and

$$[\hat{\mathsf{X}}_1,\hat{\mathsf{X}}_2] = -2\hat{\mathsf{H}}\mathrm{i}\hat{\mathsf{L}}_3\,,$$

and also due to Eqs. (3.79), $\hat{\mathsf{H}}|n,l,m\rangle = -\frac{1}{2n^2}|n,l,m\rangle$, (3.80), $\hat{\mathsf{L}}^2|n,l,m\rangle = l(l+1)|n,l,m\rangle$, and (3.81), $\hat{\mathsf{L}}_z|n,l,m\rangle = m|n,l,m\rangle$, the rhs of Eq. (4.69) can be simplified to

$$(\hat{\mathsf{X}}^2 - \mathrm{i}[\hat{\mathsf{X}}_1,\hat{\mathsf{X}}_2])|n,l,m\rangle = \frac{1}{n^2}(n^2 - l(l+1) - 1 + m)|n,l,m\rangle\,.$$

When the last result is used on the rhs of Eq. (4.69), we obtain, invoking the orthonormality of the eigenstates $|n, l, m\rangle$, the following relation for the coefficients $c_{l,n}$

$$c_{l,n}^2(2l-1)(l+m) + c_{l+1,n}^2(l+1-m)(2l+3) = \frac{1}{n^2}(n^2 - l(l+1) - 1 + m)\,.$$

In an analogical way to the consideration following Eq. (4.61), the above equation breaks into two

$$\text{0th degree in } m \qquad c_{l,n}^2(2l-1)l + c_{l+1,n}^2(l+1)(2l+3) = \frac{1}{n^2}(n^2 - l(l+1) - 1)\,,$$

$$\text{1st degree in } m \qquad c_{l,n}^2(2l-1) + c_{l+1,n}^2(-1)(2l+3) = \frac{1}{n^2}\,;$$

their solution being

$$c_{l,n} = \pm\frac{1}{n}\sqrt{\frac{n^2 - l^2}{(2l+1)(2l-1)}}\,.$$

The sign of the coefficients is not uniquely determined as there is a freedom in the choice of the phase factors of the radial hydrogen functions. We will opt for $c_{l,n} < 0$ here; this choice is consistent with that of the phase in Eq. (4.94) below. When we substitute the coefficients $c_{l,n}$ into Eq. (4.68), we obtain

$$\hat{\mathsf{X}}_3|n, l, m\rangle = -\frac{1}{n}\sqrt{n^2 - l^2}\,b_{l,m}|n, l-1, m\rangle - \frac{1}{n}\sqrt{n^2 - (l+1)^2}\,b_{l+1,m}|n, l+1, m\rangle\,, \tag{4.70}$$

where we introduced the notation

$$b_{l,m} = \sqrt{\frac{l^2 - m^2}{(2l+1)(2l-1)}}\,. \tag{4.71}$$

4.5.2 Energy Spectrum of the Hydrogen Atom

One can notice that the effect of the operator $\hat{\mathsf{X}}_3$ on the vector $|n, l, m\rangle$ is a mere shift of the orbital number l. As discussed earlier, see the discussion at the end of Sect. 4.4.5, we may require the coefficients $c_{l,n}$ be real. However, one cannot satisfy this requirement for $l > n - 1$, see Eq. (4.70). Therefore, it must hold that

$$0 = \sqrt{n^2 - (l_{\max} + 1)^2} \quad \Rightarrow \quad l_{\max} = n - 1\,. \tag{4.72}$$

The square of the angular momentum must equal at least zero, hence $l_{\min} = 0$. We can conclude therefrom that l always acquires integer values from zero to $n-1$. This finding imposes a new requirement on the principal quantum number n: it must be an integer as well. For the third time, realize that we have deduced all these results using nothing but the simple commutators!

As we commented in Sect. 3.4.3, the energy levels of an electron in the Coulomb potential are subject to the *accidental degeneracy*: the energy of the stationary states depends on neither the magnetic nor the orbital quantum number. This name might be misleading, though. As we have showed above, the degeneracy is in fact not accidental at all. On the contrary, it is related to the existence of the vector operator (4.31) which commutes with the Hamiltonian of the hydrogen atom. The existence of this operator is then related to the fact that the classical motion of a body in the Coulomb potential is restricted to an ellipse.

4.5.3 The Stark Effect

The Hamiltonian of a hydrogen-like atom placed in an external electric field of the intensity $\mathbf{E} = (0, 0, E)$ reads

$$\hat{\mathsf{H}} = \frac{\hat{\mathsf{p}}^2}{2m} - \frac{Z\alpha}{\hat{\mathsf{r}}} + eE\hat{\mathsf{z}} = m(Z\alpha)^2 \left[\hat{\mathsf{H}}_0 + \frac{\gamma_E}{Z^3}\hat{\mathsf{z}}_A\right], \quad \hat{\mathsf{H}}_0 = \frac{\hat{\mathsf{p}}_A^2}{2} - \frac{1}{\hat{\mathsf{r}}_A},$$

where we used the atomic units, $r \to r_A/(mZ\alpha)$, and introduced the notation

$$\gamma_{\mathrm{E}} = \frac{eE}{m^2\alpha^3}. \tag{4.73}$$

As we have shown in Eq. (0.6), a field of the intensity 1 V/m corresponds to $\gamma_{\mathrm{E}} = 1.945 \times 10^{-12}$. The highest intensity reached on Earth is approximately 10^7 V/m, therefore it suffices to use the first-order perturbation method for degenerate levels when studying a system under terrestrial conditions. We thus need the matrix elements

$$\langle n, l', m' | \hat{\mathsf{z}}_A | n, l, m \rangle .$$

There are a few tricks (we found them in [11]) allowing us to transform the calculation of these matrix elements to the calculation of those of the Runge-Lenz vector.

The first trick lies in the identity (prove by yourself!)

$$\hat{\boldsymbol{X}} = \frac{1}{2}\left(\hat{\boldsymbol{L}} \times \hat{\boldsymbol{p}} - \hat{\boldsymbol{p}} \times \hat{\boldsymbol{L}}\right) + \hat{\boldsymbol{n}} = -\frac{i}{2}\left[\hat{\boldsymbol{L}} \times \hat{\boldsymbol{x}} - \hat{\boldsymbol{x}} \times \hat{\boldsymbol{L}}, \hat{\mathsf{H}}_0\right] + \hat{\boldsymbol{n}}.$$

It follows clearly thereof

$$\langle n, l', m' | \hat{\boldsymbol{X}} | n, l, m \rangle = \langle n, l', m' | \hat{\boldsymbol{n}} | n, l, m \rangle .$$

The second trick constitutes the identities (prove by yourself!)

$$\hat{\mathsf{n}}_i = \hat{\mathsf{x}}_i(-\mathrm{i}[\hat{\mathsf{x}}_j\hat{\mathsf{p}}_j, \hat{\mathsf{H}}_0] - 2\hat{\mathsf{H}}_0)$$

and

$$2\mathrm{i}\hat{\mathsf{x}}_i[\hat{\mathsf{x}}_j\hat{\mathsf{p}}_j, \hat{\mathsf{H}}_0] = [2\mathrm{i}\hat{\mathsf{x}}_i\hat{\mathsf{x}}_j\hat{\mathsf{p}}_j - \mathrm{i}\hat{\mathsf{p}}_i\hat{\mathsf{x}}_j\hat{\mathsf{x}}_j + 3\hat{\mathsf{x}}_i, \hat{\mathsf{H}}_0] + \hat{\mathsf{n}}_i\,.$$

It follows from these two identities that

$$\langle n, l', m'|\hat{\boldsymbol{n}}|n, l, m\rangle = \frac{2}{3n^2}\langle n, l', m'|\hat{\boldsymbol{r}}|n, l, m\rangle\,.$$

Hence, the overall result reads

$$\langle n, l', m'|\hat{\boldsymbol{r}}|n, l, m\rangle = \frac{3n^2}{2}\langle n, l', m'|\hat{\boldsymbol{X}}|n, l, m\rangle\,.$$

Exercise 12: The Stark Effect

Determine (numerically in hertz) the splitting of the state $n = 3$ of the hydrogen atom when it is inserted into a homogeneous electric field of the intensity 2.5×10^5 V/m.

4.5.4 Radial Functions of the Hydrogen Atom

We now show how to determine the radial part of the hydrogen bound state wave function, i.e., the function R_{nl} in Eq. (3.83), $\langle \mathbf{r}|n, l, m\rangle = \psi_{nlm}(r, \vartheta, \varphi) = R_{nl}(r)Y_{lm}(\vartheta, \varphi)$. We use the action of the third component of the Runge-Lenz vector, Eq. (4.34), into which we substitute Eqs. (4.62), (4.66), (4.67), and (4.71). That is,

$$n_3|l, m\rangle = b_{lm}|l-1, m\rangle + b_{l+1,m}|l+1, m\rangle \tag{4.74}$$

and

$$\nabla_3^n|l, m\rangle = (l+1)b_{lm}|l-1, m\rangle - l b_{l+1,m}|l+1, m\rangle\,. \tag{4.75}$$

Hence, on one hand, we have

$$\langle \mathbf{r}|\hat{\mathsf{X}}_3|n,l,m\rangle = \left(\hat{\mathsf{n}}_3\left(\frac{\partial}{\partial r} - \frac{l(l+1)}{r} + 1\right) - \nabla_3^n \frac{\partial}{\partial r}\right) R_{nl}(r)Y_{lm}(\mathbf{n})$$
$$= \left((b_{lm}Y_{l-1,m} + b_{l+1,m}Y_{l+1,m}) \left(\frac{\partial}{\partial r} - \frac{l(l+1)}{r} + 1\right) \right.$$
$$\left. - \left((l+1)b_{lm}Y_{l-1,m} - lb_{l+1,m}Y_{l+1,m}\right)\frac{\partial}{\partial r}\right) R_{nl}\,.$$

On the other hand, we know from Eq. (4.70) that

$$\langle \mathbf{r}|\hat{\mathsf{X}}_3|n,l,m\rangle = -\frac{\sqrt{n^2-l^2}}{n} b_{lm}R_{n,l-1}Y_{l-1,m} - \frac{\sqrt{n^2-(l+1)^2}}{n} b_{l+1,m}R_{n,l+1}Y_{l+1,m}\,.$$

The spherical harmonics form an orthogonal basis set, thus we can compare the coefficients in front of the same Y_{lm}. This leads to a simple set of differential equations for radial functions

$$\left[\frac{\mathrm{d}}{\mathrm{d}r} + \frac{(l+1)}{r} - \frac{1}{l}\right] R_{nl} = \frac{\sqrt{n^2-l^2}}{nl} R_{n,l-1}\,, \tag{4.76}$$

$$\left[\frac{\mathrm{d}}{\mathrm{d}r} - \frac{l}{r} + \frac{1}{l+1}\right] R_{nl} = -\frac{\sqrt{n^2-(l+1)^2}}{n(l+1)} R_{n,l+1}\,. \tag{4.77}$$

From these equations, one can derive the solution in a sequence of steps. First, we set $n = l+1$. It then follows from the second equation that

$$\left[\frac{\mathrm{d}}{\mathrm{d}r} - \frac{l}{r} + \frac{1}{l+1}\right] R_{l+1,l} = 0\,.$$

One can employ the method of separation of the variables to solve this last equation:

$$R_{l+1,l}(r) = K_{(l)} r^l \mathrm{e}^{-\frac{r}{l+1}}\,, \tag{4.78}$$

and find the integration constant from the normalization condition,

$$1 = \int_0^\infty R_{l+1,l}^2 r^2 \mathrm{d}r = K_{(l)}^2 \int_0^\infty r^{2l+2} \mathrm{e}^{-\frac{2r}{l+1}} \mathrm{d}r$$
$$= K_{(l)}^2 \left(\frac{l+1}{2}\right)^{2l+3} \int_0^\infty t^{2l+2}\mathrm{e}^{-t}\mathrm{d}t = K_{(l)}^2 \left(\frac{l+1}{2}\right)^{2l+3} (2l+2)!$$

$$\Rightarrow K_{(l)} = \left(\frac{2}{l+1}\right)^{l+3/2} \frac{1}{\sqrt{(2l+2)!}}\,. \tag{4.79}$$

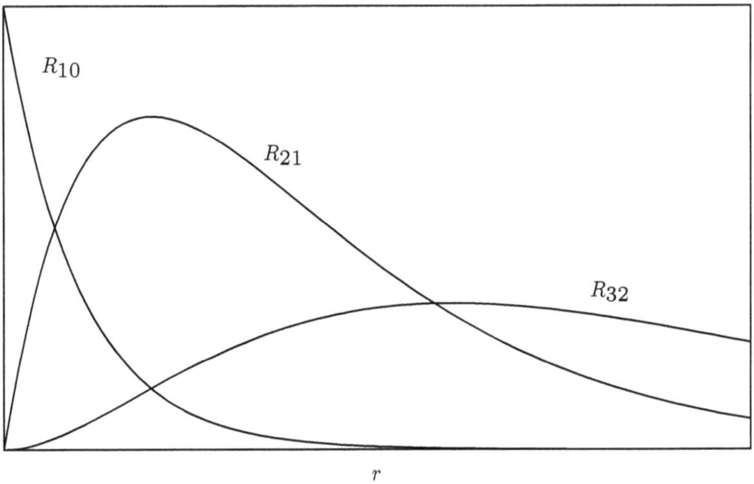

Fig. 4.2 Radial functions with the maximal angular momentum (not to scale)

This way, one obtains the radial functions without nodes (i.e., without intersections with the axis r), see Fig. 4.2. We have already encountered special cases of these functions before, Eq. (3.35) for $l = 0$ and Eq. (3.96) for $l = 1$. We then use Eq. (4.76) to express the other functions. For instance, for $n = 2$ and $l = 1$ we obtain

$$\left(-\frac{\mathrm{d}}{\mathrm{d}r} - \frac{2}{r} + 1\right) R_{21} = -\frac{\sqrt{3}}{2} R_{20}\,;$$

hence

$$R_{20} = -\frac{2}{\sqrt{3}}\frac{1}{2\sqrt{6}}\left(-3 + \frac{3r}{2}\right)\mathrm{e}^{-r/2} = \frac{1}{\sqrt{2}}\left(1 - \frac{r}{2}\right)\mathrm{e}^{-r/2}\,,$$

confront with Eq. (3.95). And so forth...

The relations (4.76) and (4.77) will come in hand in the next two chapters.

4.5.5 Parabolic Coordinates

Exercise 13: Parabolic Coordinates and Hydrogen

Find the eigenstates of the operator $\hat{X}_3$ in the subspace spanned by the states $n = 2$ and $m = 0$; denote the eigenvalue as X_3. Show that $\langle \mathbf{r}| 2X_30\rangle = f_1(\xi)f_2(\eta)$, where the so-called parabolic coordinates ξ and η are given by the expressions $z = (\xi - \eta)/2$ and $r = (\xi + \eta)/2$.

Exercise 13 relates to the following question: do we need to choose the particular commuting set of the operators $\{\hat{\mathsf{H}}, \hat{\mathsf{L}}^2, \hat{\mathsf{L}}_z\}$ when considering the hydrogen atom? Not at all, nothing prevents us from using a different complete set of commuting operators, such as $\{\hat{\mathsf{H}}, \hat{\mathsf{X}}_z, \hat{\mathsf{L}}_z\}$, which corresponds to the solution of the Schrödinger equation in the parabolic coordinates, as illustrated in the exercise. Furthermore, nothing prevents us from choosing even this complete set of commuting operators: $\{\hat{\mathsf{H}}, c_1\hat{\mathsf{L}}^2 + c_2\hat{\mathsf{X}}_z, \hat{\mathsf{L}}_z\}$, where c_1 and c_2 are arbitrary real numbers. Thus we see that there is a rich variety of possible coordinate systems in which the Schrödinger equation separates.

We will use the separation of the Schrödinger equation in the parabolic coordinates later in Sects. 6.3.2–6.3.4 to obtain the wave function of an electron ejected from the atom by a photon.

4.6 Decomposition of a Plane Wave into Spherical Waves

Finally, we focus on the simplest possible problem—a free particle,

$$\hat{\mathsf{H}} = \frac{\hat{\mathsf{p}}^2}{2}\,.$$

On one hand, a free particle is a trivial example of a spherically symmetrical potential and we can search for the common eigenvectors of the operators $\hat{\mathsf{H}}$, $\hat{\mathsf{L}}^2$ and $\hat{\mathsf{L}}_3$,

$$\hat{\mathsf{H}}|\omega, l, m\rangle = \frac{\omega^2}{2}|\omega, l, m\rangle\,. \tag{4.80}$$

In the coordinate representation, we can subsequently split the wave function into the radial and angular parts

$$\langle \mathbf{r}|\,\omega, l, m\rangle = R_l(\omega r)Y_{l,m}(\mathbf{n})\,. \tag{4.81}$$

On the other hand, for this system, and only for this system, it is $[\hat{\boldsymbol{p}}, \hat{\mathsf{H}}] = 0$. We see from Eq. (3.15) that the momentum operator is another example of the vector operator $\hat{\boldsymbol{V}}$, Eq. (4.35). Owing to $[\hat{\mathsf{p}}_1, \hat{\mathsf{p}}_2] = 0$ and $\hat{\mathsf{p}}^2 = 2\hat{\mathsf{H}}$, we obtain from Eqs. (4.59), (4.60), and (4.80)

$$\omega^2 = c_{l,\omega}^2(2l-1)(l+m) + c_{l+1,\omega}^2(l+1-m)(2l+3) \Rightarrow c_{l,\omega} = \frac{\omega}{\sqrt{(2l-1)(2l+1)}}\,.$$

Analogously to Eq. (4.70), we find

$$\hat{\mathsf{p}}_3|\omega, l, m\rangle = \omega[b_{l,m}|\omega, l-1, m\rangle + b_{l+1,m}|\omega, l+1, m\rangle]\,. \tag{4.82}$$

After the substitution of the separation (4.81) into the last equation, we obtain in the coordinate representation

$$\langle \mathbf{r}|\hat{\mathsf{p}}_3|\omega, l, m\rangle = \omega [b_{l,m} R_{l-1}(\omega r) Y_{l-1,m} + b_{l+1,m} R_{l+1}(\omega r) Y_{l+1,m}] \,.$$

However, through the direct application of the angular operators, we find from Eqs. (3.15), (4.74), and (4.75) that

$$\begin{aligned}\langle \mathbf{r}|\hat{\mathsf{p}}_3|\omega, l, m\rangle = -\mathrm{i}\Bigg[& \frac{\mathrm{d}}{\mathrm{d}r} R_l(\omega r)(b_{l,m} Y_{l-1,m} + b_{l+1,m} Y_{l+1,m}) \\ & + \frac{1}{r} R_l(\omega r)((l+1) b_{l,m} Y_{l-1,m} - l b_{l+1,m} Y_{l+1,m}) \Bigg] \,.\end{aligned}$$

Comparing of the last two equations and following the same reasoning as the one preceding Eqs. (4.76) and (4.77) lead to

$$-\mathrm{i}\left(\frac{\mathrm{d}}{\mathrm{d}r} + \frac{l+1}{r}\right) R_l(\omega r) = \omega R_{l-1}(\omega r) \tag{4.83}$$

and

$$-\mathrm{i}\left(\frac{\mathrm{d}}{\mathrm{d}r} - \frac{l}{r}\right) R_l(\omega r) = \omega R_{l+1}(\omega r) \,. \tag{4.84}$$

This pair of equations does not impose any restriction on l or ω, though. The number ω can be an arbitrary real number. For a given ω the orbital quantum number l may acquire any integer value from zero to infinity. However, if we find $R_0(\omega r)$ by solving directly the Schrödinger equation (4.80), we can use the last two equations to determine all other $R_l(\omega r)$ for $l > 0$. With the aid of Eqs. (3.21) and (4.81), we obtain from Eq. (4.80) for the radial functions with $l = 0$

$$-\left[\frac{\mathrm{d}^2}{\mathrm{d}r^2} + \frac{2}{r}\frac{\mathrm{d}}{\mathrm{d}r}\right] R_0(\omega r) = \omega^2 R_0(\omega r) \,.$$

There are two solutions to this equation: the first one is a regular one which is finite everywhere,

$$j_0(\omega r) = \frac{\sin(\omega r)}{\omega r} \,, \tag{4.85}$$

and the second one is a singular one which diverges at the origin,

$$n_0(\omega r) = \frac{\cos(\omega r)}{\omega r} \,.$$

By the substitution

$$R_l(\omega r) = \begin{cases} \mathrm{i}^l j_l(\omega r)\,, \\ \mathrm{i}^l n_l(\omega r) \end{cases} \tag{4.86}$$

into Eqs. (4.83) and (4.84), we can easily prove the functions $j_l(\omega r)$ and $n_l(\omega r)$ are real. The functions $j_l(\omega r)$ are commonly referred to as the *spherical Bessel functions*, the functions $n_l(\omega r)$ as the *spherical Neumann functions*, or the irregular spherical Bessel functions.

Now we can ask the same question as in the case of the hydrogen atom: Do we need to consider this particular set of the commuting operators, namely $\{\hat{\mathsf{H}}, \hat{\mathsf{L}}^2, \hat{\mathsf{L}}_3\}$? Of course not. Nothing prevents us from considering the complete set of the commuting operators $\{\hat{\mathsf{H}}, \hat{\mathsf{p}}_3, \hat{\mathsf{L}}_3\}$. We easily find in the Cartesian coordinates that the wave function

$$\langle \mathbf{r}|\,\omega, p_3, 0\rangle = \mathrm{e}^{\mathrm{i}\omega z} = \mathrm{e}^{\mathrm{i}\omega r\cos\vartheta} \tag{4.87}$$

is an eigenfunction of these operators. The solutions (4.81) and (4.87) are different solutions to the same Eq. (4.80) and as such they need to be related to each other. In other words, it must be possible to express the plane wave (4.87), which describes a particle with the linear momentum $p_3 = \omega$, as a linear combination of the spherical waves (4.81),

$$|\omega, p_3, 0\rangle = \sum_{l=0}^{\infty} d_l|\omega, l, 0\rangle\,. \tag{4.88}$$

After application of the operator $\hat{\mathsf{p}}_3$ to both sides of the equation, we obtain, using successively Eqs. (4.87), (4.88), and (4.82),

$$\begin{aligned}\hat{\mathsf{p}}_3|\omega, p_3, 0\rangle &= \omega|\omega, p_3, 0\rangle = \omega\sum_{l=0}^{\infty} d_l|\omega, l, 0\rangle \\ &= \omega\sum_{l=0}^{\infty} d_l\left[b_{l,0}|\omega, l-1, 0\rangle + b_{l+1,0}|\omega, l+1, 0\rangle\right]\,.\end{aligned}$$

By replacing $l \to l+1$ and $l \to l-1$ in the first and the second term, respectively, on the rhs of the last equation, by comparing the last and second to last expressions in this equation, and by invoking the orthogonality of the states $|\omega, l, m\rangle$, we arrive at the relation

$$d_l - b_{l+1,0}d_{l+1} - b_{l,0}d_{l-1} = 0\,.$$

Substituting for $b_{l,0}$ from Eq. (4.71) into the last equation leads to the solution

$$d_l = d_0\sqrt{2l+1}\,.$$

Since the plane wave (4.87) is finite everywhere, we opt for the regular solution $R_l(\omega r) = \mathrm{i}^l j_l(\omega r)$ as its radial part. The projection of Eq. (4.88) along the coordinate basis then reads:

$$\mathrm{e}^{\mathrm{i}\omega z} = \mathrm{e}^{i\omega r\cos\vartheta} = d_0 \sum_{l=0}^{\infty} \sqrt{2l+1}\,\mathrm{i}^l j_l(\omega r) Y_{l,0}(\vartheta)\,. \tag{4.89}$$

Setting $r = 0$ and using the approximate asymptotic behavior of the Bessel function near the origin $j_l(\omega r) \approx r^l$ imply

$$1 = d_0 Y_{0,0} \quad \Rightarrow \quad d_0 = \sqrt{4\pi}\,. \tag{4.90}$$

The expansion of a plane wave into spherical waves, Eq. (4.89), finds a wide application within the quantum scattering theory. One exploits it also in other parts of physics for example in the analysis of the scattering of electromagnetic waves by a conducting or dielectric sphere, see, e.g., [12]. We will return to Eq. (4.89) later on and rewrite it into a more convenient form, see Eq. (5.30).

4.7 Algebra of Radial Operators

As mentioned above in Sect. 3.2.5, the eigenstates of the hydrogen atom do not serve well when it comes to the variational calculation of more complex atoms. The reason is that the spectrum of the Hamilton operator of the hydrogen atom comprises both a discrete and a continuous part. However, we have shown for the s-states in Sect. 3.2.5 that one can use a few simple tricks to transform the original search for the eigenstates of the hydrogen atom into that for the eigenstates of the operator $\hat{\mathsf{T}}_3$, the advantage being its purely discrete spectrum. All we need to do is thus to generalize the procedure presented in Sect. 3.2.5 for the states with $l = 0$ to general states.[10]

Instead of Eq. (3.22), we need to consider Eq. (3.87)

$$\hat{\mathsf{H}}_l|\psi\rangle = \left(\frac{\hat{\mathsf{p}}_r^2}{2} + \frac{l(l+1)}{2\hat{\mathsf{r}}^2} - \frac{1}{\hat{\mathsf{r}}}\right)|\psi\rangle = -\frac{1}{2n^2}|\psi\rangle\,. \tag{4.91}$$

[10]The presentation in this section is, as in the Sect. 3.2.5, inspired by the work [3].

We substitute $r \to nr$ and multiply the whole equation by r to transform the search for the spectrum of the Hamilton operator to that of the spectrum of the generalized operator $\hat{\mathsf{T}}_3$, see Eq. (3.25),

$$\hat{\mathsf{T}}_3|\psi\rangle = n|\psi\rangle\,, \quad \hat{\mathsf{T}}_3 = \frac{\hat{\mathsf{r}}\hat{\mathsf{p}}_r^2}{2} + \frac{l(l+1)}{2\hat{\mathsf{r}}} + \frac{\hat{\mathsf{r}}}{2}\,.$$

If we change the operator $\hat{\mathsf{W}}_3$ in Eq. (3.26) to

$$\hat{\mathsf{W}}_3 = \hat{\mathsf{r}}\hat{\mathsf{p}}_r^2 + \frac{l(l+1)}{\hat{\mathsf{r}}}$$

and leave the operators $\hat{\mathsf{W}}_1 = \hat{\mathsf{r}}$ and $\hat{\mathsf{W}}_2 = \hat{\mathsf{r}}\hat{\mathsf{p}}_r$ intact, the commutation relations (3.27) do not change. It follows that neither Eqs. (3.28), (3.29), (3.30), and (3.31) change; for instance

$$[\hat{\mathsf{T}}_1, \hat{\mathsf{T}}_2] = -\mathrm{i}\hat{\mathsf{T}}_3\,, \qquad [\hat{\mathsf{T}}_2, \hat{\mathsf{T}}_3] = \mathrm{i}\hat{\mathsf{T}}_1\,, \qquad [\hat{\mathsf{T}}_3, \hat{\mathsf{T}}_1] = \mathrm{i}\hat{\mathsf{T}}_2\,. \tag{3.29}$$

Instead of Eq. (3.32), we now have (prove by yourself!)

$$\hat{\mathsf{T}}^2 = \hat{\mathsf{T}}_3^2 - \hat{\mathsf{T}}_1^2 - \hat{\mathsf{T}}_2^2 = l(l+1)\,. \tag{4.92}$$

Apart from the single sign change in the first commutator, the commutation relations (3.29) are identical to those for the angular momentum operators, Eq. (4.1). The eigenvalues and eigenstates of the operators $\hat{\mathsf{T}}_3$ and $\hat{\mathsf{T}}^2$ can be determined in the same way as those of the operators $\hat{\mathsf{J}}_3$ and $\hat{\mathsf{J}}^2$. In accordance with Eq. (4.3), we thus write

$$\hat{\mathsf{T}}^2|n,l\rangle = l(l+1)|n,l\rangle\,, \qquad \hat{\mathsf{T}}_3|n,l\rangle = n|n,l\rangle\,. \tag{4.93}$$

After the replacement of $\hat{\mathsf{J}}$ by $\hat{\mathsf{T}}$, we obtain the equation

$$\hat{\mathsf{T}}_\pm|n,l\rangle = \alpha^\pm(l,n)|n\pm 1,l\rangle\,, \quad \alpha^\pm(l,n) = \sqrt{(n\mp l)(n\pm(l+1))}\,, \tag{4.94}$$

instead of Eq. (4.8). Imposing the requirement there be a lowest possible n, which agrees with n dictating the energies in Eq. (4.91), it must hold that $\hat{\mathsf{T}}_-|l, n_{\min}\rangle = 0$, hence $n_{\min} \in \{-l, l+1\}$. Were we to set $n_{\min} = -l$ in $\alpha^+(l,n)$ we would obtain an imaginary result, which contradicts the reality of the operator $\hat{\mathsf{T}}_3$ and its eigenfunctions. Therefore, we will use $n_{\min} = l+1$ as the lowest state. This way, we easily obtain the eigenvalues of the operator $\hat{\mathsf{T}}_3$ and derive for the second time that the hydrogen spectrum acquires the form $E_n = -\frac{1}{2n^2}$, where $n = l+1+n_r$, $n_r = 0, 1, 2, \ldots$.

The projections of the eigenvectors of the operator $\hat{\mathsf{T}}_3$ onto the coordinate basis[11]

$$R^M_{n,l}(r) = \langle r|n,l\rangle$$

is almost identical to the eigenfunction of the hydrogen atom $R_{n,l}(r)$. The eigenfunctions of the operator $\hat{\mathsf{T}}_3$ differ from those of the radial Hamiltonian (4.91) only by the energy-dependent scaling of the radial coordinate $r \to nr$ and by the normalization. A slight difference lies also in the Hamiltonian (4.91) being Hermitian with respect to the scalar product with the weight r^2,

$$\langle n_1,l|\hat{\mathsf{H}}_l|n_2,l\rangle = \langle n_2,l|\hat{\mathsf{H}}_l|n_1,l\rangle$$
$$\Rightarrow \int_0^\infty r^2 R_{n_1,l}(r)\hat{\mathsf{H}}_l R_{n_2,l}(r)\,\mathrm{d}r = \int_0^\infty r^2 R_{n_2,l}(r)\hat{\mathsf{H}}_l R_{n_1,l}(r)\,\mathrm{d}r\,,$$

as can be easily proven by integration by parts. On the other hand, the operator $\hat{\mathsf{T}}_3$ is Hermitian with respect to the scalar product with the weight r,

$$\int_0^\infty r R^M_{n_1,l}(r)\hat{\mathsf{T}}_3 R^M_{n_2,l}(r)\,\mathrm{d}r = \int_0^\infty r R^M_{n_2,l}(r)\hat{\mathsf{T}}_3 R^M_{n_1,l}(r)\,\mathrm{d}r\,.$$

Hence the radial functions of the hydrogen atom form an orthonormal system with respect to the scalar product with the weight r^2,

$$\langle n_1,l|\,n_2,l\rangle = \int_0^\infty r^2 R_{n_1,l}(r)R_{n_2,l}(r)\,\mathrm{d}r = \delta_{n_1,n_2}\,, \tag{4.95}$$

whereas the eigenfunctions of the operator $\hat{\mathsf{T}}_3$ form an orthonormal system with respect to the scalar product with weight r,

$$\int_0^\infty r R^M_{n_1,l}(r)R^M_{n_2,l}(r)\,\mathrm{d}r = \delta_{n_1,n_2}\,. \tag{4.96}$$

For example, the eigenfunctions of the operator $\hat{\mathsf{T}}_3$ for $n = l+1$ read, compare with Eqs. (4.78) and (4.79),

$$R^M_{l+1,l}(r) = \frac{2}{\sqrt{(2l+1)!}}(2r)^l \mathrm{e}^{-r}\,. \tag{4.97}$$

[11]It is a common habit within literature to call these functions the *Sturmian functions*, or the *Sturmians*, see for example [2].

As we will see later, it is sometimes advantageous to introduce the *screening parameter* η

$$R^M_{l+1,l}(\eta, r) = \frac{2\eta}{\sqrt{(2l+1)!}}(2\eta r)^l \mathrm{e}^{-\eta r}\,. \tag{4.98}$$

Apart from the substitution $r \to \eta r$, the function (4.98) differs from the functions (4.97) also by the factor η which ensures correct normalization (4.96).

From the definition of the operators $\hat{\mathsf{T}}_3$ and $\hat{\mathsf{T}}_1$, see Eqs. (3.26) and (3.28), we find $\hat{\mathsf{r}} = \hat{\mathsf{T}}_3 - \hat{\mathsf{T}}_1 = \hat{\mathsf{T}}_3 - \frac{1}{2}(\hat{\mathsf{T}}_+ + \hat{\mathsf{T}}_-)$; hence

$$\begin{aligned} rR^M_{nl}(r) &= \left(\hat{\mathsf{T}}_3 - \frac{1}{2}(\hat{\mathsf{T}}_+ + \hat{\mathsf{T}}_-)\right) R^M_{nl}(r) \\ &= nR^M_{nl}(r) - \frac{1}{2}\sqrt{(n-l)(n+l+1)}R^M_{n+1,l}(r) \\ &\quad - \frac{1}{2}\sqrt{(n-l-1)(n+l)}R^M_{n-1,l}(r)\,, \end{aligned} \tag{4.99}$$

where we used Eq. (4.94). We have thus found an equation for eigenfunctions of $\hat{\mathsf{T}}_3$ operator which will prove handy in the next chapter.

As we already pointed out above, the discrete basis functions $R^M_{n,l}(r)$ differ from the hydrogen functions $R_{n,l}(r)$ only by a scale factor and a normalization constant. Let us now find the exact relation between these two functions:

$$R_{n,l}(r) = \frac{K}{n}R^M_{n,l}(r/n)\,.$$

The factor n^{-1} ensures the correct normalization of the functions $R^M_{n,l}(r/n)$, see Eq. (4.96),

$$\int_0^\infty r\left|n^{-1}R^M_{n,l}(r/n)\right|^2 \mathrm{d}r = \int_0^\infty r\left|R^M_{n,l}(r)\right|^2 \mathrm{d}r = 1\,.$$

The factor K can be determined from the requirement of the proper normalization of the functions $R_{n,l}(r)$, see Eq. (4.95),

$$\begin{aligned} \int_0^\infty r^2\left|R_{n,l}(r)\right|^2 \mathrm{d}r &= K^2 \int_0^\infty r^2\left|n^{-1}R^M_{n,l}(r/n)\right|^2 \mathrm{d}r \\ &= K^2 n^{-2} n^3 \int_0^\infty rR^M_{n,l}(r) rR^M_{n,l}(r)\mathrm{d}r \\ &= K^2 n^{-2} n^3 n = 1 \Rightarrow K = \frac{1}{n}\,. \end{aligned}$$

The third equality follows from Eqs. (4.96) and (4.99).

Substituting $r = 0$ and $l = 0$ into Eq. (4.99) yields

$$0 = nR^M_{n0}(0) - \frac{1}{2}\sqrt{n(n+1)}R^M_{n+1,0}(0) - \frac{1}{2}\sqrt{(n-1)n}R^M_{n-1,0}(0)\,. \tag{4.100}$$

We solve these equations with the initial conditions $R^M_{0,0}(0) = 0$ and $R^M_{1,0}(0) = 2$. The reader will easily verify that the solution to these equations reads

$$R^M_{n,0}(0) = 2n^{1/2}\,.$$

The above equations imply the following value of the hydrogen function at the origin

$$\psi_{n,l,m}(\mathbf{r} = 0) = \delta_{l,0}\frac{1}{\sqrt{4\pi}}R_{n,0}(0)\,, \qquad R_{n,0}(0) = \frac{2}{n^{3/2}}\,. \tag{4.101}$$

This result will come in hand several times in Chaps. 6 and 7.

4.8 Final Notes

The application of the addition of the angular momentum, Clebsch-Gordan coefficients, and selection rules of the vector operators is not restricted to atomic physics and angular momentum itself, though. If the reader is interested in the application of the presented mathematical apparatus to the internal symmetries of the elementary particles, we refer him or her to, e.g., [4, 7, 9, 14]. An elaborate theory of the composition of three and four angular momenta has been developed, see for example [6, 13].

References

1. B.G. Adams, J. Čížek, J. Paldus, Adv. Quantum Chem. **19**, 1 (1988)
2. J. Avery, *Hyperspherical Harmonics and Generalized Sturmians* (Kluwer Academic, Boston, 2000)
3. J. Čížek, J. Paldus, Int. J. Quantum Chem. **12**, 875 (1977)
4. W. Greiner, B. Müller, *Quantum Mechanics: Symmetries* (Springer, Berlin, 1997)
5. S. Hatamian et al., Phys. Rev. Lett. **58**, 1833 (1987)
6. I. Lindgren, J. Morrison, *Atomic Many-Body Theory* (Springer, Berlin, 1982)
7. H.J. Lipkin, *Lie Groups for Pedestrians* (North Holland, Amsterdam, 1965)
8. A. Messiah, *Quantum Mechanics* (North Holland, Amsterdam, 1961)
9. D.H. Perkins, *Introduction to High Energy Physics* (Addison-Wesley, Reading, 1987)
10. L. Schiff, *Quantum Mechanics* (McGraw Hill College, New York, 1968)
11. J. Schwinger, *Quantum Mechanics: Symbolism of Atomic Measurement* (Springer, Berlin, 2003)
12. J. Schwinger et al., *Classical Electrodynamics* (Perseus Books, Cambridge, 1998)
13. P.E.S. Wormer, J. Paldus, Adv. Quantum Chem. **51**, 59 (2006)
14. A. Zee, *Group Theory in a Nutshell for Physicists* (Princeton University Press, Princeton, 2016)

Chapter 5
The Helium Atom

In the previous chapters, we thoroughly investigated the simplest atom of all—the hydrogen atom. We have found that, owing to the existence of a sufficient number of integrals of motion, one can solve its energy spectrum exactly. Unfortunately, one cannot determine exactly the spectrum of helium nor of any of the heavier atoms. Nevertheless, we know from Chap. 2 that by means of the variational method we may approximate the solution to any desirable accuracy. We will show that the antisymmetry of the wave function with respect to the exchange of the electrons leads to a so-called exchange interaction. Accounting for this interaction subsequently leads to a qualitatively correct result even when only one two-electron configuration considered. This estimate can be further systematically improved by the inclusion of additional electron configurations. We will see that the symmetries of the helium atom, i.e., the existence of operators commuting with the Hamiltonian, substantially decrease the amount of configurations one needs to include in the calculations.

We also show how the variational calculation can be carried out to the very end. To do so, it will be necessary to calculate the matrix elements of the electron-electron interaction which constitutes a six-dimensional integral. With the aid of the multipole expansion, it is possible to separate the radial and angular coordinates of the electrons. We will show how properties of the spherical harmonics enable us, *always* in the case of atoms, to reduce the infinite multipole expansion to a finite number of terms. Next, we will show how to use the recurrence relations for the radial functions derived in the previous chapter for the calculation of integrals of these functions. Thus we will further develop the theme initiated in the previous chapter: the importance lies rather in the *relations* between the individual wave functions than in their specific forms. These relations then imply relations between the integrals of these functions. There are no three-particle forces acting between the electrons. Therefore, once we are able to calculate the approximate spectrum of helium, we are able to determine approximate spectra also of other more complicated atoms. Despite it all, though, the principal reason why we discuss helium to such a

J. Zamastil, J. Benda, *Quantum Mechanics and Electrodynamics*,
DOI 10.1007/978-3-319-65780-6_5

great detail is that it is the simplest physical system which can serve as an illustrative example of an efficient application of symmetries in finding approximate solution of the Schrödinger equation.

5.1 Symmetry in the Helium Atom

In Eq. (3.129), we introduce dimensionless atomic units by substituting $r \to r/(Z\alpha m_r)$ and factor out $m(Z\alpha)^2$. The Hamiltonian then acquires the form

$$\frac{\hat{\mathsf{H}}}{Z^2} = \hat{\mathsf{h}}_0(1) + \hat{\mathsf{h}}_0(2) + \frac{1}{Zr_{12}}\,. \tag{5.1}$$

where

$$\hat{\mathsf{h}}_0(a) = \frac{\hat{\mathsf{p}}_a^2}{2} - \frac{1}{r_a}\,, \qquad a = 1, 2\,. \tag{5.2}$$

Our task is to find a solution to the Schrödinger equation

$$\hat{\mathsf{H}}\psi(1,2) = E\psi(1,2)\,, \tag{5.3}$$

where the numbers in the parentheses represent a short notation for the coordinates of the respective electrons; we should have written $\mathbf{r}_1, \mathbf{r}_2$ instead of $1, 2$ to be precise.

In the case of helium, $Z = 2$, the magnitude of the electrostatic electron-electron interaction is far too large for a perturbative treatment. Moreover, the application of the perturbation method is further hindered by the fact that a simple combination of two one-electron hydrogen-like atom spectra comprises a discrete part, a continuous part and a part where these two overlap. For this reason, we will opt for the variational method to determine the energies of the stationary states.

5.1.1 *The Total Spin and the Antisymmetry of the Wave Function*

The most straightforward choice of the test function would be a product of two normalized "hydrogenic" wave functions

$$\psi(1,2) = \psi_a(\mathbf{r}_1)\psi_b(\mathbf{r}_2)\,. \tag{5.4}$$

One could certainly use this function in the calculations; however, we could expect to obtain the better estimate of the stationary state energy the more properties the test function has in common with the exact solution. In this case, the Hamiltonian (5.1) clearly does not notice the exchange of the electrons, while the wave function ψ in Eq. (5.4) does. At the same time, we can easily see that if the wave function $\psi(1,2)$ solves the Schrödinger equation (5.3), so does the function $\psi(2,1)$,

$$\hat{\mathsf{H}}\psi(2,1) = E\psi(2,1)\,. \tag{5.5}$$

Since the particle exchange operator $\hat{\mathsf{P}}_{12}$ and the Hamiltonian $\hat{\mathsf{H}}$ commute with each other, it must be possible to find their common eigenstates. What form do the common eigenstates of $\hat{\mathsf{P}}_{12}$ and $\hat{\mathsf{H}}$ and the eigenvalues of $\hat{\mathsf{P}}_{12}$ take on?

Supposing that we already have such an eigenstate

$$\hat{\mathsf{P}}_{12}|\psi\rangle = \lambda|\psi\rangle\,,$$

we apply the operator $\hat{\mathsf{P}}_{12}$ on both side of the last equation. Since $\hat{\mathsf{P}}_{12}\hat{\mathsf{P}}_{12} = 1$, we have

$$1 = \lambda^2 \quad \Rightarrow \quad \lambda = \pm 1\,.$$

We further add Eqs. (5.3) and (5.5) to each other, and also subtract one from another,

$$\hat{\mathsf{H}}(\psi(1,2) + \psi(2,1)) = E(\psi(1,2) + \psi(2,1)) \stackrel{\text{def}}{=} E\psi_S\,,$$

$$\hat{\mathsf{H}}(\psi(1,2) - \psi(2,1)) = E(\psi(1,2) - \psi(2,1)) \stackrel{\text{def}}{=} E\psi_A\,.$$

The states ψ_S and ψ_A are eigenstates of the Hamiltonian as well as eigenstates of the particle exchange operator, the corresponding eigenvalues being $+1$ and -1, respectively.

In the very beginning of this book, see Sect. 1.1, we mentioned that Eq. (1.2) holds for the scattering of two indistinguishable α-particles, see Fig. 1.3. Namely, in the case of α-particles, nature always picks the symmetrical possibility. Is this true also for electrons?

At low velocities when we can neglect the relativistic effects such as the spin-orbit interaction, the projection of the spin of an electron along an arbitrary axis is an integral of motion. Thus, if the electrons differ in projections of their spins, they are distinguishable and their scattering is given by Eq. (1.1). In the case of equal orientation of the spin projections, the experiment shows that the angular distribution of the scattered particles is given by the formula

$$\frac{\mathrm{d}\sigma}{\mathrm{d}\Omega} = |f(\vartheta) - f(\pi - \vartheta)|^2\,. \tag{5.6}$$

This means, nature selects the other, i.e., antisymmetrical, possibility when it comes to electrons.

The discussed situations of symmetrical and antisymmetrical wave functions of electrons and α-particles, respectively, are specific examples of a more general rule. This rule states that the wave function of identical particles with integer spins (*bosons*) is totally symmetric with respect to the interchange of any two particles, and that the wave function of identical particles with half-integer spins (*fermions*) is totally antisymmetric with respect to the interchange of any two particles.

However, it is necessary that this antisymmetry includes *all* coordinates of the electrons. We must therefore consider the spin of the electrons as well.[1] When we explicitly write the spin part of the wave functions, we are able to antisymmetrize even the (spatial) function ψ_S:

$$\Psi_S(1,2) \stackrel{\text{def}}{=} \frac{1}{\sqrt{2}}(\psi(1,2)+\psi(2,1))\frac{1}{\sqrt{2}}(|+\rangle_1|-\rangle_2 - |-\rangle_1|+\rangle_2)\,, \tag{5.7}$$

$$\Psi_A(1,2) \stackrel{\text{def}}{=} \frac{1}{\sqrt{2}}(\psi(1,2)-\psi(2,1))\frac{1}{\sqrt{2}}(|+\rangle_1|-\rangle_2 + |-\rangle_1|+\rangle_2)\,. \tag{5.8}$$

Note that one can use any of the three triplet states in the case of Ψ_A.

We thus see that one needs to account for both possibilities after all. The energies of states which differ only in the symmetry or antisymmetry of the spatial part of the wave function will acquire different values. After substitution for $\psi(1,2)$ from Eq. (5.4), we obtain a variational estimate

$$\begin{aligned}\frac{E_{\text{var}}}{Z^2} &= \langle\Psi(1,2)|\hat{\mathsf{H}}|\Psi(1,2)\rangle \\ &= \langle a|\hat{\mathsf{h}}_0|a\rangle + \langle b|\hat{\mathsf{h}}_0|b\rangle + \frac{1}{Z}\left(\langle a|\langle b|r_{12}^{-1}|a\rangle|b\rangle \pm \langle a|\langle b|r_{12}^{-1}|b\rangle|a\rangle\right)\,. \end{aligned}\tag{5.9}$$

Expressing the first term in the parenthesis in the coordinate representation yields

$$\begin{aligned}\langle a|\langle b|r_{12}^{-1}|a\rangle|b\rangle &= \iint \psi_a^*(\mathbf{r}_1)\psi_a(\mathbf{r}_1)\frac{1}{r_{12}}\psi_b^*(\mathbf{r}_2)\psi_b(\mathbf{r}_2)\mathrm{d}V_1\mathrm{d}V_2 \\ &= \iint \frac{\varrho_a(\mathbf{r}_1)\varrho_b(\mathbf{r}_2)}{|\mathbf{r}_1-\mathbf{r}_2|}\mathrm{d}V_1\mathrm{d}V_2\,,\end{aligned}$$

which is obviously a classical mutual potential energy of two charge distributions. We thus refer to this term as to the *Coulomb* term. The other term in the parenthesis in Eq. (5.9),

$$\langle a|\langle b|r_{12}^{-1}|b\rangle|a\rangle = \iint \psi_a^*(\mathbf{r}_1)\psi_b(\mathbf{r}_1)\frac{1}{r_{12}}\psi_b^*(\mathbf{r}_2)\psi_a(\mathbf{r}_2)\mathrm{d}V_1\mathrm{d}V_2, \tag{5.10}$$

[1] Precisely speaking, we need to consider all other internal characteristics of the particle as well. Although electrons possess no other internal properties but their spin, one would have to include also the color when discussing quarks of a given flavor, for example.

has no classical analogy and is commonly named the *exchange* term. With its aid, one can explain various phenomena, such as the covalent bond or ferromagnetism.

We thus see that in our problem the exchange interaction will lead to splitting of the singlet and triplet states. However, this splitting cannot be explained within the framework of the classical theory, and it is also the reason why Bohr-Sommerfeld theory could not elucidate spectra of many-electron atoms.

The Hamiltonian (5.1) does not depend on the spin, which might imply at first glance that one does not need to consider the spin at all. Nevertheless, we have seen that the requirement the total wave function be antisymmetric demands we include the spin, too. In case of helium-like atoms, the value of the total spin determines whether the spatial part of the wave function is symmetric or antisymmetric with respect to the exchange of two electrons. This in turn then dictates the sign of the exchange energy.

5.1.2 Where Does the Indistinguishability Come From?

One could intuitively argue for the indistinguishability of identical particles as follows. We have seen on the example of the Gaussian wave packet, considered back in Sect. 1.3.9, that the uncertainty in the determination of the position increases over time; that is, the wave packet is diffusing. It is then understandable that when we have two particles close to each other, after a very short time the uncertainty of their positions grows to such a great magnitude that we are no longer able to tell them apart.

In fact, this explanation is of greater significance that one would expect. We have seen that the indistinguishability of particles leads to the exchange contribution to energy. This exchange interaction depends on the product of the wave functions of individual particles, see Eq. (5.10). If the overlap of the functions is negligible, then also the exchange interaction does not need to be considered and the particles behave as if they were distinguishable. If two atoms are sufficiently far from each other, their electrons become practically distinguishable. However, as they come closer to a distance of 1–2 atomic radii (approximately 10^{-10} m), the wave functions start to overlap and the exchange interaction leads to the so-called covalent bond between the atoms.

5.1.3 Additional Symmetries

The reader can verify that the operator of the total orbital angular momentum commutes with the Hamiltonian (5.1),

$$\hat{\boldsymbol{L}} = \hat{\boldsymbol{L}}_1 + \hat{\boldsymbol{L}}_2 , \qquad [\hat{\boldsymbol{L}}, \hat{\mathsf{H}}] = 0 ,$$

where $\hat{\boldsymbol{L}}_1$ and $\hat{\boldsymbol{L}}_2$ represent the angular momenta of the first and second electron, respectively. The eigenvalues of this angular momentum are $0, 1, 2, \ldots$ and the corresponding many-electron states are denoted by $S, P, D, \ldots$[2] Since $\hat{\boldsymbol{L}}$ is an integral of motion, the Hamiltonian $\hat{\mathsf{H}}$ does not mix states which belong to different eigenvalues of this operator. As we will demonstrate in the next text, this fact considerably simplifies the variational calculation.

After the interchange $\mathbf{r}_1 \to -\mathbf{r}_1$ and $\mathbf{r}_2 \to -\mathbf{r}_2$, the Hamiltonian (5.1) maintains its original form. Meaning, the Hamiltonian (5.1) does not mix states with different parities. This further simplifies our calculation of the excited states, as we will discuss in detail later, namely in Sect. 5.3.8.

5.1.4 Spectroscopic Notation

The set of the operators $\{\hat{\mathsf{H}}, \hat{\mathsf{L}}^2, \hat{\mathsf{S}}^2, \hat{\Pi}, \hat{\mathsf{L}}_z, \hat{\mathsf{S}}_z\}$, where $\hat{\Pi}$ is the above introduced operator of parity, forms a complete set of commuting operators. The spectroscopic notation of their common eigenstates reads for instance $1\,^1S$, $2\,^1S$, $2\,^1P^o$, $2\,^3P^e$. The first number from the left describes the level of excitation of the particular symmetry; the singlet S-states are numbered from 1, the triplet S-states are numbered from 2, P-states from 2, D-states from 3, etc. The first upper index denotes the spin multiplicity, i.e., 1 for singlet and 3 for triplet states. The second upper index captures the parity, i.e., o labels an odd-parity state and e an even parity state. The S-states are always of even parity as we are to see later.

In summary, $1\,^1S$ marks the lowest singlet state with zero angular momentum, $2\,^1S$ the second lowest singlet state with zero angular momentum, $2\,^1P^o$ the lowest singlet state of odd parity with unit angular momentum, $2\,^3P^e$ the lowest triplet state of even parity with unit angular momentum, etc.

5.2 Variational Method with the Hartree-Fock Function

When we consider the variational function in the form (5.7) and (5.8), where $\psi(1,2)$ is of the form (5.4), we refer to it as the Hartree-Fock function adapted to the spin symmetry. In this section, we will use it to estimate the energy of the lowest levels of the helium atom.

The ground state of the hydrogen-like atom is $1s$-state. When searching for the ground state of helium we put both electrons into $1s$-orbital with opposite spins. Recalling the addition of the angular momenta, we know that the addition of two electrons both of which are in an S-state results in a combined state which is likewise an S-state.

[2] As we demonstrated in the previous chapter, a sum of two integer angular momenta gives rise to another integer angular momentum. We also showed there that both the magnetic quantum number m and the orbital quantum number l can evaluate only to integer numbers.

$$|1S(1,2)\rangle = |1s\rangle^{(1)}|1s\rangle^{(2)} \cdot \frac{1}{\sqrt{2}}\left(|+\rangle^{(1)}|-\rangle^{(2)} - |-\rangle^{(1)}|+\rangle^{(2)}\right) ,$$

where

$$\langle \mathbf{r}|\, 1s\rangle = \psi_{1s}(\mathbf{r};\alpha) = 2\alpha^{3/2}\mathrm{e}^{-\alpha r}Y_{00}(\mathbf{n}) . \tag{5.11}$$

The screening parameter α acquires the value $\alpha = 1$ in case of the hydrogen atom. Note that in case of a triplet state, the two electrons cannot both populate the $1s$ state. Here we thus see a link between the antisymmetry of the wave function and Pauli exclusion principle. If we place both electrons into the same orbital $1s$, it is then the spin part of the total wave function that must be antisymmetric. It is therefore not possible for the electrons to occupy the same orbital *and* have the same spin projection at the same time. The antisymmetry of the total wave function necessarily leads to the impossibility of describing such a state with a nontrivial wave function. The fact that two electrons cannot populate a one-electron state characterized by the same values of the quantum numbers is nothing else than the Pauli exclusion principle.

It is instructive to calculate also the lowest P-state. We will estimate its wave function as

$$|2P(1,2)\rangle = \frac{1}{\sqrt{2}}\left(|1s\rangle^{(1)}|2p\rangle^{(2)} \pm |2p\rangle^{(1)}|1s\rangle^{(2)}\right)\cdot\frac{1}{\sqrt{2}}\left(|+\rangle^{(1)}|-\rangle^{(2)} \mp |-\rangle^{(1)}|+\rangle^{(2)}\right),$$

where

$$\langle \mathbf{r}|\, 2p\rangle = \psi_{2p}(\mathbf{r};\beta) = \sqrt{\frac{\beta^5}{4!}}\, r\mathrm{e}^{-\beta r/2}Y_{1M}(\mathbf{n}) . \tag{5.12}$$

As mentioned above, the Hamiltonian (5.1) ignores the spin states, thus we will not explicitly write them henceforth. Equation (5.9) for energy can be written for both considered situations separately,

$$\frac{E_{1S}}{Z^2} = 2\langle 1s|\hat{\mathrm{h}}_0|1s\rangle + \frac{1}{Z}\langle 1s|\langle 1s|\frac{1}{r_{12}}|1s\rangle|1s\rangle , \tag{5.13}$$

$$\frac{E_{2P}}{Z^2} = \langle 1s|\hat{\mathrm{h}}_0|1s\rangle + \langle 2p|\hat{\mathrm{h}}_0|2p\rangle + \frac{1}{Z}\langle 1s|\langle 2p|\frac{1}{r_{12}}\left(|1s\rangle|2p\rangle \pm |2p\rangle|1s\rangle\right) . \tag{5.14}$$

One can easily calculate the two simple matrix elements of the Hamiltonian $\hat{\mathrm{h}}_0$,

$$\langle 1s|\hat{\mathrm{h}}_0|1s\rangle = \alpha\left(\frac{\alpha}{2} - 1\right) ,$$

$$\langle 2p|\hat{\mathrm{h}}_0|2p\rangle = \frac{\beta}{4}\left(\frac{\beta}{2} - 1\right) .$$

The remaining Coulomb and exchange terms are, on the other hand, slightly more complicated,

$$\langle 1s|\langle 2p|\frac{1}{r_{12}}\left(|1s\rangle|2p\rangle \pm |2p\rangle|1s\rangle\right) = C \pm V\,,$$

where

$$C = \int\int \psi_{1s}^*(\mathbf{r}_1)\psi_{2p}^*(\mathbf{r}_2)\frac{1}{r_{12}}\psi_{1s}(\mathbf{r}_1)\psi_{2p}(\mathbf{r}_2)\mathrm{d}^3\mathbf{r}_2\mathrm{d}^3\mathbf{r}_1\,, \tag{5.15}$$

$$V = \int\int \psi_{1s}^*(\mathbf{r}_1)\psi_{2p}^*(\mathbf{r}_2)\frac{1}{r_{12}}\psi_{2p}(\mathbf{r}_1)\psi_{1s}(\mathbf{r}_2)\mathrm{d}^3\mathbf{r}_2\mathrm{d}^3\mathbf{r}_1 \tag{5.16}$$

and

$$\frac{1}{r_{12}} = \frac{1}{\left[r_1^2 - 2\mathbf{r}_1\cdot\mathbf{r}_2 + r_2^2\right]^{1/2}}\,. \tag{5.17}$$

5.2.1 *Multipole Expansion*

We concluded the previous section by writing down the Coulomb and exchange integrals in a rather frightful form of six-dimensional integrals. We will now show that to the utmost surprise of the reader, one can evaluate them exactly. Furthermore, we will also demonstrate that the calculation itself is not in fact as difficult as one might expect at the first glance.

The hydrogen-like wave functions, for instance, ψ_{1s} and ψ_{2p}, can be split into a product of their angular and radial parts, see Eq. (3.83). One might therefore attempt a similar angular-radial decomposition for the operator r_{12}^{-1}. We will now demonstrate that one can indeed do so, though not for free. The price we have to pay is that we will be able to express r_{12}^{-1} only as an infinite series of terms. It will be possible, though, to reduce this infinite series to a finite number of terms owing to the properties of the spherical harmonics Y_{lm}.

We have already mentioned, see Sect. 3.3.1, that the Coulomb potential

$$G(\mathbf{r},\mathbf{r}') = \frac{1}{4\pi|\mathbf{r}-\mathbf{r}'|} \tag{3.40}$$

is a solution to the equation

$$-\nabla^2 G(\mathbf{r},\mathbf{r}') = \delta(\mathbf{r}-\mathbf{r}')\,. \tag{3.38}$$

By expressing the Coulomb potential with radial and angular variables split, we obtain the solution to Eq. (3.38) in the spherical coordinates.[3] Let us now consider the expansion of the function G in a complete basis of the spherical harmonics with respect to the angular variables of the position vector $\mathbf{r}$,

$$G(\mathbf{r},\mathbf{r}') = \sum_{l,m} f_{lm}(r,\mathbf{r}')Y_{lm}(\vartheta,\varphi)\,.$$

We consider the radial part $f_{lm}(r,\mathbf{r}')$ to be continuous. Later on, we show that this does not lead to a contradiction and that the singularity of the expression (3.40) at the point $\mathbf{r} = \mathbf{r}'$ does not originate in $f_{lm}(r,\mathbf{r}')$, but in the summation over l. From Eqs. (3.21) and (3.76) and the definition of the δ-function in spherical coordinates[4] we obtain

$$-\left(\frac{1}{r^2}\frac{\partial}{\partial r}\left(r^2\frac{\partial}{\partial r}\right) - \frac{\hat{\mathrm{L}}^2}{r^2}\right)G(\mathbf{r},\mathbf{r}') = \frac{1}{r^2}\delta(r-r')\frac{1}{\sin\vartheta}\delta(\vartheta-\vartheta')\delta(\varphi-\varphi')\,. \tag{5.18}$$

As our next step, we multiply the last equation by r^2 and integrate over the interval from $r'-\varepsilon$ to $r'+\varepsilon$; we obtain

$$-\int_{r'-\varepsilon}^{r'+\varepsilon}\frac{\partial}{\partial r}\left(r^2\frac{\partial G}{\partial r}\right)\mathrm{d}r + \sum_{l,m} l(l+1)Y_{lm}(\vartheta,\varphi)\int_{r'-\varepsilon}^{r'+\varepsilon} f_{lm}(r,\mathbf{r}')\mathrm{d}r$$
$$= \int_{r'-\varepsilon}^{r'+\varepsilon}\delta(r-r')\frac{1}{\sin\vartheta}\delta(\vartheta-\vartheta')\delta(\varphi-\varphi')\mathrm{d}r\,.$$

Owing to the assumed continuity of $f_{lm}(r,\mathbf{r}')$ in r, the second term on the lhs of the last equation disappears in the limit $\varepsilon \to 0$. We can deduce from the remaining parts of the equation that the function G has a jump on the spherical surface $r = r'$,

$$-\left[r^2\frac{\partial G}{\partial r}\right]_{r'-\varepsilon}^{r'+\varepsilon} = \frac{1}{\sin\vartheta}\delta(\vartheta-\vartheta')\delta(\varphi-\varphi')\,. \tag{5.19}$$

The homogeneous solutions to Eq. (5.18) can be found with the aid of Eqs. (3.84) and (3.86):

$$-\left[\frac{\mathrm{d}^2}{\mathrm{d}r^2} + \frac{2}{r}\frac{\mathrm{d}}{\mathrm{d}r} - \frac{l(l+1)}{r^2}\right]f_{lm}(r,\mathbf{r}') = 0 \Rightarrow f_{lm}(r,\mathbf{r}') \sim \begin{cases} r^l\,, \\ r^{-l-1}\,. \end{cases}$$

[3]The derivation presented henceforth is inspired by the one in [7].

[4]It holds true for the δ-function that $\int\delta^{(3)}(\mathbf{r})\mathrm{d}^3\mathbf{r} = 1$. After transformation of the differential to spherical coordinates, we have $\mathrm{d}^3\mathbf{r} = r^2\mathrm{d}r\mathrm{d}\vartheta\sin\vartheta\mathrm{d}\varphi$. Since we have $\int\delta(x)\mathrm{d}x = 1$ for the one-dimensional integrals, it must obviously be $\delta^{(3)}(\mathbf{r}) = \frac{1}{r^2}\delta(r)\frac{1}{\sin\vartheta}\delta(\vartheta)\delta(\varphi)$.

The first solution diverges for large r, while the other diverges for small r. Since we are interested in a solution which is finite for all r, and because of, see Eq. (3.40),

$$G(\mathbf{r},\mathbf{r}') = G(\mathbf{r}',\mathbf{r}) , \tag{5.20}$$

we compose the Green function as

$$G(\mathbf{r},\mathbf{r}')\Big|_{r>r'} = \sum_{l,m} C_{lm} r'^{l} r^{-l-1} Y_{lm}(\vartheta,\varphi) Y^*_{lm}(\vartheta',\varphi') , \tag{5.21}$$

$$G(\mathbf{r},\mathbf{r}')\Big|_{r<r'} = \sum_{l,m} C_{lm} r^{l} r'^{-l-1} Y_{lm}(\vartheta,\varphi) Y^*_{lm}(\vartheta',\varphi') . \tag{5.22}$$

The complex conjugation of the spherical harmonics does not violate the condition (5.20). Recalling Eq. (3.94) and realizing we are summing over all magnetic numbers m, these formulations of the Green function are symmetrical in the angular variables, even though it may not be obvious on the first sight.

Plugging Eqs. (5.21) and (5.22) into Eq. (5.19) yields

$$\sum_{l,m} (2l+1) C_{lm} Y_{lm}(\vartheta,\varphi) Y^*_{lm}(\vartheta',\varphi') = \frac{1}{\sin\vartheta} \delta(\vartheta-\vartheta')\delta(\varphi-\varphi') .$$

A comparison of the last equation with the completeness relation for the spherical harmonics

$$\sum_{l,m} |l,m\rangle\langle l,m| = 1 \Rightarrow \sum_{l,m} Y_{lm}(\vartheta,\varphi) Y^*_{lm}(\vartheta',\varphi') = \frac{1}{\sin\vartheta} \delta(\vartheta-\vartheta')\delta(\varphi-\varphi')$$

implies

$$C_{lm} = \frac{1}{2l+1} .$$

After substitution of the last equation into the two expansions of the function $G(\mathbf{r},\mathbf{r}')$, Eqs. (5.22) and (5.21), and introduction of the notation $r_> = \max\{r,r'\}$ and $r_< = \min\{r,r'\}$, we can write the resulting *multipole expansion* in a compact form

$$\frac{1}{|\mathbf{r}-\mathbf{r}'|} = \sum_{l=0}^{\infty} \frac{4\pi}{2l+1} \frac{r_<^l}{r_>^{l+1}} \sum_{m=-l}^{l} Y^*_{lm}(\vartheta',\varphi') Y_{lm}(\vartheta,\varphi) . \tag{5.23}$$

5.2.2 A Note on the Legendre Polynomials

The multipole expansion can be also derived in a more direct way. We factor out $r_>$ from Eq. (5.17) and hence write

$$(1 - 2tz + t^2)^{-1/2} = \sum_{l=0}^{\infty} P_l(z)t^l, \qquad z = \mathbf{n}_1 \cdot \mathbf{n}_2, \quad t = \frac{r_<}{r_>}, \tag{5.24}$$

where $P_l(z)$ are the so-called *Legendre polynomials*. Setting $t = 0$ in the last equation provides the first, trivial, polynomial

$$P_0(z) = 1. \tag{5.25}$$

We differentiate Eq. (5.24) with respect to t and obtain

$$(z - t)(1 - 2tz + t^2)^{-3/2} = \sum_{l=0}^{\infty} P_l(z)lt^{l-1}.$$

We multiply the last equation by the expression $(1 - 2tz + t^2)$ and manipulate the lhs and rhs into the forms

$$(z - t)(1 - 2tz + t^2)^{-1/2} = \sum_l P_l(z)t^l(z - t) = \sum_l [zP_l(z) - P_{l-1}(z)]t^l$$

and

$$\sum_l P_l(z)lt^{l-1}(1 - 2tz + t^2) = \sum_l [(l + 1)P_{l+1}(z) - 2lzP_l(z) + (l - 1)P_{l-1}(z)]t^l,$$

respectively, where we used Eq. (5.24) and shifted the summation variable. Matching now the terms with the same power of t on the rhs of the last two equations leads to recurrence relations for the Legendre polynomials

$$(l + 1)P_{l+1}(z) - z(2l + 1)P_l(z) + lP_{l-1}(z) = 0. \tag{5.26}$$

Consequently, $P_1(z) = z$, $P_2(z) = (-1 + 3z^2)/2$ and so on. When we compare Eq. (5.26) to a similar relation (4.62) for the spherical harmonics Y_{l0}, and the form of Y_{00}, (3.35), to P_0, (5.25), we obtain

$$Y_{l,0}(\vartheta) = \sqrt{\frac{2l + 1}{4\pi}} P_l(\cos\vartheta). \tag{5.27}$$

Comparison of Eqs. (5.24) and (5.23) now leads to the *addition theorem for the spherical harmonics*

$$P_l(\mathbf{n} \cdot \mathbf{n}') = \frac{4\pi}{2l + 1} \sum_{m=-l}^{l} Y^*_{lm}(\vartheta', \varphi')Y_{lm}(\vartheta, \varphi). \tag{5.28}$$

Setting $\mathbf{n} = \mathbf{n}'$ results in

$$1 = \frac{4\pi}{2l+1} \sum_{m=-l}^{l} |Y_{lm}(\vartheta, \varphi)|^2 , \tag{5.29}$$

where we used the equality

$$P_l(1) = 1 ,$$

which follows from Eqs. (5.25) and (5.26).

If we set $\mathbf{n} = \mathbf{n}'$ in Eq. (5.23), we have then with the use of Eq. (5.29) for the problematic point $r \to r'$

$$\lim_{\mathbf{r}\to\mathbf{r}'} \frac{1}{|\mathbf{r} - \mathbf{r}'|} = \frac{1}{r} \sum_{l=0}^{\infty} 1 = +\infty ,$$

in accordance with the comment preceding Eq. (5.18).

Next, after substituting Eq. (5.27) into Eq. (4.89), we arrive at the *expansion of a plane wave into spherical waves*

$$\mathrm{e}^{\mathrm{i}\omega r \cos\vartheta} = \sum_{l=0}^{\infty} (2l+1) P_l(\cos\vartheta) \mathrm{i}^l j_l(\omega r) .$$

With the aid of Eq. (5.28) and the separation $\mathbf{k} = \omega\boldsymbol{\eta}$, the last equation can be cast into the form

$$\mathrm{e}^{\mathrm{i}\mathbf{k}\cdot\mathbf{r}} = \sum_{l=0}^{\infty} (2l+1) P_l(\mathbf{n}\cdot\boldsymbol{\eta}) \mathrm{i}^l j_l(\omega r) = 4\pi \sum_{l=0}^{\infty} \mathrm{i}^l j_l(\omega r) \sum_{m=-l}^{l} Y_{lm}^*(\mathbf{n}) Y_{lm}(\boldsymbol{\eta}) . \tag{5.30}$$

On the rhs of this equation we have, finally, separated angular variables $\boldsymbol{\eta}$ and $\mathbf{n}$. We will take advantage of this decomposition later in Sects. 6.3.6 and 6.5.7.

5.2.3 Calculation of the Integrals

Now we are ready to calculate the integrals (5.15) and (5.16)! Inserting into them the multipole expansion (5.23) and the form of the wave functions (5.11) and (5.12) yields

$$C = \sum_{l=0}^{\infty} \frac{4\pi}{2l+1} \sum_{m=-l}^{l} \int Y_{00}^*(\mathbf{n}_1) Y_{lm}^*(\mathbf{n}_1) Y_{00}(\mathbf{n}_1) \mathrm{d}\Omega_1 \int Y_{1M}^*(\mathbf{n}_2) Y_{lm}(\mathbf{n}_2) Y_{1M}(\mathbf{n}_2) \mathrm{d}\Omega_2$$

$$\times \int_0^\infty r_1^2 \left(\int_0^\infty r_2^2 R_{1s}(r_1) R_{2p}(r_2) \frac{r_<^l}{r_>^{l+1}} R_{1s}(r_1) R_{2p}(r_2) \mathrm{d}r_2 \right) \mathrm{d}r_1 \,,$$

$$V = \sum_{l=0}^{\infty} \frac{4\pi}{2l+1} \sum_{m=-l}^{l} \int Y_{00}^*(\mathbf{n}_1) Y_{lm}^*(\mathbf{n}_1) Y_{1M}(\mathbf{n}_1) \mathrm{d}\Omega_1 \int Y_{1M}^*(\mathbf{n}_2) Y_{lm}(\mathbf{n}_2) Y_{00}(\mathbf{n}_2) \mathrm{d}\Omega_2$$

$$\times \int_0^\infty r_1^2 \left(\int_0^\infty r_2^2 R_{1s}(r_1) R_{2p}(r_2) \frac{r_<^l}{r_>^{l+1}} R_{2p}(r_1) R_{1s}(r_2) \mathrm{d}r_2 \right) \mathrm{d}r_1 \,.$$

Recall that Y_{00}^* is a constant function, see Eq. (3.35), therefore we factor it out in front of the angular integrals. It follows from the orthonormality of the spherical harmonics, Eq. (3.86), that the infinite summations over l and m in the Coulomb integral C reduce to a single contribution $l = m = 0$. Similarly, the infinite summations over l and m in the exchange term V reduce to a single contribution $l = 1, m = M$. We are thus left with merely the integration over the radial variables in both integrals,

$$C = \int_0^\infty r_1^2 \left(\int_0^\infty r_2^2 R_{1s}(r_1) R_{2p}(r_2) \frac{1}{r_>} R_{1s}(r_1) R_{2p}(r_2) \mathrm{d}r_2 \right) \mathrm{d}r_1 \,, \tag{5.31}$$

$$V = \frac{1}{3} \int_0^\infty r_1^2 \left(\int_0^\infty r_2^2 R_{1s}(r_1) R_{2p}(r_2) \frac{r_<}{r_>^2} R_{2p}(r_1) R_{1s}(r_2) \mathrm{d}r_2 \right) \mathrm{d}r_1 \,. \tag{5.32}$$

Let us now show how one can evaluate the following general integral:

$$I(\eta, \xi, a, b, l) = \int_0^\infty \int_0^\infty r_1^a r_2^b \mathrm{e}^{-\eta r_1} \mathrm{e}^{-\xi r_2} \frac{r_<^l}{r_>^{l+1}} \mathrm{d}r_2 \mathrm{d}r_1 \,. \tag{5.33}$$

First, we split the integration over r_2 into two regions $r_1 > r_2$ and $r_1 < r_2$. In the region $r_1 > r_2$ we obviously have $r_< = r_2$, $r_> = r_1$, while in the region $r_2 > r_1$ the opposite is true, $r_< = r_1$, $r_> = r_2$:

$$I(\eta, \xi, a, b, l) = \int_0^\infty r_1^a \mathrm{e}^{-\eta r_1} \left(\int_0^{r_1} r_2^b \mathrm{e}^{-\xi r_2} \frac{r_2^l}{r_1^{l+1}} \mathrm{d}r_2 \right) \mathrm{d}r_1$$
$$+ \int_0^\infty r_1^a \mathrm{e}^{-\eta r_1} \left(\int_{r_1}^\infty r_2^b \mathrm{e}^{-\xi r_2} \frac{r_1^l}{r_2^{l+1}} \mathrm{d}r_2 \right) \mathrm{d}r_1 \,.$$

These integrals can be most easily calculated by the differentiation of the integrals with respect to a parameter:

$$I(\eta, \xi, a, b, l) = (-1)^{a+b-1} \frac{\partial^{a-l-1}}{\partial \eta^{a-l-1}} \frac{\partial^{b+l}}{\partial \xi^{b+l}} \int_0^\infty \mathrm{e}^{-\eta r_1} \left(\int_0^{r_1} \mathrm{e}^{-\xi r_2} \mathrm{d}r_2 \right) \mathrm{d}r_1 \tag{5.34}$$

$$+(-1)^{a+b-1}\frac{\partial^{a+l}}{\partial\eta^{a+l}}\frac{\partial^{b-l-1}}{\partial\xi^{b-l-1}}\int_0^\infty \mathrm{e}^{-\eta r_1}\left(\int_{r_1}^\infty \mathrm{e}^{-\xi r_2}\mathrm{d}r_2\right)\mathrm{d}r_1$$

$$=(-1)^{a+b-1}\frac{\partial^{a-l-1}}{\partial\eta^{a-l-1}}\frac{\partial^{b+l}}{\partial\xi^{b+l}}\frac{1}{\eta(\xi+\eta)}$$

$$+(-1)^{a+b-1}\frac{\partial^{a+l}}{\partial\eta^{a+l}}\frac{\partial^{b-l-1}}{\partial\xi^{b-l-1}}\frac{1}{\xi(\xi+\eta)}$$

$$=J(\eta,\xi,a,b,l)+J(\xi,\eta,b,a,l)\,,$$

where

$$J(\eta,\xi,a,b,l)=(a-l-1)!\sum_{q=0}^{a-l-1}\frac{(b+l+q)!}{q!}\eta^{-(a-l-q)}(\eta+\xi)^{-(b+l+q+1)}\,. \quad (5.35)$$

Substitution of Eqs. (5.34) and (5.35) into Eqs. (5.31) and (5.32) yields

$$C=4\alpha^3\frac{\beta^5}{4!}I(2\alpha,\beta,2,4,0)=\frac{\alpha\beta(8\alpha^4+20\alpha^3\beta+20\alpha^2\beta^2+10\alpha\beta^3+\beta^4)}{(2\alpha+\beta)^5}\,,$$

and

$$V=4\alpha^3\frac{\beta^5}{4!}\frac{1}{3}I(\alpha+\beta/2,\alpha+\beta/2,3,3,1)=\frac{112\beta^5\alpha^3}{3(2\alpha+\beta)^7}\,.$$

Likewise, we find for the variational estimate of the ground state (5.13) that

$$\frac{E_{1S}}{Z^2}=2\alpha\left(\frac{\alpha}{2}-1\right)+\frac{1}{Z}(4\alpha^3)^2I(2\alpha,2\alpha,2,2,0)=\alpha^2-2\alpha+\frac{1}{Z}\frac{5}{8}\alpha\,.$$

5.2.4 *Optimization of the Parameters*

The value of the parameter α can be retrieved from the condition $\partial_\alpha E=0$, which reads

$$\alpha=1-\frac{1}{Z}\frac{5}{16}=\frac{27}{32}$$

in the case of helium. Using this value of α in Eq. (5.13) leads to a prediction $E_{1S}(\alpha=27/32)=-2.8476$. Had we set $\alpha=1$, the energy estimate would be $E_{1S}(\alpha=1)=-2.75$. The accurate nonrelativistic value in the approximation of an infinitely heavy nucleus reads $\bar{E}_{1S}\doteq-2.903724$ [2]. We thus see that the optimization of the parameter α decreased the relative error of the variational calculation from 5 to 2%, which is—given the simplicity of the calculation—very satisfying. In the next text, we show how to achieve an even better match.

In contrast to the case of the ground state, the solution satisfying the conditions on the energy minimum (5.14) of the state $\Psi_{2P}(1,2)$ is technically more involved. Performing the differentiations, $\partial_\alpha E_{2P} = 0$ and $\partial_\beta E_{2P} = 0$, results in a set of two nonlinear equations which cannot be solved analytically. Therefore, we use the Newton-Raphson method to find a numerical solution. We deem this method to be of enough importance as to briefly introduce it herein.

First, we will explain the method for an one-dimensional case. Suppose we look for a root of equation $f(x) = 0$ and let the value x_k be an approximate solution. This solution can be improved if we replace the function $f(x)$ in the vicinity of the point x_k by its Taylor expansion and keep only the first two terms,

$$f(x_k) + (x_{k+1} - x_k) \left.\frac{\mathrm{d}f}{\mathrm{d}x}\right|_{x=x_k} = 0\,.$$

Let x_0 be our estimated solution. From the last equation, we obtain a sequence of successively refined estimates $x_1, x_2, x_3, \ldots$ This sequence converges very fast to the correct solution if we move along an almost linear section of the graph. Unfortunately, the method performs much worse if the graph has large curvature.

A two-dimensional generalization of this method for the set of two nonlinear equations $f_1 = \partial_\alpha E_{2P}(\alpha,\beta) = 0$ and $f_2 = \partial_\beta E_{2P}(\alpha,\beta) = 0$ is the set of two linear recurrence relations

$$f_j(\alpha_k,\beta_k) + (\alpha_{k+1} - \alpha_k)\frac{\partial f_j}{\partial \alpha}(\alpha_k,\beta_k) + (\beta_{k+1} - \beta_k)\frac{\partial f_j}{\partial \beta}(\alpha_k,\beta_k) = 0\,, \qquad j = 1,2\,.$$

The choice of the initial α and β is of great importance. The first guess could be $\alpha_0 = \beta_0 = 1$, that is, we would start with the hydrogen states. A more accurate reasoning says that an electron in the state $2p$ feels the electrostatic field as if it were generated by a single proton only as the inner $1s$-electron shields one of the nuclear charges. We therefore have for the Hamiltonian (5.1) (when $Z = 2$)

$$\alpha_0 = 1\,, \qquad \beta_0 = 1 - 1/Z = 1/2\,.$$

When we carry out the numerical calculation with this initial condition, we find the solutions for the singlet state (symmetrical in the spatial part) and the triplet state (antisymmetrical in the spatial state) summarized in Table 5.1.

Table 5.1 Optimization of the parameters for the state $2P$

	Singlet	Triplet
α	1.0015	0.9955
β	0.4823	0.5445
E_{2P}	−2.12239	−2.13069
$\bar{E}_{2P}$	−2.12384	−2.13316

Table 5.2 Results of Exercise 14

	Singlet	Triplet
α	1	1
β	0.4998	0.5004
E_{3D}	−2.05555	−2.05557
$\bar{E}_{3D}$	−2.05562	−2.05564

The obtained variational energies in the table are therein compared with the exact nonrelativistic values $\bar{E}_{2P}$ [2]. Note that the result from the approximate variational method lies this time much closer to the exact value than in the case of the ground state. This is a general and easily understandable phenomenon. In the case of the same spatial distribution of electrons, as in the ground state, the electron-electron correlation is much stronger than in the case of different spatial distributions.

The electron-electron correlation is mostly characterized by the *correlation energy*. This energy is defined as a difference between the exact nonrelativistic value and the variational value obtained from the Hartree-Fock function.

Exercise 14: Nonlinear Variational Method I

Consider the function

$$|3D(1,2)\rangle = \frac{1}{\sqrt{2}}\left(|1s\rangle_1|3d\rangle_2 \pm |3d\rangle_1|1s\rangle_2\right)\cdot\frac{1}{\sqrt{2}}\left(|+\rangle_1|-\rangle_2 \mp |-\rangle_1|+\rangle_2\right),$$

where ψ_{1s} is given by Eq. (5.11) and where ψ_{3d} is given by

$$\psi_{3d}(\mathbf{r};\beta) = \sqrt{\left(\frac{2}{3}\right)^7\frac{\beta^7}{6!}}\,r^2\mathrm{e}^{-\beta r/3}\,Y_{2M}(\mathbf{n})\,.$$

Estimate the energy of the singlet and triplet states. Your results should match the numbers in Table 5.2.

One can readily see from Table 5.2 that the screening for D-states is even stronger than that for the P-states, that the difference between the result of the singlet and triplet state is smaller, and that the result from the single-configuration variational calculation likewise lies closer to the exact value. These trends continue also for other states with higher angular momenta.

Exercise 15: Nonlinear Variational Method II

Consider the function

$$|2\,^3S(1,2)\rangle = \frac{1}{\sqrt{2}}\left(|1s\rangle_1|2s\rangle_2 - |2s\rangle_1|1s\rangle_2\right)\cdot\frac{1}{\sqrt{2}}\left(|+\rangle_1|-\rangle_2 + |-\rangle_1|+\rangle_2\right),$$

where ψ_{1s} *is given by Eq. (5.11) and where* ψ_{2s} *is given by*

$$\psi_{2s}(\mathbf{r};\beta) = (c_0 + c_1 r)\mathrm{e}^{-\beta r/2}Y_{0,0}(\mathbf{n})\,. \tag{5.36}$$

Determine the coefficients c_0 *and* c_1 *using the equations*

$$\langle 1s|\,2s\rangle = 0\,, \qquad \langle 2s|\,2s\rangle = 1\,.$$

Find the variational energy E_{2s} *as a function of* α *and* β *and evaluate it for the values* $\alpha = 1, \beta = 1$ *and* $\alpha = 1, \beta = 0.5$. *Use the Newton-Raphson method for the condition on minimum to find the optimal* α *and* β *and the variational estimate of the energy. You should obtain the values listed in Table 5.3.*

Table 5.3 Results of Exercise 15

α	β	E_{2^3S}
1	1	−2.124
1	0.5	−2.157
1.0042	0.6945	−2.17195

The last row in Table 5.3 contains the optimized values. One can see that also in this case, the electron in $2s$ "feels" the charge of the nucleus as if screened by the electron in $1s$. The agreement with the exact nonrelativistic value $\bar{E}_{2^3S} = -2.175229$ [2] is again more than satisfying. One could have expected that, though, based on the discussion of the states $2\,P$. In the state $2\,^3S$, the Pauli exclusion principle forces the electrons to occupy a different spatial distribution and the correlation energy thus reaches substantially lower values than for the ground state.

5.3 Variational Method: Configuration Interaction

In the previous section, we showed that one can obtain a very good estimate of the exact energies of the ground and excited states of helium already when using very simple test functions. In this section, we will see how one can further systematically improve this estimate. We will demonstrate this method on the ground state where the error of the calculation with a Hartree-Fock function does not go below 2%.

The substitution $r \to r\eta$ in Eq. (5.1) transforms the Schrödinger equation into the form

$$\left(-\frac{\nabla_1^2}{2} - \frac{\eta}{r_1} + \frac{1}{2} - \frac{\nabla_2^2}{2} - \frac{\eta}{r_2} + \frac{1}{2} + \frac{\eta}{Zr_{12}}\right)\psi = \frac{\Delta E \eta^2}{Z^2}\psi\,, \tag{5.37}$$

where η is a nonlinear variational parameter and ΔE the energy arising from the interaction of electrons:

$$E_{\text{var}}(\eta) = -2\frac{Z^2}{2\eta^2} + \Delta E\,.$$

We now solve Eq. (5.37) by the optimization of the linear parameters. We expand the sought wave function ψ into a complete basis, see Eq. (2.4). The optimization of the expansion coefficients then leads to the generalized eigenvalue problem (2.6).

We can set the parameter η equal to one or—better— to an optimized value which we find as follows. For a given value of the nonlinear parameter η, we optimize N linear parameters c_i. We thus find ΔE as well as the variational energy estimate $E_{\text{var}}(\eta)$. Subsequently, we choose a different value of the parameter $\eta = \eta'$, run the optimization process, and obtain a new energy estimate $E_{\text{var}}(\eta')$. This way, we find the value of the parameter η which minimizes the variational energy for the given number N of the basis functions. In the below calculation, we found the optimal value of η by numerical means, see Exercise 16.

This method of solution, namely the decomposition of the sought many-particle eigenstate into a many-particle complete basis, see Eq. (2.4), is commonly referred to as the *configuration interaction* method.

5.3.1 Adaptation of the Basis to Symmetry

We can consider the basis functions $|j\rangle$ in Eq. (2.4) in the form of antisymmetrized products of the hydrogen-like wave functions (omitting the spin parts)

$$|j\rangle = \left|n_{1j}, l_{1j}, m_{1j}\right\rangle^{(1)} \left|n_{2j}, l_{2j}, m_{2j}\right\rangle^{(2)} \pm \left|n_{2j}, l_{2j}, m_{2j}\right\rangle^{(1)} \left|n_{1j}, l_{1j}, m_{1j}\right\rangle^{(2)}.$$

We thus see that there is a set of six one-electron quantum numbers $n_{1j}, l_{1j}, m_{1j}, n_{2j}, l_{2j}, m_{2j}$ to describe every single electron configuration $|j\rangle$. Unfortunately, even for small values of n_{1j} and n_{2j}, one needs to consider a great number of such configurations. Recall that for a given principal quantum number n the orbital quantum number l acquires the values $l = 0, 1, \ldots, n-1$, and that for a given l the magnetic quantum number equals $m = -l, -l+1, \ldots, l-1, l$. One can thus readily imagine that this is definitely not a way to go, even with powerful computers at one's disposal. As we demonstrate below, integrals of motion substantially reduce the number of configurations that need to be considered. As in the previous calculation with the Hartree-Fock function, we know that the ground state is characterized by the total $l = 0$. From Sect. 4.2, we also know that to compose $l = 0$ from l_1 and l_2, it must be $|l_1 - l_2| \leq l = 0$, hence $l_1 = l_2$. The basis states adapted to the symmetry, from which we choose only those with their total angular momentum equal to zero, are

$$|j\rangle = \frac{1}{\sqrt{2}}\left(|n_{1j}, l_j\rangle^{(1)}|n_{2j}, l_j\rangle^{(2)} \sum_{i=-l_j}^{l_j}(l_j, i, l_j, -i|0, 0)|l_j, i\rangle^{(1)}|l_j, -i\rangle^{(2)}\right. \tag{5.38}$$

$$\left.\pm|n_{2j}, l_j\rangle^{(1)}|n_{1j}, l_j\rangle^{(2)} \sum_{i=-l_j}^{l_j}(l_j, -i, l_j, i|0, 0)|l_j, -i\rangle^{(1)}|l_j, i\rangle^{(2)}\right),$$

where the one-particle states $|n, l\rangle$ and $|l, m\rangle$ are given by Eqs. (4.93) and (4.3), respectively. The plus and minus signs refer to singlets and triplets, respectively. We thus see that, within the variational calculation of the ground state, only three quantum numbers (as opposed to the initial six) determine unambiguously each of the configurations adapted to the symmetry. Moreover, owing to the antisymmetry of the full state, it suffices to consider merely the cases $n_{1j} \leq n_{2j}$. Setting $j_1 = l_j$, $j_2 = l_j$, $j = 0$ and $m = 0$ in Eq. (4.20) results in a simple expression for the needed Clebsch-Gordan coefficients, namely

$$(l_j, i, l_j, -i|0, 0) = (-1)^{i+l_j}\frac{1}{\sqrt{2l_j + 1}},$$

which can be then used to simplify the basis functions (5.38) to

$$|j\rangle = \frac{1}{\sqrt{2}}\sum_{i=-l_j}^{l_j}\frac{(-1)^{i+l_j}}{\sqrt{2l_j + 1}}\left(|n_{1j}, l_j, i\rangle^{(1)}|n_{2j}, l_j, -i\rangle^{(2)} \pm |n_{2j}, l_j, -i\rangle^{(1)}|n_{1j}, l_j, i\rangle^{(2)}\right).$$

In the coordinate representation, the states $|j\rangle$ acquire the form

$$\langle \mathbf{r}_1, \mathbf{r}_2|j\rangle = \frac{1}{\sqrt{2}}\sum_{i=-l_j}^{l_j}\frac{(-1)^{i+l_j}}{\sqrt{2l_j + 1}}\left(\varphi_{1j}^{(1)}\varphi_{2j}^{(2)} \pm \varphi_{2j}^{(1)}\varphi_{1j}^{(2)}\right),$$

where

$$\varphi_{1j}(\mathbf{r}) = R^M_{n_{1j}l_j}(r)Y_{l_j i}(\mathbf{n}), \qquad \varphi_{2j}(\mathbf{r}) = R^M_{n_{2j}l_j}(r)Y_{l_j,-i}(\mathbf{n}), \tag{5.39}$$

$$\varphi_{1j}^{(1)} = \varphi_{1j}(\mathbf{r}_1)\,, \qquad \varphi_{1j}^{(2)} = \varphi_{1j}(\mathbf{r}_2)\,,$$

and so on. Keep in mind that this basis is not orthonormal, therefore the overlap matrix $S_{kj} = \langle k|j\rangle$ is not diagonal.

We now need to find the matrix elements $\langle k|\hat{\mathsf{H}}|j\rangle$, which consists of three steps: calculation of two matrix elements of simple hydrogen Hamiltonians and then of the matrix elements

$$\langle k|\frac{1}{r_{12}}|j\rangle = \frac{1}{2}\sum_{p=-l_k}^{l_k}\sum_{i=-l_j}^{l_j}\frac{(-1)^{i+l_j+p+l_k}}{\sqrt{(2l_k+1)(2l_j+1)}}$$
$$\times \int_1\int_2 \left(\varphi_{1k}^{(1)*}\varphi_{2k}^{(2)*} \pm \varphi_{2k}^{(1)*}\varphi_{1k}^{(2)*}\right)\frac{1}{r_{12}}\left(\varphi_{1j}^{(1)}\varphi_{2j}^{(2)} \pm \varphi_{2j}^{(1)}\varphi_{1j}^{(2)}\right)\mathrm{d}^3\hat{\boldsymbol{r}}_1\,\mathrm{d}^3\hat{\boldsymbol{r}}_2\,.$$

After distribution of the parentheses, we obtain the integrand as

$$\left(\varphi_{1k}^{(1)*}\varphi_{2k}^{(2)*}\frac{1}{r_{12}}\varphi_{1j}^{(1)}\varphi_{2j}^{(2)} + \varphi_{2k}^{(1)*}\varphi_{1k}^{(2)*}\frac{1}{r_{12}}\varphi_{2j}^{(1)}\varphi_{1j}^{(2)}\right)$$
$$\pm\left(\varphi_{2k}^{(1)*}\varphi_{1k}^{(2)*}\frac{1}{r_{12}}\varphi_{1j}^{(1)}\varphi_{2j}^{(2)} + \varphi_{1k}^{(1)*}\varphi_{2k}^{(2)*}\frac{1}{r_{12}}\varphi_{2j}^{(1)}\varphi_{1j}^{(2)}\right),$$

which can be rearranged by changing the integration variables in the second term in each of the parentheses (one can do so due to the symmetry in the variables $\mathbf{r}_1$ and $\mathbf{r}_2$) to

$$2\varphi_{1j}^{(1)}\varphi_{2j}^{(2)}\frac{1}{r_{12}}\left(\varphi_{1k}^{(1)*}\varphi_{2k}^{(2)*} \pm \varphi_{2k}^{(1)*}\varphi_{1k}^{(2)*}\right).$$

Using the multipole expansion (5.23), changing to spherical coordinates in integrations over $\mathbf{r}_1$ and $\mathbf{r}_2$, and substituting from Eq. (5.39) yield the final (for the time being) form

$$\langle k|\frac{1}{r_{12}}|j\rangle = \sum_{l=0}^{\infty}\sum_{p=-l_k}^{l_k}\sum_{i=-l_j}^{l_j}\sum_{m=-l}^{l}\frac{(-1)^{i+l_j+p+l_k}}{\sqrt{(2l_k+1)(2l_j+1)}} \tag{5.40}$$
$$\times\int\int r_1^2 r_2^2\frac{r_<^l}{r_>^{l+1}}R_{n_{1j}l_j}^{M(1)}R_{n_{2j}l_j}^{M(2)}\left(R_{n_{1k}l_k}^{M(1)}R_{n_{2k}l_k}^{M(2)} \pm R_{n_{2k}l_k}^{M(1)}R_{n_{1k}l_k}^{M(2)}\right)\mathrm{d}r_1\mathrm{d}r_2$$
$$\times\frac{4\pi}{2l+1}\int Y_{lm}^{(1)*}Y_{l_k,-p}^{(1)*}Y_{l_ji}^{(1)}\,\mathrm{d}\Omega_1\int Y_{lm}^{(2)}Y_{l_kp}^{(2)*}Y_{l_j,-i}^{(2)}\mathrm{d}\Omega_2\,.$$

5.3.2 Angular Integration: The Wigner-Eckart Theorem

Four summations, one of which is infinite, and three (though only) two-dimensional integrals are still quite worrying. Fortunately, there is an interesting rule which allows us to calculate the integrals from the properties of the spherical harmonics. This rule is widely known as the *Wigner-Eckart theorem*,[5] and states

$$Y_{l_1,i}(\mathbf{n})Y_{l_2,m-i}(\mathbf{n}) = \sum_l k_l Y_{lm}(\mathbf{n})\,. \tag{5.41}$$

In words: A product of two spherical harmonics can be expressed as a linear combination of other spherical harmonics. The sought 3-Y integrals are then indeed simple to evaluate,

$$\int Y^*_{l'm}Y_{l_1,i}Y_{l_2,m-i}\mathrm{d}\Omega = \sum_l k_l \int Y^*_{l'm}Y_{lm}\mathrm{d}\Omega = \sum_l k_l\delta_{ll'} = k_{l'}\,. \tag{5.42}$$

We will take a minor detour to determine the coefficients k_l. First, we find the coefficients of the inverse problem

$$Y_{lm}(\mathbf{n}) = \sum_i a_i Y_{l_1,i}(\mathbf{n})Y_{l_2,m-i}(\mathbf{n})\,.$$

We apply the operator $\hat{\mathsf{L}}^2$ to this equation and obtain, according to the product rule for differentiation,

$$\begin{aligned}\hat{\mathsf{L}}^2 Y_{lm} &= l(l+1)Y_{lm}\\ &= \sum_i a_i \left\{(\hat{\mathsf{L}}^2 Y_{l_1,i})Y_{l_2,m-i} + Y_{l_1,i}(\hat{\mathsf{L}}^2 Y_{l_2,m-i}) + 2\left(\hat{\mathsf{L}}_j Y_{l_1,i}\right)\left(\hat{\mathsf{L}}_j Y_{l_2,m-i}\right)\right\}.\end{aligned}$$

By comparison of this equation to Eq. (4.19), one comes to the conclusion that the equation for the coefficients a_i is *identical* to Eq. (4.20) for the Clebsch-Gordan coefficients c_i! The coefficients a_i and c_i may still differ by a normalization condition, though. (And they indeed do.) Therefore we write

$$K_{l_1,l_2,l}Y_{lm}(\mathbf{n}) = \sum_i (l_1, i, l_2, m-i|l, m)Y_{l_1,i}(\mathbf{n})Y_{l_2,m-i}(\mathbf{n})\,, \tag{5.43}$$

where K is a number that ensures the correct normalization. The spherical harmonics are normalized by the condition (3.86), though there is no guarantee that the rhs will have the same normalization.

[5] For a different presentation see for example [3, 4].

Example: Consider $l_1 = l_2 = 1$, hence the integral (5.42) differs from zero only for $l = 0, 1, 2$. According to the Wigner-Eckart theorem, it should be

$$\frac{1}{\sqrt{3}}(Y_{1,-1}Y_{11} - Y_{10}Y_{10} + Y_{11}Y_{1,-1}) = -\frac{\sqrt{3}}{4\pi} \sim Y_{00} = \frac{1}{\sqrt{4\pi}},$$

$$\frac{1}{\sqrt{2}}(Y_{1,-1}Y_{11} - Y_{11}Y_{1,-1}) = 0 \sim Y_{10} = \sqrt{\frac{3}{4\pi}}\cos\vartheta\,,$$

$$\frac{1}{\sqrt{6}}(Y_{1,-1}Y_{11} + 2Y_{10}Y_{10} + Y_{11}Y_{1,-1}) = \frac{\sqrt{6}}{8\pi}(3\cos^2\vartheta - 1) \sim Y_{20}$$

$$= -\sqrt{\frac{5}{16\pi}}(1 - 3\cos^2\vartheta)\,,$$

where we substituted from Eqs. (3.91), (3.92) and (3.93) on the lhs, and from Eqs. (3.35) and (4.65) on the rhs. In this case, the theorem seems to hold.

Since the number K is independent of angles, we can calculate it for one particular direction which simplifies the equation, for example $\vartheta = 0$. It holds that[6]

$$Y_{lm}(\vartheta = 0, \varphi) = \sqrt{\frac{2l+1}{4\pi}}\delta_{m0}\,,$$

therefore there will be only one term left from the sum on the rhs of Eq. (5.43), resulting in

$$K_{l_1,l_2,l}\sqrt{\frac{2l+1}{4\pi}} = (l_1, 0, l_2, 0|l, 0)\sqrt{\frac{2l_1+1}{4\pi}}\sqrt{\frac{2l_2+1}{4\pi}}$$

$$\Rightarrow K_{l_1,l_2,l} = \sqrt{\frac{(2l_1+1)(2l_2+1)}{(2l+1)4\pi}}(l_1, 0, l_2, 0|l, 0)\,.$$

We have shown before, see Sect. 4.2, that the transform via the CG coefficients is orthogonal; therefore the inverse relation to Eq. (5.43), Eq. (5.41), can be written as[7]

$$Y_{l_1,i}(\mathbf{n})Y_{l_2,m-i}(\mathbf{n}) = \sum_l K_{l_1,l_2,l}(l_1, i, l_2, m-i|l, m)Y_{lm}(\mathbf{n})\,.$$

[6]This can be deduced from Eqs. (4.62), (4.63), (4.64).

[7]Confront with Eqs. (4.17) and (4.22).

We find the value of the angular integral from the last two equations and the orthonormality relations (3.86)

$$\int Y^*_{l,m}(\mathbf{n})Y_{l_1,i}(\mathbf{n})Y_{l_2,m-i}(\mathbf{n})\,\mathrm{d}\Omega$$
$$= (l_1,i,l_2,m-i|l,m)(l_1,0,l_2,0|l,0)\frac{1}{\sqrt{4\pi}}\sqrt{\frac{(2l_1+1)(2l_2+1)}{2l+1}}\,. \tag{5.44}$$

When dealing with the complex conjugation, we can use the even/odd property of the spherical harmonics in m, Eq. (3.94), $Y_{l,-m} = (-1)^m Y^*_{l,m}$.

Finally, the angular part of the matrix element $\langle k|r_{12}^{-1}|j\rangle$ can be rewritten as

$$\frac{4\pi}{2l+1}\sum_{m=-l}^{l}\sum_{i=-l_j}^{l_j}\sum_{p=-l_k}^{l_k}\frac{(-1)^{i+l_j+l_k+p}}{\sqrt{(2l_j+1)(2l_k+1)}}$$
$$\times\,(-1)^p\int Y^*_{lm}Y_{l_ji}Y_{l_kp}\mathrm{d}\Omega\cdot(-1)^{m+p}\int Y^*_{l,-m}Y_{l_j,-i}Y_{l_k,-p}\mathrm{d}\Omega$$
$$=\frac{4\pi}{2l+1}\sum_{m=-l}^{l}\sum_{i=-l_j}^{l_j}\sum_{p=-l_k}^{l_k}\frac{(-1)^{l_j+l_k}}{\sqrt{(2l_j+1)(2l_k+1)}}$$
$$\times\,\delta_{m,i+p}\frac{1}{4\pi}\frac{(2l_j+1)(2l_k+1)}{2l+1}(l_j,0,l_k,0|l,0)^2$$
$$\times\,(l_j,i,l_k,m-i|l,m)(l_j,-i,l_k,-m+i|l,-m)\,.$$

The last two CG coefficients in the product are the same, hence the sum over i equals 1 due to the normalization of the coefficients, see Eq. (4.21). The Kronecker symbol cancels the summation over p. There are no other terms dependent on m, therefore the sum over m yields $2l+1$. We can then write the integration over angular variables in a very compact form

$$\langle k|\frac{1}{r_{12}}|j\rangle = \sum_{\substack{l=|l_j-l_k|\\ \text{by two}}}^{l_j+l_k}\frac{\sqrt{(2l_j+1)(2l_k+1)}}{2l+1}(-1)^{l_j+l_k}(l_j,0,l_k,0|l,0)^2 \tag{5.45}$$
$$\times\int_1\int_2 r_1^2r_2^2\frac{r_<^l}{r_>^{l+1}}R^{M(1)}_{n_{1j}l_j}R^{M(2)}_{n_{2j}l_j}\left(R^{M(1)}_{n_{1k}l_k}R^{M(2)}_{n_{2k}l_k}\pm R^{M(1)}_{n_{2k}l_k}R^{M(2)}_{n_{1k}l_k}\right)\mathrm{d}r_1\mathrm{d}r_2\,,$$

which—to remind ourselves—holds true for S-states. We can follow the same steps also for other states than the S-states, the resulting expression being then generally more complicated, though. We have thus seen how powerful the Wigner-Eckart theorem is: four summations in Eq. (5.40), one of them being infinite, are reduced to merely one finite summation in Eq. (5.45). Moreover, it transforms the analytical task of calculating integrals of three spherical harmonics to an algebraical task of calculating the CG coefficients.

5.3.3 Angular Integration: Calculation of Reduced Matrix Elements

The only part of the angular integration we are still left to find is the so-called *reduced matrix elements*

$$
\begin{aligned}
(l_1, l_2||l) &= \int Y_{l,0}(\mathbf{n}) Y_{l_1,0}(\mathbf{n}) Y_{l_2,0}(\mathbf{n})\, \mathrm{d}\Omega \\
&= (l_1, 0, l_2, 0|l, 0)^2 \frac{1}{\sqrt{4\pi}} \sqrt{\frac{(2l_1+1)(2l_2+1)}{2l+1}} .
\end{aligned}
$$

In fact, there is nothing to calculate anymore as the needed CG coefficients can be retrieved from the relation (4.20). However, for the fun of it and for better understanding, we will now present a different way of calculating the coefficients $(l_1, l_2||l)$. It follows from Eq. (5.41)

$$
Y_{l_1,0}(\mathbf{n}) Y_{l_2,0}(\mathbf{n}) = \sum_{l=|l_1-l_2|}^{l_1+l_2} (l_1, l_2||l) Y_{l,0}(\mathbf{n}) . \tag{5.46}
$$

After applying the operator $\hat{\mathsf{n}}_3$ to this equation, we have from Eq. (4.62)

$$
\begin{aligned}
&\left(\frac{l_1}{\sqrt{(2l_1-1)(2l_1+1)}} Y_{l_1-1,0} + \frac{l_1+1}{\sqrt{(2l_1+1)(2l_1+3)}} Y_{l_1+1,0} \right) Y_{l_2,0} \\
&= \sum_{l=|l_1-l_2|}^{l_1+l_2} (l_1, l_2||l) \left(\frac{l}{\sqrt{(2l-1)(2l+1)}} Y_{l-1,0} + \frac{l+1}{\sqrt{(2l+1)(2l+3)}} Y_{l+1,0} \right) .
\end{aligned}
$$

Multiplying by $Y_{l',0}$ and integrating over $\mathrm{d}\Omega$ lead together with the orthonormality of the spherical harmonics, Eq. (3.86), to

$$
\begin{aligned}
\frac{l_1+1}{\sqrt{(2l_1+1)(2l_1+3)}} (l_1+1, l_2||l) &= -\frac{l_1}{\sqrt{(2l_1-1)(2l_1+1)}} (l_1-1, l_2||l) \\
&+ \frac{l}{\sqrt{(2l-1)(2l+1)}} (l_1, l_2||l-1) + \frac{l+1}{\sqrt{(2l+1)(2l+3)}} (l_1, l_2||l+1) .
\end{aligned} \tag{5.47}
$$

These equations are solved with the condition

$$
(l_1, l_2||l) = 0 ,
$$

when

$$l_1 < 0, l_2 < 0, l < 0, l < |l_1 - l_2|, l > l_1 + l_2 .$$

Exploiting Eq. (5.47), we lower l_1 to zero. Subsequently, we swap l_1 and l_2 and lower l_2 to zero. The reader can calculate the matrix element $(0, 0||l)$ on his or her own. We thus see that one can use Eqs. (4.62)–(4.64) to calculate the CG coefficients! The reader can easily verify from Eq. (5.47) that

$$(-1)^{l_1+l_2+l} = -1 \Rightarrow (l_1, l_2||l) = 0 . \tag{5.48}$$

This is the reason the summation over l in Eq. (5.45) goes by two.

5.3.4 Calculation of the One-Electron Matrix Elements

The one-electron matrix elements in Eq. (5.37) can be calculated algebraically. For instance,

$$\langle k| \left(-\frac{\nabla_1^2}{2} - \frac{\eta}{r_1} + \frac{1}{2} \right) |j\rangle = \frac{1}{2} \sum_{p=-l_k}^{l_k} \sum_{i=-l_j}^{l_j} \frac{(-1)^{i+l_j+p+l_k}}{\sqrt{(2l_k+1)(2l_j+1)}}$$
$$\times \int_1 \int_2 \left(\varphi_{1k}^{(1)*} \varphi_{2k}^{(2)*} \pm \varphi_{2k}^{(1)*} \varphi_{1k}^{(2)*} \right) \left(-\frac{\nabla_1^2}{2} - \frac{\eta}{r_1} + \frac{1}{2} \right)$$
$$\left(\varphi_{1j}^{(1)} \varphi_{2j}^{(2)} \pm \varphi_{2j}^{(1)} \varphi_{1j}^{(2)} \right) \mathrm{d}^3\mathbf{r}_1 \mathrm{d}^3\mathbf{r}_2 .$$

This expression has a structure of the type

$$(\langle a|^1 \langle b|^2 \pm \langle b|^1 \langle a|^2) \hat{O}_1 \hat{O}_2 (|a\rangle^1 |b\rangle^2 \pm |b\rangle^1 |a\rangle^2)$$
$$= \langle a|\hat{O}_1|a\rangle \langle b|\hat{O}_2|b\rangle + \langle a|\hat{O}_2|a\rangle \langle b|\hat{O}_1|b\rangle \pm (\langle b|\hat{O}_1|a\rangle \langle a|\hat{O}_2|b\rangle$$
$$+ \langle a|\hat{O}_1|b\rangle \langle b|\hat{O}_2|a\rangle) ,$$

where the operators $\hat{O}_1$ and $\hat{O}_2$ act only on the state of the first and second electron, respectively. For this reason, we analyze only the expression

$$\sum_{p=-l_k}^{l_k} \sum_{i=-l_j}^{l_j} \frac{(-1)^{i+l_j+p+l_k}}{\sqrt{(2l_k+1)(2l_j+1)}} \int_1 \int_2 \varphi_{1k}^{(1)*} \varphi_{2k}^{(2)*} \left(-\frac{\nabla_1^2}{2} - \frac{\eta}{r_1} + \frac{1}{2} \right) \varphi_{1j}^{(1)} \varphi_{2j}^{(2)} \mathrm{d}^3\mathbf{r}_1 \mathrm{d}^3\mathbf{r}_2 .$$

Using Eq. (5.39), changing to the spherical coordinates and exploiting the orthonormality of the spherical harmonics, Eq. (3.86), and the relation for the radial functions, Eq. (4.99), the last integral can be cast into the form

$$
\begin{aligned}
&\delta_{l_k,l_j}\int_0^\infty rR^M_{n_{1k},l_k}(r)\left(\frac{r}{2}\left(\hat{\mathsf{p}}_r^2+\frac{l(l+1)}{r^2}+1\right)-\eta\right)R^M_{n_{1j},l_j}(r)\mathrm{d}r\\
&\quad\times\int_0^\infty rR^M_{n_{2k},l_k}(r)rR^M_{n_{2j},l_j}(r)\mathrm{d}r\\
&=\delta_{l_k,l_j}\int_0^\infty rR^M_{n_{1k},l_k}(r)\left(\hat{\mathsf{T}}_3-\eta\right)R^M_{n_{1j},l_j}(r)\mathrm{d}r\\
&\quad\times\int_0^\infty rR^M_{n_{2k},l_k}(r)\left(\hat{\mathsf{T}}_3-(\hat{\mathsf{T}}_++\hat{\mathsf{T}}_-)/2\right)R^M_{n_{2j},l_j}(r)\mathrm{d}r\\
&=\delta_{l_k,l_j}\delta_{n_{1k},n_{1j}}\left(n_{1j}-\eta\right)\left(\delta_{n_{2k},n_{2j}}n_{2j}\right.\\
&\quad-\frac{\sqrt{(n_{2j}-l_j)(n_{2j}+l_j+1)}}{2}\delta_{n_{2k},n_{2j}+1}\\
&\quad\left.-\frac{\sqrt{(n_{2j}+l_j)(n_{2j}-l_j-1)}}{2}\delta_{n_{2k},n_{2j}-1}\right).
\end{aligned}
$$

5.3.5 Radial Integrations

The calculation of the matrix elements of the last term in the Hamiltonian in Eq. (5.37) is slightly more complicated. We introduce the following notation for the radial integrals in Eq. (5.45)

$$
(l_4,l_3,l_2,l_1)=(l_4,l_3,l_2,l_1)^++(l_4,l_3,l_2,l_1)^-,
$$

$$
\begin{aligned}
(l_4,l_3,l_2,l_1)^-\overset{\text{def}}{=}&\int_0^\infty \mathrm{d}r_1 r_1^{2-l-1}R^M_{n_4l_4}(\xi_4,r_1)R^M_{n_3l_3}(\xi_3,r_1)\\
&\times\int_0^{r_1}\mathrm{d}r_2 r_2^{2+l}R^M_{n_2l_2}(\xi_2,r_2)R^M_{n_1l_1}(\xi_1,r_2)
\end{aligned}
\tag{5.49}
$$

and

$$
\begin{aligned}
(l_4,l_3,l_2,l_1)^+\overset{\text{def}}{=}&\int_0^\infty \mathrm{d}r_1 r_1^{2+l}R^M_{n_4l_4}(\xi_4,r_1)R^M_{n_3l_3}(\xi_3,r_1)\\
&\times\int_{r_1}^{\infty}\mathrm{d}r_2 r_2^{2-l-1}R^M_{n_2l_2}(\xi_2,r_2)R^M_{n_1l_1}(\xi_1,r_2).
\end{aligned}
\tag{5.50}
$$

In the next text, we consider a general case with arbitrary screening parameters ξ_1, ξ_2, ξ_3, and ξ_4, see Eq. (4.98). One can return back to the original integral (5.45) by setting

$$
\xi_1=\xi_2=\xi_3=\xi_4=1,\qquad l_4=l_3,\qquad l_2=l_1.
$$

We start from Eqs. (4.76) and (4.77) where we substitute $r \to n\xi r$

$$\frac{\mathrm{d}}{\mathrm{d}r} R^M_{n,l} = \left(-\frac{l+1}{r} + \frac{n\xi}{l}\right) R^M_{n,l} + \frac{\xi\sqrt{n^2-l^2}}{l} R^M_{n,l-1}\,, \tag{5.51}$$

$$\frac{\mathrm{d}}{\mathrm{d}r} R^M_{n,l} = -\left(-\frac{l}{r} + \frac{n\xi}{l+1}\right) R^M_{n,l} - \frac{\xi\sqrt{n^2-(l+1)^2}}{l+1} R^M_{n,l+1}\,. \tag{5.52}$$

Subtracting the last two equations one from another yields

$$r^{-1} R^M_{n,l} = \frac{\xi}{l(l+1)(2l+1)} \Big[+l\sqrt{(n+l+1)(n-l-1)} R^M_{n,l+1} + (l+1)\sqrt{(n+l)(n-l)} R^M_{n,l-1} + (2l+1) n R^M_{n,l} \Big]\,. \tag{5.53}$$

The problem when evaluating the integrals (5.49) and (5.50) is that they generally comprise the functions R^M_{nl} which feature $(n-l-1)$ nodes, i.e., they cross the r-axis $(n-l-1)$ times. From a numerical point of view, a straightforward calculation of such integrals is troublesome since a frequent sign change results in subtracting numbers of a comparable size, hence an operation leading to the so-called numerical instabilities. These instabilities arise from the fact that we represent numbers using a finite number of decimal (or binary) places in numerical calculations. When we, for example, work within the single precision (8 decimal digits) and subtract two numbers which do not differ before the twelfth decimal place, the result is due to the rounding errors almost always completely off. Thus, our strategy will be as follows. We know how to calculate the integrals (5.49) and (5.50) for $l_4 = n_4 - 1$, $l_3 = n_3 - 1$, $l_2 = n_2 - 1$, and $l_1 = n_1 - 1$, see Eqs. (5.33), (5.34), and (5.35). We will subsequently attempt to transform every radial function with $n-l-1$ nodes in the integrals (5.49) and (5.50) to a function without any nodes by using Eqs. (5.51), (5.52), and (5.53). So let us get started!

Multiplying Eq. (5.51) by the function $R^M_{n_0,l_0} r^{p+2} f$, where f is an arbitrary function of r, we obtain

$$-\frac{\xi\sqrt{n^2-l^2}}{l} R^M_{n,l-1} R^M_{n_0,l_0} r^{p+2} f = -\frac{\mathrm{d}}{\mathrm{d}r}\left(R^M_{n,l}\right) R^M_{n_0,l_0} r^{p+2} f + \left(-\frac{l+1}{r} + \frac{n\xi}{l}\right) R^M_{n,l} R^M_{n_0,l_0} r^{p+2} f\,. \tag{5.54}$$

From the Leibnitz rule for differentiation of a product, we have

$$\frac{\mathrm{d}}{\mathrm{d}r}\left(R^M_{n,l}\right) R^M_{n_0,l_0} r^{p+2} f = \frac{\mathrm{d}}{\mathrm{d}r}\left(R^M_{n,l} R^M_{n_0,l_0} r^{p+2} f\right) - \frac{\mathrm{d}}{\mathrm{d}r}\left(R^M_{n_0,l_0}\right) R^M_{n,l} r^{p+2} f - \frac{\mathrm{d}}{\mathrm{d}r}\left(r^{p+2} f\right) R^M_{n,l} R^M_{n_0,l_0}\,.$$

We now substitute this equation into the previous one and use Eq. (5.52) with $n = n_0$, $l = l_0$, $\xi = \xi_0$ for the calculation of $\frac{\mathrm{d}}{\mathrm{d}r}\left(R^M_{n_0,l_0}\right)$. We can eliminate the multiplication by r^{-1} with the aid of Eq. (5.53). With a little effort, we then obtain from Eq. (5.54) a new equation

$$\begin{aligned}
&-\frac{\xi\sqrt{n^2-l^2}}{l}R^M_{n,l-1}R^M_{n_0,l_0}r^{p+2}f\left(-1+\frac{l+1-l_0-p-2}{2l+1}\right)=\frac{\mathrm{d}}{\mathrm{d}r}\left(R^M_{n,l}R^M_{n_0,l_0}r^{p+2}f\right)\\
&+\left(\frac{\xi_0 n_0}{l_0+1}-\frac{\xi n}{l}+\frac{\xi n(l+1-l_0-p-2)}{l(l+1)}\right)R^M_{n,l}R^M_{n_0,l_0}r^{p+2}f\\
&+\frac{\xi(l+1-l_0-p-2)}{2l+1}\frac{\sqrt{n^2-(l+1)^2}}{l+1}R^M_{n,l+1}R^M_{n_0,l_0}r^{p+2}f\\
&+\frac{\xi_0\sqrt{n_0^2-(l_0+1)^2}}{l_0+1}R^M_{n,l}R^M_{n_0,l_0+1}r^{p+2}f-\frac{\mathrm{d}f}{\mathrm{d}r}R^M_{n,l}R^M_{n_0,l_0}r^{p+2}\,. \qquad (5.55)
\end{aligned}$$

We then substitute into the last equation

$$f=1\,,\quad l=l_2\,,\quad n=n_2\,,\quad \xi=\xi_2\,,\quad l_0=l_1\,,$$
$$n_0=n_1\,,\quad \xi_0=\xi_1\,,\quad p=l_\pm\,,\quad r=r_2\,,$$

where

$$l_+=-l-1\,,\qquad l_-=l\,,$$

integrate the equation from r_1 to ∞ for l_+ and from 0 do r_1 for l_- and finally multiply both sides of the equation by

$$r_1^{2-l_\pm}R^M_{n_4,l_4}(\xi_4,r_1)R^M_{n_3,l_3}(\xi_3,r_1)\,.$$

We subsequently integrate over r_1 from 0 to infinity to obtain recurrence relations for the integrals $(l_4,l_3,l_2,l_1)^\pm$, which raise the quantum number l_2,

$$\begin{aligned}
&-(l_4,l_3,l_2-1,l_1)^\pm\frac{\xi_2\sqrt{n_2^2-l_2^2}}{l_2}\left(-1+\frac{l_2+1-l_1-2-l_\pm}{2l_2+1}\right)\\
&=\mp[l_4,l_3,l_2,l_1]+\left(\frac{\xi_1 n_1}{l_1+1}-\frac{\xi_2 n_2}{l_2}+\frac{\xi_2 n_2(l_2+1-l_1-l_\pm-2)}{l_2(l_2+1)}\right)(l_4,l_3,l_2,l_1)^\pm\\
&+\frac{\xi_2(l_2+1-l_1-l_\pm-2)}{2l_2+1}\frac{\sqrt{n_2^2-(l_2+1)^2}}{l_2+1}(l_4,l_3,l_2+1,l_1)^\pm
\end{aligned}$$

$$+\frac{\xi_1\sqrt{n_1^2-(l_1+1)^2}}{l_1+1}(l_4,l_3,l_2,l_1+1)^{\pm}\,, \tag{5.56}$$

where

$$[l_4,l_3,l_2,l_1] \stackrel{\text{def}}{=} \int_0^\infty r^3 R^M_{n_4 l_4}(\xi_4,r)R^M_{n_3 l_3}(\xi_3,r)R^M_{n_2 l_2}(\xi_2,r)R^M_{n_1 l_1}(\xi_1,r)\,\mathrm{d}r\,.$$

Substituting into Eq. (5.55) for

$$f(r_1)=\int_{d_\pm}^{h^\pm} r_2^{2+l_\pm} R^M_{n_2,l_2}(\xi_2,r_2)R^M_{n_1,l_1}(\xi_1,r_2)\,\mathrm{d}r_2\,,$$

where

$$d_\pm=\begin{cases} r_1\,,\\ 0\,,\end{cases} \qquad h_\pm=\begin{cases} \infty\,,\\ r_1\,,\end{cases}$$

$$l=l_4\,,\quad n=n_4\,,\quad \xi=\xi_4\,,\quad l_0=l_3\,,\quad n_0=n_3\,,\quad \xi_0=\xi_3\,,\quad p=l_\mp\,,\quad r=r_1\,,$$

and integrating over r_1 from 0 to infinity yields recurrence relations for the integrals $(l_4,l_3,l_2,l_1)^\pm$ which raise the quantum number l_4:

$$\begin{aligned}
&-(l_4-1,l_3,l_2,l_1)^{\pm}\frac{\xi_4\sqrt{n_4^2-l_4^2}}{l_4}\left(-1+\frac{l_4+1-l_3-2-l_\mp}{2l_4+1}\right)\\
&=\pm[l_4,l_3,l_2,l_1]+\left(\frac{\xi_3 n_3}{l_3+1}-\frac{\xi_4 n_4}{l_4}+\frac{\xi_4 n_4(l_4+1-l_3-l_\mp-2)}{l_4(l_4+1)}\right)(l_4,l_3,l_2,l_1)^{\pm}\\
&\quad+\frac{\xi_4(l_4+1-l_3-l_\mp-2)}{2l_4+1}\frac{\sqrt{n_4^2-(l_4+1)^2}}{l_4+1}(l_4+1,l_3,l_2,l_1)^{\pm}\\
&\quad+\frac{\xi_3\sqrt{n_3^2-(l_3+1)^2}}{l_3+1}(l_4,l_3+1,l_2,l_1)^{\pm}\,.
\end{aligned} \tag{5.57}$$

Finally, we substitute into Eq. (5.55) for

$$f(r)=R^M_{n_2,l_2}(\xi_2,r)R^M_{n_1,l_1}(\xi_1,r)\,,$$

$$l=l_4\,,\quad n=n_4\,,\quad \xi=\xi_4\,,\quad l_0=l_3\,,\quad n_0=n_3\,,\quad \xi_0=\xi_3\,,\quad p=1$$

and integrate the whole equation over r from 0 to infinity, and thus obtain recurrence relations for one-dimensional integrals $[l_4,l_3,l_2,l_1]$ which raise the quantum number l_4,

$$
\begin{aligned}
&- [l_4-1,l_3,l_2,l_1]\frac{\xi_4\sqrt{n_4^2-l_4^2}}{l_4}\left(-1+\frac{l_4-l_1-l_2-l_3-2}{2l_4+1}\right) \\
&= \left(\frac{\xi_3 n_3}{l_3+1}+\frac{\xi_2 n_2}{l_2+1}+\frac{\xi_1 n_1}{l_1+1}-\frac{\xi_4 n_4}{l_4}+\frac{\xi_4 n_4(l_4-l_3-l_2-l_1-2)}{l_4(l_4+1)}\right)[l_4,l_3,l_2,l_1] \\
&+ \frac{\xi_4(l_4-l_3-l_2-l_1-2)}{2l_4+1}\frac{\sqrt{n_4^2-(l_4+1)^2}}{l_4+1}[l_4+1,l_3,l_2,l_1] \\
&+ \frac{\xi_3\sqrt{n_3^2-(l_3+1)^2}}{l_3+1}[l_4,l_3+1,l_2,l_1]-\frac{\xi_2\sqrt{n_2^2-(l_2+1)^2}}{l_2+1}[l_4,l_3,l_2+1,l_1] \\
&+ \frac{\xi_1\sqrt{n_1^2-(l_1+1)^2}}{l_1+1}[l_4,l_3,l_2,l_1+1]\,. \qquad (5.58)
\end{aligned}
$$

Equations (5.56), (5.57), and (5.58) are solved with the condition

$$
(l_4,l_3,l_2,l_1)^{\pm}=0\,, \qquad [l_4,l_3,l_2,l_1]=0\,,
$$

when

$$
\begin{gathered}
l_1>n_1-1\,,\quad l_2>n_2-1\,,\quad l_3>n_3-1\,,\quad l_4>n_4-1\,,\\
l_1<0\,,\quad l_2<0\,,\quad l_3<0 \quad\text{or}\quad l_4<0\,.
\end{gathered}
$$

The algorithm for calculation of the integrals $(l_4,l_3,l_2,l_1)^{\pm}$ is the following. We raise l_2 to n_2-1 using Eq. (5.56), then swap l_1 and l_2 and raise l_1 to n_1-1. Next, we raise l_4 to n_4-1 using Eq. (5.57), then swap l_4 and l_3 and raise l_3 to n_3-1. Subsequently, we raise l_4 to n_4-1 using Eq. (5.58), then swap l_4 and l_3 and raise l_3 to n_3-1, then swap l_3 and l_2 and raise l_2 to n_2-1. Finally, we swap l_2 and l_1 and raise l_1 to n_1-1. As noted above, the integrals $(n_4-1,n_3-1,n_2-1,n_1-1)^{\pm}$ are already given, see Eq. (5.33), (5.34) and (5.35). The reader can calculate the one-dimensional integrals $[n_4-1,n_3-1,n_2-1,n_1-1]$ by himself.

This procedure may appear much worse than it actually is. In fact, it is quite simple[8] to program the relations (5.56), (5.57), and (5.58) in a form of recursive procedures.

[8] So simple that the even authors managed it.

5.3.6 Convergence of the Variational Method

We have thus finally developed all of the required tools for the calculation of the elements of the Hamiltonian and the overlap matrices. All that is still to be done is to assign a set of three quantum numbers $\{n_{1i}, n_{2i}, l_i\}$ to each of the basis states $|i\rangle$. One may choose from several approaches, we will opt for the one listed in the table below. We create state shells with $n_{12} = n_{1i} + n_{2i}$ where $n_{1i} \leq n_{2i}$, and let l_i run from 0 to $n_{1i} - 1$ for every value of n_{1i} and n_{2i}. The obtained variational energies are presented in Table 5.5. N denotes the total number of states used in the calculation, see Eq. (2.4) and Table 5.4.

So far, we always restricted ourselves to finding the variational estimate of the ground state of given symmetry. However, the method can be applied to excited states of a given symmetry as well. If we are interested in the first excited singlet S-state, we optimize the parameter η so as the energy of the *second smallest* eigenvalue is as small as possible. The results are, again, summarized in Table 5.5. The exact nonrelativistic values were adopted from [2].

Exercise 16: Helium Ground State

Using a programming language of your choice, program the variational calculation of the ground state of helium atom. Use built-in procedures available in the chosen language to find the eigenvalues. You will do fine already with a simple—non-generalized—eigenvalue problem as the overlap matrix is invertible. For the optimization of the parameter η search the internet (e.g.) for the method of golden section search.

Table 5.4 Assignment of quantum numbers to the basis states

i	n_{1i}	n_{2i}	l_i	n_{12}
1	1	1	0	2
2	1	2	0	3
3	1	3	0	4
4	2	2	0	4
5	2	2	1	4
6	1	4	0	5
7	2	3	0	5
8	2	3	1	5
9	1	5	0	6
10	2	4	0	6
11	2	4	1	6
12	3	3	0	6
13	3	3	1	6
14	3	3	2	6

5.3.7 Comparison with the Experiment

To ascertain the difference of our calculations from the correct values as well as to assess the therein neglected effects, namely the relativistic and magnetic effects and nuclear motion discussed in Sects. 3.6.2 and 3.6.3, we compare the prediction for the frequency corresponding to the transition between the lowest singlet S-states of helium obtained from the variational calculation with 70 functions, see Table 5.5,

$$\nu_{\text{var70}}(2\,{}^1S - 1\,{}^1S) = 2R_\infty c\,(2.903127 - 2.145694) = 4.983670 \times 10^{15}\text{Hz}\,,$$

and also from the exact nonrelativistic limit with nuclear motion neglected, i.e., from the exact solution of the Schrödinger equation (5.3),

$$\nu_{\text{nr}}(2\,{}^1S - 1\,{}^1S) = 2R_\infty c\,(2.903724 - 2.145974) = 4.985755 \times 10^{15}\text{Hz}\,,$$

with the experimental value [1]

$$\nu_{\text{exp}}(2\,{}^1S - 1\,{}^1S) = 4.984872315(48) \times 10^{15}\text{Hz}\,.$$

It follows from this comparison that not only the fundamental principles, but also the approximate methods of quantum mechanics are correct. In addition, one can notice that a significant improvement of the variational calculation does not lead to a better match to the experiment. The reason is obvious: the Schrödinger equation (5.3) does not capture all phenomena occurring in the real helium atom. Note that the relativistic and magnetic effects discussed in Sect. 3.6.2 lead to the fine structure of the helium energy levels. For instance, the state $2\,{}^3P^o$ splits into three $2\,{}^3P^o_J$, where $J = 0, 1, 2$ denotes three possible total angular momenta of electrons which arise

Table 5.5 Variational energies of the two lowest singlet helium S-states

		$1\,{}^1S$		$2\,{}^1S$	
n_{12}	N	η	$E_{\text{var}}(\eta)$	η	$E_{\text{var}}(\eta)$
2	1	1.185	−2.847656		
3	2	1.185	−2.847656	1.659	−1.452885
4	5	0.971	−2.895444	1.875	−2.031260
5	8	0.940	−2.897109	1.439	−2.049180
6	14	0.796	−2.900714	1.833	−2.134045
7	20	0.760	−2.901452	1.607	−2.134897
8	30	0.682	−2.902341	1.902	−2.144425
9	40	0.648	−2.902654	1.674	−2.144749
10	55	0.595	−2.902975	1.788	−2.145552
11	70	0.566	−2.903127	1.568	−2.145694
∞	∞		−2.903724		−2.145974

from the composition of the total orbital angular momentum $L = 1$ of the electrons and their total spin angular momentum $S = 1$.[9]

5.3.8 A Note on the Parity

One can follow the same reasoning for states of other symmetries as well. In case of a general state with its total angular momentum l (and the total $m = 0$), we consider the following symmetry-adapted basis states

$$|j\rangle = \frac{1}{\sqrt{2}}\left(|n_{1j}, l_{1j}\rangle^{(1)}|n_{2j}, l_{2j}\rangle^{(2)} \sum_{i=-l_{1j}}^{l_{1j}} (l_{1j}, i, l_{2j}, -i|l, 0)|l_{1j}, i\rangle^{(1)}|l_{2j}, -i\rangle^{(2)} \right. \tag{5.59}$$

$$\left. \pm|n_{2j}, l_{2j}\rangle^{(1)}|n_{1j}, l_{1j}\rangle^{(2)} \sum_{i=-l_{1j}}^{l_{1j}} (l_{2j}, -i, l_{1j}, i|l, 0)|l_{2j}, -i\rangle^{(1)}|l_{1j}, i\rangle^{(2)}\right),$$

instead of those given by Eq. (5.38). For example, for P-states, i.e., states with the total angular momentum $l = 1$, we find from the condition $|l_{1j} - l_{2j}| \leq 1$ that either $l_{2j} = l_{1j} + 1$, or $l_{2j} = l_{1j}$. Therefore, we compose the P-state from pairs of states in Table 5.6.

From Eq. (5.44) and the condition (5.48), we derive that $\langle j|\hat{\mathsf{H}}|k\rangle = 0$ when

$$(-1)^{l_{1j}+l_{2j}} \neq (-1)^{l_{1k}+l_{2k}}.$$

In other words, the P-states with $l_{2j} = l_{1j} + 1$ and the P-states with $l_{2j} = l_{1j}$ do not interact with each other; the matrix elements between these two classes of states equal zero.

This finding relates to the parity operator. It follows from Eqs. (4.62), (4.63), and (4.64) that spherical harmonics obey the relation

$$Y_{l,m}(-\mathbf{n}) = (-1)^l Y_{l,m}(\mathbf{n}). \tag{5.60}$$

Table 5.6 One electron orbital quantum numbers used for composition of odd- and even-parity P-states in Eq. (5.59)

Even parity		Odd parity	
l_{1j}	l_{2j}	l_{1j}	l_{2j}
0	1	1	1
1	2	2	2
2	3	3	3
Etc.			

[9]Confront with the fine structure of positronium, see Exercise 10 in Sect. 4.4.9.

When the coordinates are inverted, $\mathbf{r}_1 \to -\mathbf{r}_1$, $\mathbf{r}_2 \to -\mathbf{r}_2$, the basis states (5.59) behave as

$$\langle -\mathbf{r}_1, -\mathbf{r}_2 | j \rangle = (-1)^{l_{1j}+l_{2j}} \langle \mathbf{r}_1, \mathbf{r}_2 | j \rangle \,.$$

One can thus readily see that the P-states with $l_{2j} = l_{1j} + 1$ are of odd parity, while the P-states with $l_{2j} = l_{1j}$ of even parity. Owing to the fact that parity is an integral of motion, matrix elements of the Hamiltonian between these two classes of states equal zero.

5.3.9 A Note on Complex Atoms

Following the same procedure that led to the Hamiltonian (5.1), one can derive for instance for the Hamiltonian of lithium-like atom in nonrelativistic approximation with neglect of the nuclear motion

$$\frac{\hat{\mathsf{H}}}{Z^2} = \left(\frac{1}{2}\hat{\mathsf{p}}_1^2 - \frac{1}{r_1}\right) + \left(\frac{1}{2}\hat{\mathsf{p}}_2^2 - \frac{1}{r_2}\right) + \left(\frac{1}{2}\hat{\mathsf{p}}_3^2 - \frac{1}{r_3}\right) + \frac{1}{Z}\left(\frac{1}{r_{12}} + \frac{1}{r_{23}} + \frac{1}{r_{31}}\right). \tag{5.61}$$

The variational calculation is then similar to the one for helium-like atoms. For example, we opt for the ground-state test function in the form of a so-called *Slater determinant*; that is, the determinant of the matrix

$$\frac{1}{\sqrt{3!}} \begin{pmatrix} |1s\rangle^{(1)}|+\rangle^{(1)} & |1s\rangle^{(1)}|-\rangle^{(1)} & |2s\rangle^{(1)}|+\rangle^{(1)} \\ |1s\rangle^{(2)}|+\rangle^{(2)} & |1s\rangle^{(2)}|-\rangle^{(2)} & |2s\rangle^{(2)}|+\rangle^{(2)} \\ |1s\rangle^{(3)}|+\rangle^{(3)} & |1s\rangle^{(3)}|-\rangle^{(3)} & |2s\rangle^{(3)}|+\rangle^{(3)} \end{pmatrix},$$

where the one-electron orbitals $\langle \mathbf{r} |\, 1s \rangle$ and $\langle \mathbf{r} |\, 2s \rangle$ are given by Eqs. (5.11) and (5.36). This choice ensures that the wave function is totally antisymmetric with respect to the interchange of any two variables. The variational estimate can be systematically improved by the expansion of the wave function into a linear combination of Slater determinants, i.e., the determinants of matrices

$$\frac{1}{\sqrt{3!}} \begin{pmatrix} |1i\rangle^{(1)}|+\rangle^{(1)} & |2i\rangle^{(1)}|-\rangle^{(1)} & |3i\rangle^{(1)}|+\rangle^{(1)} \\ |1i\rangle^{(2)}|+\rangle^{(2)} & |2i\rangle^{(2)}|-\rangle^{(2)} & |3i\rangle^{(2)}|+\rangle^{(2)} \\ |1i\rangle^{(3)}|+\rangle^{(3)} & |2i\rangle^{(3)}|-\rangle^{(3)} & |3i\rangle^{(3)}|+\rangle^{(3)} \end{pmatrix}, \tag{5.62}$$

where $\varphi_{1i} = \langle \mathbf{r} |\, 1i \rangle$, etc., stand for one-electron orbitals (5.39). Unfortunately, these basis states are generally not the eigenstates of the total $\hat{\mathsf{S}}^2$ nor of the total $\hat{\mathsf{L}}^2$. In case we aim for the configuration interaction method, it proves advantageous to find appropriate linear combinations of these states (5.62) so that they become the eigenstates. However, it is anything but trivial to satisfy the requirement the basis states be both totally antisymmetric with respect to the interchange of an arbitrary pair of electrons, *and* they be the eigenstates of the total $\hat{\mathsf{S}}^2$ and $\hat{\mathsf{L}}^2$. If the reader is interested in such a procedure, we refer him or her to, e.g., [3]. Note also that it shows advantageous to exploit the formalism of the second quantization, described in Sect. 6.6, when calculating the matrix elements of the Hamiltonian (5.61) between the states (5.62).

5.4 Final Notes

We have devoted considerable attention to the method of employing the variational function constituted by one-electron functions when searching for a sufficiently accurate energy estimate. Note, though, that generally for *S*-states and especially for the ground state, one usually cannot obtain this way results with the relative error being significantly smaller than one part in 10^5. The reason is that the wave function in the form (5.38) does not accurately describe the behavior of the exact wave function in the vicinity of the point where the inter-electron distance r_{12} approaches zero. This point $r_{12} = 0$ as well as those $r_1 = 0$ and $r_2 = 0$ are of great importance, though, since the potential energy diverges there, see Eq. (5.37). If the wave function ψ is to be continuous everywhere and solve the Schrödinger equation, the kinetic energy must diverge at these points as well, though with an opposite sign to cancel out the potential term. To accurately capture the behavior of the wave function in the vicinity of $r_{12} = 0$, one needs to switch to the collective coordinates of the electrons. The reader can find a clear treatment of behavior of the wave function in the vicinity of points with diverging potential, i.e., points $r_1 = 0$, $r_2 = 0$, $r_{12} = 0$, and references to the literature that uses this wave function for construction of the variational function in [5]. The wave functions dependent on the collective coordinates of the electrons are termed as *explicitly correlated* functions. With a suitable choice of these functions, one can reach estimates of a significantly greater accuracy than one needs for the inclusion of the relativistic and other corrections, and for comparison with experiment, see, e.g., [2].

The disadvantage of the collective electron coordinates is that their generalization to more than two-electron atoms is very complicated, and to more than four-electron atoms practically inapplicable. Moreover, the relative accuracy of 1 part in 10^5 suffices well for most applications. On the other hand, the method based on the search for the wave function in the form of a linear combination of products of one-electron functions allows for highly accurate calculations of atoms with a large number of electrons. For instance, the accuracy reached for the cesium atom (55 electrons) amounts to less than 1% [6].

We have shown that from the relations for the radial functions, Eqs. (5.51) and (5.52), obtained from the conservation of the Runge-Lenz vector, one can derive relations for the integrals of these functions, Eqs. (5.56) and (5.57). The relations (5.51) and (5.52) can be further combined with the relations for the radial functions arising from the algebra of the radial operators. From the relations for radial functions built in this way, it is then possible to derive more relations for the integrals (5.49) and (5.50). If the reader is interested therein, we refer him or her to [8].

References

1. S.D. Bergeson et al., Phys. Rev. Lett. **80**, 3475 (1998)
2. G.W.F. Drake, ed., *Springer Handbook of Atomic, Molecular and Optical Physics* (Springer, Berlin, 2006)
3. I. Lindgren, J. Morrison, *Atomic Many-Body Theory* (Springer, Berlin, 1982)
4. A. Messiah, *Quantum Mechanics* (North Holland, Amsterdam, 1961)
5. C.R. Myers et al., Phys. Rev. A **44**, 5537 (1991)
6. S.G. Porsev et al., Phys. Rev. D **82**, 036008 (2010)
7. J. Schwinger et al., *Classical Electrodynamics* (Perseus Books, Cambridge, 1998)
8. J. Zamastil, F. Vinette, M. Šimánek, Phys. Rev. A **75**, 022506 (2007)

Chapter 6
Dynamics: The Nonrelativistic Theory

In this chapter, we introduce the basic principles and applications of quantum electrodynamics while describing the motion of a particle within the nonrelativistic approximation. We start by showing how classical electrodynamics may be cast into the Hamilton formalism, the subsequent transition from classical to quantum electrodynamics being then straightforward. Next, we focus on the quantization of a free electromagnetic field wherein the abstract solution for harmonic oscillator based on the introduction of non-Hermitian ladder operators plays a decisive role.

The problem of mutual interaction of atoms and a quantized electromagnetic field is complicated and in general unsolved. Owing to the fact that the coupling between the charged particles and the quantized electromagnetic field is proportional to the fine structure constant $\alpha \simeq 1/137.036$, for most, though not all, practical purposes it is sufficient to consider this interaction in a perturbative manner. The first order of the perturbation method suffices to describe spontaneous emission and photoelectric effect, while the second order is needed to capture photon-atom scattering. At this order, one obtains also the so-called *virtual processes*: processes when the radiated photons are not registered in a detector, but absorbed by one of the involved particles instead, i.e., either by the same particle that radiated them or by another charged particle.

The exchange of one photon between two charged particles may be described with sufficient accuracy within the nonrelativistic approximation. In case the retardation is neglected, we arrive at the magnetic interactions discussed in Chap. 3. We will show in the following text how to take into account the effect of retardation.

There are two additional virtual processes which are of the same importance as the retardation effect in one-photon exchange, namely the self-energy effect and two-photon exchange. The former comprises an emission and subsequent absorption of a photon by the same particle and the latter is an exchange of two photons between two charged particles. One can deal with these two processes within the nonrelativistic approximation only in the region of low frequencies of virtual photons. In order

J. Zamastil, J. Benda, *Quantum Mechanics and Electrodynamics*,
DOI 10.1007/978-3-319-65780-6_6

to develop a more complete description of the phenomena, we need to include the relativistic theory, as we will do in the next chapter.

To conclude this chapter, we will show that electrons can be viewed as quanta of an electron field, and emphasize the advantages of such a view.

6.1 Quantization of the Electromagnetic Field

6.1.1 *Why Quantize?*

Until now, we considered the electromagnetic (EM) field as a static quantity without its own dynamics, which is clearly an approximation. The EM field should be viewed as a dynamic and evolving quantity with its own energy, momentum, and angular momentum instead. The universality of the quantum-mechanical laws consequently leads to the quantization of these characteristics. Furthermore, there is now a large experimental evidence of the EM field behaving in accordance with quantum-mechanical laws. The classical experiments include for instance the black-body radiation, photoelectric effect, Raman and Compton scattering, and so on.

6.1.2 *How to Quantize?*

If we know the classical behavior of a given physical system, we find its quantum behavior by the application of the procedure of canonical quantization, confront with Sect. 1.3.2.[1] First, we determine the classical Hamiltonian H and the canonical coordinates q_k and momenta p_k, i.e., quantities obeying the Hamilton canonical equations

$$\frac{\partial H}{\partial q_k} = -\dot{p}_k\,, \qquad \frac{\partial H}{\partial p_k} = \dot{q}_k\,. \tag{6.1}$$

To obtain the quantum Hamiltonian, we substitute in the classical Hamiltonian for the number-valued time-dependent functions $q_k(t)$ and $p_k(t)$ the time-independent operators $\hat{\mathsf{q}}_k$ and $\hat{\mathsf{p}}_k$ obeying the canonical commutation relation

$$[\hat{\mathsf{p}}_k, \hat{\mathsf{q}}_j] = -\mathrm{i}\delta_{kj}\,. \tag{6.2}$$

Once we have the Hamilton operator, the remaining task is "merely" to solve the Schrödinger equation and to determine the observable consequences of the theory.

[1] The exposition given in this Sect. 6.1 is inspired by that given in [12].

6.1.3 Classical Electrodynamics in Conventional Formalism

What is classical electrodynamics? It is a theory describing the interaction between charged particles. This interaction is captured by two sets of equations.

The Maxwell equations constitute the first set. Given the densities of electrical charge and current, these equations determine the electric and magnetic fields, **E** and **B**,

$$\nabla \times \mathbf{E} + \frac{\partial \mathbf{B}}{\partial t} = 0\,, \qquad \nabla \cdot \mathbf{B} = 0\,, \tag{6.3}$$

$$\nabla \times \mathbf{B} - \frac{\partial \mathbf{E}}{\partial t} = \mathbf{j}\,, \qquad \nabla \cdot \mathbf{E} = \rho\,. \tag{6.4}$$

The total densities of charge and current on the rhs of the equations are given as a sum of the densities of charge and current of elementary particles which we deem as structureless and dimensionless points,

$$\rho(\mathbf{r}, t) = \sum_{n=1}^{N} e_n \delta(\mathbf{r} - \mathbf{q}^{(n)}(t))\,, \qquad \mathbf{j}(\mathbf{r}, t) = \sum_{n=1}^{N} e_n \dot{\mathbf{q}}^{(n)}(t) \delta(\mathbf{r} - \mathbf{q}^{(n)}(t))\,. \tag{6.5}$$

Here, N denotes the number of the charged particles "in the Universe" and e_n represents their charges.

The Newton equation with the Lorentz force forms the second set. Given the electric and magnetic fields, **E** and **B**, this equation determines the motion of charged particles, namely their coordinates and momenta, $\mathbf{q}^{(n)}$ and $\dot{\mathbf{q}}^{(n)}$, respectively,

$$m\ddot{\mathbf{q}}^{(n)} = e_n(\mathbf{E} + \dot{\mathbf{q}}^{(n)} \times \mathbf{B})\,, \tag{6.6}$$

where **E** and **B** are evaluated at the point $\mathbf{q}^{(n)}$. Once the coordinates and momenta of the particles are known, Eq. (6.5) yields charge and current densities.

Why do we describe the electromagnetic interaction in this very way, namely via the electromagnetic field? Why do not we solve the Maxwell equations (6.3) and (6.4) and insert for **E** and **B** into the Newton equation (6.6)? We would thus find the action of a single charge, albeit with retardation, on another charge. Hence, one would not need to consider the EM field at all. The reason why we bear the notion of the EM field is that the EM field possesses its own energy, momentum, and angular momentum. Were the notion of EM field eliminated and were we to keep track of the motion of charged particles, energy, momentum, and angular momentum would not be locally conserved quantities.

Within the classical theory, it is *advantageous*, although not at all necessary, to introduce the scalar and vector electromagnetic potentials φ and **A**, respectively

$$\mathbf{E} = -\frac{\partial \mathbf{A}}{\partial t} - \nabla\varphi\,, \qquad \mathbf{B} = \nabla \times \mathbf{A}\,. \tag{6.7}$$

Exploiting the identities of vector analysis, $\nabla \cdot (\nabla \times \mathbf{A}) = 0$ and $\nabla \times (\nabla \varphi) = 0$, the first pair of the Maxwell equations is solved regardless of the functional form of φ and $\mathbf{A}$. If we insert (6.7) into the second pair (6.4), we obtain

$$\left(\frac{\partial^2}{\partial t^2} - \nabla^2\right)\mathbf{A} + \nabla\left(\nabla \cdot \mathbf{A} + \frac{\partial \varphi}{\partial t}\right) = \mathbf{j}\,, \qquad -\nabla^2\varphi - \frac{\partial}{\partial t}\nabla \cdot \mathbf{A} = \rho\,, \tag{6.8}$$

where we used the identity $\nabla \times (\nabla \times \mathbf{A}) = \nabla\nabla \cdot \mathbf{A} - \nabla^2 \mathbf{A}$.

On the other hand, it is *necessary* to introduce these potentials upon transition to the quantum theory. There are two reasons. Firstly, from a formal point of view, their introduction is requisite for the Hamiltonian formulation of classical electrodynamics. Secondly, from a physical point of view, the Bohm-Aharonov experiment (see, e.g., [8]) shows that the potentials $\mathbf{A}$ and φ appear to be "more fundamental" (in a sense discussed in [8]) than the electric and magnetic fields $\mathbf{E}$ and $\mathbf{B}$. Apparently, the Hamiltonian formalism must be the right one to proceed with as it enforces the introduction of these potentials.

6.1.4 Gauge Invariance and Number of Degrees of Freedom

Equations (6.8) determine the electromagnetic potentials but for a *gauge transformation*

$$\mathbf{A} = \mathbf{A}' + \nabla \chi\,, \qquad \varphi = \varphi' - \frac{\partial \chi}{\partial t}\,, \tag{6.9}$$

where χ is an arbitrary "reasonable" function of space and time. Likewise, irrespective of whether the original or primed potentials are used, one obtains the same electric and magnetic fields $\mathbf{E}$ and $\mathbf{B}$ from Eq. (6.7). This gauge invariance results in a mismatch between the number of functions used for the mathematical description of the EM field and the number of physical degrees of freedom of the EM field. Namely, there are four functions (4=3+1, 3 components of the vector $\mathbf{A}$ + 1 scalar φ) used for the mathematical description. On the other hand, as is widely known, the EM field features only two polarizations. This is of an immediate concern for us; only the dynamical, evolving part of the field is to be quantized. This part has to be determined prior to the quantization, otherwise we would run into problems; for the static part of the field the associated canonical momentum is identically zero. Before we proceed further, we need to determine the dynamic and the static parts of the field.

6.1.5 Coulomb Gauge

The simplest way to determine the dynamic part of the field is to use the Coulomb gauge[2]

$$\nabla \cdot \mathbf{A} = 0 . \tag{6.10}$$

In this gauge, the Maxwell equations (6.8) take the form

$$\left(\frac{\partial^2}{\partial t^2} - \nabla^2\right) \mathbf{A}(\mathbf{r}, t) = \mathbf{j}(\mathbf{r}, t) - \nabla \frac{\partial \varphi}{\partial t}(\mathbf{r}, t) , \qquad -\nabla^2 \varphi(\mathbf{r}, t) = \rho(\mathbf{r}, t) . \tag{6.11}$$

The equation for the scalar potential has a character of a static problem, hence may be solved "once and for all":

$$\varphi(\mathbf{r}, t) = \frac{1}{4\pi} \int \frac{\rho(\mathbf{r}', t)}{|\mathbf{r} - \mathbf{r}'|} \, \mathrm{d}^3\mathbf{r}' = \frac{1}{4\pi} \sum_{n=1}^{N} \frac{e_n}{|\mathbf{r} - \mathbf{q}^{(n)}(t)|} , \tag{6.12}$$

where we used Eq. (6.5) in the second equality. This way, the scalar potential can be completely eliminated from the theory. There is one constraint (6.10) for the three remaining components of the vector potential, which yields the two independent physical degrees of freedom mentioned above.

In order to better understand the static and the dynamic part of the field within the Coulomb gauge, we now decompose the electric field into its longitudinal and transverse parts:

$$\mathbf{E} = \mathbf{E}_{||} + \mathbf{E}_{\perp} .$$

The longitudinal part has zero curl, while the transverse part has zero divergence,

$$\nabla \times \mathbf{E}_{||} = 0 , \quad \nabla \cdot \mathbf{E}_{\perp} = 0 .$$

If we decompose the field into Fourier components

$$\mathbf{E}(\mathbf{r}, t) = \frac{1}{(2\pi)^{3/2}} \int \mathbf{E}(\mathbf{k}, t) \mathrm{e}^{\mathrm{i}\mathbf{k}\cdot\mathbf{r}} \, \mathrm{d}^3\mathbf{k} ,$$

the above equations take the form

$$\mathbf{k} \times \mathbf{E}_{||} = 0 , \qquad \mathbf{k} \cdot \mathbf{E}_{\perp} = 0 .$$

[2]Mr. Coulomb died 25 years prior to the birth of Mr. Maxwell, but the reader will certainly not be confused.

Thus the assignments “longitudinal” and “transverse” should be clear. It should be now also clear, see Eqs. (6.7) and (6.10), that the scalar potential φ describes the longitudinal part, $\mathbf{E}_{||} = -\nabla\varphi$, and the vector potential $\mathbf{A}$ the transverse part, $\mathbf{E}_{\perp} = -\frac{\partial \mathbf{A}}{\partial t}$. The magnetic field $\mathbf{B}$ is purely transverse, see Eq. (6.3). The longitudinal part describes the instantaneous, electrostatic part of the interaction and carries no momentum. See the second equation of (6.11): if this equation is satisfied at an initial time t_0, it is automatically satisfied at an arbitrary instant later. On the other hand, the transverse part of the field describes the retarded part of the interaction which carries momentum and thus has to be quantized.

This a posteriori justifies the way we have been proceeding so far. If we restrict ourselves to the electrostatic part of the interaction, the scalar potential can be taken in the same form as in classical electrodynamics. If the motion of the particles is slow enough, contribution of the retarded part of the interaction is relatively small in comparison with the instantaneous part, and hence can be neglected within the first approximation; such is the case of light atoms as hydrogen, helium, and so on.

6.1.6 Hamiltonian of Free Electromagnetic Field

Let us first consider the case of a free field, $\rho = 0, \mathbf{j} = 0$; then

$$\varphi = 0\,, \qquad \left(\frac{\partial^2}{\partial t^2} - \nabla^2\right)\mathbf{A} = 0\,, \qquad \nabla\cdot\mathbf{A} = 0\,. \tag{6.13}$$

We decompose the vector potential into *modes*,

$$\mathbf{A}(\mathbf{r}, t) = \sum_{\tau} q_{\tau}(t)\mathbf{T}_{\tau}(\mathbf{r})\,, \tag{6.14}$$

where $\mathbf{T}_{\tau}(\mathbf{r})$ are real functions of spatial variables obeying equations

$$-\nabla^2\mathbf{T}_{\tau} = \omega_{\tau}^2\mathbf{T}_{\tau}\,, \qquad \nabla\cdot\mathbf{T}_{\tau} = 0\,; \tag{6.15}$$

$q_{\tau}(t)$ are coefficients of the expansion of the vector potential into these functions for given time t.

The summation in Eq. (6.14) runs over all solutions to Eq. (6.15), and changes into an integral in the limit of infinite space. To avoid, at least for now, the integrations, we assume that the field is periodic, with period L,

$$\mathbf{A}(x, y, z, t) = \mathbf{A}(x + L, y, z, t) = \mathbf{A}(x, y + L, z, t) = \mathbf{A}(x, y, z + L, t)\,. \tag{6.16}$$

This means that we can restrict our further considerations to a box with dimensions $L \times L \times L$. Once we complete the procedure of canonical quantization, we easily take the limit $L \to \infty$. The Laplace operator is Hermitian, hence the functions $\mathbf{T}_{\tau}$ are orthonormal:

$$\int_{-L/2}^{L/2} \mathrm{d}x \int_{-L/2}^{L/2} \mathrm{d}y \int_{-L/2}^{L/2} \mathrm{d}z\, \mathbf{T}_\tau \cdot \mathbf{T}_{\tau'} = \delta_{\tau\tau'} \,. \tag{6.17}$$

We insert Eqs. (6.14) and (6.15) into Eq. (6.13), take the scalar product of both sides with $\mathbf{T}_{\tau'}$, and integrate over the whole space. Owing to the orthonormality relations (6.17), we have for all τ

$$\ddot{q}_\tau + \omega_\tau^2 q_\tau = 0 \,. \tag{6.18}$$

We obtained an "equation of motion" for the coefficients of the expansion (6.14). This equation matches that for the harmonic oscillator oscillating with frequency ω_τ. We know, though, how to express the dynamics of the harmonic oscillator in the framework of the Hamilton formalism. The Hamiltonian of the harmonic oscillator has the form (do you remember?)

$$H_\tau = \frac{p_\tau^2}{2} + \frac{1}{2}\omega_\tau^2 q_\tau^2 \,, \tag{6.19}$$

where q_τ and p_τ are canonical coordinates and momenta, respectively. From the Hamilton canonical equations (6.1) (where we substitute the index τ for the index k), we obtain Eq. (6.18). To determine the dynamics of the complete field, we need to sum H_τ into the *Hamiltonian of a free EM field*

$$H_{\mathrm{EM}} = \sum_\tau H_\tau \,. \tag{6.20}$$

6.1.7 Classical Electrodynamics in Hamiltonian Formalism

The Hamiltonian of N charged particles and an EM field comprises a sum of the Hamiltonian of the EM field, H_{EM}, the particle Hamiltonian and the interaction Hamiltonian. The particle Hamiltonian consists of a sum of kinetic energies $\sum_j \frac{\hat{\mathbf{p}}^{(j)}\cdot\hat{\mathbf{p}}^{(j)}}{2m_j}$ and electrostatic potential energies $\frac{1}{2}\sum_{l\neq j}\frac{1}{4\pi}\frac{e_j e_l}{|\mathbf{q}^{(j)}-\mathbf{q}^{(l)}|}$. To obtain the interaction between charged particles and EM field it suffices to replace

$$\mathbf{p}^{(j)} \to \mathbf{p}^{(j)} - e_j \mathbf{A}^{(j)} \,. \tag{6.21}$$

Here, $\mathbf{A}^{(j)} = \mathbf{A}(\mathbf{q}^{(j)})$ is the value of the vector potential at the location of the j-th particle. Thus, the desired Hamiltonian has the form

$$H = H_{\mathrm{EM}} + \sum_j \frac{\left(\mathbf{p}^{(j)} - e_j\mathbf{A}^{(j)}\right)\cdot\left(\mathbf{p}^{(j)} - e_j\mathbf{A}^{(j)}\right)}{2m_j} + \frac{1}{2}\sum_{l\neq j}\frac{1}{4\pi}\frac{e_j e_l}{|\mathbf{q}^{(j)} - \mathbf{q}^{(l)}|} \,. \tag{6.22}$$

Let us now show that this Hamiltonian contains classical electrodynamics. That is, if we write the Hamilton canonical equations for the field coordinates and momenta

$$\frac{\partial H}{\partial q_\tau} = -\dot{p}_\tau\,, \qquad \frac{\partial H}{\partial p_\tau} = \dot{q}_\tau\,, \tag{6.23}$$

we arrive at the Maxwell equations. Similarly, if we write the Hamilton canonical equations for particle coordinates and momenta

$$\frac{\partial H}{\partial q_k^{(j)}} = -\dot{p}_k^{(j)}\,, \qquad \frac{\partial H}{\partial p_k^{(j)}} = \dot{q}_k^{(j)}\,, \tag{6.24}$$

we obtain the Newton equation with the Lorentz force. The notation $q_k^{(j)}$ means that we have in mind the k-th component of the radius vector of the j-th particle and so on.

Let us start with the second assertion and differentiate

$$\dot{q}_k^{(j)} = \frac{\partial H}{\partial p_k^{(j)}} = \frac{p_k^{(j)} - e_j A_k^{(j)}}{m_j}\,, \tag{6.25}$$

$$\dot{p}_k^{(j)} = -\frac{\partial H}{\partial q_k^{(j)}} = -\frac{p_n^{(j)} - e_j A_n^{(j)}}{m_j}(-e_j)\frac{\partial A_n^{(j)}}{\partial q_k^{(j)}} - e_j \frac{\partial \varphi^{(j)}}{\partial q_k^{(j)}}\,, \tag{6.26}$$

where the scalar potential acting on the j-th particle is a sum of electrostatic interactions from other particles

$$\varphi^{(j)} = \sum_{l \neq j} \frac{1}{4\pi} \frac{e_l}{|\mathbf{q}^{(j)} - \mathbf{q}^{(l)}|}\,.$$

Substituting Eq. (6.25) for $\frac{p_n^{(j)} - e_j A_n^{(j)}}{m_j}$ on the rhs of Eq. (6.26), we obtain

$$\dot{p}_k^{(j)} = e_j \left[\dot{q}_n^{(j)} \frac{\partial A_n^{(j)}}{\partial q_k^{(j)}} - \frac{\partial \varphi^{(j)}}{\partial q_k^{(j)}} \right]. \tag{6.27}$$

This equation does not resemble the Newton equation (6.6), yet. We need to recognize that the lhs of the Newton equation features the time derivation of the physical momentum $m\dot{\mathbf{q}}$. The relation between the physical ($m\dot{\mathbf{q}}$) and canonical ($\mathbf{p}$) momenta is given by Eq. (6.25). By differentiating Eq. (6.25) with respect to time and substituting

$$\frac{\mathrm{d}A_k^{(j)}}{\mathrm{d}t} = \frac{\partial A_k^{(j)}}{\partial t} + \frac{\partial A_k^{(j)}}{\partial q_n^{(j)}} \dot{q}_n^{(j)}$$

for the total time derivative of the vector potential and Eq. (6.27) for $\dot{p}_k^{(j)}$, we obtain

$$m_j\ddot{q}_k^{(j)} = \dot{p}_k^{(j)} - e_j\frac{\mathrm{d}A_k^{(j)}}{\mathrm{d}t} = e_j\left[\dot{q}_n^{(j)}\frac{\partial A_n^{(j)}}{\partial q_k^{(j)}} - \dot{q}_n^{(j)}\frac{\partial A_k^{(j)}}{\partial q_n^{(j)}} - \frac{\partial A_k^{(j)}}{\partial t} - \frac{\partial \varphi^{(j)}}{\partial q_k^{(j)}}\right].$$

Now, if we express the fields in the Newton equation (6.6) in terms of potentials via Eq. (6.7), we arrive at the last equation.

One can prove the first assertion in a similar manner; first we differentiate

$$\dot{q}_\tau = \frac{\partial H}{\partial p_\tau} = p_\tau\,,$$

$$\dot{p}_\tau = \ddot{q}_\tau = -\frac{\partial H}{\partial q_\tau} = -\omega_\tau^2 q_\tau - \sum_j \frac{p_k^{(j)} - e_j A_k^{(j)}}{m_j}(-e_j)\frac{\partial A_k^{(j)}}{\partial q_\tau} = -\omega_\tau^2 q_\tau + \sum_j e_j\dot{q}_k^{(j)}T_{\tau k}^{(j)}$$

$$\Rightarrow \ddot{q}_\tau + \omega_\tau^2 q_\tau = \sum_j e_j\dot{\mathbf{q}}^{(j)}\cdot\mathbf{T}_\tau^{(j)}\,, \tag{6.28}$$

where in the last equality on the second row we inserted from Eqs. (6.14) and (6.25). As our next step, we show that the last equation is the vector Maxwell equation (6.11).We substitute Eq. (6.14) for the vector potential into the last mentioned equation

$$\sum_{\tau'}\left(\ddot{q}_{\tau'} + \omega_{\tau'}^2 q_{\tau'}\right)\mathbf{T}_{\tau'} = \mathbf{j} - \frac{\partial\nabla\varphi}{\partial t}\,.$$

We then take the scalar product of both sides with vector $\mathbf{T}_\tau$ and integrate over the box. Owing to the orthogonality of vectors $\mathbf{T}_\tau$, the sum on the lhs reduces to the one term with $\tau' = \tau$,

$$\ddot{q}_\tau + \omega_\tau^2 q_\tau = \int \mathbf{j}\cdot\mathbf{T}_\tau \mathrm{d}^3\mathbf{r} - \int \tfrac{\partial\nabla\varphi}{\partial t}\cdot\mathbf{T}_\tau \mathrm{d}^3\mathbf{r} = \int \mathbf{j}\cdot\mathbf{T}_\tau \mathrm{d}^3\mathbf{r} - \oint \tfrac{\partial\varphi}{\partial t}\mathbf{T}_\tau\cdot\mathrm{d}\mathbf{S} + \int \tfrac{\partial\varphi}{\partial t}\nabla\cdot\mathbf{T}_\tau \mathrm{d}^3\mathbf{r}\,.$$

Let us analyze the above expressions from the end: the last integral vanishes because of the condition (6.13) for basis vectors $\mathbf{T}_\tau$. The second to last integral may be written in the form $\frac{\mathrm{d}}{\mathrm{d}t}\oint\varphi\mathbf{T}_\tau\cdot\mathrm{d}\mathbf{S}$. For a finite size of the box, this integral does not vanish since we imposed periodic boundary conditions on the transverse part of the EM field, but not on the longitudinal part. However, as we are about to show, it is justifiable to neglect this integral in the limit of the infinite box. In this limit, we have $\varphi(\mathbf{r}) = \varphi(-\mathbf{r})$ on the boundary of the box, and also $\mathbf{T}_\tau(\mathbf{r}) = \mathbf{T}_\tau(-\mathbf{r})$ as follows from the periodicity of the vector potential on the boundary of the box. Thus, the arguments of the integral at the points $\mathbf{r}$ a $-\mathbf{r}$ are identical, but the differential $\mathrm{d}\mathbf{S}$ differs in the sign at these points. Hence, the contribution from each of the pairs of points $\mathbf{r}$ and $-\mathbf{r}$ cancels out. If we substitute Eq. (6.5) for the charge density current $\mathbf{j}$ in the first term on the rhs of the last equation, we finally arrive at Eq. (6.28).

Exercise 17: Energy of EM Field and Charged Particles

Using Eqs. (6.3), (6.4) and (6.6), show that the expression

$$E = \frac{1}{2}\int \mathrm{d}^3\mathbf{r}\,(\mathbf{E}\cdot\mathbf{E}+\mathbf{B}\cdot\mathbf{B}) + \frac{1}{2}\sum_{j=1}^{N} m_j \dot{\mathbf{q}}^{(j)}\cdot\dot{\mathbf{q}}^{(j)}$$

is the total energy of the system of N charged particles and an EM field, i.e., $\frac{\mathrm{d}E}{\mathrm{d}t}=0$. In addition, using Eqs. (6.7), (6.12), (6.14), and (6.25), show that this expression matches the expression (6.22).

6.1.8 Polarization

So far, we did not specify the functions $\mathbf{T}_\tau$ introduced by Eq. (6.15) with the condition (6.16). The simplest solution are standing waves

$$\mathbf{T}_\tau = \begin{cases} \mathbf{T}_{\tau_c} = \boldsymbol{\varepsilon}^{(\tau)}\sqrt{\frac{2}{L^3}}\cos\mathbf{k}_\tau\cdot\mathbf{r}\,, \\ \mathbf{T}_{\tau_s} = \boldsymbol{\varepsilon}^{(\tau)}\sqrt{\frac{2}{L^3}}\sin\mathbf{k}_\tau\cdot\mathbf{r}\,, \end{cases}$$

where $\boldsymbol{\varepsilon}^{(\tau)}$ are the polarization vectors and $\mathbf{k}_\tau$ the wave vectors which are given as

$$\mathbf{k}_\tau = \frac{2\pi}{L}\mathbf{n}_\tau\,, \qquad \mathbf{k}_\tau\cdot\mathbf{k}_\tau = \omega_\tau^2\,, \tag{6.29}$$

where $\mathbf{n}_\tau$ represent triples of integers, i.e., $(n_\tau)_{x,y,z} = 0, \pm1, \pm2, \ldots$ The functions $\cos(\mathbf{k}_\tau\cdot\mathbf{r})$ and $\sin(\mathbf{k}_\tau\cdot\mathbf{r})$ are even and odd, respectively; the functions $\cos(-\mathbf{k}_\tau\cdot\mathbf{r})$ and $\sin(-\mathbf{k}_\tau\cdot\mathbf{r})$ are thus linearly dependent thereon. Namely, the function $\mathbf{T}_{-\tau}$ belonging to the vector $-\mathbf{k}_\tau$ is linearly dependent on the function $\mathbf{T}_\tau$ belonging to vector $\mathbf{k}_\tau$. Thus, one of them needs to be excluded in order to avoid overcompleteness.

With this choice of the functions $\mathbf{T}_\tau$, the condition $\nabla\cdot\mathbf{T}_\tau = 0$ leads to the requirement

$$\mathbf{k}_\tau\cdot\boldsymbol{\varepsilon}^{(\tau)} = 0\,. \tag{6.30}$$

In the coordinate system where

$$\mathbf{k}_\tau = (0, 0, \omega_\tau)\,, \tag{6.31}$$

the two linearly independent solutions to Eq. (6.30) are

$$\boldsymbol{\varepsilon}^{(1)} = (1,0,0)\,, \qquad \boldsymbol{\varepsilon}^{(2)} = (0,1,0)\,.$$

In the following sections, we will often need the following summation over polarizations

$$\sum_{\lambda=1}^{2} \varepsilon_i^{(\lambda)} \varepsilon_j^{(\lambda)} = \delta_{ij} - \frac{(\mathbf{k}_\tau)_i (\mathbf{k}_\tau)_j}{\omega_\tau^2}\,. \tag{6.32}$$

The simplest way to check this identity is true is to opt for a coordinate system where (6.31) holds. The rhs of Eq. (6.32) is a projector onto the plane perpendicular to the vector $\mathbf{k}_\tau$

$$\left(\delta_{ij} - \frac{(\mathbf{k}_\tau)_i (\mathbf{k}_\tau)_j}{\omega_\tau^2}\right)(\mathbf{k}_\tau)_i = 0\,, \tag{6.33}$$

where we used the second equation in (6.29). As you may recall, we already encountered this projector in the Sect. 3.6.1, see Eq. (3.117).

6.1.9 Quantized Electromagnetic Field

To proceed to quantum electrodynamics, we impose the canonical commutation relations on the canonical coordinates and momenta of the EM field,

$$[\hat{\mathsf{p}}_\tau, \hat{\mathsf{q}}_{\tau'}] = -\mathrm{i}\delta_{\tau\tau'}\,, \tag{6.34}$$

and of particles

$$[\hat{\mathsf{p}}_k^{(j)}, \hat{\mathsf{q}}_m^{(l)}] = -\mathrm{i}\delta_{km}\delta_{jl}\,. \tag{6.35}$$

The Hamiltonian (6.22), where we substitute the corresponding operators for canonical coordinates and momenta, is then the correct Hamiltonian of quantum electrodynamics within the applicability of the nonrelativistic description on the motion of particles.

As mentioned above, the Hamiltonian of a free EM field comprises a sum of Hamiltonians of harmonic oscillators. In Sect. 1.3.3, we learned how to determine the spectrum of a harmonic oscillator. The key step is to introduce the ladder operators

$$\hat{\mathsf{a}}_\tau = \frac{1}{\sqrt{2}}\left(\sqrt{\omega_\tau}\hat{\mathsf{q}}_\tau + \frac{\mathrm{i}\hat{\mathsf{p}}_\tau}{\sqrt{\omega_\tau}}\right) \tag{6.36}$$

and their Hermitian conjugates $\hat{a}_\tau^+$. It follows from Eqs. (6.34) and (6.36) that

$$[\hat{a}_\tau, \hat{a}_{\tau'}^+] = \delta_{\tau,\tau'} . \tag{6.37}$$

By inverting Eq. (6.36) and its Hermitian conjugate, we can express canonical coordinates and momenta in terms of the ladder operators. Consequently, we are able to express also the Hamiltonian, vector potential, electric and magnetic fields in terms of these operators:

$$\hat{H}_{EM} = \sum_\tau \frac{\hat{p}_\tau^2}{2} + \frac{1}{2}\omega_\tau^2 \hat{q}_\tau^2 = \sum_\tau \omega_\tau \left(\hat{a}_\tau^+ \hat{a}_\tau + \frac{1}{2} \right), \tag{6.38}$$

$$\hat{\boldsymbol{A}} = \sum_\tau \hat{q}_\tau \mathbf{T}_\tau = \sum_\tau \frac{1}{\sqrt{2\omega_\tau}} (\hat{a}_\tau + \hat{a}_\tau^+) \mathbf{T}_\tau , \tag{6.39}$$

$$\mathbf{E}_\perp = -\frac{\partial}{\partial t}\mathbf{A} = -\sum_\tau \frac{dq_\tau}{dt}\mathbf{T}_\tau = -\sum_\tau p_\tau \mathbf{T}_\tau$$

$$\Rightarrow \hat{\boldsymbol{E}}_\perp = -\sum_\tau \hat{p}_\tau \mathbf{T}_\tau = \mathrm{i}\sum_\tau \sqrt{\frac{\omega_\tau}{2}} (\hat{a}_\tau - \hat{a}_\tau^+) \mathbf{T}_\tau , \tag{6.40}$$

$$\hat{\boldsymbol{B}} = \nabla \times \hat{\boldsymbol{A}} = \sum_\tau \hat{q}_\tau \nabla \times \mathbf{T}_\tau = \sum_\tau \frac{1}{\sqrt{2\omega_\tau}} (\hat{a}_\tau + \hat{a}_\tau^+) \nabla \times \mathbf{T}_\tau . \tag{6.41}$$

The ground state (vacuum) of EM field is determined by

$$\hat{a}_\tau |0\rangle = 0 \tag{6.42}$$

for all τ. All possible excited states are then obtained by the action of the operators $\hat{O} = \hat{a}_{\tau_1}^+ \hat{a}_{\tau_2}^+ \dots \hat{a}_{\tau_n}^+$, $n \in \mathbb{N}$, thereon.

6.1.10 Transition to the Complex Basis

Having derived Eqs. (6.38)–(6.41), we successfully solved the problem of quantizing the EM field. However, for the purpose of the following calculations, it is more advantageous to work with running waves rather than with the standing ones. Let us therefore change to the complex basis

$$\mathbf{T}_\tau = \frac{\boldsymbol{\varepsilon}^{(\tau)}}{\sqrt{L^3}} \mathrm{e}^{\mathrm{i}\mathbf{k}_\tau \cdot \mathbf{r}} = \frac{1}{\sqrt{2}} (\mathbf{T}_{\tau_c} + \mathrm{i}\mathbf{T}_{\tau_s}) . \tag{6.43}$$

After complex conjugation we obtain

$$\mathbf{T}_\tau^* = \mathbf{T}_{-\tau}\,.$$

The orthonormality relations (6.17) now take the form

$$\int \mathbf{T}_{\tau'}^*(\mathbf{r}) \cdot \mathbf{T}_\tau(\mathbf{r})\, \mathrm{d}^3\mathbf{r} = \delta_{\tau\tau'}\,. \tag{6.44}$$

For now, we denote the operator annihilating odd (sin) states as $\hat{\mathrm{a}}_{\tau_s}$ and similarly the operator annihilating even (cos) states as $\hat{\mathrm{a}}_{\tau_c}$. We also rewrite Eq. (6.39) in a more detailed form

$$\begin{aligned}\hat{\boldsymbol{A}} &= \sum_{\tau>0} \frac{1}{\sqrt{2\omega_\tau}} \left[(\hat{\mathrm{a}}_{\tau_c} + \hat{\mathrm{a}}_{\tau_c}^+)\mathbf{T}_{\tau_c} + (\hat{\mathrm{a}}_{\tau_s} + \hat{\mathrm{a}}_{\tau_s}^+)\mathbf{T}_{\tau_s}\right] \\ &= \sum_{\tau>0} \frac{1}{\sqrt{2\omega_\tau}} \left[\mathbf{T}_\tau \frac{1}{\sqrt{2}} \left(\hat{\mathrm{a}}_{\tau_c} - \mathrm{i}\hat{\mathrm{a}}_{\tau_s} + (\hat{\mathrm{a}}_{\tau_c} + \mathrm{i}\hat{\mathrm{a}}_{\tau_s})^+\right)\right. \\ &\quad \left. + \mathbf{T}_\tau^* \frac{1}{\sqrt{2}} \left(\hat{\mathrm{a}}_{\tau_c} + \mathrm{i}\hat{\mathrm{a}}_{\tau_s} + (\hat{\mathrm{a}}_{\tau_c} - \mathrm{i}\hat{\mathrm{a}}_{\tau_s})^+\right)\right],\end{aligned} \tag{6.45}$$

where we used Eq. (6.43). The restriction $\tau > 0$ means that half of the triples $\mathbf{n}_\tau$ is excluded in Eq. (6.45); namely those triples that one can obtain from the remaining terms by the replacement $\mathbf{n}_\tau \to -\mathbf{n}_\tau$, see discussion below Eq. (6.29). We would like to replace the operators creating and annihilating the standing waves by those creating and annihilating the running waves. Considering the equations

$$\langle 0|\hat{\boldsymbol{A}}\hat{\mathrm{a}}_{\tau_c}^+|0\rangle = \frac{1}{\sqrt{2\omega_\tau}}\mathbf{T}_{\tau_c}, \qquad \langle 0|\hat{\boldsymbol{A}}\hat{\mathrm{a}}_{\tau_s}^+|0\rangle = \frac{1}{\sqrt{2\omega_\tau}}\mathbf{T}_{\tau_s}\,,$$

and

$$\langle 0|\hat{\boldsymbol{A}}\frac{1}{\sqrt{2}}(\hat{\mathrm{a}}_{\tau_c}^+ \pm \mathrm{i}\hat{\mathrm{a}}_{\tau_s}^+)|0\rangle = \frac{1}{\sqrt{2\omega_\tau}}\mathbf{T}_{\pm\tau}\,,$$

it is obvious that the operators

$$\hat{\mathrm{a}}_{\pm\tau}^+ = \frac{1}{\sqrt{2}}(\hat{\mathrm{a}}_{\tau_c}^+ \pm \mathrm{i}\hat{\mathrm{a}}_{\tau_s}^+), \quad \hat{\mathrm{a}}_{\pm\tau} = \frac{1}{\sqrt{2}}(\hat{\mathrm{a}}_{\tau_c} \mp \mathrm{i}\hat{\mathrm{a}}_{\tau_s}) \tag{6.46}$$

are the sought creation and annihilation operators for running waves. By substituting these relations into Eq. (6.45), we arrive at

$$\hat{\boldsymbol{A}} = \sum_{\tau>0} \frac{1}{\sqrt{2\omega_\tau}}[(\hat{\mathrm{a}}_\tau + \hat{\mathrm{a}}_{-\tau}^+)\mathbf{T}_\tau + (\hat{\mathrm{a}}_{-\tau} + \hat{\mathrm{a}}_\tau^+)\mathbf{T}_{-\tau}] = \sum_{\tau} \frac{1}{\sqrt{2\omega_\tau}}(\hat{\mathrm{a}}_\tau\mathbf{T}_\tau + \hat{\mathrm{a}}_\tau^+\mathbf{T}_\tau^*)\,, \tag{6.47}$$

where τ in the last expression runs without restriction, that is, $\mathbf{n}_\tau$ in Eq. (6.29) are now triples of integers without any restriction. Likewise, after inverting the relations (6.46),

$$\hat{a}_{\tau_c} = \frac{1}{\sqrt{2}}(\hat{a}_\tau + \hat{a}_{-\tau}), \qquad \hat{a}_{\tau_s} = \frac{i}{\sqrt{2}}(\hat{a}_\tau - \hat{a}_{-\tau}),$$

we find for the Hamilton operator from Eq. (6.38) that

$$\begin{aligned}\hat{H} &= \sum_{\tau>0} \omega_\tau \left(\hat{a}^+_{\tau_c}\hat{a}_{\tau_c} + \frac{1}{2} + \hat{a}^+_{\tau_s}\hat{a}_{\tau_s} + \frac{1}{2} \right) \\ &= \sum_{\tau>0} \omega_\tau \left(\hat{a}^+_{\tau}\hat{a}_{\tau} + \frac{1}{2} + \hat{a}^+_{-\tau}\hat{a}_{-\tau} + \frac{1}{2} \right) = \sum_{\tau} \omega_\tau \left(\hat{a}^+_{\tau}\hat{a}_{\tau} + \frac{1}{2} \right).\end{aligned} \tag{6.48}$$

6.1.11 Transition to the Continuous Basis

Finally, let us take the limit $L \to \infty$. The basis states $\mathbf{T}$ are then normalized to the δ-function

$$\mathbf{T}^{(\lambda)}_{\mathbf{k}}(\mathbf{r}) = \frac{\boldsymbol{\varepsilon}^{(\lambda)}}{\sqrt{(2\pi)^3}} e^{i\mathbf{k}\cdot\mathbf{r}}. \tag{6.49}$$

This is, but for the unit polarization vector, a three-dimensional generalization of (1.95). The Hamiltonian, vector potential, electric and magnetic fields consequently take the final forms

$$\hat{H}_{EM} = \int \sum_{\lambda=1}^{2} \omega \left(\hat{a}^+(\mathbf{k},\lambda)\hat{a}(\mathbf{k},\lambda) + \frac{1}{2} \right) d^3\mathbf{k}, \tag{6.50}$$

$$\hat{\boldsymbol{A}} = \frac{1}{\sqrt{(2\pi)^3}} \int \frac{1}{\sqrt{2\omega}} \sum_{\lambda=1}^{2} \boldsymbol{\varepsilon}^{(\lambda)} \left(\hat{a}(\mathbf{k},\lambda)e^{i\mathbf{k}\cdot\mathbf{r}} + \hat{a}^+(\mathbf{k},\lambda)e^{-i\mathbf{k}\cdot\mathbf{r}} \right) d^3\mathbf{k}, \tag{6.51}$$

$$\hat{\boldsymbol{E}}_\perp = \frac{i}{\sqrt{(2\pi)^3}} \int \sqrt{\frac{\omega}{2}} \sum_{\lambda=1}^{2} \boldsymbol{\varepsilon}^{(\lambda)} \left(\hat{a}(\mathbf{k},\lambda)e^{i\mathbf{k}\cdot\mathbf{r}} - \hat{a}^+(\mathbf{k},\lambda)e^{-i\mathbf{k}\cdot\mathbf{r}} \right) d^3\mathbf{k} \tag{6.52}$$

and

$$\hat{\boldsymbol{B}} = \frac{i}{\sqrt{(2\pi)^3}} \int \frac{1}{\sqrt{2\omega}} \sum_{\lambda=1}^{2} (\mathbf{k}\times\boldsymbol{\varepsilon}^{(\lambda)}) \left(\hat{a}(\mathbf{k},\lambda)e^{i\mathbf{k}\cdot\mathbf{r}} - \hat{a}^+(\mathbf{k},\lambda)e^{-i\mathbf{k}\cdot\mathbf{r}} \right) d^3\mathbf{k}. \tag{6.53}$$

The continuous version of the commutation relations (6.37) then reads

$$[\hat{a}(\mathbf{k}',\lambda'),\hat{a}^+(\mathbf{k},\lambda)] = \delta^{(3)}(\mathbf{k}-\mathbf{k}')\delta_{\lambda\lambda'}\,. \tag{6.54}$$

Finally, the continuous version of the summation over polarizations, Eq. (6.32), is given as

$$\sum_{\lambda=1}^{2} \varepsilon_i^{(\lambda)}\varepsilon_j^{(\lambda)} = \delta_{ij} - \eta_i\eta_j = P_{ij}\,, \tag{6.55}$$

where $\boldsymbol{\eta}$ is a unit vector in the direction of the wave vector $\mathbf{k}$

$$\mathbf{k} = \omega\boldsymbol{\eta}\,. \tag{6.56}$$

6.1.12 States of the Field

The ground state (vacuum) is determined by the continuous version of Eq. (6.42),

$$\hat{a}(\mathbf{k},\lambda)|0\rangle = 0\,, \tag{6.57}$$

which holds for all possible $\mathbf{k}$ and λ. To find the energy of the ground state, we act on it with the Hamiltonian

$$\hat{H}_{\mathrm{EM}|0\rangle} = \int\sum_{\lambda=1}^{2}\omega\left(\hat{a}^+(\mathbf{k},\lambda)\hat{a}(\mathbf{k},\lambda)+\frac{1}{2}\right)\mathrm{d}^3\mathbf{k}|0\rangle = \sum_{\lambda=1}^{2}\int\frac{\omega}{2}\mathrm{d}^3\mathbf{k}\,|0\rangle \stackrel{\text{def}}{=} E_0|0\rangle\,. \tag{6.58}$$

Likewise, we have for the energy of an excited state

$$\begin{aligned}\hat{H}^+_{\mathrm{EM}\hat{a}}(\mathbf{k}',\lambda')|0\rangle &= \int\sum_{\lambda=1}^{2}\omega\left(\hat{a}^+(\mathbf{k},\lambda)\hat{a}(\mathbf{k},\lambda)+\frac{1}{2}\right)\mathrm{d}^3\mathbf{k}\,\hat{a}^+(\mathbf{k}',\lambda')|0\rangle \\ &= \int\sum_{\lambda=1}^{2}\omega\left(\hat{a}^+(\mathbf{k},\lambda)\delta(\mathbf{k}-\mathbf{k}')\delta_{\lambda\lambda'}+\frac{1}{2}\hat{a}^+(\mathbf{k}',\lambda')\right)\mathrm{d}^3\mathbf{k}\,|0\rangle \\ &= (E_0+\omega')\hat{a}^+(\mathbf{k}',\lambda')|0\rangle\,. \end{aligned} \tag{6.59}$$

Clearly, the energy is higher by ω'—there is a photon there! Proceeding in the same manner, we easily find

$$\hat{H}_{\mathrm{EM}}\hat{a}^+(\mathbf{k}',\lambda')\hat{a}^+(\mathbf{k}'',\lambda'')|0\rangle = (E_0+\omega'+\omega'')\hat{a}^+(\mathbf{k}',\lambda')\hat{a}^+(\mathbf{k}'',\lambda'')|0\rangle\,. \tag{6.60}$$

Up to now, the creation and annihilation operators $\hat{a}^+(\mathbf{k},\lambda)$ and $\hat{a}(\mathbf{k},\lambda)$ could have been viewed merely as useful mathematical tools. However, they have now received a clear physical meaning.[3] The symbol $\hat{a}^+(\mathbf{k},\lambda)$ denotes the creation operator of a photon with the wave vector $\mathbf{k}$ and polarization λ and $\hat{a}(\mathbf{k},\lambda)$ represents the corresponding annihilation operator.

One can readily see from Eq. (6.58) that the energy of the vacuum is infinite. However, one can also notice from Eqs. (6.59) and (6.60) that all excited states are shifted by this very energy. Since we can measure only energy differences,[4] we can subtract the vacuum energy from all of the states, and speak of the Hamilton operator (6.50) as in the so-called *normally ordered form*

$$:\hat{\mathsf{H}}_{\mathrm{EM}}: = \hat{\mathsf{H}}_{\mathrm{EM}} - \langle 0|\hat{\mathsf{H}}_{\mathrm{EM}|0\rangle} = \int \sum_{\lambda=1}^{2} \omega \hat{a}^+(\mathbf{k},\lambda)\hat{a}(\mathbf{k},\lambda)\mathrm{d}^3\mathbf{k}\,. \tag{6.61}$$

One last note: owing to the commutation relation

$$[\hat{a}^+(\mathbf{k}',\lambda'), \hat{a}^+(\mathbf{k}'',\lambda'')] = 0\,, \tag{6.62}$$

interchange of the first and the second photon in Eq. (6.60) leads to the same state. That is, quanta of EM field (i.e., photons) are bosons.

6.2 Spontaneous Emission

The quantum version of the Hamiltonian (6.22) can be easily brought into the form

$$\hat{\mathsf{H}} = \hat{\mathsf{H}}_0 + \hat{\mathsf{H}}_{\mathrm{int}}\,, \qquad \hat{\mathsf{H}}_0 = :\hat{\mathsf{H}}_{\mathrm{EM}}: + \hat{\mathsf{H}}_{\mathrm{at}}\,, \tag{6.63}$$

$$\hat{\mathsf{H}}_{\mathrm{at}} = \sum_j \frac{\hat{\boldsymbol{p}}^{(j)} \cdot \hat{\boldsymbol{p}}^{(j)}}{2m_j} + \frac{1}{4\pi}\frac{1}{2}\sum_{m\neq n} \frac{e_m e_n}{|\hat{\boldsymbol{r}}^{(m)} - \hat{\boldsymbol{r}}^{(n)}|}\,, \tag{6.64}$$

$$\hat{\mathsf{H}}_{\mathrm{int}} = e\hat{\mathsf{H}}_1 + e^2\hat{\mathsf{H}}_2 = \sum_j \left(-\frac{e_j}{m_j}\hat{\boldsymbol{A}}^{(j)} \cdot \hat{\boldsymbol{p}}^{(j)} + \frac{e_j^2}{2m_j}\hat{\boldsymbol{A}}^{(j)} \cdot \hat{\boldsymbol{A}}^{(j)} \right)\,, \tag{6.65}$$

---

[3] One usually claims that the physical meaning is carried by the Hermitian operators, whereas the non-Hermitian ones are merely useful mathematical tools. However, we see that in the case of the quantized EM field, the situation is in fact the very opposite: while the canonical coordinates and momenta bear no obvious physical meaning, their non-Hermitian combinations $\hat{a}$ and $\hat{a}^+$ do.

[4] Strictly speaking, this is true only if the gravitation is excluded. Let us also note that this energy of the vacuum represents a part of the cosmological constant problem, see for instance [16].

where we used the Coulomb gauge condition (6.10), $[\hat{\mathsf{p}}_i, \hat{\mathsf{A}}_i] = -\mathrm{i}\nabla \cdot \hat{\boldsymbol{A}} = 0$. Unfortunately, we are not able to find the exact eigenvalues and eigenstates of the complete Hamiltonian. Therefore, in the spirit of the perturbation method, we have split the Hamiltonian into two parts: $\hat{\mathsf{H}}_0$ that we are able to solve exactly, and $\hat{\mathsf{H}}_{\mathrm{int}}$ that we are unable to solve exactly, but which does not markedly influence the solution obtained for $\hat{\mathsf{H}}_0$. In case of slowly moving particles, the electrostatic interaction is so strong that one cannot treat it perturbatively and we have to include it in the "free" Hamiltonian $\hat{\mathsf{H}}_0$. Therefore, we need to know the spectrum of the atomic Hamiltonian (6.64), not of a free-particle Hamiltonian! However, for atoms with more than one electron, we are not able to solve the spectrum of the atomic Hamiltonian (6.64) exactly. Nevertheless, as we showed in Chap. 5, we are able to approach the exact solution to any proximity, at least in principle. Note that the use of the Coulomb gauge substantially simplifies this problem. In case of other than Coulomb gauges the distinction between the electrostatic and delayed interaction is obscured.

6.2.1 Interaction Representation

We have found the Hamilton operator; how do we relate it to what experimenters measure?

Let us start with a general situation. We assume the system is in the initial state $|I\rangle$ at the initial time t_0. The probability of finding this system in the final state $|F\rangle$ any later in time t is—according to the basic principles of quantum mechanics—given by

$$P_{I\to F} = |\langle F|\,\psi(t)\rangle|^2 \,. \tag{6.66}$$

Here, $\psi(t)$ is a solution to the time-dependent Schrödinger equation

$$\mathrm{i}\frac{\mathrm{d}|\psi\rangle}{\mathrm{d}t} = \hat{\mathsf{H}}|\psi\rangle \,, \tag{6.67}$$

with the initial condition

$$|\psi(t = t_0)\rangle = |I\rangle \,. \tag{6.68}$$

We search for the solution to the Schrödinger equation in the form of a time-dependent linear combination of the stationary states of the Hamiltonian $\hat{\mathsf{H}}_0$

$$|\psi(t)\rangle = \sum_n c_n(t)|n\rangle \mathrm{e}^{-\mathrm{i}E_n t} \,, \tag{6.69}$$

where

$$\hat{\mathsf{H}}_0|n\rangle = E_n|n\rangle\,. \tag{6.70}$$

After insertion of Eq. (6.69) into Eq. (6.67), we obtain

$$\sum_n (\mathrm{i}\dot{c}_n + E_n c_n)|n\rangle \mathrm{e}^{-\mathrm{i}E_n t} = \sum_n (\hat{\mathsf{H}}_0 + \hat{\mathsf{H}}_{\mathrm{int}})c_n|n\rangle \mathrm{e}^{-\mathrm{i}E_n t}\,.$$

We multiply this equation from the left by the bra-vector $\langle F|$ and by the number $\mathrm{e}^{\mathrm{i}E_F t}$. Owing to the orthonormality of the states $|n\rangle$ and Eq. (6.70), we find the Schrödinger equation in the *interaction representation*

$$\mathrm{i}\dot{c}_F = \sum_n (\hat{\mathsf{H}}_{\mathrm{int}})_{Fn} c_n \mathrm{e}^{\mathrm{i}\omega_{Fn} t}\,, \tag{6.71}$$

where we introduced the transition frequency

$$\omega_{Fn} = E_F - E_n\,. \tag{6.72}$$

We now solve Eq. (6.71) with the initial condition

$$c_n(t_0) = \delta_{nI}\,, \tag{6.73}$$

which follows from Eqs. (6.68) and (6.69) and from the orthonormality of the states $|n\rangle$. Insertion of Eq. (6.69) into Eq. (6.66) results in, again with the use of the orthonormality of the states $|n\rangle$,

$$P_{I\to F} = |c_F(t)|^2\,. \tag{6.74}$$

6.2.2 Time-Dependent Perturbation Method and the Fermi Golden Rule

Note that our considerations were exact so far.[5] Now we use the perturbation method. We consider the interaction Hamiltonian (6.65) as $\hat{\mathsf{H}}_{\mathrm{int}}$. Since this Hamiltonian is proportional to the elementary charge e, it is reasonable to expand the coefficients c_F into a series in e,

$$c_F(t) = c_F^{(0)}(t) + e c_F^{(1)}(t) + e^2 c_F^{(2)}(t) + \dots\,, \tag{6.75}$$

[5]The only assumption we have made is that the initial and final states can be identified with the eigenstates of the free Hamiltonian. This assumption holds approximately, not exactly, see Sect. 6.5.3, and relates to the issue of a so-called "renormalization" which we will discuss in great detail in Chap. 7.

and substitute this expansion into Eq. (6.71). Comparison of the terms proportional to e^0 on both sides results in

$$\dot{c}_F^{(0)} = 0 \quad \Rightarrow \quad c_F^{(0)} = \delta_{FI}\,, \tag{6.76}$$

where the implication follows from Eq. (6.73). Next, by comparing the terms proportional to e^1 on both sides, we find

$$\dot{c}_F^{(1)} = -\mathrm{i}\sum_n (\hat{\mathsf{H}}_1)_{Fn} c_n^{(0)} \mathrm{e}^{\mathrm{i}\omega_{Fn}t} = -\mathrm{i}(\hat{\mathsf{H}}_1)_{FI}\mathrm{e}^{\mathrm{i}\omega_{FI}t}\,,$$

where we used Eq. (6.76) in the last step. Integration of the last equation together with the initial condition $c_F^{(1)}(t_0) = 0$ yields

$$c_F^{(1)} = -\mathrm{i}(\hat{\mathsf{H}}_1)_{FI}\int_{t_0}^{t} \mathrm{e}^{\mathrm{i}\omega_{FI}t'}\,\mathrm{d}t'\,.$$

By inserting the expansion (6.75) into Eq. (6.74), we arrive at the first approximation for $F \neq I$

$$P_{I\to F}(t) \simeq |ec_F^{(1)}|^2 = |e(\hat{\mathsf{H}}_1)_{FI}|^2 \left| \int_{t_0}^{t} \mathrm{e}^{\mathrm{i}\omega_{FI}t'}\,\mathrm{d}t' \right|^2 . \tag{6.77}$$

One can find that for large times it holds that

$$\lim_{t-t_0\to\infty} \left| \int_{t_0}^{t} \mathrm{e}^{\mathrm{i}\omega_{FI}t'}\,\mathrm{d}t' \right|^2 = \lim_{t-t_0\to\infty} \frac{2(1-\cos\omega_{FI}(t-t_0))}{\omega_{FI}^2} = 2\pi(t-t_0)\delta(\omega_{FI})\,. \tag{6.78}$$

To accustom to this equation, it helps to draw a graph of the functional dependence of the lhs on ω_{FI} for different values of the parameter $(t-t_0) = 10^{-1}, 1, 10, 10^2$, and so on. The factor $2\pi(t-t_0)$ on the rhs stems from the requirement that integration over all possible values of ω_{FI} produces the same number on both sides

$$\int_{-\infty}^{\infty} \frac{2(1-\cos\omega_{FI}(t-t_0))}{\omega_{FI}^2}\,\mathrm{d}\omega_{FI} = 2\pi(t-t_0)\,.$$

When studying decays, the experimenters do not measure directly the probability of decay, but rather the probability of decay per unit time, called the *transition rate* w_{IF}

$$w_{I\to F} \stackrel{\text{def}}{=} \lim_{t-t_0\to\infty} \frac{P_{I\to F}(t)}{t-t_0} = 2\pi\delta(\omega_{FI})\,|e(\hat{\mathsf{H}}_1)_{FI}|^2\,. \tag{6.79}$$

Here, we used Eqs. (6.77) and (6.78). The last equation is called the *Fermi golden rule* and bears the following meaning. Firstly, this rule states that in the limit of infinite times, the energy of the final state equals exactly the energy of the initial state. For instance, we show in a moment that in case of spontaneous emission, the δ-function is merely an expression of the Bohr quantization condition, namely that the energy of the radiated photon equals the difference between the energies of the initial and final atomic states. The Fermi rule thus determines the *position* of atomic spectral lines. Secondly, the Fermi rule states that the probability of a given process, for instance of the spontaneous emission, depends on the value of the squared matrix element $(\hat{\mathsf{H}}_1)_{FI}$. Thus, the Fermi rule determines the *intensity* of atomic spectral lines as well.

There is no need to fear the formidably looking δ-function in Eq. (6.79), though. Whenever we can make use of this equation, we necessarily integrate over a continuous infinity of final states, as we will see later in the text.

6.2.3 Elimination of the Field Operators

We now focus on the process of spontaneous emission, that is, on the change of the state of an atom accompanied by a radiation of a photon with the wave vector $\mathbf{k}$ and polarization $\boldsymbol{\varepsilon}^{(\lambda)}$. The initial and final states are in this case

$$|I\rangle = |I_{\rm at}\rangle|0\rangle\,, \qquad |F\rangle = |F_{\rm at}\rangle\hat{\mathsf{a}}^+(\mathbf{k},\lambda)|0\rangle\,. \tag{6.80}$$

The eigenvalues of the Hamiltonian $\hat{\mathsf{H}}_0$, Eq. (6.63), are

$$\hat{\mathsf{H}}_0|I\rangle = E_I^{\rm at}|I\rangle \quad \text{and} \quad \hat{\mathsf{H}}_0|F\rangle = (E_F^{\rm at}+\omega)|F\rangle\,.$$

The difference between the energies of the initial and final states reads

$$\omega_{IF} = E_I^{\rm at} - E_F^{\rm at} - \omega\,.$$

There is a continuous infinity of the final photon states. Thus, we rewrite the Fermi rule, Eq. (6.79), into the differential form

$$\mathrm{d}w_{I\to F} = 2\pi\delta(E_I^{\rm at} - E_F^{\rm at} - \omega)|e(\hat{\mathsf{H}}_1)_{IF}|^2\mathrm{d}^3\mathbf{k}\,. \tag{6.81}$$

We need to calculate the matrix element of $\hat{\mathsf{H}}_1$. Therefore, we substitute for $\hat{\mathsf{H}}_1$ from Eq. (6.65), where we substitute for the quantized field from Eq. (6.51), and arrive at

$$e(\hat{\mathsf{H}}_1)_{IF} = -\sum_j \frac{e_j}{m_j}\langle 0|\langle I_{\rm at}|\left(\frac{1}{(2\pi)^{3/2}}\int\sum_{\lambda'=1}^{2}\boldsymbol{\varepsilon}^{(\lambda')}\left(\hat{\mathsf{a}}(\mathbf{k}',\lambda')\mathrm{e}^{\mathrm{i}\mathbf{k}'\cdot\hat{\boldsymbol{r}}^{(j)}}\right.\right.$$
$$\left.\left.+\,\hat{\mathsf{a}}^+(\mathbf{k}',\lambda')\mathrm{e}^{-\mathrm{i}\mathbf{k}'\cdot\hat{\boldsymbol{r}}^{(j)}}\right)\frac{\mathrm{d}^3\mathbf{k}'}{\sqrt{2\omega'}}\right)\cdot\hat{\boldsymbol{p}}^{(j)}\hat{\mathsf{a}}^+(\mathbf{k},\lambda)|F_{\rm at}\rangle|0\rangle\,.$$

All expressions of the type $\langle 0|\hat{a}^+\hat{a}^+|0\rangle$ vanish and only the first term in the bracket contributes to the value. Using Eqs. (6.54) and (6.57), we find for the sought matrix element that

$$e(\hat{H}_1)_{IF} = -\sum_j \frac{e_j}{m_j}\langle I_{at}|\left(\frac{1}{(2\pi)^{3/2}}\int\sum_{\lambda'=1}^{2}\boldsymbol{\varepsilon}^{(\lambda')}\delta(\mathbf{k}'-\mathbf{k})\delta_{\lambda,\lambda'}e^{i\mathbf{k}'\cdot\hat{\boldsymbol{r}}^{(j)}}\frac{d^3\mathbf{k}'}{\sqrt{2\omega'}}\right)\cdot\hat{\boldsymbol{p}}^{(j)}|F_{at}\rangle \tag{6.82}$$

$$= -\frac{1}{(2\pi)^{3/2}}\sum_j \frac{e_j}{m_j}\frac{\boldsymbol{\varepsilon}^{(\lambda)}}{\sqrt{2\omega}}\cdot\langle I_{at}|e^{i\mathbf{k}\cdot\hat{\boldsymbol{r}}^{(j)}}\hat{\boldsymbol{p}}^{(j)}|F_{at}\rangle .$$

As our next step, it is advantageous to change in Eq. (6.81) in the k-space from the Cartesian to spherical coordinates, Eq. (6.56). We thus rewrite the differential $d^3\mathbf{k}$ in Eq. (6.81) as $\omega^2 d\omega d\Omega_k$, in an analogy to $d^3\mathbf{r} = r^2 dr d\Omega$ in the ordinary space. Owing to the presence of the δ-function in Eq. (6.81), the integration over frequencies of the radiated photon is trivial: its effect is to replace ω by $E_I^{at} - E_F^{at}$. Finally, we may ignore the interaction of the EM field with the nucleus since the interaction term contains the factor $1/m_n$ and a typical nucleus is at least by three orders of magnitude heavier than electrons. Thus, we can set $m_j = m$, $e_j = -e$, where m and $-e$ are the electron mass and charge. Insertion of Eq. (6.82) into Eq. (6.81) finally yields

$$dw_{I\to F} = e^2 2\pi\frac{(E_I^{at}-E_F^{at})}{2}\frac{1}{(2\pi)^3}\frac{1}{m^2}\left|\boldsymbol{\varepsilon}^{(\lambda)}\cdot\sum_j\langle I_{at}|e^{i\mathbf{k}\cdot\hat{\boldsymbol{r}}^{(j)}}\hat{\boldsymbol{p}}^{(j)}|F_{at}\rangle\right|^2 d\Omega_k .$$

6.2.4 *Electric Dipole Radiation*

Let us make the transition to the dimensionless atomic units in the last equation by substituting

$$\hat{\boldsymbol{r}} = \frac{\hat{\boldsymbol{r}}_A}{mZ\alpha}, \quad \hat{\boldsymbol{p}} = (mZ\alpha)\hat{\boldsymbol{p}}_A \tag{6.83}$$

and

$$E_I^{at} - E_F^{at} = m(Z\alpha)^2\omega_{IF}^{at}, \quad \mathbf{k} = m(Z\alpha)^2\mathbf{k}_A, \quad |\mathbf{k}_A| = \omega_{IF}^{at}; \tag{6.84}$$

we obtain

$$dw_{I\to F} = \frac{\alpha}{2\pi}\omega_{IF}^{at}m(Z\alpha)^4\left|\boldsymbol{\varepsilon}^{(\lambda)}\cdot\sum_j\langle I_{at}|e^{Z\alpha i\mathbf{k}_A\cdot\hat{\boldsymbol{r}}_A^{(j)}}\hat{\boldsymbol{p}}_A^{(j)}|F_{at}\rangle\right|^2 d\Omega_k . \tag{6.85}$$

The exponent $Z\alpha\mathbf{k}_A\cdot\mathbf{r}_A$ is a small number since all the quantities therein reach values which are in atomic units of the order of unity and α is significantly smaller than 1. Therefore, the whole exponential function can be replaced by unity

$$\mathrm{e}^{Z\alpha \mathrm{i}\mathbf{k}_A\cdot\hat{\boldsymbol{r}}_A^{(j)}} \simeq 1\,. \tag{6.86}$$

This approximation is commonly termed as the *dipole approximation* and is appropriate in nonrelativistic calculations (such as those we are performing now), since—as we know from Sect. 3.5—the ratio of the relativistic corrections to the nonrelativistic energy is of the order $(Z\alpha)^2$. Thus, *the same conditions hold for the validity of both the nonrelativistic and dipole approximations.* The usual argument for the validity of the dipole approximation is that the wavelength of the EM radiation has to be substantially larger than the typical dimensions of the object the EM field is interacting with. Although this argument is more general (except for the case of atoms when both of them are the same), the one given here is somewhat cleaner and makes the connection with the nonrelativistic approximation transparent.

Within this approximation, we can write Eq. (6.85) in a significantly simpler form

$$\mathrm{d}w_{I\to F} = \frac{\alpha}{2\pi}\omega_{IF}^{\mathrm{at}} m(Z\alpha)^4 \left|\boldsymbol{\varepsilon}^{(\lambda)}\cdot\left(\hat{\boldsymbol{p}}_A\right)_{IF}\right|^2 \mathrm{d}\Omega_k\,. \tag{6.87}$$

Here, we introduced the total linear momentum operator

$$\hat{\boldsymbol{p}} = \sum_j \hat{\boldsymbol{p}}^{(j)}; \tag{6.88}$$

where the sum over j runs over the electrons. We can further recast Eq. (6.87) into a more commonly used form. The Hamiltonian (6.64) complies with the identity, see Eqs. (1.52) and (1.54),

$$\hat{\mathsf{p}}_i = -\mathrm{i}m[\hat{\mathsf{x}}_i, \hat{\mathsf{H}}_{\mathrm{at}}]\,, \tag{6.89}$$

where we introduced the total radius vector operator

$$\hat{\boldsymbol{r}} = \sum_j \hat{\boldsymbol{r}}^{(j)}. \tag{6.90}$$

Owing to Eq. (6.89), there is a simple relation between the matrix elements of the total momentum and the matrix elements of the total radius vector, namely

$$(\hat{\boldsymbol{p}})_{IF} = \langle I_{\mathrm{at}}|\hat{\boldsymbol{p}}|F_{\mathrm{at}}\rangle = -\mathrm{i}m\langle I_{\mathrm{at}}|[\hat{\boldsymbol{r}},\hat{\mathsf{H}}_{\mathrm{at}}]|F_{\mathrm{at}}\rangle = \mathrm{i}m(E_I^{\mathrm{at}} - E_F^{\mathrm{at}})(\hat{\boldsymbol{r}})_{IF}\,. \tag{6.91}$$

In atomic units, we easily derive from Eqs. (6.83), (6.84), and (6.91)

$$(\hat{\boldsymbol{p}}_A)_{IF} = \mathrm{i}\omega_{IF}^{\mathrm{at}}(\hat{\boldsymbol{r}}_A)_{IF}\,.$$

Thus, Eq. (6.87) can be rewritten as

$$\mathrm{d}w_{I\to F} = \frac{\alpha}{2\pi}(\omega_{IF}^{\mathrm{at}})^3 m(Z\alpha)^4 \left|\boldsymbol{\varepsilon}^{(\lambda)}\cdot(\hat{\boldsymbol{r}}_A)_{IF}\right|^2 \mathrm{d}\Omega_k\,. \tag{6.92}$$

The advantage of the latter form over the former lies in the calculation of the matrix element of a coordinate, which is usually simpler than that of a momentum.

Alkali metal atoms feature only one valence electron, therefore it is a reasonable approximation to assume that only this valence electron is responsible for the transitions in such atoms corresponding to the radiation within the visible part of the spectra. The electron moves in an effective potential which is to a satisfactory approximation spherically symmetric. We can then assign the quantum numbers n_I, l_I, m_I and n_F, l_F, m_F to the initial and final states of the atom, respectively. With the exception of hydrogen, n_I and n_F are not natural numbers, however, it does not matter for now. Recalling Chap. 4, relations (4.62), (4.63), and (4.64), we immediately see that the matrix element $(\hat{\boldsymbol{r}})_{IF}$ vanishes unless $l_F = l_I \pm 1$ and $m_F = m_I$ or $m_F = m_I \pm 1$. In other words, or rather with more words, the electron in an s-state can relax only to a p-state, from a p-state only to an s- or a d-state, and so on. These rules are called the *selection rules for dipole radiation*.

6.2.5 Polarization and Angular Distribution of the Radiated Photons

Let us now examine how the angular distribution of the circularly polarized radiated photons varies with the magnetic quantum number of the valence electron. For this purpose, we rewrite the scalar product of the polarization and radius vectors as

$$\boldsymbol{\varepsilon}^{(\lambda)}\cdot(\hat{\boldsymbol{r}}_A)_{IF} = \frac{1}{2}\varepsilon_+^{(\lambda)}(\hat{\mathrm{x}}_-)_{IF} + \frac{1}{2}\varepsilon_-^{(\lambda)}(\hat{\mathrm{x}}_+)_{IF} + \varepsilon_3^{(\lambda)}(\hat{\mathrm{x}}_3)_{IF} \tag{6.93}$$

and express vectors $\mathbf{k}$, $\boldsymbol{\varepsilon}^{(1)}$, and $\boldsymbol{\varepsilon}^{(2)}$ in the spherical coordinates in the k space

$$\mathbf{k} = \omega\boldsymbol{\eta}, \qquad \boldsymbol{\eta} = (\sin\vartheta\cos\varphi, \sin\vartheta\sin\varphi, \cos\vartheta)\,,$$
$$\boldsymbol{\varepsilon}^{(1)} = (\cos\vartheta\cos\varphi, \cos\vartheta\sin\varphi, -\sin\vartheta)$$

and

$$\boldsymbol{\varepsilon}^{(2)} = (\sin\varphi, -\cos\varphi, 0)\,.$$

One can easily verify that $\boldsymbol{\varepsilon}^{(1)}\cdot\boldsymbol{\varepsilon}^{(2)} = \boldsymbol{\varepsilon}^{(1)}\cdot\boldsymbol{\eta} = \boldsymbol{\varepsilon}^{(2)}\cdot\boldsymbol{\eta} = 0$. A circularly polarized wave (left- or right-handed) is given as the following combination of the linearly polarized waves

$$\boldsymbol{\varepsilon}^{(L)} = \frac{1}{\sqrt{2}}(\boldsymbol{\varepsilon}^{(1)} + \mathrm{i}\boldsymbol{\varepsilon}^{(2)}), \quad \boldsymbol{\varepsilon}^{(R)} = \frac{1}{\sqrt{2}}(\boldsymbol{\varepsilon}^{(1)} - \mathrm{i}\boldsymbol{\varepsilon}^{(2)}).$$

For definiteness, we henceforth consider the right-handed photon only. From the last four equations, we derive

$$\boldsymbol{\varepsilon}^{(R)} = \frac{1}{\sqrt{2}}(\boldsymbol{\varepsilon}^{(1)} - \mathrm{i}\boldsymbol{\varepsilon}^{(2)}) = \frac{1}{\sqrt{2}}(\cos\vartheta\cos\varphi - \mathrm{i}\sin\varphi, \cos\vartheta\sin\varphi + \mathrm{i}\cos\varphi, -\sin\vartheta).$$

We now calculate the probability that an atom undergoes a transition from the state with m_I to the state $m_F = m_I - 1$ and radiates a right-handed photon with the wave vector $\mathbf{k} = \omega\boldsymbol{\eta}$. By inserting the last equation into Eqs. (6.92) and (6.93), we obtain by means of selection rules (4.62), (4.63), and (4.64)[6]

$$\mathrm{d}w_{I\to F} = \frac{\alpha}{2\pi} m(Z\alpha)^4(\omega_{IF}^{\mathrm{at}})^3 \frac{1}{4} |\langle I_{\mathrm{at}}|\hat{\mathbf{x}}_{+}|F_{\mathrm{at}}\rangle|^2 \left|\frac{\cos\vartheta + 1}{2}\right|^2 \mathrm{d}\Omega_k. \tag{6.94}$$

What is the physical interpretation of this result? One can readily notice that the probability of photon emission reaches its maximum for $\vartheta = 0$, that is, for the direction of the z-axis. Recalling that the magnetic quantum number is the projection of the electron angular momentum with respect to this axis, we readily see that in this case, the photon carries away the amount of the angular momentum (with respect to the z-axis) lost by the atom. On the other hand, the expression (6.94) vanishes for $\vartheta = \pi$; that is no photon can go in this direction when the atom undergoes the transition from the state with m_I to the state $m_I - 1$. Why? The law of conservation of angular momentum prohibits that. Bearing this in mind, we rewrite Eq. (6.94) as

$$\begin{aligned}\frac{\mathrm{d}w_{I\to F}}{\mathrm{d}\Omega_k} &= |\langle \mathrm{at}| +\boldsymbol{\eta}\rangle|^2 = |\langle \mathrm{at}| +z\rangle\langle +z| +\boldsymbol{\eta}\rangle + \langle \mathrm{at}| -z\rangle\langle -z| +\boldsymbol{\eta}\rangle|^2 \\ &= |\langle \mathrm{at}| +z\rangle|^2 |\langle +z| +\boldsymbol{\eta}\rangle|^2 .\end{aligned}$$

Here, $\langle \mathrm{at}| +\boldsymbol{\eta}\rangle$ denotes the probability amplitude of the process when the atom radiates a right-handed photon in the direction $\boldsymbol{\eta}$ and the atom angular momentum with respect to the z-axis decreases by 1, $\langle \mathrm{at}| +z\rangle$ denotes the amplitude that the photon is radiated in the direction of the z-axis and so on. In the second equality, we used the decomposition of unity in the photon spin states. The second term does not contribute, though, since—as noted a few lines above—$\langle \mathrm{at}| -z\rangle$ vanishes. Obviously,

$$|\langle +z| +\boldsymbol{\eta}\rangle|^2 = \left|\frac{\cos\vartheta + 1}{2}\right|^2$$

and $|\langle \mathrm{at}| +z\rangle|^2$ is obtained by comparison of the last two equations with Eq. (6.94).

[6] Recall that $|I_{\mathrm{at}}\rangle = |n_I, l_I, m_I\rangle$ and $|F_{\mathrm{at}}\rangle = |n_F, l_F, m_I - 1\rangle$.

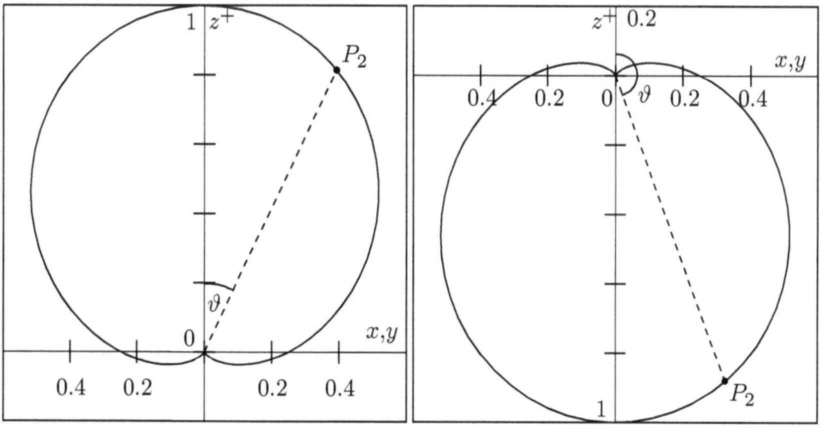

Fig. 6.1 Angular distribution of circularly polarized photons for the transition involving a decrease in atom angular momentum with respect to the z-axis by 1: On the *left* for the right-handed photon and on the *right* for the left-handed photon

Thus, the sought probability can be expressed as a product of the probability that atom radiates a photon in the direction of z-axis ($P_1 = |\langle \mathrm{at}|+z\rangle|^2$) and the probability that a photon with the spin projection $+1$ along the z-axis will be found with the spin projection $+1$ into the direction of $\boldsymbol{\eta}$ ($P_2 = |\langle +z|+\boldsymbol{\eta}\rangle|^2$).

The aim of this discussion is to conclude that the angular distribution of the emitted photons can be deduced from the conservation of angular momentum and the result of Exercise 5. Polar radiation diagrams for the right- and left-handed photons are shown in Fig. 6.1.

If we do not seek the polarizations of the photon and ask merely for their angular distribution, we can sum over the polarizations. By means of Eq. (6.55), we obtain from Eq. (6.92)

$$\mathrm{d}w_{I\to F} = \frac{\alpha}{2\pi}(\omega_{IF}^{\mathrm{at}})^3 m(Z\alpha)^4(\delta_{ij} - \eta_i\eta_j)\langle F_{\mathrm{at}}|\hat{\mathrm{x}}_i|I_{\mathrm{at}}\rangle\langle F_{\mathrm{at}}|\hat{\mathrm{x}}_j|I_{\mathrm{at}}\rangle \mathrm{d}\Omega_k\,. \tag{6.95}$$

The reader will surely be able to sketch the corresponding polar diagram.

6.2.6 *Lifetime of States*

And if the angular distribution is of no interest to us and we merely seek the probability of an atom radiating a photon? We then clearly need to integrate over all directions. The only photon-direction-dependent variable in Eq. (6.95) is the vector $\boldsymbol{\eta}$. Therefore, by means of the integral, see Eq. (3.47),

$$\int \eta_i\eta_j \mathrm{d}\Omega_k = \frac{4}{3}\pi\delta_{ij}\,, \tag{6.96}$$

we find that

$$w_{I\to F} = \frac{4}{3} m\alpha (Z\alpha)^4 (\omega_{IF}^{\rm at})^3 |\langle I_{\rm at}|\hat{\boldsymbol{r}}_A|F_{\rm at}\rangle|^2 = \frac{4}{3} m\alpha (Z\alpha)^4 (\omega_{IF}^{\rm at}) |\langle I_{\rm at}|\hat{\boldsymbol{p}}_A|F_{\rm at}\rangle|^2 . \tag{6.97}$$

The total transition rate Γ_I at a time t is given as the ratio of the rate of change in the number of atoms $N(t)$ which are in the state $I_{\rm at}$ at the time t to this number $N(t)$

$$-\frac{\frac{dN(t)}{dt}}{N(t)} = \Gamma_I \Rightarrow N(t) = N(t=0) \mathrm{e}^{-\Gamma_I t} . \tag{6.98}$$

As you may recall, this equation resembles the Rutherford formula for the radioactive decay, see Sect. 1.1. Therefore, the lifetime τ_I of the initial atomic state equals the inverse value of total transition rate Γ_I. We can obtain the latter from Eq. (6.97) by summing over all possible final atomic states

$$\tau_I = \Gamma_I^{-1} = \left(\sum_{E_F^{\rm at} < E_I^{\rm at}} w_{I\to F} \right)^{-1} . \tag{6.99}$$

Again, if we restrict ourselves to alkali metal atoms, we can use the relations (4.62), (4.63), and (4.64) in the calculation of the matrix element in Eq. (6.97), and sum over all possible magnetic quantum numbers m_F of the final atomic state. We thus obtain after angular integration in Eq. (6.97) (check it!)

$$\sum_{m_F} \langle l_I, m_I | n_i | l_F, m_F \rangle \langle l_F, m_F | n_i | l_I, m_I \rangle = \delta_{l_F, l_I - 1} \frac{l_I}{2l_I + 1} + \delta_{l_F, l_I + 1} \frac{l_I + 1}{2l_I + 1} . \tag{6.100}$$

Note that the result is independent of the initial quantum number m_I; this should not be surprising.

To convert the transition rate from electronvolts to hertz

$$w(\mathrm{eV}) \to w(\mathrm{Hz}) ,$$

it suffices to replace

$$m \to \frac{2R_\infty c}{\alpha^2} 2\pi . \tag{6.101}$$

Had this reasoning been too fast, the reader can find another way of unit conversion in Eq. (8).

Table 6.1 Wavelengths corresponding to colors of visible light

Color	Wavelength [nm]
Violet	380–435
Blue	435–500
Cyan	500–520
Green	520–565
Yellow	565–590
Orange	590–625
Red	625–740

By means of Table 6.1[7] and what he learned so far, the reader can attempt the following exercise.

Exercise 18: Hydrogen Spectral Lines

What colors can we observe in the hydrogen spectrum? What are the corresponding transitions? What are the ratios of the intensities of the individual lines?
Hint: Determine the principal and orbital quantum numbers of the initial and final states. You easily find that for a given n_I and n_F, there are more choices of l_I. For the sake of simplicity, assume that for a given n_I there is a uniform distribution of the states with different l_I and m_I. When calculating the ratio of intensities, calculate the ratios of expressions

$$\overline{w}\,(n_I \to n_F) = \frac{1}{n_I^2} \sum_{l_I=0}^{n_I-1} (2l_I + 1) w(n_I, l_I \to n_F, l_F)\,.$$

The value of the speed of light can be found in Eq. (3). If the reader is not in the mood to calculate the needed radial integrals, he can consult Sect. 6.4.6.

6.2.7 Circular States and Connection with Classical Theory

Now, we apply the results obtained so far to the so-called *circular states*, that is, states with $l = n-1$. Little examination of Eqs. (4.62)–(4.64) together with Eq. (4.72) leads to the conclusion that a circular state with $l_I = n_I - 1$ necessarily ends up in another

[7]The distinction between individual colors is not sharp, of course. Similarly, the exact wavelength range for visible light is a subject to an endless debate. The reader should thus not be surprised to find slightly different values in the literature.

circular state $l_F = n_F - 1$ after a spontaneous emission. Inserting Eq. (6.100) into Eq. (6.97) we obtain

$$\begin{aligned} \Gamma &= \frac{4}{3} m\alpha (Z\alpha)^4 \frac{1}{2^3} \left(\frac{1}{l^2} - \frac{1}{(l+1)^2} \right)^3 \frac{l}{2l+1} |\langle l, l-1 | \hat{\mathrm{r}}_A | l+1, l \rangle|^2 \\ &= \frac{2}{3} m\alpha (Z\alpha)^4 \frac{1}{n^5} \frac{\left(1 - \frac{1}{n}\right)^{2n-2}}{\left(1 - \frac{1}{2n}\right)^{4n-1}}, \end{aligned} \tag{6.102}$$

where $l = l_I$ and $n = l + 1$, and where we used Eqs. (4.78) and (4.79) in the last step.

It is highly instructive to compare this result with the quasi-classical one [3]. The classical formula (due to Larmor) for the radiated power in the dipole approximation reads (see, e.g., [13])

$$P_{\mathrm{cl}} = \frac{2}{3} \alpha |a|^2 . \tag{6.103}$$

Now, we insert a for acceleration from the Newton equation

$$a = \frac{Z\alpha}{m} \left\langle \frac{1}{r^2} \right\rangle = m(Z\alpha)^3 \left\langle \frac{1}{r_A^2} \right\rangle . \tag{6.104}$$

For average inverse distance squared $\langle r_A^{-2} \rangle$, we find for the circular states, see Eqs. (4.78) and (4.79),

$$\langle r_A^{-2} \rangle = \langle n, n-1 | \hat{\mathrm{r}}_A^{-2} | n, n-1 \rangle = \frac{1}{n^3 \left(n - \frac{1}{2}\right)} . \tag{6.105}$$

If we take the ratio of the radiated power (that is, the radiated energy per unit time) to the radiated energy, we arrive at the classical prediction for the total transition probability

$$\Gamma_{\mathrm{cl}} = \frac{P_{\mathrm{cl}}}{\Delta E},$$

where

$$\Delta E = m(Z\alpha)^2 \Delta\varepsilon , \qquad \Delta\varepsilon = \frac{1}{2} \left(-\frac{1}{n^2} + \frac{1}{(n-1)^2} \right) .$$

Last five equations yield

$$\Gamma_{\mathrm{cl}} = \frac{2}{3} m\alpha (Z\alpha)^4 \frac{1}{n^5} \frac{\left(1 - \frac{1}{n}\right)^2}{\left(1 - \frac{1}{2n}\right)^3} .$$

Comparing Eq. (6.102) with the last one, we see that in the limit of large quantum numbers $n \to \infty$, quantum and quasi-classical predictions coincide.

The expressions (6.103) and (6.104) are usually regarded as a proof that classical physics is incapable of explaining the stability of an atom. Within the framework of classical physics, an electron must orbit the nucleus so that the centrifugal force balances the attractive electrostatic force stemming from the nucleus. However, as the electron orbits, it radiates EM waves, hence loses its energy and falls to the nucleus.

However, the failure of classical physics to explain the stability of an atom is somewhat more subtle. Classical physics cannot—not even qualitatively—elucidate the stability of the ground state. On the other hand, it explains fairly well the instability of states with nonzero angular momentum; moreover, within the limit of large quantum numbers, the explanation is even quantitatively correct. Thus, it is obvious that the spontaneous emission is nothing but a quantum counterpart of the classical phenomenon that an accelerated charged particle emits EM waves. From this point of view, one can also readily understand that the lifetime of circular states increases with increasing l, see Eq. (6.102). The reason is: with increasing angular momentum, their average distance from the nucleus increases as well, see Eq. (6.105). However, as the distance grows, the electrostatic attraction weakens, see Eq. (6.104). With a decrease of attraction, the acceleration decreases, too, and a decrease in acceleration results in a decrease of the radiation of an EM wave.

To conclude this section, note that the validity of the expression (6.102) is not restricted to hydrogen-like atoms only. There are states above the ground level which one can describe within a reasonable approximation as one-electron excitations of the ground-state configuration. For instance, we saw in the last chapter that the helium ground-state configuration lies close to $\{1s, 1s\}$, i.e., both electrons are in an s-state; the pertinent orbital being approximately hydrogen-like, see Eq. (5.11). One-electron excitations with valence electrons in circular states are then $\{1s, 2p\}$, $\{1s, 3d\}$ and so on. If we follow these one-electron excitations high enough, they begin to resemble those of hydrogen. Intuitively, if one electron is highly excited while all of the other ones remains "as low as possible," the excited electron is "far" from the nucleus and hence feels a potential which is in a reasonable approximation Coulombic with an effective charge close to one as the remaining $Z - 1$ "lower" electrons screen the charge of the nucleus Z. One can notice this screening already in case of the lowest P- and D-states in helium, see Sect. 5.2.4, and it further grows with increasing angular momentum of the valence electron. Therefore, Eq. (6.102) holds for circular states of an *arbitrary* atom if n is sufficiently large. Highly excited states that we can describe well as one-electron excitations from the ground configuration are commonly termed as the *Rydberg* states, and comprise highly excited states of all atoms which lie below the ionization threshold. Above the ionization threshold, one of the electrons lies in the continuous part of the spectrum; that is, the atom is ionized. One could similarly consider excited states which are well describable as two-electron excitations from the ground configuration. In case of helium, we find such states already above the ionization threshold.

6.2.8 Forbidden Transitions

It follows from the selection rules for dipole radiation that some of the transitions are forbidden. The prime example is the process $2s \to 1s+\gamma$ in the hydrogen which is forbidden in all orders of $Z\alpha$ for the interaction Hamiltonian (6.65). Let us now show why. Note, though, that the proof applies to all transitions between arbitrary s-states.

We let the initial and final states be s-states that is, $|I_{\rm at}\rangle = |n_I, 0, 0\rangle$ and $|F_{\rm at}\rangle = |n_F, 0, 0\rangle$. In case of hydrogen, Eq. (6.85) reduces to

$$\frac{\mathrm{d}w}{\mathrm{d}\Omega} = m\frac{\alpha}{2\pi}(Z\alpha)^4 \omega_{IF}^{\rm at} |\langle F_{\rm at}| \mathrm{e}^{\mathrm{i}Z\alpha \mathbf{k}\cdot\hat{\mathbf{r}}_A} \hat{\boldsymbol{p}}_A \cdot \boldsymbol{\varepsilon} |I_{\rm at}\rangle|^2 .$$

Firstly, recall Eq. (3.15), $\hat{\boldsymbol{p}}_A \to -\mathrm{i}\left(\mathbf{n}\frac{\partial}{\partial r} + \frac{\nabla^{(n)}}{r}\right)$. The action of the second term in the bracket on the s-states vanishes. Secondly, we expand the exponential

$$\begin{aligned}\langle F_{\rm at}| \mathrm{e}^{\mathrm{i}Z\alpha \mathbf{k}\cdot\hat{\mathbf{r}}_A} \hat{\boldsymbol{p}}_A \cdot \boldsymbol{\varepsilon} |I_{\rm at}\rangle &= -\frac{\mathrm{i}}{4\pi}\int_0^\infty R_{n_F 0}\frac{\partial}{\partial r}R_{n_I 0}\left(\int\left(1 + Z\alpha \mathrm{i} r k_j n_j\right.\right.\\ &\quad \left. - \frac{1}{2}(Z\alpha)^2 r^2 k_j k_l n_i n_j n_l - \frac{\mathrm{i}}{6}(Z\alpha)^3 r^3 k_j k_l k_m n_j n_l n_m + \ldots\right)\\ &\quad \left.\times \mathbf{n}\cdot\boldsymbol{\varepsilon}\,\mathrm{d}\Omega\right) r^2 \mathrm{d}r .\end{aligned}$$

To proceed further, it suffices to calculate the integrals of the type

$$\int \underbrace{n_i n_j \ldots}_{N\text{-times}} \mathrm{d}\Omega .$$

These integrals vanish for N odd; for N even they are proportional to the products of Kronecker symbols. For instance, Eq. (3.47) holds for $N = 2$,

$$\int n_i n_j n_k n_l \mathrm{d}\Omega = \frac{4\pi}{15}(\delta_{ij}\delta_{kl} + \delta_{ik}\delta_{jl} + \delta_{il}\delta_{jk}) \tag{6.106}$$

for $N = 4$, and so on. Thus, upon integration over angles, the individual terms of the expansion either vanish or the scalar product $\mathbf{k}\cdot\boldsymbol{\varepsilon}$ appears. However, this product vanishes as well, see Eq. (6.30).

The lifetime of the $2s$-state is not infinite, of course. There are two additional mechanisms how the transition may proceed if the interaction (6.65) "does not work." Either the magnetic interaction of the spin of the electron with the quantized magnetic field operates, as we will show further, or the atom radiates two photons. In the latter case, the atom and the EM field firstly "borrow energy from vacuum for the moment": the atom radiates photon and goes to a linear superposition of

p-states, all of which lie higher than the initial s-state. Secondly, the atom and EM field "return the energy back": the atom radiates an additional photon and relaxes from the superposition to the ground state. To describe this process, one needs to opt for the second order of the perturbation method. We learn how to go beyond the first order of the perturbation method later, see Sect. 6.4.

In case of hydrogen, the dominant mechanism of the $2s - 1s$ transition is a two-photon radiation. The lifetime of the $2s$-state equals, see, e.g., [5],

$$\tau_{2s} \simeq 0.12\ \text{s}\,.$$

For comparison, the lifetime of the $2p$-state is $\tau_{2p} \simeq 0.16 \times 10^{-8}$ s, see (6.102). On the atomic scale, the lifetime of the $2s$-state is enormous, hence these states are called *metastable*.

In case of helium, the dominant mechanism is the interaction of the spin with the magnetic field. Lifetime of the metastable $2\,^3S$ state is 7870 s (!), see [10].

6.2.9 Radiation Associated with a Change of Spin

In the nonrelativistic limit, the interaction of the electron with the EM field contains the term $-\frac{e}{m}\hat{\boldsymbol{S}}\cdot\mathbf{B}$, see Eqs. (3.50) and (3.51). As the interaction Hamiltonian (6.65) features no such term, we should add one into it:

$$\hat{\mathsf{H}}_1^{\text{mag}} = -\sum_j \frac{e_j}{m_j}\hat{\boldsymbol{S}}^{(j)}\cdot\hat{\boldsymbol{B}}^{(j)}\,,$$

where $\hat{\boldsymbol{B}}$ is given by Eq. (6.53). If the selection rules for dipole radiation allow the transition in question, one may neglect this spin interaction.

Exercise 19: Lifetime of the $1\,^3s$ State

Find the lifetime of the $1\,^3s$ state.
Hint: Write the Fermi golden rule (6.79) in the form

$$dw_{I\to F} = \sum_\lambda 2\pi\delta(\omega_{IF})|(\hat{\mathsf{H}}_1^{\text{mag}})_{IF}|^2 \mathrm{d}^3\mathbf{k},$$

and take

$$|I\rangle = |1s\rangle\frac{1}{\sqrt{2}}\left(|+\rangle_e|-\rangle_p + |-\rangle_e|+\rangle_p\right)|0\rangle$$

as the initial state,

$$|F\rangle = |1s\rangle \frac{1}{\sqrt{2}} \left(|+\rangle_e |-\rangle_p - |-\rangle_e |+\rangle_p \right) \hat{\mathsf{a}}^+(\mathbf{k}, \lambda)|0\rangle$$

as the final state, and

$$\hat{\mathsf{H}}_1^{\mathrm{mag}} \simeq -\frac{e}{m_e} \hat{\boldsymbol{S}}_e \cdot \hat{\boldsymbol{B}}$$

as the interaction Hamiltonian. Clearly, $\omega_{IF} = E_I^{\mathrm{at}} - E_F^{\mathrm{at}} - \omega$*, where insert from Eq. (3.58) for* $E_I^{\mathrm{at}} - E_F^{\mathrm{at}}$*. Invoke the dipole approximation, in this case called the magnetic dipole approximation.*

You should reach the value

$$\tau_{1\,{}^3s} = 3.4 \times 10^{14}\,\mathrm{s} \simeq 10^7 \text{ years.}$$

In this case, the lifetime is large not only on the atomic scales.

6.3 Photoelectric Effect

6.3.1 Introductory Notes

In case of photon absorption, the initial and final states read

$$|I\rangle = |I_{\mathrm{at}}\rangle \hat{\mathsf{a}}^+(\mathbf{k}, \lambda)|0\rangle\,, \qquad |F\rangle = |F_{\mathrm{at}}\rangle|0\rangle\,. \tag{6.107}$$

Since we now must sum over all final atomic states, we write the Fermi golden rule (6.79) in the form

$$w_{I\to F} = \sum_{\mathrm{at}} 2\pi\delta(E_I^{\mathrm{at}} - E_F^{\mathrm{at}} + \omega)|e(\hat{\mathsf{H}}_1)_{IF}|^2\,. \tag{6.108}$$

For instance in the case of hydrogen, we have

$$\sum_{\mathrm{at}} \to \sum_{l_F=0}^{\infty} \sum_{m_F=-l_F}^{l_F} \left(\sum_{n_F=l_F+1}^{\infty} + \int_0^{\infty} \mathrm{d}k_e \right)\,.$$

One can readily recognize the necessity to distinguish between the “final” states lying in the discrete part and those in the continuous part of the spectrum.

If the atom goes to the discrete part of the spectrum, it sooner or later radiates a photon and relaxes to a state of lower energy. In other words, the discrete excited states of the atom are not the true final states; hence the Fermi golden rule (6.79) cannot be directly applied to them. Nevertheless, we can use the Fermi rule provided we seek the probability of a photon absorption succeeded by a photon emission. We will return to this problem in the next section.

We will consider only the transition to the continuous part of the spectrum for now. We write the Fermi golden rule in the differential form

$$\mathrm{d}w_{I\to F} = \sum_{l_F, m_F} 2\pi\delta(E_I^{\mathrm{at}} - E_F^{\mathrm{at}} + \omega)|e(\hat{\mathsf{H}}_1)_{IF}|^2 \mathrm{d}k_e\,. \tag{6.109}$$

Owing to the δ-function, the integration over electron wave numbers k_e is trivial. The energy of the final atomic state is then given as a sum of the energies of the incident photon and of the initial atomic state

$$E_F^{\mathrm{at}} = \omega + E_I^{\mathrm{at}} \Rightarrow m(Z\alpha)^2\frac{k_e^2}{2} = \omega - \frac{m(Z\alpha)^2}{2}\,, \tag{6.110}$$

which is exactly the *Einstein equation for photoeffect*. We restricted ourselves to the case of hydrogen in the second equality, and we also assumed that the initial state is the ground state.

Note that if the energy of the incident photon is too large, the energy of the outgoing electron acquires large values as well, hence relativistic description may be in place. For validity of the nonrelativistic approximation, the kinetic energy of the outgoing electron must be significantly smaller than its rest energy

$$m(Z\alpha)^2\frac{k_e^2}{2} \ll m \quad \Rightarrow \quad k_e \ll (Z\alpha)^{-1}\,.$$

Insertion of this requirement into Eq. (6.110) then leads to the condition

$$\omega \ll m\,.$$

The rest mass of the electron is about 0.511 MeV while the ionization energy of the hydrogen atom in its ground state equals approximately $m(Z\alpha)^2/2 \simeq 13.6\,\mathrm{eV}$. Hence, the theory we are about to outline holds for incident photon energies within the range 10^1–10^5 eV.

The matrix element $(\hat{\mathsf{H}}_1)_{FI}$ enters Fermi rule only in the form $|(\hat{\mathsf{H}}_1)_{FI}|^2$. Equation (6.82) can be thus used for the initial and final states (6.107) as well. After invoking the dipole approximation,[8] changing to the atomic units, restricting to the hydrogen-

[8] As already emphasized in the previous section, the dipole approximation can be used whenever the nonrelativistic approximation has been invoked.

like atoms and taking advantage of the fact that $\hat{\boldsymbol{p}}_A|I_{\mathrm{at}}\rangle = \mathrm{i}\hat{\boldsymbol{n}}|I_{\mathrm{at}}\rangle$ for the ground state, we arrive at the expression

$$|e(\hat{\mathsf{H}}_1)_{FI}|^2 = \frac{(mZ\alpha)^2 e^2}{(2\pi)^3 2\omega m^2} |\boldsymbol{\varepsilon}^{(\lambda)} \cdot \langle F_{\mathrm{at}}|\hat{\boldsymbol{n}}|I_{\mathrm{at}}\rangle|^2 . \tag{6.111}$$

Equation (6.109) is the correct one for the transition rate from the initial atomic state to the state when the atom is ionized and has definite value of the orbital and angular quantum number. For example, the hydrogen atom initially in the ground state ends up, owing to the selection rules for dipole radiation, in one of its p-states.

Note, though, that we have not described the photoeffect yet. What experimenters detect is a *current* of outgoing electrons. However, states with a definite value of angular momentum do not create a current one could register in the detector. Such states describe an electron, whether in the discrete or in the continuous part of the spectrum, which "circulates" around the nucleus, and thus the average distance of the electron from the nucleus does not change. If we wish to describe the electrons that are "going away" from the nucleus, we must require the electron be described by a plane wave in the limit of large distances from the nucleus. Formally, the wave function of the final atomic state ψ must comply with the Schrödinger equation

$$\left[-\frac{\nabla^2}{2} - \frac{1}{r}\right]\psi = \frac{k_e^2}{2}\psi \tag{6.112}$$

with the boundary condition

$$\psi(r \to \infty) \longrightarrow \frac{\mathrm{e}^{\mathrm{i}\mathbf{k}_e \cdot \mathbf{r}}}{(2\pi)^{3/2}} . \tag{6.113}$$

The above considerations can be formulated more precisely. We start from the time-dependent Schrödinger equation for one particle moving under the influence of an external potential V

$$\mathrm{i}\frac{\partial \psi}{\partial t} = \left[-\frac{\nabla^2}{2} + V(\mathbf{r})\right]\psi . \tag{6.114}$$

We find the complex conjugate of the last equation and multiply it from the left by ψ, and subsequently subtract therefrom equation Eq. (6.114) multiplied from the left by ψ^*. We thus obtain

$$\mathrm{i}(\psi^*\frac{\partial \psi}{\partial t} + \psi\frac{\partial \psi^*}{\partial t}) = -\psi^*\frac{\nabla^2}{2}\psi + \psi\frac{\nabla^2}{2}\psi^* .$$

Exploiting the knowledge of differentiating a product of functions, we manipulate the last equation into the form

$$\mathrm{i}\frac{\partial|\psi|^2}{\partial t} = \frac{1}{2}\frac{\partial}{\partial x_i}\left(\psi\frac{\partial}{\partial x_i}\psi^* - \psi^*\frac{\partial}{\partial x_i}\psi\right).$$

We now integrate over a space region and use Gauss's theorem to obtain

$$-\frac{\mathrm{d}}{\mathrm{d}t}\int|\psi|^2\,\mathrm{d}^3\mathbf{r} = \oint\frac{1}{2\mathrm{i}}\left[\psi^*\nabla\psi - \psi\nabla\psi^*\right]\cdot\mathrm{d}\mathbf{S}. \tag{6.115}$$

One can interpret this equation as a continuity equation for the probability density: a decrease of probability per unit time inside the volume of the region, i.e., the lhs of the equation, equals the probability current over the surface of the region, i.e., the rhs of the equation. That is, shall the electrons "move away" from the nucleus, their probability density inside a sphere centered at the nucleus and of radius which well exceeds the "atomic radius" (that is, approaching infinity from the mathematical point of view) must decrease. Hence, we must describe the electrons with such a wave function that leads to a nonvanishing probability current density

$$\mathbf{j} = \frac{1}{2\mathrm{i}}\left[\psi^*\nabla\psi - \psi\nabla\psi^*\right] \tag{6.116}$$

through the surface of the sphere, radius of which goes to infinity. One can easily see that should the wave function lead to nonvanishing current density at all, it must be complex. Wave functions with a definite value of angular momentum are complex only in the variable φ; thus they lead to a nonvanishing current density in the direction along the changing coordinate φ. These states "circulate" around the nucleus and their current density through the spherical surface vanishes. On the other hand, the wave functions obeying condition (6.113) lead to nonvanishing current density through the spherical surface in the direction of the vector $\mathbf{k}_e$. To see this, insert Eq. (6.113) into Eq. (6.116).

The solution to Eq. (6.112) with the boundary condition (6.113) is most easily found in the parabolic coordinates. However, prior to our turning to this, we find a relation between such a solution and an experimentally measurable quantity. We rewrite Eq. (6.108) into the differential form

$$\mathrm{d}w_{I\to F} = 2\pi\delta\left(-m(Z\alpha)^2\frac{1+k_e^2}{2} + \omega\right)|e(\hat{\mathrm{H}}_1)_{IF}|^2\mathrm{d}^3\mathbf{k}_e\,, \tag{6.117}$$

and then change to the spherical coordinates in the space of the electron final states; consequently $\mathrm{d}^3\mathbf{k}_e = k_e^2\mathrm{d}k_e\mathrm{d}\Omega$. The δ-function yields a condition for $m(Z\alpha)^2k_e^2/2$, not for k_e though. To turn it into the condition for the latter, we write

$$\mathrm{d}\left(\frac{k_e^2}{2}m(Z\alpha)^2\right) = \frac{\mathrm{d}\left(\frac{k_e^2}{2}m(Z\alpha)^2\right)}{\mathrm{d}k_e}\mathrm{d}k_e = k_e m(Z\alpha)^2\mathrm{d}k_e\,.$$

We then find from Eq. (6.117)

$$\frac{\mathrm{d}w_{I\to F}}{\mathrm{d}\Omega} = \frac{2\pi k_e}{m(Z\alpha)^2}|e(\hat{\mathsf{H}}_1)_{IF}|^2 , \tag{6.118}$$

with k_e given by Eq. (6.110).

The probability of the atom ionization depends on the intensity of the incident radiation which can be freely tuned by the experimenter. To remove this freedom, we divide the transition rate by the incident probability current density per unit area

$$\sigma_{I\to F} = \frac{w_{I\to F}}{j_I} . \tag{6.119}$$

The quantity σ is called the *cross-section* of the reaction and depends only on the "physics" of the scattering process, not on the "initial conditions."

Exercise 20: The Average of the Poynting Vector

Check that the average of the energy density current of an EM field (i.e., the Poynting vector) in a one-photon state equals

$$\langle 0|\hat{\mathsf{a}}(\mathbf{k},\lambda)(\hat{\boldsymbol{E}} \times \hat{\boldsymbol{B}})\hat{\mathsf{a}}^{+}(\mathbf{k},\lambda)|0\rangle = \frac{\mathbf{k}}{(2\pi)^3} .$$

Hint: The operators $\hat{\mathbf{E}}$ *and* $\hat{\mathbf{B}}$ *are given by Eqs. (6.52) and (6.53), respectively.*

The incident current density in the photoelectric experiment is given by the power of incident radiation per unit area (i.e., the magnitude of the Poynting vector) divided by the energy of the incident radiation. It follows from Exercise 20 that the incident current density equals

$$j_I = \frac{1}{(2\pi)^3} . \tag{6.120}$$

After inserting Eqs. (6.111) and (6.110) into Eq. (6.118) and inserting the resulting equation together with Eq. (6.120) into Eq. (6.119), we obtain

$$\frac{\mathrm{d}\sigma}{\mathrm{d}\Omega} = \frac{\alpha(2\pi)^3}{\pi(mZ\alpha)^2}\frac{k_e}{k_e^2+1}|\boldsymbol{\varepsilon}^{(\lambda)} \cdot \langle F_{\mathrm{at}}|\hat{\boldsymbol{n}}|I_{\mathrm{at}}\rangle|^2 . \tag{6.121}$$

Once we find the wave function of the final state and calculate the corresponding matrix element, this equation yields predictions about the outcome of the photoelectric experiment which go beyond the Einstein relation (6.110). For instance, given the polarization and the wave vector of the incident photon, what is the angular and energetic distribution of the outgoing electrons?

The following three sections are devoted to the calculation of the squared matrix element $|\boldsymbol{\varepsilon}^{(\lambda)} \cdot \langle F_{\mathrm{at}}|\hat{\boldsymbol{n}}|I_{\mathrm{at}}\rangle|^2$. An impatient reader may skip directly to the resulting Eq. (6.152).

6.3.2 Parabolic Coordinates

As mentioned above, the solution to Eq. (6.112) with the boundary condition (6.113) is relatively easy to find using the *parabolic coordinates*. Recall, see Sect. 4.5.5, Exercise 13, that solving the Schrödinger equation (6.112) in parabolic coordinates corresponds to searching for common eigenvectors of the operators $\hat{\mathsf{H}}$, $\hat{\mathsf{X}}_z$ and $\hat{\mathsf{L}}_z$.

The parabolic coordinates are defined by the relations

$$\xi = r + z\,, \qquad \eta = r - z\,, \qquad \varphi = \operatorname{arctg}\frac{y}{x}\,, \qquad r = \sqrt{x^2 + y^2 + z^2}\,, \tag{6.122}$$

where the ranges of the individual coordinates are

$$\xi \in (0, \infty)\,, \qquad \eta \in (0, \infty)\,, \qquad \varphi \in (0, 2\pi)\,. \tag{6.123}$$

The inverse transformation has the form

$$x = \sqrt{\xi\eta}\cos\varphi\,, \qquad y = \sqrt{\xi\eta}\sin\varphi\,, \qquad z = \frac{1}{2}(\xi - \eta)\,. \tag{6.124}$$

Differential volume element is

$$\mathrm{d}V = \mathrm{d}^3\mathbf{r} = \frac{1}{4}(\xi + \eta)\,\mathrm{d}\xi\,\mathrm{d}\eta\,\mathrm{d}\varphi\,. \tag{6.125}$$

Laplace operator in parabolic coordinates takes the form

$$\nabla^2 = \frac{4}{\xi + \eta}\left[\frac{\partial}{\partial\xi}\left(\xi\frac{\partial}{\partial\xi}\right) + \frac{\partial}{\partial\eta}\left(\eta\frac{\partial}{\partial\eta}\right)\right] + \frac{1}{\eta\xi}\frac{\partial^2}{\partial\varphi^2}\,. \tag{6.126}$$

The proofs of these statements are elementary, though a bit involved.

In the following text, we will need only an axially symmetric solution, i.e., one which does not depend on the angle φ. After substituting $\psi = \psi(\xi, \eta)$ into the Schrödinger equation (6.112), $\left[-\frac{\nabla^2}{2} - \frac{1}{r}\right]\psi = \frac{k_e^2}{2}\psi$, we obtain

$$\left[\frac{\partial}{\partial\xi}\left(\xi\frac{\partial}{\partial\xi}\right) + \frac{\partial}{\partial\eta}\left(\eta\frac{\partial}{\partial\eta}\right) + 1\right]\psi = -\frac{\xi + \eta}{4}k_e^2\psi\,. \tag{6.127}$$

6.3.3 Wave Functions of the Continuous Spectrum

For large distances from the nucleus, the wave function of the outgoing electron has to obey the boundary condition (6.113). As the reader will shortly see, we can substantially simplify our calculation if we place the z-axis in the direction of the wave vector $\mathbf{k}_e$ of the outgoing electron. The condition (6.113) then reads

$$\psi(r \to +\infty) \to \frac{\mathrm{e}^{\mathrm{i}k_e z}}{(2\pi)^{3/2}} = \frac{\mathrm{e}^{\mathrm{i}k_e(\xi-\eta)/2}}{(2\pi)^{3/2}} . \tag{6.128}$$

For an electron leaving the atom along the positive direction of the z-axis, i.e., for $z \to \infty$, we easily find that $r \to \infty$ and consequently $\xi \to \infty$. On the other hand, the coordinate η could still be finite in this limit, see Fig. 6.2. Therefore, shall Eq. (6.128) hold, we need to search for a solution to Eq. (6.127) of the form

$$\psi = \frac{\mathrm{e}^{-\mathrm{i}k_e\eta/2}}{(2\pi)^{3/2}} \phi(\xi) . \tag{6.129}$$

By inserting Eq. (6.129) into Eq. (6.127), we obtain an equation for the function $\phi(\xi)$, namely

$$\left[\frac{\partial}{\partial\xi}\left(\xi\frac{\partial}{\partial\xi}\right) + 1 - \frac{\mathrm{i}k_e}{2} + \frac{k_e^2}{4}\xi\right]\phi = 0 . \tag{6.130}$$

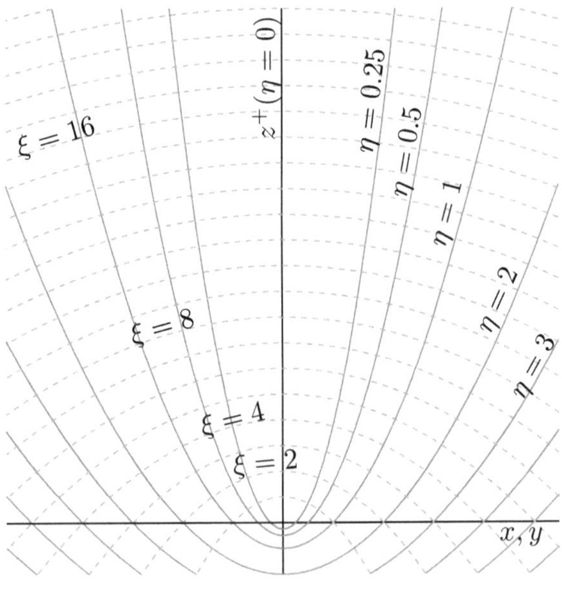

Fig. 6.2 Parabolic coordinates. The *dashed lines* are the lines of constant ξ, the *continuous ones* of constant η

The corresponding boundary condition follows from the insertion of Eq. (6.129) into Eq. (6.128)

$$\phi(\xi \to +\infty) \to \mathrm{e}^{\mathrm{i}k_e\xi/2} . \tag{6.131}$$

We thus search for a solution to Eq. (6.130) in the form

$$\phi(\xi) = C_{k_e}\mathrm{e}^{\mathrm{i}k_e\xi/2} f(-\mathrm{i}k_e\xi) , \tag{6.132}$$

where C_{k_e} is a normalization constant which we will determine so that it fulfills the condition (6.128). By inserting Eq. (6.132) into Eq. (6.130), we arrive at equation for the function $f(z)$

$$\left(z\frac{\mathrm{d}^2}{\mathrm{d}z^2} + (b-z)\frac{\mathrm{d}}{\mathrm{d}z} - a \right) f(z) = 0 , \tag{6.133}$$

where

$$a = -\frac{\mathrm{i}}{k_e} , \quad b = 1 , \quad z = -\mathrm{i}k_e\xi . \tag{6.134}$$

The least difficult way to find a solution to Eq. (6.133) is to opt for the method of Laplace transform.[9] We will now proceed by writing down a solution and subsequently showing that it is indeed the solution[10]:

$$f(z) = \int_C \frac{\mathrm{d}t}{2\pi\mathrm{i}} \mathrm{e}^t t^{a-b} (t-z)^{-a} . \tag{6.135}$$

We insert this function into Eq. (6.133) to find

$$\int_C \frac{\mathrm{d}t}{2\pi\mathrm{i}} \frac{\mathrm{d}}{\mathrm{d}t} V(t) = 0 , \qquad V(t) = \mathrm{e}^t t^{a-b+1} (t-z)^{-1-a} .$$

Hence, C is an arbitrary curve in the complex t plane with the only restriction the value of the function $V(t)$ in initial and final points be the same.

Let us define an analytic continuation of a real function $\psi(t)$ of a real variable t to complex values of t on the *principal branch of the Riemann surface* as follows. The functional relation $\psi = \psi(t)$ maintains its form and the complex values of t are given as

$$t = |t| \exp\{\mathrm{i} \arg t\} , \quad \arg t \in (-\pi, \pi) . \tag{6.136}$$

---

[9]The following exposition concerning the wave functions of the continuous part of the spectrum and transitions from the discrete to continuous parts is inspired by the exposition given in [11].

[10]A mathematician would most certainly not proceed this way. However, for the purpose of this book, we do not wish to systematically explain the method of solution of the second order linear differential equations nor the method of the Laplace transform.

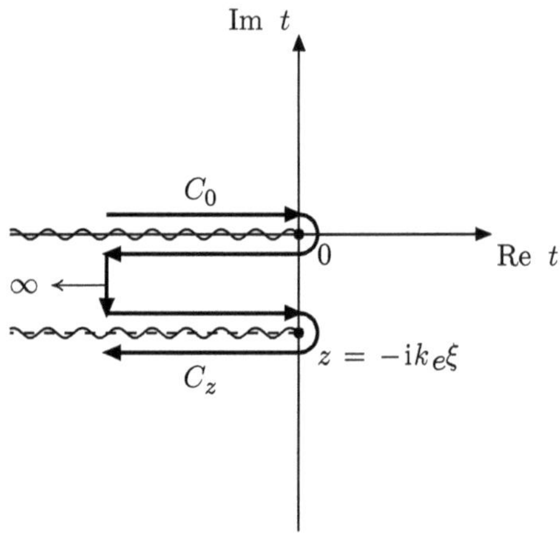

Fig. 6.3 Integration contour for the integral (6.135)

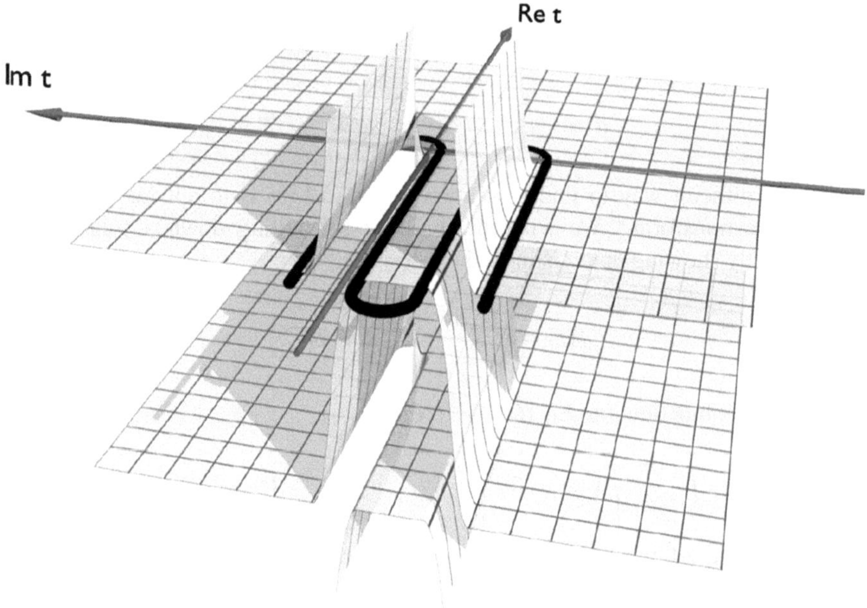

Fig. 6.4 The first two sheets of the Riemann surface of the integrand (6.135); the surface continues above and below

When the negative real axis is approached from the upper half of the complex plane, $t = |t| \exp\{i\pi\}$, we generally obtain different values of the function ψ from those if the negative real axis is approached from the lower half of the complex plane, $t = |t| \exp\{-i\pi\}$. That is, the function ψ has a discontinuity on the negative real axis. We consider the function $t^{1/2}$ for example: we obtain either $+i|t|^{1/2}$ or $-i|t|^{1/2}$ for negative t. We term the point $t = 0$ as a *branch point*, and in general, there may be more than one of such points.

In case of the integrand of the function f, Eq. (6.135), there are two branch points $t = 0$ and $t = z = -ik_e\xi$ on the principal branch of the Riemannn surface. Therefore, we want to choose the curve C in such a way that it does not leave the principal branch of the Riemann surface. Hence we have to avoid the discontinuities stemming from the points $t = 0$ and $t = z = -ik_e\xi$. One possible choice of the curve C is to compose it from three connected parts: a part C_0 where $t \in (-\infty - \mathrm{i}\varepsilon, 0 - \mathrm{i}\varepsilon)$, then a semi-circle around the origin with radius ε and $t \in (0 + \mathrm{i}\varepsilon, -\infty + \mathrm{i}\varepsilon)$, a part C_∞ connecting the points $-\infty + \mathrm{i}\varepsilon$ and $-\infty + z - \mathrm{i}\varepsilon$ and a part C_z where $t \in (z - \infty - \mathrm{i}\varepsilon, z - \mathrm{i}\varepsilon)$, then a semi-circle around z with radius ε and $t \in (z + \mathrm{i}\varepsilon, z - \infty + \mathrm{i}\varepsilon)$; see Figs. 6.3 and 6.4. The contribution from the part C_∞ vanishes, hence

$$f(z) = \int_{C_0} + \int_{C_z} \left[\frac{\mathrm{d}t}{2\pi\mathrm{i}} \mathrm{e}^t t^{a-b} (t-z)^{-a} \right] .$$

Substitution $t \to t + z$ along the part C_z transforms the function $f(z)$ into the form

$$f(z) = \int_{C_0} \frac{\mathrm{d}t}{2\pi\mathrm{i}} \left[\mathrm{e}^t t^{a-b} (t-z)^{-a} + \mathrm{e}^{t+z} (t+z)^{a-b} t^{-a} \right] .$$

For $z \to \infty$, one has

$$f(z \to \infty) \to (-z)^{-a} \int_{C_0} \frac{\mathrm{d}t}{2\pi\mathrm{i}} \mathrm{e}^t t^{a-b} + \mathrm{e}^z (z)^{a-b} \int_{C_0} \frac{\mathrm{d}t}{2\pi\mathrm{i}} \mathrm{e}^t t^{-a} .$$

By means of the formula for Γ function[11]

$$\frac{1}{\Gamma(c)} = \int_{C_0} \frac{\mathrm{d}t}{2\pi\mathrm{i}} \mathrm{e}^t t^{-c} , \tag{6.137}$$

we find that

$$f(z \to \infty) \to \frac{(-z)^{-a}}{\Gamma(b-a)} + \frac{\mathrm{e}^z (z)^{a-b}}{\Gamma(a)} . \tag{6.138}$$

[11]This is because

$$\int_{C_0} \frac{\mathrm{d}t}{2\pi\mathrm{i}} \mathrm{e}^t t^{-c} = \frac{1}{2\pi\mathrm{i}} \left[\int_0^\infty \mathrm{d}x \mathrm{e}^{-x} (x\mathrm{e}^{-\mathrm{i}\pi})^{-c} - \int_0^\infty \mathrm{d}x \mathrm{e}^{-x} (x\mathrm{e}^{\mathrm{i}\pi})^{-c} \right]$$
$$= \frac{1}{\pi} \int_0^\infty \mathrm{d}x \mathrm{e}^{-x} x^{-c} \frac{\mathrm{e}^{\mathrm{i}\pi c} - \mathrm{e}^{-\mathrm{i}\pi c}}{2\mathrm{i}} = \frac{1}{\pi} \Gamma(1-c) \sin(\pi c) = \frac{1}{\Gamma(c)} ,$$

where we used the defining formula for Γ-function, $\Gamma(z) = \int_0^\infty \mathrm{d}x \mathrm{e}^{-x} x^{z-1}$, and Eq. (6.155) in the last two steps.

After substituting into the last equation for z, a, and b, see Eq. (6.134), we have

$$f(\xi \to +\infty) \to \frac{(\mathrm{i}k_e\xi)^{\mathrm{i}/k_e}}{\Gamma(1+\mathrm{i}/k_e)} + \frac{\mathrm{e}^{-\mathrm{i}k_e\xi}(-\mathrm{i}k_e\xi)^{-\mathrm{i}/k_e-1}}{\Gamma(-\mathrm{i}/k_e)} \to \frac{(\mathrm{i}k_e\xi)^{\mathrm{i}/k_e}}{\Gamma(1+\mathrm{i}/k_e)} . \tag{6.139}$$

The numerator can be rewritten as a product of an absolute value and a phase factor,

$$(\mathrm{i}k_e\xi)^{\mathrm{i}/k_e} = \mathrm{e}^{\mathrm{i}\,\mathrm{Ln}(\mathrm{i}k_e\xi)/k_e} = \mathrm{e}^{-\pi/(2k_e)}\mathrm{e}^{+\mathrm{i}\ln|k_e\xi|/k_e} ,$$

where we took the value of the logarithm on the principal branch of the Riemann surface, $\mathrm{Ln}\,\mathrm{i} = \mathrm{i}\frac{\pi}{2}$. For large ξ, $\ln|k_e\xi|/k_e$ is negligible in comparison to $k_e\xi/2$, thus the function (6.132), obeying the condition (6.131), acquires the form

$$\phi(\xi) = \mathrm{e}^{\mathrm{i}k_e\xi/2}\mathrm{e}^{\pi/(2k_e)}\Gamma(1+\mathrm{i}/k_e)F(-\mathrm{i}/k_e, 1, -\mathrm{i}k_e\xi) . \tag{6.140}$$

Here, $F(a, 1, z)$ is a special case of a *confluent hypergeometric function*

$$F(a,b,z) = \Gamma(b)f(z) = \Gamma(b)\int_{C_0}\frac{\mathrm{d}t}{2\pi\mathrm{i}}\mathrm{e}^t t^{a-b}(t-z)^{-a} , \tag{6.141}$$

where the factor $\Gamma(b)$ is chosen so that, see Eq. (6.137),

$$F(a,b,0) = 1 .$$

By substituting the Taylor expansion of the function $F(a,b,z)$ around $z = 0$ into Eq. (6.133), we easily find

$$F(a,b,z) = 1 + \frac{a}{b}z + \frac{a(a+1)}{b(b+1)}\frac{z^2}{2!} + \frac{a(a+1)(a+2)}{b(b+1)(b+2)}\frac{z^3}{3!} + \dots \tag{6.142}$$

6.3.4 *Transition from the Discrete to Continuous Part of the Spectrum I*

Let us now calculate the following integral

$$I_s = \int \frac{\mathrm{d}^3\mathbf{r}}{r}\psi^*_{\mathbf{k}_e}(\mathbf{r})\mathrm{e}^{-\mathrm{i}\mathbf{k}\cdot\mathbf{r}}\psi_0(\nu\mathbf{r}) , \tag{6.143}$$

where

$$\psi_0(\nu\mathbf{r}) = \frac{\mathrm{e}^{-\nu r}}{\sqrt{\pi}} = \frac{\mathrm{e}^{-\nu(\xi+\eta)/2}}{\sqrt{\pi}} \tag{6.144}$$

is for $\nu = 1$ the wave function of the hydrogen ground state and $\psi_{\mathbf{k}_e}$ is the wave function of the outgoing electron obeying condition (6.113), explicit expression of which we will write down in a moment. The sought matrix element in Eq. (6.121) is obtained from this integral by differentiating with respect to the parameter **k**.

The integral (6.143) parametrically depends on the vectors **k** and $\mathbf{k}_e$. We place the vector $\mathbf{k}_e$ in the direction of z-axis and the vector **k** in the xz plane

$$\mathbf{k}_e = (0, 0, k_e)\,, \quad \mathbf{k} = (\omega \sin\vartheta, 0, \omega\cos\vartheta)\,.$$

Here, ϑ is the angle between the vectors **k** and $\mathbf{k}_e$:

$$\mathbf{k}_e \cdot \mathbf{k} = k_e\omega\cos\vartheta\,. \tag{6.145}$$

With this choice of the vector $\mathbf{k}_e$, the function $\psi_{\mathbf{k}_e}$ appearing in Eq. (6.143) acquires the following form in the parabolic coordinates[12]

$$\psi_{\mathbf{k}_e}(\xi,\eta) = \frac{e^{\pi/(2k_e)}\Gamma(1+i/k_e)}{(2\pi)^{3/2}} e^{ik_e(\xi-\eta)/2} F(-i/k_e, 1, -ik_e\xi)\,, \tag{6.146}$$

as follows from Eqs. (6.129) and (6.140). In literature, this form is commonly called the *Coulomb function.*

Similarly, the plane wave appearing in Eq. (6.143) has the form

$$e^{-i\mathbf{k}\cdot\mathbf{r}} = e^{-i\left[\omega\sin\vartheta(\xi\eta)^{1/2}\cos\varphi + \omega\cos\vartheta(\xi-\eta)/2\right]}\,.$$

After substituting these functions, the differential (6.125) and the boundaries (6.123) into the integral (6.143), we arrive at the expression

$$I_s = \frac{1}{2}\int_0^\infty d\xi \int_0^\infty d\eta \int_0^{2\pi} d\varphi \left(e^{-\nu(\xi+\eta)/2} e^{-i\left[\omega\sin\vartheta(\xi\eta)^{1/2}\cos\varphi + \omega\cos\vartheta(\xi-\eta)/2\right]}\right.$$
$$\left.\times\, e^{-ik_e(\xi-\eta)/2} F(i/k_e, 1, ik_e\xi)\right)\left[\frac{e^{\pi/(2k_e)}\Gamma(1-i/k_e)}{(2\pi)^{3/2}\sqrt{\pi}}\right]. \tag{+}$$

[12] In literature, for example in the already mentioned course [11], one can find the notation $\psi_{\mathbf{k}_e}^{(-)}$ for this function. The sign is related to the choice of the boundary condition; here *the outgoing plane wave*, see Eq. (6.113). The plane wave yields a positive value of the probability current $J = \oint \mathbf{j}\cdot d\mathbf{S}$, where $\mathbf{j}$ is given by Eq. (6.116). Similarly, one can encounter the function $\psi_{\mathbf{k}_e}^{(+)}$ with the boundary condition of an *ingoing plane wave*. These functions are nearly identical; they differ here and there merely by signs. For instance, it holds that $\psi_{\mathbf{k}_e}^{(-)} = \psi_{-\mathbf{k}_e}^{(+)*}$.

We firstly focus on the integration over the variables η and φ:

$$\frac{1}{2}\int_0^{2\pi} \mathrm{d}\varphi \int_0^{\infty} \mathrm{d}\eta \exp\left\{-[\nu - \mathrm{i}(\omega\cos\vartheta + k_e)]\frac{\eta}{2} - \mathrm{i}\omega\sin\vartheta\cos\varphi(\xi\eta)^{1/2}\right\}. \tag{6.147}$$

To calculate this integral, it is advantageous to substitute

$$u = \sqrt{\eta}\cos\varphi,\ v = \sqrt{\eta}\sin\varphi \Rightarrow \mathrm{d}u\mathrm{d}v = 2^{-1}\mathrm{d}\eta\mathrm{d}\varphi\,. \tag{6.148}$$

The integral then splits up into two one-dimensional integrals over u and v

$$\begin{aligned}(6.147) &= \int_{-\infty}^{\infty} \mathrm{d}u \exp\left\{-[\nu - \mathrm{i}(\omega\cos\vartheta + k_e)]\frac{u^2}{2} - \mathrm{i}\omega\sin\vartheta\,\xi^{1/2}u\right\} \\ &\quad\times \int_{-\infty}^{\infty} \mathrm{d}v \exp\left\{-[\nu - \mathrm{i}(\omega\cos\vartheta + k_e)]\frac{v^2}{2}\right\} \\ &= \exp\left\{-\frac{\omega^2\sin^2\vartheta}{\nu - \mathrm{i}(\omega\cos\vartheta + k_e)}\frac{\xi}{2}\right\}\left[\frac{2\pi}{\nu - \mathrm{i}(\omega\cos\vartheta + k_e)}\right], \end{aligned} \tag{++}$$

where we used formula (1.96) for integration of the Gauss function. The remaining integral over ξ is calculated by means of the formula (as we will derive in a moment)

$$\int_0^{\infty} \mathrm{e}^{-\lambda\xi} F(\mathrm{i}/k_e, 1, \mathrm{i}k_e\xi)\,\mathrm{d}\xi = \lambda^{\mathrm{i}/k_e - 1}(\lambda - \mathrm{i}k_e)^{-\mathrm{i}/k_e}\,, \tag{6.149}$$

where

$$2\lambda = \frac{\omega^2\sin^2\vartheta}{\nu - \mathrm{i}(\omega\cos\vartheta + k_e)} + \nu + \mathrm{i}(\omega\cos\vartheta + k_e) = \frac{\nu^2 + |\mathbf{k_e} + \mathbf{k}|^2}{\nu - \mathrm{i}(k_e + \omega\cos\vartheta)}\,.$$

In the last equality, we substituted from Eq. (6.145). After substituting for λ and performing several manipulations, we finally obtain

$$(6.149) = \left[2\frac{\nu - \mathrm{i}(k_e + \omega\cos\vartheta)}{|\mathbf{k}_e + \mathbf{k}|^2 + \nu^2}\left(\frac{\nu^2 + |\mathbf{k}_e + \mathbf{k}|^2}{\omega^2 + (\nu - \mathrm{i}k_e)^2}\right)^{\mathrm{i}/k_e}\right]. \tag{+++}$$

The formula (6.149) is a special case of the formula

$$\begin{aligned}\frac{1}{\Gamma(b)}\int_0^{\infty} \mathrm{e}^{-\lambda z}z^{b-1}F(a,b,z)\mathrm{d}z &= \int_C \frac{\mathrm{d}t}{2\pi\mathrm{i}}t^{a-b}(t-1)^{-a}\left(\int_0^{\infty}\mathrm{e}^{-(\lambda-t)z}\mathrm{d}z\right) \\ &= \int_C \frac{\mathrm{d}t}{2\pi\mathrm{i}}t^{a-b}(t-1)^{-a}\frac{1}{\lambda - t} = \lambda^{a-b}(\lambda-1)^{-a}\,. \end{aligned} \tag{6.150}$$

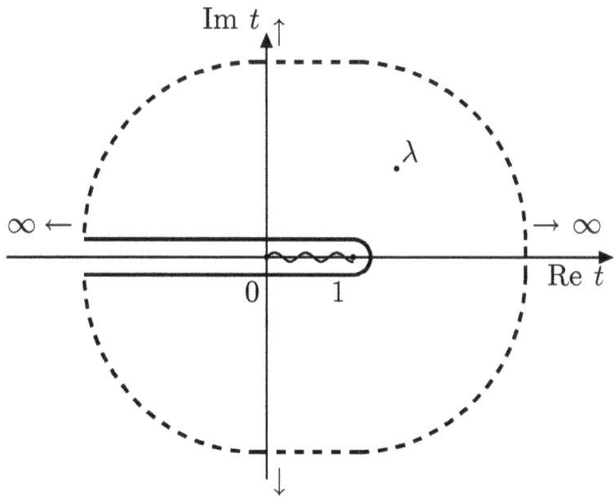

Fig. 6.5 Integration contour for the integral (6.150). The *continuous line* denotes the part of the integration contour necessary for the definition of the hypergeometric function. *The dashed line* denotes the part of the contour necessary for the use of the residue theorem. On this line, the integrand asymptotically vanishes

In the first equality, we substituted $t \to tz$ to manipulate the function $f(z)$, Eq. (6.135), into the form

$$f(z) = z^{1-b} \int_C \frac{\mathrm{d}t}{2\pi \mathrm{i}} \mathrm{e}^{tz} t^{a-b} (t-1)^{-a} ,$$

where the curve C now goes from $-\infty$ closely below the real axis, encircles the point $z = 1$ and returns back to $-\infty$ tightly above the real axis. The discontinuities from branch points $z = 0$ and $z = 1$ for negative z in this case cancel out each other (owing to the opposite exponents) and only the cut between them remains, see Fig. 6.5. The topology of the Riemann surface is displayed in Fig. 6.6. The second equality in Eq. (6.150) follows from trivial integration over z. In the last equality in Eq. (6.150), we used the residue theorem as follows. In the third form of Eq. (6.150), the integrand behaves for large t as $|t|^{-(b+1)}$. Thus for $b \geq 1$, the integral over a large circle vanishes in the limit of an infinite radius. From the residue theorem, the integral along the curve C consequently equals $2\pi \mathrm{i}$ times the residue at the pole $t = \lambda$.

After multiplying the expressions in the square brackets in Eqs. (+), (++), and (+++), we finally obtain for the total integral I_s, Eq. (6.143),

$$I_s = \frac{\mathrm{e}^{\pi/(2k_e)} \Gamma(1 - \mathrm{i}/k_e) 2^{1/2}}{\pi} \frac{1}{|\mathbf{k}_e + \mathbf{k}|^2 + \nu^2} \left(\frac{\nu^2 + |\mathbf{k}_e + \mathbf{k}|^2}{\omega^2 + (\nu - \mathrm{i}k_e)^2} \right)^{\mathrm{i}/k_e} . \tag{6.151}$$

6.3.5 *Angular and Energy Distribution of Outgoing Electrons*

Returning now back to the photoelectric effect, we obtain the sought matrix element in Eq. (6.121) from the last equation by differentiation with respect to a parameter

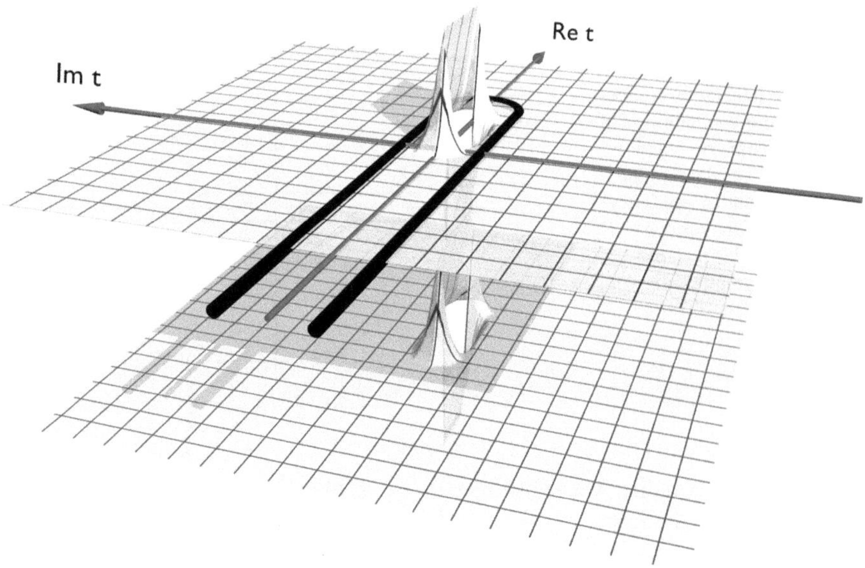

Fig. 6.6 The first two sheets of the Riemann surface of the integrand (6.150). The surface continues above and below

$$\left|\boldsymbol{\varepsilon}^{(\nu)}\cdot\langle F_{\rm at}|\hat{\boldsymbol{n}}|I_{\rm at}\rangle\right|^2=\left|\varepsilon_i^{(\nu)}\left(\frac{\partial}{\partial k_i}I_s\right)_{\substack{\mathbf{k}=0\\ \nu=1}}\right|^2$$
$$=\left|\boldsymbol{\varepsilon}^{(\nu)}\cdot\boldsymbol{\eta}\right|^2\frac{8}{(1+k_e^2)^3}\left(\frac{1+\mathrm{i}k_e}{1-\mathrm{i}k_e}\right)^{2\mathrm{i}/k_e}\frac{\mathrm{e}^{\pi/k_e}\left|\Gamma\left(1-\frac{\mathrm{i}}{k_e}\right)\right|^2}{\pi^2}\,. \tag{6.152}$$

Here, we introduced spherical coordinates for the electron wave vector

$$\mathbf{k}_e=k_e\boldsymbol{\eta}\,.$$

In the first equality, we inserted Eq. (6.146) for $\langle\mathbf{r}|F_{\rm at}\rangle=\psi_{\mathbf{k}_e}(\mathbf{r})$ and Eq. (6.144) for $\langle\mathbf{r}|I_{\rm at}\rangle=\psi_0(\mathbf{r})$. One can see from Eq. (6.152) that the most probable direction of the outgoing electron matches that of the polarization vector of the incident photon. However, we could have guessed this result beforehand, without any calculation, couldn't we?

After inserting Eq. (6.152) into Eq. (6.121), integrating over directions of outgoing electrons, using Eq. (6.96) and relations

$$\boldsymbol{\varepsilon}^{(\lambda)}\cdot\boldsymbol{\varepsilon}^{(\lambda)}=1$$

and

$$\frac{k_e}{2\pi}\left(\frac{1+\mathrm{i}k_e}{1-\mathrm{i}k_e}\right)^{2\mathrm{i}/k_e} \mathrm{e}^{\pi/k_e}\left|\Gamma\left(1-\frac{\mathrm{i}}{k_e}\right)\right|^2 = \frac{\mathrm{e}^{-\frac{4}{k_e}\arctan k_e}}{1-\mathrm{e}^{-2\pi/k_e}}, \tag{6.153}$$

we finally find the total cross-section for photoelectric effect

$$\sigma = \frac{\alpha}{(mZ\alpha)^2}\frac{2^9\pi^2}{3}\frac{1}{(1+k_e^2)^4}\frac{\mathrm{e}^{-(4\arctan k_e)/k_e}}{1-\mathrm{e}^{-2\pi/k_e}}. \tag{6.154}$$

In Eq. (6.153), we used the formulas for the Γ-function

$$\Gamma(z)\Gamma(1-z) = \frac{\pi}{\sin(\pi z)}, \qquad \Gamma(z+1) = z\Gamma(z). \tag{6.155}$$

The expression (6.154) is given in terms of the magnitude of the electron wave vector k_e. This quantity is related to the frequency ω, which experimenters tune, of the incident photon via the Einstein equation (6.110).

The function on the rhs of Eq. (6.154) reaches its maximum for $k_e \to 0$; with increasing k_e it plummets to zero. The probability of the photoeffect is the largest if the frequency of the incident photons is slightly above the lowest possible frequency necessary for the effect, i.e., slightly above the ionization threshold. For this particular case, the expression (6.153) equals e^{-4}; hence

$$\sigma_{k_e\to 0} \to \frac{\alpha}{(mZ\alpha)^2}\frac{2^9\pi^2}{3\mathrm{e}^4} \to 0.630\times 10^{-21}\,\mathrm{m}^2, \tag{6.156}$$

where the conversion from $(\mathrm{eV})^{-2}$ to m^2 was outlined in Eq. (5).

Note, though, that we could have expected this result (6.156) as well. The total cross-section has the dimension of an area, therefore we can perceive it as an area "seen" by the incident photon. The "atom radius," called the *Bohr radius*, is given as

$$a_0 = \frac{1}{m_e\alpha} \to 0.529\times 10^{-10}\,\mathrm{m}, \tag{6.157}$$

hence the magnitude of the "atom area" equals approximately $10^{-20}\,\mathrm{m}^2$. We need to multiply this area by the coupling between the electron and the photon, $\alpha \simeq 10^{-2}$. We thus obtain the order estimate $10^{-22}\,\mathrm{m}^2$ which lies close to the exact result (6.156).

With increasing k_e, the wave function of continuous spectrum becomes more and more oscillatory. Consecutively, the overlap integral between the wave functions of the ground state and of the continuous spectrum approaches zero. For large k_e, the expression (6.153) approaches $k_e/(2\pi)$; hence

$$\sigma_{k_e \to \infty} \to \frac{\alpha}{(mZ\alpha)^2} \frac{2^8 \pi}{3k_e^7} = Z^5 \frac{\alpha}{(m\alpha)^2} \frac{2^8 \pi}{3} \left(\frac{m\alpha^2}{2\omega} \right)^{7/2} , \tag{6.158}$$

where we inserted $k_e \simeq (Z\alpha)^{-1}(2\omega/m)^{1/2}$ in the last equality, as follows from Einstein equation (6.110) for large k_e. For larger frequencies of the incident photon, the dependence of the photoeffect goes with the fifth power of the nuclear charge Z (compare with the dependence Z^{-2} in Eq. (6.156)!), and with the $7/2$-th power of the ratio of the hydrogen ionization energy $m\alpha^2/2$ and photon energy ω.

6.3.6 Excitation of an Atom by an Electron Impact

We have already learned enough to be able to estimate the probability of an atom excitation by an electron impact. For the sake of simplicity, we restrict ourselves to the hydrogen atom. Subsequent generalization to more complex atoms is, at least in principle, simple. We assume that the hydrogen initial state is its ground state and that the energy of the incident electron falls within the range $(10 - 10^5)$ eV. On one hand, this range allows the process of the atom excitation to happen at all, and on the other hand, it allows us to treat the problem within the nonrelativistic approximation. The overlap between the wave functions of the free and bound electron is small, hence we will treat the electrons as distinguishable.[13]

As usual, we perform the calculation in the atomic units.

1. We split the Hamiltonian as follows:

$$\hat{\mathsf{H}} = \hat{\mathsf{H}}_0 + \hat{\mathsf{H}}_1 , \quad \hat{\mathsf{H}}_0 = \hat{\mathsf{H}}_0^{(1)} + \hat{\mathsf{H}}_0^{(2)} ,$$

$$\hat{\mathsf{H}}_0^{(1)} = \frac{\hat{\mathsf{p}}_1^2}{2} - \frac{1}{\hat{\mathsf{r}}_1} , \quad \hat{\mathsf{H}}_0^{(2)} = \frac{\hat{\mathsf{p}}_2^2}{2} , \quad \hat{\mathsf{H}}_1 = \frac{1}{\hat{\mathsf{r}}_{12}} - \frac{1}{\hat{\mathsf{r}}_2} .$$

   Here, $\hat{\mathsf{H}}_0^{(1)}$ is the hydrogen Hamiltonian, $\hat{\mathsf{H}}_0^{(2)}$ is a Hamiltonian of a free electron and $\hat{\mathsf{H}}_1$ describes their mutual electrostatic interaction.
2. After inserting these relations into the Fermi golden rule (6.79), we obtain[14]

$$\mathrm{d}w_{I \to F} = 2\pi \delta \left(\frac{k_F^2 - k_I^2}{2} - \omega_{IF}^{at} \right) \sum_{l,m} \left| (\hat{\mathsf{H}}_1)_{IF} \right|^2 \mathrm{d}^3 \mathbf{k}_F , \tag{6.159}$$

[13] De Broglie electron wavelength is related to the electron energy via the formula $\lambda = 2\pi/p = 2\pi/\sqrt{2mE}$. By means of Eqs. (1) and (5), this translates to $\lambda \approx \frac{10^{-9}}{E(\mathrm{eV})}$ [m]. Since the "size" of an atom is of the order 10^{-10} m, we see that the overlap is in fact negligible only for energies above 10^2 eV.

[14] The use of the Fermi golden rule for scattering problems is usually referred to as the *first Born approximation*.

where the initial and final states are

$$|I\rangle = |1,0,0\rangle_1 |\mathbf{k}_I\rangle_2 \quad |F\rangle = |n,l,m\rangle_1 |\mathbf{k}_F\rangle_2$$

and ω_{IF}^{at} is the difference (6.84) between the atom final and initial energies (in atomic units). To eliminate the δ-function, we write

$$\mathrm{d}^3\mathbf{k}_F = k_F \mathrm{d}\left(\frac{k_F^2}{2}\right) \mathrm{d}\Omega\,. \tag{6.160}$$

3. Substituting

$$\langle \mathbf{r}|\,\mathbf{k}_I\rangle = \frac{\mathrm{e}^{\mathrm{i}\mathbf{k}_I\cdot\mathbf{r}}}{(2\pi)^{3/2}}$$

into Eq. (6.116) yields

$$j_I = \frac{k_I}{(2\pi)^3}\,. \tag{6.161}$$

4. After substituting Eqs. (6.159), (6.160) and (6.161) into Eq. (6.119), we find

$$\frac{\mathrm{d}\sigma}{\mathrm{d}\Omega} = (2\pi)^4 \frac{k_F}{k_I} \sum_{l,m} \left|(\hat{\mathsf{H}}_1)_{IF}\right|^2\,. \tag{6.162}$$

Here, k_F is determined by the energy conservation following from the δ-function in Eq. (6.159)

$$k_F^2 = k_I^2 + 2\omega_{IF}^{\text{at}}\,.$$

5. To calculate $(\hat{\mathsf{H}}_1)_{IF}$, we evaluate first

$$\begin{aligned} \langle \mathbf{k}_I|_2 \left(\frac{1}{\hat{\mathsf{r}}_{12}} - \frac{1}{\hat{\mathsf{r}}_2}\right) |\mathbf{k}_F\rangle_2 &= \int \frac{\mathrm{e}^{\mathrm{i}(\mathbf{k}_F-\mathbf{k}_I)\cdot\mathbf{r}}}{(2\pi)^3} \left(\frac{1}{|\mathbf{r}-\mathbf{r}_1|} - \frac{1}{r}\right) \mathrm{d}^3\mathbf{r} \\ &= \frac{4\pi}{(2\pi)^3} \frac{1}{q^2} \left(\mathrm{e}^{\mathrm{i}\mathbf{q}\cdot\mathbf{r}_1} - 1\right)\,, \end{aligned} \tag{6.163}$$

where we used Eq. (3.114) and denoted

$$\mathbf{q} = \mathbf{k}_F - \mathbf{k}_I\,. \tag{6.164}$$

For obvious reasons, $\mathbf{q}$ is referred to as the *exchanged momentum*.

6. What now remains to calculate is the matrix element

$$(\hat{\mathsf{H}}_1)_{IF} = \frac{4\pi}{(2\pi)^3}\frac{1}{q^2}\langle 1,0,0|\left(\mathrm{e}^{\mathrm{i}\mathbf{q}\cdot\hat{r}} - 1\right)|n,l,m\rangle\,.$$

Owing to the orthogonality of the hydrogen states, the second term in the brackets vanishes. We now use the decomposition of a plane wave into spherical waves, Eq. (5.30). Owing to the orthogonality of the spherical harmonics, the angular integration is trivial and we arrive at the expression

$$(\hat{\mathsf{H}}_1)_{IF} = \frac{(4\pi)^{3/2}}{(2\pi)^3}\frac{1}{q^2}Y_{lm}(\boldsymbol{\eta})\mathrm{i}^l\langle 1,0|j_l(q\hat{\mathrm{r}})|n,l\rangle\,, \tag{6.165}$$

where $\mathbf{q} = q\boldsymbol{\eta}$.

7. Substitution of the last equation into Eq. (6.162) leads to

$$\sum_{m=-l}^{l}|Y_{lm}(\boldsymbol{\eta})|^2 = \frac{2l+1}{4\pi}\,, \tag{6.166}$$

where we used Eq. (5.29) for $\mathbf{n} = \boldsymbol{\eta}$,

8. The rhs of Eq. (6.162) now depends on the angle ϑ between the directions of the momentum of the ingoing and outgoing electrons. Squaring Eq. (6.164), we have

$$q^2 = k_I^2 + k_F^2 - 2k_Ik_F\cos\vartheta\,.$$

We can now simplify our further calculation by the transformation of the integral over the angle ϑ to an integral over the magnitude of the exchanged momentum q. We thus obtain from the last equation

$$\mathrm{d}\Omega = 2\pi\mathrm{d}\vartheta\sin\vartheta = \frac{2\pi}{k_Ik_F}q\,\mathrm{d}q\,. \tag{6.167}$$

9. From Eqs. (6.162), (6.165), (6.166), and (6.167), we find for the total cross-section

$$\sigma = \frac{8\pi}{k_I^2}\sum_{l=0}^{n-1}(2l+1)\int\limits_{|k_F-k_I|}^{k_F+k_I}\frac{|\langle 1,0|j_l(q\hat{\mathrm{r}})|n,l\rangle|^2}{q^4}q\,\mathrm{d}q\,. \tag{6.168}$$

For excitation to $2s$ and $2p$ states, one can easily evaluate the remaining radial integrals[15]

[15] Use Eqs. (3.35), (3.95), (3.96), (4.84), (4.85) and (4.86).

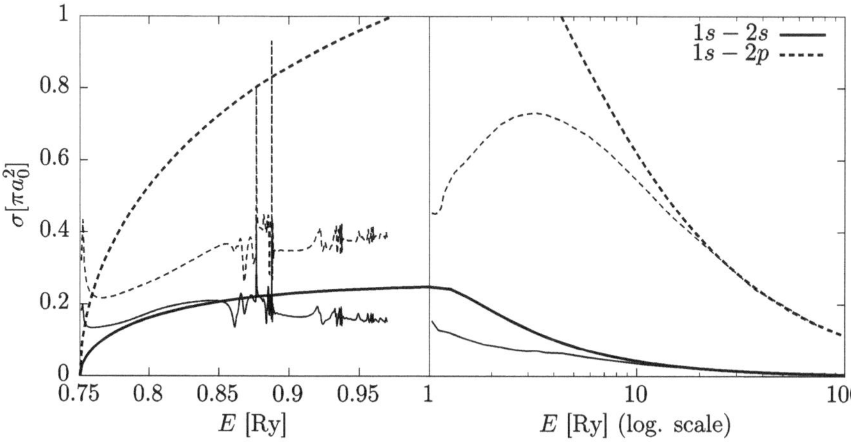

Fig. 6.7 The *bold lines* denote the dependence of the total cross-section (6.168) on the incident electron energy $E = k_I^2$. The *thin lines* denote results of more accurate methods that solve the Schrödinger equation non-perturbatively. The accurate data on the *left and right panel* are adopted from [1] and [2], respectively. The value $E = 1\,\text{Ry} \doteq 13.6\,\text{eV}$ is the ionization threshold; i.e., it is the minimal value necessary for the ionization of hydrogen by the incident electron. The region directly below the ionization threshold is very difficult to calculate due to the large number of closely lying hydrogen energy levels. The total cross-section is in the units of πa_0^2, where a_0 is the Bohr radius (6.157)

$$|F_{1s2s}| = \left|\langle 1,0|j_0(q\hat{r})|2,0\rangle\right| = \frac{2^{17/2}q^2}{(4q^2+9)^3},$$

$$|F_{1s2p}| = \left|\langle 1,0|j_1(q\hat{r})|2,1\rangle\right| = \frac{2^{15/2}\sqrt{3}q}{(4q^2+9)^3},$$

where the quantities F are called the *form-factors*. Finally, we integrate over q. The result, however, cannot be cast into a well-arranged compact form, therefore we rather display it graphically in Fig. 6.7.

In Fig. 6.7, we compare the results of our calculation (bold lines) with those of more accurate ones to assess the use of the first order of the perturbation method. One can readily see that we overshoot in the region of low energies and reach a satisfactory agreement only in the region above approximately 20 eV of the incident electron energy. In addition, the reader surely does not overlook that this approach is not able to capture the *resonances* visible on the curves of the cross-section below the ionization threshold. These resonances appear when an electron "does not simply fly by" the hydrogen atom, but creates the unstable excited state of H^- instead.[16] One can also notice that the cross-section of the reaction $1s$-$2p$ is always higher than that of the reaction $1s$-$2s$.

[16] This kind of resonance is called the Feshbach resonance.

Resonances in electron-atom scattering experiments are generally non-perturbative phenomena and Lippmann-Schwinger equation is usually solved for their theoretical description. In the next section, we will turn our attention to its derivation.

6.4 Photon-Atom Scattering

Having discussed the spontaneous emission and the photoelectric effect, we now turn our attention to the next simplest quantum electrodynamic process: photon-atom scattering. At time t_0, an experimenter sends a photon with a wave vector $\mathbf{k}_1$ and polarization $\boldsymbol{\varepsilon}^{(1)}$ towards an atom; the atom is in the ground state. In later time t, the experimenter detects a photon with a wave vector $\mathbf{k}_2$ and polarization $\boldsymbol{\varepsilon}^{(2)}$. This is, of course, an idealization of a real experiment. The real experiment involves a very large number of photons and very large number of atoms. However, unless special care is taken for collective effects to appear, such as lasers, slowing the light and so on, our idealization is sufficient for the correct description of the real situation.

Photons possess no charge; hence the interaction between a photon and an atom takes place on the scale of an atom, i.e., 10^{-10} m, which is by ten orders of magnitude smaller than the typical distance between the emission and absorption of photons by the experimenter. Thus, the interaction time is about ten orders of magnitude smaller than the typical time span between the emission and absorption of the photons.[17]

The expression (6.74), $P_{I\to F} = |c_F(t)|^2$, gives the probability that a system prepared at time t_0 in the state $|I\rangle$, will be found at later time t in the state $|F\rangle$. As follows from our previous discussion, we are allowed to take the limit $t - t_0 \to \infty$. We obtain the so-called *S-matrix* elements, see for instance [14],

$$P_{I\to F} = |S_{FI}|^2 = |c_F(t - t_0 \to +\infty)|^2 \,. \tag{6.169}$$

[17]Strictly speaking, this is not entirely correct. Firstly, a photon with a definite wave vector $\mathbf{k}_1$ is completely delocalized in space and time. One thus cannot talk about the times of emission, interaction and absorption. Nonetheless, taking the appropriate linear combination of states with wave vectors close to $\mathbf{k}_1$, we can obtain a state localized in space and time. This also corresponds to the photons the experimenter is able to create. Secondly, the interaction between the atom and the EM field can never be "switched off," see Sect. 6.5. Even if no photon is present, the atom still interacts with fluctuations of the EM field. However, we ignore both reservations. The first one is merely a technicality; when considering the appropriate linear combination, we arrive at the very same result, but complicate the calculation leading to it. The second one leads only to a small correction to the result, see the discussion in Sect. 6.4.3.

Recall that the coefficients c_F are a solution to Eq. (6.71) with the initial condition (6.73), $c_n(t_0) = \delta_{nI}$. Thus, the initial state $|I\rangle$ enters into the coefficients through the initial condition.

Now, recall that the Hamiltonian has the form $\hat{\mathsf{H}} = \hat{\mathsf{H}}_0 + \hat{\mathsf{H}}_{\text{int}}$, see Eq. (6.63), where $\hat{\mathsf{H}}_0$ describes dynamics of a free EM field and a free atom, Eqs. (6.61) and (6.64), and $\hat{\mathsf{H}}_{\text{int}}$ describes the interaction of the atom and EM field, (6.65). The Hamiltonian does not depend directly on time, therefore we attempt to eliminate the time dependence completely. This approach is advantageous in nonrelativistic theory where the time is the same in all inertial systems.

6.4.1 *Lippmann-Schwinger Equation*

Firstly, we transform the differential equation (6.71) with the initial condition (6.73) to the integral equation

$$c_q(t) = \delta_{qI} - \mathrm{i} \sum_m (\hat{\mathsf{H}}_{\text{int}})_{qm} \int_{t_0}^{t} \mathrm{e}^{\mathrm{i}\omega_{qm}t'} c_m(t')\, \mathrm{d}t' \,. \tag{6.170}$$

Let us isolate the case of no scattering

$$S_{FI} = \delta_{FI} + T_{FI} \tag{6.171}$$

from matrix S_{FI}. For elements of the so-called *T-matrix*, see [14], we find after inserting Eqs. (6.170) and (6.171) into Eq. (6.169)

$$T_{FI} = -\mathrm{i} d_F(t_1 - t_0 \to \infty)\,, \tag{6.172}$$

where

$$d_n = \sum_q (\hat{\mathsf{H}}_{\text{int}})_{nq} \int_{t_0}^{t_1} \mathrm{e}^{\mathrm{i}\omega_{Fq}t'} c_q(t')\, \mathrm{d}t' \,. \tag{6.173}$$

Note that for a general coefficient d_n, $\omega_{Fq} = E_F - E_q$ appears in the argument of the exponential function, where E_F denotes the energy of the final state. If we perform the indicated limit $t_1 - t_0 \to \infty$, we easily obtain in Eqs. (6.173) and (6.170) an undefined integral of an infinitely oscillating exponential function. To avoid it, we add an infinitesimal imaginary part to the transition frequencies

$$\mathrm{i}\omega_{Fq}t \to \mathrm{i}\omega_{Fq}t - \varepsilon|t| \,. \tag{6.174}$$

Later on, in Sect. 7.4.1, we will analyze the physical significance of this replacement in detail. In terms of mathematics, it is clear—for an arbitrarily small positive ε we transform the integral (6.173) to a convergent one.

Let us now try to find an equation for the coefficients d_n. We substitute the formula (6.170) for c_q on the rhs of Eq. (6.173); we find

$$d_n = (\hat{\mathsf{H}}_{\rm int})_{nI} \int_{t_0}^{t_1} \mathrm{d}t\, \mathrm{e}^{\mathrm{i}\omega_{FI}t}$$
$$- \mathrm{i} \sum_q (\hat{\mathsf{H}}_{\rm int})_{nq} \sum_m (\hat{\mathsf{H}}_{\rm int})_{qm} \int_{t_0}^{t_1} \mathrm{d}t\, \mathrm{e}^{\mathrm{i}\omega_{Fq}t} \int_{t_0}^{t} \mathrm{d}t'\, \mathrm{e}^{\mathrm{i}\omega_{qm}t'} c_m(t') \,.$$

The last term features an integral of a two-dimensional function $f(t,t') = \mathrm{e}^{\mathrm{i}\omega_{Fq}t}\mathrm{e}^{\mathrm{i}\omega_{qm}t'} c_m(t')$ over the triangle $t \in (t_0, t_1)$, $t' \in (t_0, t)$. The reader will be surely able to sketch the corresponding picture. Owing to the replacement (6.174), the function $f(t,t')$ is damped when t or t' approach $\pm\infty$. Hence we can exchange the order of the integrations while preserving the integration region

$$\int_{t_0}^{t_1} \mathrm{d}t\, \mathrm{e}^{\mathrm{i}\omega_{Fq}t} \int_{t_0}^{t} \mathrm{d}t'\, \mathrm{e}^{\mathrm{i}\omega_{qm}t'} c_m(t') = \int_{t_0}^{t_1} \mathrm{d}t'\, \mathrm{e}^{\mathrm{i}\omega_{qm}t'} c_m(t') \int_{t'}^{t_1} \mathrm{d}t\, \mathrm{e}^{\mathrm{i}\omega_{Fq}t}$$

and subsequently shift the inner integration variable $t \to t' + t$:

$$\int_{t_0}^{t_1} \mathrm{d}t'\, \mathrm{e}^{\mathrm{i}\omega_{qm}t'} c_m(t') \int_{t'}^{t_1} \mathrm{d}t\, \mathrm{e}^{\mathrm{i}\omega_{Fq}t} = \int_{t_0}^{t_1} \mathrm{d}t' \int_{0}^{t_1-t'} \mathrm{d}t\, \mathrm{e}^{\mathrm{i}\omega_{Fq}(t+t')}\mathrm{e}^{\mathrm{i}\omega_{qm}t'} c_m(t') \,.$$

Now, bearing in mind the replacement (6.174), the region of very large t' contributes little, and we obtain appreciable contribution only for $t' \ll +\infty$. Consecutively, in the limit $t_1 \to +\infty$, we can replace in the upper limit of the inner integral $t_1 - t'$ by t_1. The double integral then factorizes into a product of two one-dimensional integrals. By means of the definition (6.173) and the last three equations, we have

$$d_n = (\hat{\mathsf{H}}_{\rm int})_{nI} \int_{t_0}^{t_1} \mathrm{d}t\, \mathrm{e}^{\mathrm{i}\omega_{FI}t} - \mathrm{i} \sum_q (\hat{\mathsf{H}}_{\rm int})_{nq} d_q \int_{0}^{t_1} \mathrm{d}t\, \mathrm{e}^{\mathrm{i}(\omega_{Fq}+\mathrm{i}\varepsilon)t} \,.$$

There is still a large number of unused letters, so let us pick one and denote

$$d_n = \int_{t_0}^{t_1} \mathrm{d}t\, \mathrm{e}^{\mathrm{i}\omega_{FI}t} b_n \,. \tag{6.175}$$

In the last term of the above expression, we perform integration over t for $t_1 \to \infty$. For *time-independent* coefficients b_F, we obtain using $E_I = E_F$ (see below) the *Lippmann-Schwinger equation*

$$b_n = (\hat{\mathsf{H}}_{\rm int})_{nI} - \sum_q \frac{(\hat{\mathsf{H}}_{\rm int})_{nq} b_q}{\omega_{qI} - \mathrm{i}\varepsilon} \,. \tag{6.176}$$

Let us emphasize once again that the summation over q has to be understood symbolically; it is a "summation" over the complete spectrum of the operator $\hat{\mathsf{H}}_0$. For the transition probability, we derive from Eqs. (6.169), (6.171), (6.172), and (6.175)

$$P_{I\to F\neq I} = |T_{FI}|^2 = |b_F|^2 \lim_{t_1-t_0\to\infty} \left| \int_{t_0}^{t_1} \mathrm{d}t \, \mathrm{e}^{\mathrm{i}\omega_{FI}t} \right|^2$$

By means of Eq. (6.78), $\lim_{t_1-t_0\to\infty} \left| \int_{t_0}^{t_1} \mathrm{e}^{\mathrm{i}\omega_{FI}t'} \mathrm{d}t' \right|^2 = 2\pi(t_1-t_0)\delta(\omega_{FI})$, we obtain a generalization of the Fermi golden rule, Eq. (6.79), for the transition rate in the form of the non-perturbative expression

$$w_{I\to F} = 2\pi\delta(\omega_{IF})|b_F|^2 \,, \tag{6.177}$$

where the coefficients b_F solve Eq. (6.176).

6.4.2 Elimination of Field Operators

As noted above, when considering the photon-atom scattering, the initial and final states are

$$|I\rangle = |I_{\mathrm{at}}\rangle \hat{\mathsf{a}}^+(\mathbf{k}_1, \lambda_1)|0\rangle \,, \qquad |F\rangle = |F_{\mathrm{at}}\rangle \hat{\mathsf{a}}^+(\mathbf{k}_2, \lambda_2)|0\rangle \,; \tag{6.178}$$

with energies

$$\hat{\mathsf{H}}_0|I\rangle = (E_I^{\mathrm{at}} + \omega_1)|I\rangle \,, \quad \hat{\mathsf{H}}_0|F\rangle = (E_F^{\mathrm{at}} + \omega_2)|F\rangle \,. \tag{6.179}$$

In the first and second orders of e, we derive from Eqs. (6.176) and (6.65)

$$b_F^{(1)} = (\hat{\mathsf{H}}_1)_{FI} \,, \tag{6.180}$$

$$b_F^{(2)} = (\hat{\mathsf{H}}_2)_{FI} - \sum_n (\hat{\mathsf{H}}_1)_{Fn} \frac{1}{\omega_{nI}} (\hat{\mathsf{H}}_1)_{nI} \,. \tag{6.181}$$

The matrix element $(\hat{\mathsf{H}}_1)_{FI}$ vanishes in this case

$$\sum_j \langle F_{\mathrm{at}}|\langle 0|\hat{\mathsf{a}}(\mathbf{k}_2, \lambda_2) \Bigg(-\frac{1}{m_j} \int \frac{\mathrm{d}^3\mathbf{k}}{(2\pi)^{3/2}} \frac{1}{\sqrt{2\omega}} \sum_\lambda \boldsymbol{\varepsilon}^{(\lambda)} \cdot \hat{\mathbf{p}}^{(j)}$$
$$\times \left(\hat{\mathsf{a}}(\mathbf{k}, \lambda) \mathrm{e}^{\mathrm{i}\mathbf{k}\cdot\hat{\mathbf{r}}^{(j)}} + \hat{\mathsf{a}}^+(\mathbf{k}, \lambda) \mathrm{e}^{-\mathrm{i}\mathbf{k}\cdot\hat{\mathbf{r}}^{(j)}} \right) \Bigg) \hat{\mathsf{a}}^+(\mathbf{k}_1, \lambda_1)|0\rangle|I_{\mathrm{at}}\rangle = 0 \,.$$

Obviously, the initial and final states are one-photon states, and the vector potential $\hat{\boldsymbol{A}}$ is a linear combination of annihilation and creation operators. Within the first order of the perturbation method, we are able to describe an emission or an absorption of a photon, i.e., a creation or a destruction of a photon. However, to describe scattering, we need to first annihilate the initial photon and subsequently create the final one, or vice versa. In both cases, we thus need employ the second order of the perturbation method.

Let us start with the first term on the rhs of Eq. (6.181), $(\hat{\mathsf{H}}_2)_{FI}$, see Eqs. (6.51) and (6.65),

$$\begin{aligned}(\hat{\mathsf{H}}_2)_{FI} = \sum_j \frac{1}{(2\pi)^3}\langle F_{\mathrm{at}}|\langle 0|\hat{\mathsf{a}}(\mathbf{k}_2,\lambda_2)\frac{e_j^2}{2m_j}\int\frac{\mathrm{d}^3\mathbf{k}}{\sqrt{2\omega}}\int\frac{\mathrm{d}^3\mathbf{k}'}{\sqrt{2\omega'}}\sum_{\lambda,\lambda'}\boldsymbol{\varepsilon}^{(\lambda)}\cdot\boldsymbol{\varepsilon}^{(\lambda')} \\ \times\left(\hat{\mathsf{a}}(\mathbf{k},\lambda)\mathrm{e}^{\mathrm{i}\mathbf{k}\cdot\hat{\boldsymbol{r}}^{(j)}} + \hat{\mathsf{a}}^+(\mathbf{k},\lambda)\mathrm{e}^{-\mathrm{i}\mathbf{k}\cdot\hat{\boldsymbol{r}}^{(j)}}\right) \\ \times\left(\hat{\mathsf{a}}(\mathbf{k}',\lambda')\mathrm{e}^{\mathrm{i}\mathbf{k}'\cdot\hat{\boldsymbol{r}}^{(j)}} + \hat{\mathsf{a}}^+(\mathbf{k}',\lambda')\mathrm{e}^{-\mathrm{i}\mathbf{k}'\cdot\hat{\boldsymbol{r}}^{(j)}}\right)\hat{\mathsf{a}}^+(\mathbf{k}_1,\lambda_1)|0\rangle\,|I_{\mathrm{at}}\rangle\,. \end{aligned} \tag{6.182}$$

From the expression

$$\begin{aligned}\langle 0|\hat{\mathsf{a}}(\mathbf{k}_2,\lambda_2)\left(\hat{\mathsf{a}}(\mathbf{k},\lambda)\mathrm{e}^{\mathrm{i}\mathbf{k}\cdot\hat{\boldsymbol{r}}^{(j)}} + \hat{\mathsf{a}}^+(\mathbf{k},\lambda)\mathrm{e}^{-\mathrm{i}\mathbf{k}\cdot\hat{\boldsymbol{r}}^{(j)}}\right) \\ \times\left(\hat{\mathsf{a}}(\mathbf{k}',\lambda')\mathrm{e}^{\mathrm{i}\mathbf{k}'\cdot\hat{\boldsymbol{r}}^{(j)}} + \hat{\mathsf{a}}^+(\mathbf{k}',\lambda')\mathrm{e}^{-\mathrm{i}\mathbf{k}'\cdot\hat{\boldsymbol{r}}^{(j)}}\right)\hat{\mathsf{a}}^+(\mathbf{k}_1,\lambda_1)|0\rangle\,,\end{aligned}$$

only the following commutators remain

$$[\hat{\mathsf{a}}(\mathbf{k}_2,\lambda_2),\hat{\mathsf{a}}^+(\mathbf{k},\lambda)][\hat{\mathsf{a}}(\mathbf{k}',\lambda'),\hat{\mathsf{a}}^+(\mathbf{k}_1,\lambda_1)]\mathrm{e}^{-\mathrm{i}(\mathbf{k}-\mathbf{k}')\cdot\hat{\boldsymbol{r}}^{(j)}} \tag{6.183}$$

and

$$[\hat{\mathsf{a}}(\mathbf{k}_2,\lambda_2),\hat{\mathsf{a}}^+(\mathbf{k}',\lambda')][\hat{\mathsf{a}}(\mathbf{k},\lambda),\hat{\mathsf{a}}^+(\mathbf{k}_1,\lambda_1)]\mathrm{e}^{\mathrm{i}(\mathbf{k}-\mathbf{k}')\cdot\hat{\boldsymbol{r}}^{(j)}}\,. \tag{6.184}$$

Owing to the commutation relation (6.54), $[\hat{\mathsf{a}}(\mathbf{k},\lambda),\hat{\mathsf{a}}^+(\mathbf{k}',\lambda')] = \delta_{\lambda\lambda'}\delta(\mathbf{k}-\mathbf{k}')$, the integration over wave vectors is trivial

$$\begin{aligned}(\hat{\mathsf{H}}_2)_{FI} &= \frac{1}{(2\pi)^3}\sum_j\langle F_{\mathrm{at}}|\frac{e_j^2}{2m_j}\int\frac{\mathrm{d}^3\mathbf{k}}{\sqrt{2\omega}}\int\frac{\mathrm{d}^3\mathbf{k}'}{\sqrt{2\omega'}}\sum_{\lambda,\lambda'}\boldsymbol{\varepsilon}^{(\lambda)}\cdot\boldsymbol{\varepsilon}^{(\lambda')} \\ &\quad\times\left(\delta_{\lambda,\lambda_2}\delta(\mathbf{k}_2-\mathbf{k})\cdot\delta_{\lambda',\lambda_1}\delta(\mathbf{k}_1-\mathbf{k}')\mathrm{e}^{-\mathrm{i}(\mathbf{k}-\mathbf{k}')\cdot\hat{\boldsymbol{r}}^{(j)}}\right. \\ &\quad\left.+\ \delta_{\lambda,\lambda_1}\delta(\mathbf{k}-\mathbf{k}_1)\cdot\delta_{\lambda',\lambda_2}\delta(\mathbf{k}'-\mathbf{k}_2)\mathrm{e}^{\mathrm{i}(\mathbf{k}-\mathbf{k}')\cdot\hat{\boldsymbol{r}}^{(j)}}\right)|I_{\mathrm{at}}\rangle \\ &= \frac{1}{(2\pi)^3}\sum_j\frac{e_j^2}{m_j}\boldsymbol{\varepsilon}^{(1)}\cdot\boldsymbol{\varepsilon}^{(2)}\frac{1}{2\sqrt{\omega_1\omega_2}}\langle F_{\mathrm{at}}|\exp\left[\mathrm{i}(\mathbf{k}_1-\mathbf{k}_2)\cdot\hat{\boldsymbol{r}}^{(j)}\right]|I_{\mathrm{at}}\rangle\,.\end{aligned} \tag{6.185}$$

In our next step, we turn to the second term on the rhs of Eq. (6.181). Prior to any manipulations, note that the sum over n can be rewritten in a more abstract way

$$\sum_n (\hat{\mathsf{H}}_1)_{Fn} \frac{1}{\omega_{nI}} (\hat{\mathsf{H}}_1)_{nI} = \sum_n \langle F|\hat{\mathsf{H}}_1 \left(|n\rangle \frac{1}{E_n - E_I} \langle n| \right) \hat{\mathsf{H}}_1 |I\rangle = \langle F|\hat{\mathsf{H}}_1 \frac{1}{\hat{\mathsf{H}}_0 - E_I} \hat{\mathsf{H}}_1 |I\rangle \,. \tag{6.186}$$

In the second equality, we used spectral decomposition of the operator $(\hat{\mathsf{H}}_0 - E_I)^{-1}$. We manipulate the second term from Eq. (6.181) as follows, see again Eqs. (6.51) and (6.65),

$$\begin{aligned} \sum_n (\hat{\mathsf{H}}_1)_{Fn} \frac{1}{\omega_{np}} (\hat{\mathsf{H}}_1)_{nI} = \sum_{j,l} \Bigg\{ & \frac{1}{(2\pi)^3} \langle 0|\hat{\mathsf{a}}(\mathbf{k}_2, \lambda_2) \langle F_{\text{at}}| \left(\frac{-e_j}{m_j}\right) \int \frac{\mathrm{d}^3\mathbf{k}}{\sqrt{2\omega}} \sum_\lambda \boldsymbol{\varepsilon}^{(\lambda)} \cdot \hat{\boldsymbol{p}}^{(j)} \\ & \times \left(\hat{\mathsf{a}}(\mathbf{k}, \lambda) \mathrm{e}^{\mathrm{i}\mathbf{k}\cdot\hat{\boldsymbol{r}}^{(j)}} + \hat{\mathsf{a}}^+(\mathbf{k}, \lambda) \mathrm{e}^{-\mathrm{i}\mathbf{k}\cdot\hat{\boldsymbol{r}}^{(j)}} \right) \\ & \times \frac{1}{\hat{\mathsf{H}}_{\text{at}} + \hat{\mathsf{H}}_{\text{EM}} - E_I} \left(\frac{-e_l}{m_l}\right) \int \frac{\mathrm{d}^3\mathbf{k}'}{\sqrt{2\omega'}} \sum_{\lambda'} \boldsymbol{\varepsilon}^{(\lambda')} \cdot \hat{\boldsymbol{p}}^{(l)} \\ & \times \left(\hat{\mathsf{a}}(\mathbf{k}', \lambda') \mathrm{e}^{\mathrm{i}\mathbf{k}'\cdot\hat{\boldsymbol{r}}^{(l)}} + \hat{\mathsf{a}}^+(\mathbf{k}', \lambda') \mathrm{e}^{-\mathrm{i}\mathbf{k}'\cdot\hat{\boldsymbol{r}}^{(l)}} \right) |I_{\text{at}}\rangle \hat{\mathsf{a}}^+(\mathbf{k}_1, \lambda_1)|0\rangle \Bigg\} \,. \end{aligned} \tag{6.187}$$

Nonvanishing contribution stems again only from the commutators (6.183) and (6.184). Owing to the commutation relations (6.54), the integration over the wave vectors is again trivial. To determine the action of $(\hat{\mathsf{H}}_{\text{at}} + \hat{\mathsf{H}}_{\text{EM}} - E_I)^{-1}$ on the intermediate states, we use Eqs. (6.58) and (6.60). From the last equation, we obtain

$$\begin{aligned} (6.187) = & \frac{1}{(2\pi)^3} \sum_{j,l} \left(\frac{e_j}{m_j}\right) \left(\frac{e_l}{m_l}\right) \int \frac{\mathrm{d}^3\mathbf{k}}{\sqrt{2\omega}} \sum_\lambda \int \frac{\mathrm{d}^3\mathbf{k}'}{\sqrt{2\omega'}} \sum_{\lambda'} \langle F_{\text{at}}| \boldsymbol{\varepsilon}^{(\lambda)} \cdot \hat{\boldsymbol{p}}^{(j)} \\ & \times \Bigg\{ \mathrm{e}^{-\mathrm{i}\mathbf{k}\cdot\hat{\boldsymbol{r}}^{(j)}} \, \delta(\mathbf{k} - \mathbf{k}_2) \delta_{\lambda,\lambda_2} \frac{1}{\hat{\mathsf{H}}_{\text{at}} - E_I} \, \delta(\mathbf{k}' - \mathbf{k}_1) \delta_{\lambda',\lambda_1} \mathrm{e}^{\mathrm{i}\mathbf{k}'\cdot\hat{\boldsymbol{r}}^{(l)}} \\ & + \, \mathrm{e}^{+\mathrm{i}\mathbf{k}\cdot\hat{\boldsymbol{r}}^{(j)}} \, \delta(\mathbf{k}_1 - \mathbf{k}) \delta_{\lambda,\lambda_1} \frac{1}{\hat{\mathsf{H}}_{\text{at}} + \omega_1 + \omega_2 - E_I} \, \delta(\mathbf{k}' - \mathbf{k}_2) \delta_{\lambda',\lambda_2} \mathrm{e}^{-\mathrm{i}\mathbf{k}'\cdot\hat{\boldsymbol{r}}^{(l)}} \Bigg\} \\ & \times \boldsymbol{\varepsilon}^{(\lambda')} \cdot \hat{\boldsymbol{p}}^{(l)} |I_{\text{at}}\rangle = \sum_{j,l} \Bigg\{ \frac{e_j e_l}{(2\pi)^3 m_j m_l} \frac{1}{2\sqrt{\omega_1 \omega_2}} \\ & \times \langle F_{\text{at}}| \Bigg(\boldsymbol{\varepsilon}^{(2)} \cdot \hat{\boldsymbol{p}}^{(j)} \, \mathrm{e}^{-\mathrm{i}\mathbf{k}_2\cdot\hat{\boldsymbol{r}}^{(j)}} \frac{1}{\hat{\mathsf{H}}_{\text{at}} - E_I^{\text{at}} - \omega_1} \boldsymbol{\varepsilon}^{(1)} \cdot \hat{\boldsymbol{p}}^{(l)} \, \mathrm{e}^{\mathrm{i}\mathbf{k}_1\cdot\hat{\boldsymbol{r}}^{(l)}} \\ & + \, \boldsymbol{\varepsilon}^{(1)} \cdot \hat{\boldsymbol{p}}^{(j)} \, \mathrm{e}^{\mathrm{i}\mathbf{k}_1\cdot\hat{\boldsymbol{r}}^{(j)}} \frac{1}{\hat{\mathsf{H}}_{\text{at}} + \omega_2 - E_I^{\text{at}}} \boldsymbol{\varepsilon}^{(2)} \cdot \hat{\boldsymbol{p}}^{(l)} \, \mathrm{e}^{-\mathrm{i}\mathbf{k}_2\cdot\hat{\boldsymbol{r}}^{(l)}} \Bigg) |I_{\text{at}}\rangle \Bigg\} \,, \end{aligned} \tag{6.188}$$

where we substituted from Eq. (6.179) in the last equality.

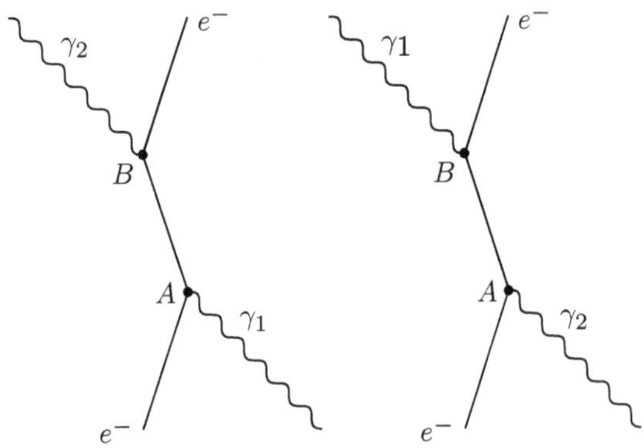

Fig. 6.8 Photon-atom scattering is a sum of two amplitudes. The *left diagram* depicts an atom absorbing the photon γ_1 first and then radiating the photon γ_2. The *right diagram* depicts an atom radiating the photon γ_2 first and then absorbing the photon γ_1

We have thus obtained the scattering amplitude as a sum of two terms, and Fig. 6.8 illustrates their meaning nicely. This figure is our first encounter with Feynman diagrams. For now, we can view them simply as "pictures" illustrating the physical situation in question. However, their significance goes far beyond that, see Sect. 7.4.10.

The first one corresponds to the process when the atom absorbs the incident photon and changes to a *virtual state* which comprises a linear combination of the stationary states of the atom. This becomes obvious if we substitute the spectral decomposition $\frac{1}{\hat{\mathsf{H}}_{\text{at}}-E_I^{\text{at}}-\omega} = \sum_n |n_{\text{at}}\rangle \frac{1}{E_n^{\text{at}}-E_I^{\text{at}}-\omega}\langle n_{\text{at}}|$ in the last equation. Once in the virtual state, the atom emits the final photon and relaxes to its final state. The second term, on the other hand, corresponds to the process when the atom first emits the final photon, goes to the virtual state and subsequently absorbs the incident photon. The principle of addition of amplitudes is again on the scene: we are not able to distinguish whether the photon-atom scattering proceeds via the first or second mechanism discussed above. Therefore, we must sum over the amplitudes of the both possibilities.

It follows from this picture why the denominators of the individual terms differ: in the first process, there is only the atom in the virtual state, hence the denominator reads $(\hat{\mathsf{H}}_{\text{at}} - E_I)$. In the second process, there are two photons and the atom in the virtual state, hence the denominator is $(\hat{\mathsf{H}}_{\text{at}} + \omega_1 + \omega_2 - E_I)$.

The total scattering amplitude is a sum of Eqs. (6.185) and (6.188). We invoke the dipole approximation (6.86), neglect the interaction of the EM field with the nucleus, i.e., set $m_j = m$, and use the definition of the operator of total momentum, Eq. (6.88), to obtain for the scattering amplitude

$$b_F \simeq \frac{e^2}{m(2\pi)^3} \frac{1}{2\sqrt{\omega_1\omega_2}} \boldsymbol{\varepsilon}_i^{(1)} \boldsymbol{\varepsilon}_j^{(2)} M_{ij} \tag{6.189}$$

where we introduced the dimensionless amplitude

$$M_{ij} = \langle F_{\rm at}| \left(\delta_{ij} - \frac{1}{m} \left[\hat{\mathsf{p}}_i \frac{1}{\hat{\mathsf{H}}_{\rm at} - E_I^{\rm at} - \omega_1} \hat{\mathsf{p}}_j + \hat{\mathsf{p}}_j \frac{1}{\hat{\mathsf{H}}_{\rm at} + \omega_2 - E_I^{\rm at}} \hat{\mathsf{p}}_i \right] \right) |I_{\rm at}\rangle \, . \tag{6.190}$$

For differential cross-section of the photon-atom scattering, we find from Eqs. (6.177), (6.119), and (6.120)

$$\mathrm{d}\sigma = (2\pi)^4 \delta(\omega_2 - \omega_1 + E_F^{\rm at} - E_I^{\rm at}) |b_F|^2 \mathrm{d}^3\mathbf{k}_2 \, . \tag{6.191}$$

After substituting for b_F from Eq. (6.189), we have

$$\mathrm{d}\sigma = \frac{\omega_2}{\omega_1} \left(\frac{\alpha}{m} \right)^2 |\boldsymbol{\varepsilon}_i^{(1)} M_{ij} \boldsymbol{\varepsilon}_j^{(2)}|^2 \mathrm{d}\Omega_2 \, . \tag{6.192}$$

In literature, one often encounters the last equation in a different form. It follows from Eq. (6.89) that

$$\langle F_{\rm at}|\hat{\mathsf{p}}_i = -\mathrm{i}m\langle F_{\rm at}|\hat{\mathsf{x}}_i(\hat{\mathsf{H}}_{\rm at} - E_F^{\rm at}) \, , \qquad \hat{\mathsf{p}}_i|I_{\rm at}\rangle = \mathrm{i}m(\hat{\mathsf{H}}_{\rm at} - E_I^{\rm at})\hat{\mathsf{x}}_i|I_{\rm at}\rangle \, .$$

By repetitive use of the last equality together with conservation of energy

$$\omega_1 + E_I^{\rm at} = \omega_2 + E_F^{\rm at} \, , \tag{6.193}$$

we manipulate Eq. (6.190) as follows

$$\begin{aligned}
&\langle F_{\rm at}| \left(\hat{\mathsf{p}}_i \frac{1}{\hat{\mathsf{H}}_{\rm at} - E_I^{\rm at} - \omega_1} \hat{\mathsf{p}}_j + \hat{\mathsf{p}}_j \frac{1}{\hat{\mathsf{H}}_{\rm at} + \omega_2 - E_I^{\rm at}} \hat{\mathsf{p}}_i \right) |I_{\rm at}\rangle \\
&= \mathrm{i}m\langle F_{\rm at}| \left\{ [\hat{\mathsf{p}}_i, \hat{\mathsf{x}}_j] + \omega_1 \left(\hat{\mathsf{p}}_i \frac{1}{\hat{\mathsf{H}}_{\rm at} - E_I^{\rm at} - \omega_1} \hat{\mathsf{x}}_j + \hat{\mathsf{x}}_j \frac{1}{\hat{\mathsf{H}}_{\rm at} + \omega_1 - E_F^{\rm at}} \hat{\mathsf{p}}_i \right) \right\} |I_{\rm at}\rangle \\
&= \mathrm{i}m\langle F_{\rm at}| \left\{ [\hat{\mathsf{p}}_i, \hat{\mathsf{x}}_j] + (-\mathrm{i})m\omega_1 \left[[\hat{\mathsf{x}}_i, \hat{\mathsf{x}}_j] \right.\right. \\
&\left.\left. + \, \omega_2 \left(\hat{\mathsf{x}}_i \frac{1}{\hat{\mathsf{H}}_{\rm at} - E_I^{\rm at} - \omega_1} \hat{\mathsf{x}}_j + \hat{\mathsf{x}}_j \frac{1}{\hat{\mathsf{H}}_{\rm at} + \omega_2 - E_I^{\rm at}} \hat{\mathsf{x}}_i \right) \right] \right\} |I_{\rm at}\rangle \, .
\end{aligned}$$

After evaluating the commutators $[\hat{\mathsf{p}}_i, \hat{\mathsf{x}}_j] = -\mathrm{i}\delta_{ij}$ and $[\hat{\mathsf{x}}_i, \hat{\mathsf{x}}_j] = 0$ and substituting the last equation into Eq. (6.192), we can write an alternative form of Eq. (6.191)

$$\mathrm{d}\sigma = \alpha^2 \omega_1 \omega_2^3 |\boldsymbol{\varepsilon}_i^{(1)} \beta_{ij} \boldsymbol{\varepsilon}_j^{(2)}|^2 \mathrm{d}\Omega_2 \, . \tag{6.194}$$

Here, β_{ij} equals

$$\beta_{ij} = \langle F_{\rm at}| \left(\hat{\mathsf{x}}_i \frac{1}{\hat{\mathsf{H}}_{\rm at} - E_I^{\rm at} - \omega_1} \hat{\mathsf{x}}_j + \hat{\mathsf{x}}_j \frac{1}{\hat{\mathsf{H}}_{\rm at} + \omega_2 - E_I^{\rm at}} \hat{\mathsf{x}}_i \right) |I_{\rm at}\rangle$$
$$= \sum_n \left(\frac{(\hat{\mathsf{x}}_i)_{Fn} (\hat{\mathsf{x}}_j)_{nI}}{E_n^{\rm at} - E_I^{\rm at} - \omega_1} + \frac{(\hat{\mathsf{x}}_j)_{Fn} (\hat{\mathsf{x}}_i)_{nI}}{E_n^{\rm at} - E_I^{\rm at} + \omega_2} \right). \tag{6.195}$$

The last equation can be cast into the form

$$\beta_{ij} = \alpha_{ij}(-\omega_1) + \alpha_{ij}(\omega_2)\,,$$

where

$$\alpha_{ij}(\omega) = \sum_n \frac{(\hat{\mathsf{x}}_i)_{Fn} (\hat{\mathsf{x}}_j)_{nI}}{E_n^{\rm at} - E_I^{\rm at} + \omega} \tag{6.196}$$

is called the atom *polarizability tensor*.

6.4.3 Rayleigh, Raman, and Resonance Scattering

The case of elastic scattering, i.e., $\omega_2 = \omega_1$, is referred to as the *Rayleigh scattering*. As one can see from Eq. (6.194), the cross-section depends on the fourth power of the frequency of the incident radiation. This dependence can explain why the sky is blue and the sunset reddish. The higher the frequency, the greater the scattering; thus out of the white light reaching the Earth from the Sun, the blue light features the highest frequency, hence scatters the most. In the latter case, the sunlight that reaches Earth in the evening is depleted of the scattered blue component, and thus we observe the complementary colors, that is, red and yellow.

In classical electrodynamics, see, e.g., [8, 13], the Rayleigh scattering is explained by a model where the electrons in the atoms behave as harmonic oscillators. Despite this model not being qualitatively misleading, it still is only a model and there is no classical theory that would be able to derive such a model. Thus the derivation presented herein might well have been the first thorough one the reader encountered.

The case of inelastic scattering, i.e., $\omega_2 \neq \omega_1$ is referred to as the *Raman scattering*. This is a purely quantum effect with no classical analog.

The case when frequency of the incident radiation ω_1 lies close to one of the atomic frequencies $E_n^{\rm at} - E_I^{\rm at}$ is called the *resonance scattering*. As seen from Eq. (6.195), the scattering amplitude diverges in this case. However, this divergence can be removed if we recognize that the excited states are not the stationary states, see Sect. 6.2. This means, the evolution of the excited "stationary" states

$$|\psi_n(t)\rangle = |\psi_n(0)\rangle e^{-iE_n t},$$

where E_n is real, is not entirely correct. A more accurate description is to add an imaginary part to the energy,

$$E_n = \Re E_n - \frac{1}{2}\mathrm{i}\Gamma_n\,, \tag{6.197}$$

where Γ_n is given by Eq. (6.99). The lifetime of the state is then $\tau = \Gamma_n^{-1}$,

$$\langle \psi_n(t)|\,\psi_n(t)\rangle = \mathrm{e}^{-\Gamma_n t}\,,$$

compare with Eq. (6.98). If we make replacement (6.197) in Eq. (6.195) we find that close to the resonance $\omega_1 \approx E_n^{\mathrm{at}} - E_I^{\mathrm{at}}$, the cross-section grows up, though remains finite.

Are our considerations correct? We asked for the probability of the photon-atom scattering and used the first two orders of the perturbation method to find an answer. Yet once we have reached it, we were forced to further modify it by taking into account that the lifetime of excited states is finite.

The perturbation method encounters difficulties every time two energies of the unperturbed system lie close to each other. Formally, this occurs when one of the frequencies ω_{nI} in the Lippmann-Schwinger equation (6.176) approaches zero. This problem is not restricted to dynamics, though. The perturbation method for stationary states runs into the very same difficulties if in Eq. (2.16), $E^{(2)} = -\sum_{n\neq N} \frac{|(\hat{\mathsf{H}}_1)_{Nn}|^2}{E_n^{(0)} - E_N^{(0)}}$, one of $E_n^{(0)}$ is close to $E_N^{(0)}$. Fortunately, we know from Chap. 2 that on the subspace of degenerate (or quasi-degenerate) energies we have to treat the perturbation non-perturbatively. Bearing this in mind, we rewrite the Lippmann-Schwinger equation (6.176) into a different form. If we continue in iterations (6.180) and (6.181), and search for a solution to Eq. (6.176) in the form of the series

$$b_n = \sum_l b_n^{(l)}\,,$$

where individual terms are obtained by iterations of Eq. (6.176)

$$b_n^{(l)} = (\hat{\mathsf{H}}_{\mathrm{int}})_{nI} - \sum_q \frac{(\hat{\mathsf{H}}_{\mathrm{int}})_{nq} b_q^{(l-1)}}{\omega_{qI} - \mathrm{i}\varepsilon}\,,\; b_n^{(0)} = 0\,,$$

we obtain the following series for the transition amplitude

$$\begin{aligned} b_F &= (\hat{\mathsf{H}}_{\mathrm{int}})_{FI} - \sum_q \frac{(\hat{\mathsf{H}}_{\mathrm{int}})_{Fq}(\hat{\mathsf{H}}_{\mathrm{int}})_{qI}}{\omega_{qI} - \mathrm{i}\varepsilon} + \sum_{q,n} \frac{(\hat{\mathsf{H}}_{\mathrm{int}})_{Fq}(\hat{\mathsf{H}}_{\mathrm{int}})_{qn}(\hat{\mathsf{H}}_{\mathrm{int}})_{nI}}{(\omega_{qI} - \mathrm{i}\varepsilon)(\omega_{nI} - \mathrm{i}\varepsilon)} + \ldots \\ &= \langle F|\hat{\mathsf{H}}_{\mathrm{int}} - \hat{\mathsf{H}}_{\mathrm{int}} \frac{1}{\hat{\mathsf{H}}_0 - E_I - \mathrm{i}\varepsilon} \hat{\mathsf{H}}_{\mathrm{int}} + \hat{\mathsf{H}}_{\mathrm{int}} \frac{1}{\hat{\mathsf{H}}_0 - E_I - \mathrm{i}\varepsilon} \hat{\mathsf{H}}_{\mathrm{int}} \frac{1}{\hat{\mathsf{H}}_0 - E_I - \mathrm{i}\varepsilon} \hat{\mathsf{H}}_{\mathrm{int}} + \ldots |I\rangle. \end{aligned}$$

By means of the expansion

$$\frac{1}{\hat{\mathsf{A}}-\hat{\mathsf{B}}}=\frac{1}{\hat{\mathsf{A}}}+\frac{1}{\hat{\mathsf{A}}}\hat{\mathsf{B}}\frac{1}{\hat{\mathsf{A}}}+\frac{1}{\hat{\mathsf{A}}}\hat{\mathsf{B}}\frac{1}{\hat{\mathsf{A}}}\hat{\mathsf{B}}\frac{1}{\hat{\mathsf{A}}}+\dots, \tag{6.198}$$

valid for two arbitrary invertible operators $\hat{\mathsf{A}}$ and $\hat{\mathsf{B}}$, the Lippmann-Schwinger equation can be manipulated into the form

$$b_F=\langle F|\left[\hat{\mathsf{H}}_{\text{int}}-\hat{\mathsf{H}}_{\text{int}}\frac{1}{\hat{\mathsf{H}}_0+\hat{\mathsf{H}}_{\text{int}}-E_I-\mathrm{i}\varepsilon}\hat{\mathsf{H}}_{\text{int}}\right]|I\rangle. \tag{6.199}$$

In case of the resonance scattering, the discrete state $|n_{\text{at}}\rangle|0\rangle$ lies inside the continuum of the states $|I_{\text{at}}\rangle\hat{\mathsf{a}}^+(\mathbf{k},\lambda)|0\rangle$, $|\mathbf{k}|\in(0,\infty)$, $E_I^{\text{at}}<E_n^{\text{at}}$. We show later in Sect. 6.5 that when taking into account the interaction between the discrete state $|n_{\text{at}}\rangle|0\rangle$ and continuum of the states $|I_{\text{at}}\rangle\hat{\mathsf{a}}^+(\mathbf{k},\lambda)|0\rangle$, the real part of the energy E_n^{at} is slightly shifted and the imaginary part $-\mathrm{i}\Gamma_n/2$ appears. The above replacement (6.197) is thus equivalent to the approximation where $\frac{1}{\hat{\mathsf{H}}_0+\hat{\mathsf{H}}_{\text{int}}-E_I-\mathrm{i}\varepsilon}$ in Eq. (6.199) is replaced by $\sum_n\frac{|n_{\text{at}}\rangle\langle n_{\text{at}}|}{E_n^{\text{at}}-E_I-\mathrm{i}\Gamma_n/2}$.[18]

Exercise 21: Rabi Oscillation II

It follows from above discussion of the resonance scattering that in such a case the following simplification holds. Firstly, it suffices to consider only the mode corresponding to the frequency ω_c out of the infinite number of modes of the EM field. Secondly, it suffices to consider merely the two metastable states, $|-\rangle$ and $|+\rangle$, of the infinite number of atom states, which energy difference ω_a lies close to ω_c.

To be able to choose only one mode of the EM field, there must be a countable number of them. In practice, this is achieved by confining the EM field inside a cavity and requiring the EM field vanish on the boundary of the cavity. Namely, we require $\mathbf{A}\left(x,y,-\frac{L}{2}\right)=\mathbf{A}\left(x,y,\frac{L}{2}\right)=0$, instead of (6.16), in the simplest instance. Furthermore, we demand the vector $\mathbf{k}$, (6.29), have nonzero the z-component only. A cavity constituted by two mirrors with a very high reflectivity serves as an excellent example of a cavity meeting these requirements [3], and highly excited circular

[18] Having developed our considerations to such an extent, we should also mention that for the very same reason of interaction of the discrete state with continuum of the states, the physical initial and final states do not match exactly the eigenstates (6.178) of the "free" Hamiltonian. If we take the interaction into account, the energy of atom initial state E_I^{at} is also slightly shifted, see Sect. 6.5. For a more systematic treatment, we refer the reader to [4].

states of alkali-metal atoms, $|+\rangle = |l+1, l, l\rangle$ *a* $|-\rangle = |l, l-1, l-1\rangle$, *where* $l \simeq 50$, *are excellent examples of metastable atomic states [3]. By adjusting the mirror distance L at the order of centimeters, the lowest mode of EM field with nonzero energy will lie close to* ω_a *[3].*
For individual terms of the total Hamiltonian (6.63)

$$\hat{\mathsf{H}} = \hat{\mathsf{H}}_{\text{at}} + :\hat{\mathsf{H}}_{\text{EM}}: + \hat{\mathsf{H}}_{\text{int}}\,,$$

we can write on such a restricted state space

$$\hat{\mathsf{H}}_{\text{at}} \simeq \omega_a \hat{\mathsf{S}}_z, \qquad :\hat{\mathsf{H}}_{\text{EM}}: \simeq \omega_c \hat{\mathsf{a}}^+ \hat{\mathsf{a}}$$

and

$$\begin{aligned}
\hat{\mathsf{H}}_{\text{int}} &\simeq -\frac{e}{m}\hat{\boldsymbol{A}}(0)\cdot\hat{\boldsymbol{p}} \\
&\simeq -\frac{e}{m}\frac{1}{(2\pi)^{3/2}(2\omega_c)^{1/2}}(\hat{\mathsf{a}}+\hat{\mathsf{a}}^+)\left[\frac{1}{2}(\varepsilon_+\hat{\mathsf{p}}_- + \varepsilon_-\hat{\mathsf{p}}_+) + \varepsilon_3\hat{\mathsf{p}}_3\right] \\
&\simeq -\frac{e}{2m}\frac{1}{(2\pi)^{3/2}(2\omega_c)^{1/2}}(\hat{\mathsf{a}}+\hat{\mathsf{a}}^+)(\hat{\mathsf{p}}_+ + \hat{\mathsf{p}}_-) \\
&\to \Omega(\hat{\mathsf{a}}^+ + \hat{\mathsf{a}})(\hat{\mathsf{S}}_+ + \hat{\mathsf{S}}_-)
\end{aligned}$$

$$\Rightarrow \hat{\mathsf{H}}_{\text{int}} \simeq \Omega(\hat{\mathsf{a}}^+\hat{\mathsf{S}}_- + \hat{\mathsf{a}}\hat{\mathsf{S}}_+)\,.$$

How did the spin operators appear in $\hat{\mathsf{H}}_{\text{at}}$ *and* $\hat{\mathsf{H}}_{\text{int}}$*? Every two-level system can be formally viewed as a spin 1/2 state, see, e.g., [7]. The eigenvalues of the operator* $\omega_a\hat{\mathsf{S}}_z$ *are* $\pm\omega_a/2$. *This means that we only shifted the energies of the atomic levels so as the zero-energy level lies in the middle between the energies of the states* $|+\rangle$ *and* $|-\rangle$. *Furthermore, we invoked the dipole approximation in the above expression for* $\hat{\mathsf{H}}_{\text{int}}$. *For the sake of simplicity, we assumed that the EM field is polarized in the x-direction, i.e.,* $\boldsymbol{\varepsilon} = (1,0,0)$*; hence* $\varepsilon_+ = \varepsilon_- = 1$ *and* $\varepsilon_3 = 0$. *The replacement of* $\hat{\mathsf{p}}_-$ *and* $\hat{\mathsf{p}}_+$ *by the spin operators* $\hat{\mathsf{S}}_-$ *and* $\hat{\mathsf{S}}_+$ *means that in the subspace of states* $|l+1, l, l\rangle$ *and* $|l, l-1, l-1\rangle$, *the operators* $\hat{\mathsf{p}}_-$ *and* $\hat{\mathsf{p}}_+$ *act similarly as* $\hat{\mathsf{S}}_-$ *and* $\hat{\mathsf{S}}_+$ *on the spin states* $|+\rangle$ *and* $|-\rangle$. *Finally, in the last equality of the last equation, we invoked the so-called rotating wave approximation. Within the second order of the perturbation method, this approximation corresponds to neglecting the contribution of the second process in Fig. 6.8 in comparison with the first one. It clearly follows from Eq. (6.195) that this omission is justifiable in the case of resonance scattering.*

Let the system be in the state $|-\rangle|n+1\rangle$ at time $t=0$, where the vector $|n+1\rangle$ describes a state when $n+1$ photons are present in the one mode of EM field; that is : $\hat{\mathsf{H}}_{\mathrm{EM}} : |n+1\rangle = \omega_c \hat{\mathsf{a}}^{+}\hat{\mathsf{a}}|n+1\rangle = \omega_c(n+1)|n+1\rangle$. What is the probability that the system will be found in the state $|+\rangle|n\rangle$ at time $t>0$?
Hint: Find the eigenvectors and eigenvalues of the Hamiltonian $\hat{\mathsf{H}}$. Search for the eigenvectors in the form

$$|\psi\rangle = c_{-}|-\rangle|n+1\rangle + c_{+}|+\rangle|n\rangle\,.$$

6.4.4 Averaging and Summing over Polarizations and Angles

The expressions (6.192) and (6.194) contain more information than we sometimes wish to exploit. Firstly, the polarization of neither the incident photon beams (which are generally unpolarized) nor of the scattered photons are frequently of interest. Secondly, the knowledge of the total cross-section often suffices and we take no interest in the specific angular distribution of the scattered photons. Thirdly, the atoms themselves are randomly oriented in the sample, hence we should average over these orientations, which is equivalent to averaging over the directions of the incident photons, though.

In case we do not measure the polarization of the scattered photons, the summation over the polarizations of the scattered photons leads to the replacement, see (6.55),

$$\varepsilon_i^{(2)}\varepsilon_j^{(2)} \to \delta_{ij} - \eta_i^{(2)}\eta_j^{(2)}\,. \tag{6.200}$$

In case the incident photons are unpolarized, averaging over the polarizations of the incident photons results in replacing

$$\varepsilon_k^{(1)}\varepsilon_l^{(1)} \to \frac{1}{2}\left(\delta_{kl} - \eta_k^{(1)}\eta_l^{(1)}\right)\,. \tag{6.201}$$

After inserting these replacements into Eq. (6.192), we obtain

$$\mathrm{d}\sigma = \frac{\omega_2}{\omega_1}\left(\frac{\alpha}{m}\right)^2 M_{ik}M_{jl}^{*}\left(\delta_{ij} - \eta_i^{(2)}\eta_j^{(2)}\right)\frac{1}{2}\left(\delta_{kl} - \eta_k^{(1)}\eta_l^{(1)}\right)\mathrm{d}\Omega_2\,.$$

Averaging over the directions of the incident photon and integration over the directions of the scattered photon leads to

$$\overline{\sigma} = \frac{1}{4\pi}\int \mathrm{d}\Omega_1\sigma = \frac{\omega_2}{\omega_1}\left(\frac{\alpha}{m}\right)^2 M_{ik}M_{jl}^{*}\frac{2}{3}4\pi\,\delta_{ij}\frac{1}{2}\frac{2}{3}\delta_{kl} = \frac{\omega_2}{\omega_1}\left(\frac{\alpha}{m}\right)^2 M_{ik}M_{ik}^{*}\frac{8\pi}{9}\,,$$

where we once again used the formula (6.96). Furthermore, if the initial and final states of the atom are spherically symmetric then[19]

$$M_{ik} = \delta_{ik} M,$$

where[20]

$$M = \langle F_{\rm at}| \left(1 - \frac{1}{3m} \hat{\mathsf{p}}_i \left[\frac{1}{\hat{\mathsf{H}}_{\rm at} - E_I^{\rm at} - \omega_1} + \frac{1}{\hat{\mathsf{H}}_{\rm at} + \omega_2 - E_I^{\rm at}} \right] \hat{\mathsf{p}}_i \right) |I_{\rm at}\rangle \,. \tag{6.202}$$

For the total cross-section we thus have

$$\overline{\sigma} = \frac{\omega_2}{\omega_1} \left(\frac{\alpha}{m} \right)^2 |M|^2 \frac{8\pi}{3}. \tag{6.203}$$

We show in the following text, at least for the simplest case of photon-hydrogen scattering, how one can calculate such a type of expressions.

6.4.5 *Calculation of Expressions Containing a Function of the Hamilton Operator*

In order to calculate the scattering amplitude (6.202) in the case of hydrogen, we first make transition to the atomic units, (6.83),

$$\hat{\mathsf{H}}_{\rm at} - E_I^{\rm at} = m(Z\alpha)^2(\hat{\mathsf{h}}_0 - \varepsilon_0)\,, \qquad \hat{\mathsf{h}}_0 = \frac{\hat{\boldsymbol{p}}_A \cdot \hat{\boldsymbol{p}}_A}{2} - \frac{1}{\hat{\mathsf{r}}_A}\,, \qquad \varepsilon_0 = -\frac{1}{2N^2}\,. \tag{6.204}$$

[19] For instance for the hydrogen atom, if $|I_{\rm at}\rangle = |n_I, 0, 0\rangle$ and $|F_{\rm at}\rangle = |n_F, 0, 0\rangle$ this equality follows from Eqs. (4.50)–(4.52).

[20] One third appears in Eq. (6.202) for the following reason. In Eq. (6.190), we have schematically

$$M_{ij} = \delta_{ij} - Q_{ij}\,.$$

Setting $i = j$ in this equation leads to (note that the Einstein summation convention is used)

$$M_{ii} = 3 - Q_{ii}\,.$$

Setting $i = k$ in $M_{ik} = \delta_{ik} M$ leads to

$$M_{ii} = 3M\,.$$

From the last two equations, we find

$$M = 1 - \frac{1}{3} Q_{ii}\,.$$

Using spectral decomposition of the hydrogen Hamiltonian (6.204), we have for a general function f

$$\langle N', L, M|(\hat{\mathsf{p}}_A)_i f(\hat{\mathsf{h}}_0 - \varepsilon_0)(\hat{\mathsf{p}}_A)_i|N, L, M\rangle = \sum_{l=0}^{\infty}\sum_{m=-l}^{l} \tag{6.205}$$

$$\left\{\int_0^{\infty} \mathrm{d}k_e \langle N', L, M|(\hat{\mathsf{p}}_A)_i|k_e, l, m\rangle\langle k_e, l, m|(\hat{\mathsf{p}}_A)_i|N, L, M\rangle f\left(x = \frac{k_e^2}{2} + \frac{1}{2N^2}\right)\right.$$

$$\left. + \sum_{n=l+1}^{\infty} \langle N', L, M|(\hat{\mathsf{p}}_A)_i|n, l, m\rangle\langle n, l, m|(\hat{\mathsf{p}}_A)_i|N, L, M\rangle f\left(x = \frac{1}{2N^2} - \frac{1}{2n^2}\right)\right\}.$$

Here, $|N, L, M\rangle$ and $|N', L, M\rangle$ are the eigenvectors corresponding to the eigenvalues $-1/(2N)^2$ and $-1/(2N')^2$, respectively. The scattering amplitude (6.202) is a special case of the last equation for

$$f(x) = -\frac{1}{3}\left(\frac{1}{x - k_1/2} + \frac{1}{x + k_2/2}\right), \tag{6.206}$$

where

$$k_{1,2} = \frac{2\omega_{1,2}}{m(Z\alpha)^2}. \tag{6.207}$$

For Rayleigh scattering $N = N' = 1$, $L = M = 0$, $k_2 = k_1$; for Raman scattering, where the final state is 2*s*, $N = 1$, $L = M = 0$, $N' = 2$, $k_2 = k_1 - 3/4$. The last equality follows from the conservation of energy (6.193).

Using Eqs. (3.15), (4.66), (4.67), (4.76), (4.77), and (6.100), we can perform integrations over angular variables, summations over m and l, and the action of the momentum operator:

$$\sum_{l=0}^{\infty}\sum_{m=-l}^{l} \langle N', L, M|(\hat{\mathsf{p}}_A)_i|n, l, m\rangle\langle n, l, m|(\hat{\mathsf{p}}_A)_i|N, L, M\rangle$$

$$= \frac{1}{2L+1}\left(\frac{\langle N', L|n, L-1\rangle\langle n, L-1|N, L\rangle + c^-}{L}\right.$$

$$\left. + \frac{\langle N', L|n, L+1\rangle\langle n, L+1|N, L\rangle + c^+}{L+1}\right), \tag{6.208}$$

where

$$\begin{aligned} c^+ &= \delta_{n,N'} \frac{\sqrt{N'^2 - (L+1)^2}}{N'} \langle N', L+1 | N, L \rangle \\ &+ \delta_{n,N} \frac{\sqrt{N^2 - (L+1)^2}}{N} \langle N', L | N, L+1 \rangle \\ &+ \delta_{n,N} \delta_{n,N'} \frac{N^2 - (L+1)^2}{N^2} \end{aligned}$$

and

$$\begin{aligned} c^- &= \delta_{n,N'} \frac{\sqrt{N'^2 - L^2}}{N'} \langle N', L-1 | N, L \rangle \\ &+ \delta_{n,N} \frac{\sqrt{N^2 - L^2}}{N} \langle N', L | N, L-1 \rangle \\ &+ \delta_{n,N} \delta_{n,N'} \frac{N^2 - L^2}{N^2}. \end{aligned}$$

In the case $L = 0$, the first term on the rhs of Eq. (6.208) does not appear. Obviously, for the continuous part of the spectrum, it is necessary only to replace n by k_e; $c^{\pm}$ vanish in this case.

Note that only radial integrals appear on the rhs of Eq. (6.208). We show in the next section how to calculate them.

6.4.6 *Transition from the Discrete to the Continuous Part of the Spectrum II*

This section is a continuation of Sects. 6.3.3 and 6.3.4. If the reader skipped these sections, he shall skip this one as well.

If the hydrogen atom is initially in the ground state, we need to know the radial parts of the p-state wave functions for both the discrete and continuous parts of the spectrum, see Eqs. (6.205) and (6.208). The high symmetry of the hydrogen atom provides several simplifications, though. By means of Eq. (4.77) for the discrete part and equation

$$\left(\frac{\mathrm{d}}{\mathrm{d}r} - \frac{l}{r} + \frac{1}{l+1} \right) R_l(k_e, r) = -\frac{\sqrt{1 + k_e^2 (l+1)^2}}{(l+1)} R_{l+1}(k_e, r) \tag{6.209}$$

for the continuous part of the spectrum, the radial parts of the states with higher angular momenta can be derived from the radial parts of the states of lower angular momenta. Equation (6.209) is obtained from Eq. (4.77) by the replacement

$$n \to -\mathrm{i}/k_e\,, \tag{6.210}$$

compare Eqs. (3.82) and (6.112). In particular, one finds from Eq. (6.209) for $l = 0$

$$\left(\frac{\mathrm{d}}{\mathrm{d}r} + 1\right) R_0(k_e, r) = -\sqrt{1 + k_e^2} R_1(k_e, r)\,. \tag{6.211}$$

The integrals appearing on the rhs of Eq. (6.208) are special cases of general integrals of the type

$$I_{l,p}^s(\nu, k_e) = \int_0^\infty \mathrm{d}r r^{2+p} \mathrm{e}^{-\nu r} R_l(k_e, r)\,, \quad I_{l,p}^d(\nu, n) = \int_0^\infty \mathrm{d}r r^{2+p} \mathrm{e}^{-\nu r} R_{nl}(r)\,, \tag{6.212}$$

for instance,

$$\langle 1, 0 |\, k_e, 1\rangle = 2 I_{1,0}^s(1, k_e)\,, \qquad \langle 2, 0 |\, k_e, 1\rangle = \frac{1}{\sqrt{2}} I_{1,0}^s\left(\frac{1}{2}, k_e\right).$$

In the second equation, we exploited the orthogonality of p-states and the fact that the radial function of the $2s$-state can be written, see Eq. (3.95), as $R_{2s}(r) = \frac{1}{\sqrt{2}}\mathrm{e}^{-r/2} - \sqrt{3} R_{2p}(r)$.

After inserting Eq. (6.211) into the integral (6.212) for $l = 1$ and $p = 0$ and integrating by parts, we obtain

$$I_{1,0}^s = \frac{1}{\sqrt{1 + k_e^2}} \left[2 + (\nu + 1)\frac{\partial}{\partial \nu}\right] I_{0,-1}^s\,. \tag{6.213}$$

This equation transforms the integration of the p-states to the integration of the s-states.

One can obtain the Schrödinger equation for the continuous spectrum of the s-state from Eq. (3.22) by replacement (6.210)

$$\left[-\frac{1}{2}\left(\frac{\mathrm{d}^2}{\mathrm{d}r^2} + \frac{2}{r}\frac{\mathrm{d}}{\mathrm{d}r}\right) - \frac{1}{r}\right] R_0(k_e, r) = \frac{k_e^2}{2} R_0(k_e, r)\,. \tag{6.214}$$

Substitution

$$R_0(k_e, r) = C_{k_e} \mathrm{e}^{-\mathrm{i}k_e r} f\,(2\mathrm{i}k_e r) \tag{6.215}$$

transforms Eq. (6.214) into Eq. (6.133) for the function $f(z)$, $\left(z\frac{d^2}{dz^2} + (b-z)\frac{d}{dz} - a\right) f(z) = 0$, where now

$$a = \frac{i}{k_e} + 1\,, \quad b = 2\,, \quad z = 2ik_e r\,. \tag{6.216}$$

Hence we know, see Eq. (6.135), $f(z) = \int_C \frac{dt}{2\pi i} e^t t^{a-b}(t-z)^{-a}$, that the solution (6.215) to Eq. (6.214) takes the form

$$\begin{aligned} R_0(k_e, r) &= C_{k_e} e^{-ik_e r} \int_C \frac{dt}{2\pi i} e^t t^{i/k_e - 1}(t - 2ik_e r)^{-i/k_e - 1} \\ &= C_{k_e} e^{-ik_e r} F\left(\frac{i}{k_e} + 1, 2, 2ik_e r\right), \end{aligned} \tag{6.217}$$

where the normalization constant

$$C_{k_e} = \sqrt{\frac{2}{\pi}} e^{\pi/(2k_e)} \left|\Gamma\left(1 - \frac{i}{k_e}\right)\right| k_e \tag{6.218}$$

was determined from the form of the asymptotic behavior of the wave function for large distances from the nucleus,

$$R_0(k_e, r) \to \frac{C_{k_e} e^{-\pi/(2k_e)}}{k_e r |\Gamma(1 - i/k_e)|} \sin\left[k_e r + \frac{\ln(2k_e r)}{k_e} + \arg\Gamma\left(1 - \frac{i}{k_e}\right)\right],$$

obtained from substituting Eq. (6.216) into Eq. (6.138), $f(z \to \infty) \to \frac{(-z)^{-a}}{\Gamma(b-a)} + \frac{e^z (z)^{a-b}}{\Gamma(a)}$, and the requirement the function $R_0(k_e, r)$ approach the wave function of a free particle in the s-state, Eq. (4.85), for large r,

$$R_0(k_e, r) \to \sqrt{\frac{2}{\pi}} \frac{\sin(k_e r)}{r}\,. \tag{6.219}$$

Factor $\sqrt{2/\pi}$ on the rhs guarantees the correct normalization of the radial wave function:

$$\int_0^\infty dr r^2 R_0(k_e, r) R_0(k'_e, r) = \frac{2}{\pi} \int_0^\infty dr \sin(k_e r) \sin(k'_e r) = \delta(k_e - k'_e)\,. \tag{6.220}$$

Since the norm of the function $R_0(k_e, r)$ is infinite, its different behavior from the free-particle solution for small r is inessential for its normalization.

The solution for the discrete spectrum

$$R_{n,0}(r) = C_n e^{-r/n} \int_C \frac{dt}{2\pi i} e^t t^{-n-1} \left(t - \frac{2r}{n}\right)^{n-1} = C_n e^{-r/n} F(-n+1, 2, 2r/n) \tag{6.221}$$

is obtained from the solution for the continuous spectrum by the replacement inverse to (6.210), namely

$$k_e \to -i/n \,. \tag{6.222}$$

Contrary to the continuum states, the discrete states can be normalized to unity. Hence, it is not possible to find the normalization constant C_n from the normalization constant C_{k_e} by the substitution (6.222). We do not have to calculate the normalization integral, though. One can obtain the correct normalization from the requirement $R_{n,0}(0) = 2/n^{3/2}$, see Eq. (4.101). Obviously, see Eqs. (6.141), (6.142), and (6.221),

$$C_n = \frac{2}{n^{3/2}} \,. \tag{6.223}$$

From Eqs. (6.150), $\frac{1}{\Gamma(b)} \int_0^\infty e^{-\lambda z} z^{b-1} F(a, b, z) dz = \lambda^{a-b}(\lambda - 1)^{-a}$, and (6.216), we have

$$I^s_{0,-1}(\nu, k_e) = \frac{C_{k_e}}{\nu^2 + k_e^2} \left(\frac{\nu + ik_e}{\nu - ik_e}\right)^{i/k_e} \,. \tag{6.224}$$

By inserting this expression into Eq. (6.213), we find the sought integrals

$$I^s_{1,0}(\nu, k_e) = \frac{2\sqrt{1 + k_e^2}}{(\nu^2 + k_e^2)^2} \left(\frac{\nu + ik_e}{\nu - ik_e}\right)^{i/k_e} C_{k_e} \,. \tag{6.225}$$

By replacement (6.222), we obtain from the last equation

$$I^d_{1,0}(\nu, n) = \frac{2\sqrt{1 - \frac{1}{n^2}}}{(\nu^2 - 1/n^2)^2} \left(\frac{\nu + 1/n}{\nu - 1/n}\right)^{-n} C_n \,. \tag{6.226}$$

6.4.7 *Photon-Hydrogen Scattering*

By inserting Eqs. (6.225), (6.226), and (6.208) into Eq. (6.205) for the function (6.206), we obtain the results for Rayleigh and Raman photon-hydrogen scattering. In particular, for Rayleigh scattering we find

$$\langle 1,0,0|(\hat{\mathsf{p}}_A)_i f(\hat{\mathsf{h}}_0 - \varepsilon_0)(\hat{\mathsf{p}}_A)_i|1,0,0\rangle$$

$$= \int_0^{\infty} \mathrm{d}k_e \frac{32k_e^2}{(1+k_e^2)^3}\left(\frac{1+\mathrm{i}k_e}{1-\mathrm{i}k_e}\right)^{2\mathrm{i}/k_e} \frac{\mathrm{e}^{\pi/k_e}\left|\Gamma\left(1-\frac{\mathrm{i}}{k_e}\right)\right|^2}{\pi} f\left(x=\frac{k_e^2+1}{2}\right)$$

$$+\sum_{n=2}^{\infty}\frac{8^2}{n^3}\frac{1}{\left(1-\frac{1}{n^2}\right)^3}\left(\frac{1-\frac{1}{n}}{1+\frac{1}{n}}\right)^{2n} f\left(x=\frac{n^2-1}{2n^2}\right), \tag{6.227}$$

where the function f is taken in the form

$$f(x) = 1 - \frac{1}{3}\left(\frac{1}{x-k_1/2} + \frac{1}{x+k_2/2}\right).$$

We added unity from Eq. (6.202) to the expression (6.206), which is justifiable since the summation rule

$$\left\langle (\hat{\mathsf{p}}_A)_i(\hat{\mathsf{p}}_A)_i\right\rangle = \left\langle \hat{\mathsf{n}}_i\hat{\mathsf{n}}_i\right\rangle = 1 \tag{6.228}$$

holds for the ground state. This rule follows from the decomposition of unity and the form of the wave function for the ground state, Eq. (3.35).

Expression (6.227) has to be calculated numerically. The reader will not be surprised to learn that the part containing the contribution of the continuous part of the spectrum could have been obtained from Eq. (6.152) by the replacement $\left|\boldsymbol{\varepsilon}^{(\nu)}\cdot\boldsymbol{\eta}\right|^2 \to 1$. The differential of the integration over all states is in this case $\mathrm{d}^3\mathbf{k}_e = \mathrm{d}k_e k_e^2 \mathrm{d}\Omega$. Integration over angles is trivial and produces the factor 4π.

The results for the Rayleigh and Raman scattering are listed in Tables 6.2 and 6.3.

There are a few features worth of attention in these tables. For low frequencies, i.e., low in comparison with the differences of atomic energy levels, the cross-section of the Rayleigh scattering indeed goes with fourth power of frequency of incident photon. For low frequencies, we can thus neglect ω_1 in comparison with $E_n^{\mathrm{at}} - E_I^{\mathrm{at}}$ in Eq. (6.195). The polarizability tensor (6.196) is consequently frequency-independent and equals the static polarizability tensor $\alpha_{ij}(\omega = 0)$. Furthermore, the contribution of the continuous part increases for low frequencies and nearly cancels the contribution of the discrete part of the hydrogen spectrum. This cancellation further increases with the decreasing frequency of the incident photon.

Furthermore, these tables nicely illustrate when one should talk in terms of scattering and when in terms of absorption and subsequent emission. If the frequency of the incident photon approaches one of the atomic characteristic frequencies, for instance if k_1 approaches $3/4 = 0.75$ or $8/9 \doteq 0.8888\ldots$, the contribution of the states $n = 2$ or $n = 3$ greatly exceeds those of the other virtual states. One can thus see that close to the resonance, it is reasonable to consider only one of the infinitely many virtual states. In such a case, use of the terms absorption and subsequent emission is more appropriate.

Table 6.2 Rayleigh scattering on hydrogen

k_1	$n = 2$	$n = 3$	$n \geq 4$	$k_e \in \mathbb{R}^+$	M	$\overline{\sigma}[m^2]$
0.01	−0.182	−0.026	−0.020	0.228	−0.0001	0.843×10^{-36}
0.1	−0.190	−0.027	−0.020	0.226	−0.0114	0.869×10^{-32}
0.745	−31.085	−0.213	−0.126	0.038	−31.386	0.656×10^{-25}
0.749	−155.944	−0.220	−0.129	0.034	−156.259	0.162×10^{-23}
0.751	156.204	−0.224	−0.131	0.033	155.882	0.162×10^{-23}
0.755	31.345	−0.231	−0.134	0.029	31.009	0.641×10^{-25}
0.82	2.364	−0.478	−0.213	−0.044	1.629	0.177×10^{-27}
0.883	1.312	−5.937	−0.436	−0.158	−5.219	0.181×10^{-26}
0.887	1.278	−18.579	−0.466	−0.168	−17.935	0.214×10^{-25}
0.889	1.262	316.439	−0.483	−0.173	317.045	0.670×10^{-23}
0.893	1.231	8.585	−0.520	−0.183	9.112	0.553×10^{-26}

The first column displays the energy values of the incident photon in units of the Rydberg constant, see Eq. (6.207). The second column presents the contribution to the amplitude M, Eq. (6.202), from the virtual state $n = 2$, the third column shows the contribution from the virtual state $n = 3$, the fourth column displays the contribution from other discrete states, and the fifth column gives the contribution of the continuum spectrum. The sixth column indicates the total amplitude M, and finally the seventh column displays total cross-section (6.203)

Table 6.3 Raman scattering $1s \rightarrow 2s$ on hydrogen

k_1	$n = 3$	$n \geq 4$	$k_e \in \mathbb{R}^+$	M	$\overline{\sigma}[m^2]$
0.751	−0.273	−0.160	−0.408	−0.841	0.627×10^{-31}
0.755	−0.280	−0.163	−0.410	−0.852	0.321×10^{-30}
0.82	−0.507	−0.226	−0.456	−1.188	0.803×10^{-29}
0.883	−5.565	−0.409	−0.530	−6.504	0.424×10^{-27}
0.887	−17.282	−0.434	−0.537	−18.252	0.343×10^{-26}
0.889	293.220	−0.447	−0.540	292.232	0.890×10^{-24}
0.893	7.894	−0.478	−0.547	6.869	0.503×10^{-27}

In contrast to the Rayleigh scattering, there is no contribution from the virtual state $n = 2$ in the case of the Raman scattering

Finally, let us note that the Raman scattering is typically by an order of magnitude smaller in comparison to the Rayleigh scattering, including when in the vicinity of resonance.

Exercise 22: Photon-Hydrogen Scattering

In a programming language of your choice, though with a built-in procedure for numerical integration, write the formula (6.227) for the function (6.206). You can cope with the infinite summation in the following

way. To $n = 200$, for instance, you sum the series exactly. For n from 201 to infinity, use (an indeed nice example of the usefulness of calculus of limits!)

$$\frac{8^2}{n^3}\frac{1}{\left(1-\frac{1}{n^2}\right)^3}\left(\frac{1-\frac{1}{n}}{1+\frac{1}{n}}\right)^{2n} f\left(x=\frac{n^2-1}{2n^2}\right) \simeq \frac{64\mathrm{e}^{-4}f(x=1/2)}{n^3}(1+O(1/n)).$$

Using programming languages such as Maple *or* Mathematica*, you easily find that the sum $\sum_{n_0}^{\infty} n^{-3}$ can be expressed in terms of special functions.*
Try to reproduce the numbers displayed in Table 6.2, as well as those in Table 6.3. When facing the latter, use Eqs. (6.205), (6.206), (6.208), (6.225), and (6.226).

6.4.8 Thomson Scattering

Prior to our conclusion of this section, we consider the case of the photon scattering on a free electron. We change to the coordinate system where the electron is at rest, $E_I^{\mathrm{at}} = 0$. The action of the momentum operator $\hat{\boldsymbol{p}}$ on $|I_{\mathrm{at}}\rangle$ thus yields zero and merely the first term in the expression (6.190) remains. For a long-wave EM radiation (and only for such Eq. (6.190) is valid at all), we can neglect the change in electron energy resulting from the scattering, i.e., $E_F^{\mathrm{at}} \simeq 0$. Then from Eq. (6.193), we have $\omega_2 \simeq \omega_1$. Insertion of Eq. (6.189) into Eq. (6.192) leads to

$$\mathrm{d}\sigma = \left(\frac{\alpha}{m_e}\right)^2 |\boldsymbol{\varepsilon}^{(1)} \cdot \boldsymbol{\varepsilon}^{(2)}|^2 \mathrm{d}\Omega_2 \,. \tag{6.229}$$

Substitution of the replacements (6.200) and (6.201) into Eq. (6.229) yields

$$\mathrm{d}\sigma = \left(\frac{\alpha}{m_e}\right)^2 \frac{1}{2}(1+\cos^2\vartheta)\sin\vartheta\,\mathrm{d}\vartheta\,\mathrm{d}\varphi \,, \tag{6.230}$$

where $\cos\vartheta = \boldsymbol{\eta}^{(1)} \cdot \boldsymbol{\eta}^{(2)}$ and where we have chosen the z-axis in the direction of the incident photon. Finally, we integrate over the directions of the scattered photon. We thus find for the total cross-section

$$\sigma = \left(\frac{\alpha}{m_e}\right)^2 \frac{8\pi}{3} \to 0.665 \times 10^{-28}\ \mathrm{m}^2 \,. \tag{6.231}$$

The last three equations were derived at the turn of the twentieth century, long time prior to the discovery of quantum electrodynamics by J.J. Thomson. The classical derivation of these formulas can be found in, e.g., [13]. The quantity

$$r_0 = \frac{\alpha}{m_e} \rightarrow 2.82 \times 10^{-15}\ \text{m} \tag{6.232}$$

is called the *classical electron radius*, the only reason being it appears in formula (6.231). The large difference between the photoelectric cross-section (6.156) and the Thomson cross-section (6.231) stems from the large difference between the Bohr radius (6.157) and the classical electron radius (6.232). The reader should be warned, though. As it is clear from above discussion of the resonance scattering, the cross-section does not necessarily correspond to the radius of the target particle.

It is worth noting that the Thomson scattering finds its place in the latest development of cosmology. Once the temperature of our Universe fell below 10^9 K, there was not enough energy for the creation of electron-positron pairs. The temperature further decreased and as it dropped below 10^3 K, the EM radiation ceased to scatter on the matter. Between these two moments, the Universe was composed of merely four components: photons, neutrinos, dark matter and baryon plasma which constituted free electrons, positive ions and a small number of light atoms. The photon-electron scattering during this time period, Eq. (6.229) serving well for its description, changed the photon polarization, which we can measure today in the fluctuations of the cosmic microwave background. The reader can learn more about this interesting subject in e.g. [15].

6.5 Virtual Processes

6.5.1 Introductory Notes

The nonrelativistic approximation to quantum electrodynamics has, not surprisingly, a limited domain of validity. As already mentioned, the error caused by its use is of the order of $Z(\alpha)^2$ as we do not consider the spin-orbit interaction and additional relativistic corrections. The interpretation of this restriction is clear: the charge of the nucleus Z must be sufficiently small for the use of the nonrelativistic approximation. Roughly speaking, in case of light atoms, the "speed of the orbiting" electron must be much smaller in comparison with the speed of light.

However, nonrelativistic quantum electrodynamics fails to describe correctly even the light atoms, including hydrogen, once we take into account the so-called *virtual processes*. So far, we always tacitly assumed that a photon radiated by an electron will be registered in a detector. However, quantum electrodynamics, even in its nonrelativistic approximation, predicts the possibility that the electron absorbs the radiated photon back, see Fig. 6.9!

Fig. 6.9 Electron acting on itself

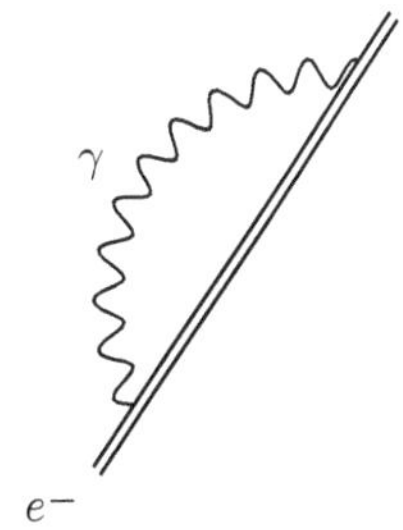

As we will show in a moment, quantum mechanics forces us to integrate over all possible wave vectors of the virtual photon. The recoil momentum is imparted on the electron, but there is no bound on the momentum of virtual photon, hence there is a contribution from the electron states with very high momenta. The nonrelativistic approximation is clearly inapplicable to these states. Hence, if we wish to understand the spectra of even the lightest atoms, we need to learn relativistic quantum electrodynamics first. We emphasize that this *self-energy effect* is relatively small, though not too much. We will show later in the text that the absolute value of this effect in case of hydrogen is by merely an order of magnitude smaller than the most important relativistic corrections, and is in fact greater than the spin-spin interaction.

6.5.2 Lamb-Retherford Experiment

There are two vector operators, namely the angular momentum and the Runge-Lenz vector, commuting with the nonrelativistic hydrogen Hamiltonian. Thus the hydrogen energy spectrum depends merely on the principal quantum number n, as we already analyzed in depth in Chap. 4. The reader convinced himself in Exercises 7 and 8 on the particular example of the states with $n = 2$ that the relativistic corrections remove this degeneracy, although only partially. Once relativistic corrections are taken into account, the energy depends also on the total angular momentum j. However, for a fixed j, it does not depend on the orbital angular momentum l. In particular, the states $2p_{1/2}$ and $2p_{3/2}$ are split by the relativistic corrections, but the degeneracy of the states $2p_{1/2}$ and $2s$ persists.

In 1947, Willis Lamb and Robert Retherford performed a beautiful experiment and showed that the states $2p_{1/2}$ and $2s$ are in fact split apart as well, the measured difference being about 1058 MHz. Victor Weisskopf and Julian S. Schwinger suggested immediately that this splitting stems from the self-energy effect. From that time on, the difference between the prediction given by the Dirac equation and the real experiment is referred to as the *Lamb shift*.

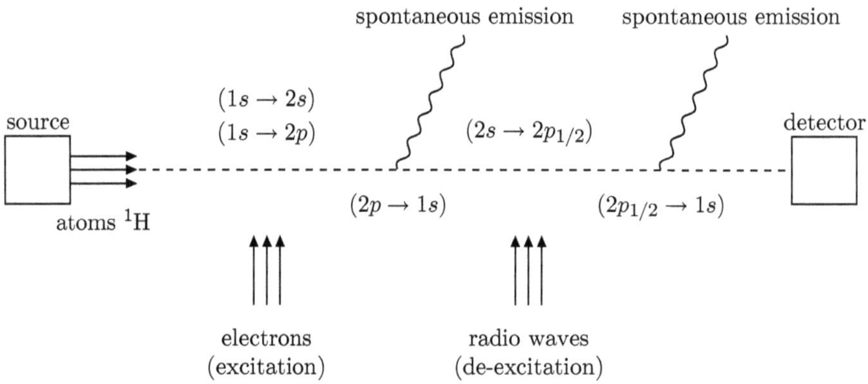

Fig. 6.10 A schematic setup of the Lamb-Retherford experiment

The Lamb-Retherford experiment is based on knowledge we have already adopted, thus its brief description is in place. Figure 6.10 captures the experimental setup. Atomic hydrogen is sent to the apparatus where it is first irradiated by electrons. This in turn leads to the excitation of the atoms to the states with $n = 2$, the number of s- and p-states being similar, see Fig. 6.7. As we already know, the lifetime of the $2s$-state is enormous in comparison to that of the $2p$-state. Hence, after having covered a certain distance, all excited atoms are in the $2s$-state. Subsequently, the atoms are irradiated by EM radio waves and finally they impact on the electrode. In case the atoms remain in the $2s$-state after being irradiated, they possess enough energy when they hit the electrode to produce a secondary current. In case the radio waves are tuned to the energy difference between the $2s$ and $2p_{1/2}$ levels, the atoms make transition to the $2p_{1/2}$ state and subsequently relax to the ground state by spontaneous emission prior to hitting the electrode. Thus no secondary current is produced. Therefore, the correct frequency is determined by a sudden decrease of the secondary current, see Fig. 6.11. The ingenious principles of this experiment are still used these days.

6.5.3 Self-energy: Bethe Estimate

Prior to our dive into the relativistic theory, we attempt to make an estimate of the effect within the nonrelativistic approximation that is able to describe correctly the part of the effect related to the emission and absorption of long-wavelength photons. As we will see, this turns out to be the most significant part of the effect. We will follow the steps of Hans Bethe who managed to do it back in 1947.

As it is clear from Fig. 6.9, the atom interacts twice with the EM field. Let us thus apply the perturbation method to the second order in e^2,

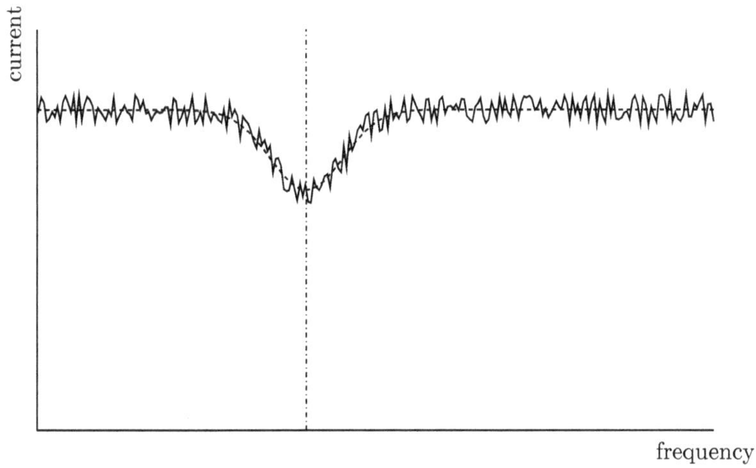

Fig. 6.11 Record of the secondary current in the Lamb-Retherford experiment

$$
\begin{aligned}
E^{(2)} &= \langle\psi_0|e^2\hat{\mathsf{H}}_2|\psi_0\rangle - \sum_n \langle\psi_0|e\hat{\mathsf{H}}_1|n\rangle \frac{1}{E_n - E_0}\langle n|e\hat{\mathsf{H}}_1|\psi_0\rangle \\
&= \langle\psi_0|e^2\hat{\mathsf{H}}_2|\psi_0\rangle - \langle\psi_0|e\hat{\mathsf{H}}_1 \frac{1}{\hat{\mathsf{H}}_0 - E_0} e\hat{\mathsf{H}}_1|\psi_0\rangle ,
\end{aligned}
$$

where $\hat{\mathsf{H}}_0$ is given as the sum of Eqs. (6.61) and (6.64), and $\hat{\mathsf{H}}_1$ by Eq. (6.65). The first term on the rhs is not necessary to consider as it only shifts all energy levels by the same infinite value. The reference state $|\psi_0\rangle = \left|\psi_{\mathrm{at}}^0\right\rangle|0\rangle$ describes the atom in one of its stationary states and the EM field in its ground state. Substitution of Eq. (6.51) for the vector potential into Eq. (6.65) for the interaction Hamiltonian $\hat{\mathsf{H}}_1$ leads to

$$
\begin{aligned}
E^{(2)} &= -\frac{1}{(2\pi)^3}\frac{e^2}{m^2}\int \frac{d^3\mathbf{k}}{\sqrt{2\omega}}\int \frac{d^3\mathbf{k}'}{\sqrt{2\omega'}}\sum_{\lambda\lambda'}\left\langle\psi_{\mathrm{at}}^0\right|\langle 0|\boldsymbol{\varepsilon}^{(\lambda)}\cdot\hat{\boldsymbol{p}} \\
&\times \left(\hat{\mathsf{a}}(\mathbf{k},\lambda)e^{i\mathbf{k}\cdot\hat{\boldsymbol{r}}} + \hat{\mathsf{a}}^+(\mathbf{k},\lambda)e^{-i\mathbf{k}\cdot\hat{\boldsymbol{r}}}\right)\frac{1}{\hat{\mathsf{H}}_{\mathrm{EM}} + \hat{\mathsf{H}}_{\mathrm{at}} - E_0} \\
&\times \left(\hat{\mathsf{a}}(\mathbf{k}',\lambda')e^{i\mathbf{k}'\cdot\hat{\boldsymbol{r}}} + \hat{\mathsf{a}}^+(\mathbf{k}',\lambda')e^{-i\mathbf{k}'\cdot\hat{\boldsymbol{r}}}\right)\boldsymbol{\varepsilon}^{(\lambda')}\cdot\hat{\boldsymbol{p}}\,|0\rangle\left|\psi_{\mathrm{at}}^0\right\rangle .
\end{aligned}
$$

Elimination of the field operators produces as usually $\delta(\mathbf{k}-\mathbf{k}')\delta_{\lambda\lambda'}$. This in turn eliminates the integration over wave vectors $\mathbf{k}'$ and summation over polarizations λ'. Next, we invoke the dipole approximation $\exp(i\mathbf{k}\cdot\mathbf{r}) \simeq 1$, as we are interested in the long-wavelength photons only; we obtain

$$
E^{(2)} = -\frac{e^2}{(2\pi)^3 m^2}\int \frac{d^3\mathbf{k}}{2\omega}\sum_\lambda \left\langle\psi_{\mathrm{at}}^0\right|\boldsymbol{\varepsilon}^{(\lambda)}\cdot\hat{\boldsymbol{p}}\frac{1}{\omega + \hat{\mathsf{H}}_{\mathrm{at}} - E_0}\boldsymbol{\varepsilon}^{(\lambda)}\cdot\hat{\boldsymbol{p}}\left|\psi_{\mathrm{at}}^0\right\rangle .
$$

We write the wave vector in spherical coordinates in the k-space, $\mathrm{d}^3\mathbf{k} = \omega^2 \mathrm{d}\omega \mathrm{d}\Omega$, and restrict the integration to the domain $\omega < \Lambda \simeq m$ where the nonrelativistic and dipole approximations hold. We then sum over the polarizations, Eq. (6.55):

$$E^{(2)} = -\frac{1}{(2\pi)^3}\frac{e^2}{m^2}\int_0^{\Lambda}\mathrm{d}\omega\frac{\omega}{2}\int \mathrm{d}\Omega\,(\delta_{ij} - \eta_i\eta_j)\langle\psi_{\mathrm{at}}^0|\hat{\mathsf{p}}_i\frac{1}{\omega + \hat{\mathsf{H}}_{\mathrm{at}} - E_0}\hat{\mathsf{p}}_j|\psi_{\mathrm{at}}^0\rangle\,.$$

We have already performed the integration over angles several times, see Eq. (6.96). Finally, we integrate over the frequencies of the virtual photon (we add and subtract $\hat{\mathsf{H}}_{\mathrm{at}} - E_0$ in the numerator),

$$E^{(2)} = -\frac{2}{3}\frac{\alpha}{\pi m^2}\left(\Lambda\langle\psi_{\mathrm{at}}^0|\hat{\mathsf{p}}^2|\psi_{\mathrm{at}}^0\rangle - \langle\psi_{\mathrm{at}}^0|\hat{\mathsf{p}}_i(\hat{\mathsf{H}}_{\mathrm{at}} - E_0)\left\{\ln\frac{\Lambda}{\hat{\mathsf{H}}_{\mathrm{at}} - E_0}\right\}\hat{\mathsf{p}}_i|\psi_{\mathrm{at}}^0\rangle\right)\,, \tag{6.233}$$

where in the upper bound of the integration, we neglected $(\hat{\mathsf{H}}_{\mathrm{at}} - E_0)$ in comparison with Λ.

We now focus on the result of the last equation for the specific case of a free electron. The momentum and Hamilton operators commute, hence $|\psi_{\mathrm{at}}\rangle = |\mathbf{p}\rangle$, and the second term with the logarithm yields zero; then

$$E_{\mathrm{EM}}^{(2)} = -\frac{2}{3}\frac{\alpha}{\pi}\frac{p^2}{m^2}\Lambda\,. \tag{6.234}$$

What is the physical meaning of this correction to the energy of a free particle? It has to be a correction to the inertial mass of the particle. We express the observable mass m_{exp} of the electron as a sum of the mass m of "bare" electron and of the electromagnetic mass Δm

$$m_{\mathrm{exp}} = m + \Delta m\,, \qquad m \gg \Delta m\,. \tag{6.235}$$

The "bare" electron is a hypothetical electron which is not surrounded by an EM field. The electromagnetic mass is the part of the mass corresponding to the energy (6.234).

We substitute the last equation into the Hamilton operator for the bare electron

$$\hat{\mathsf{H}}_{\mathrm{at}} = \frac{\hat{\mathsf{p}}^2}{2m} = \frac{\hat{\mathsf{p}}^2}{2(m_{\mathrm{exp}} - \Delta m)} \simeq \frac{\hat{\mathsf{p}}^2}{2m_{\mathrm{exp}}} + \frac{\hat{\mathsf{p}}^2}{2m_{\mathrm{exp}}^2}\Delta m\,.$$

The observable part of the self-energy effect is the sum of Eq. (6.233) and the change of the energy due to the electromagnetic mass, i.e., the second term in the last equation,

$$E_{\mathrm{exp}}^{(2)} = E^{(2)} + \langle\psi_{\mathrm{at}}^0|\frac{\hat{\mathsf{p}}^2}{2m_{\mathrm{exp}}^2}|\psi_{\mathrm{at}}^0\rangle\Delta m\,. \tag{6.236}$$

In case of a free electron, there should be no observable consequence, $E^{(2)}_{\text{exp}} = 0$; hence

$$E^{(2)}_{\text{EM}} + \frac{p^2}{2m^2_{\text{exp}}} \Delta m = 0 \, .$$

After substituting from Eq. (6.234) for the first term, we obtain

$$\Delta m = \frac{4}{3} \frac{\alpha}{\pi} \Lambda \frac{m^2_{\text{exp}}}{m^2} = \frac{4}{3} \frac{\alpha}{\pi} \Lambda \left(1 + \frac{\Delta m}{m} \right)^2 \simeq \frac{4}{3} \frac{\alpha}{\pi} \Lambda \, .$$

In the second equality we substituted from Eq. (6.235) and in the last one we neglected $\frac{\Delta m}{m}$ with respect to 1 since Δm is proportional to α.

Having determined Δm in this way, the first term in Eq. (6.233) cancels with the second term in Eq. (6.236) for bound electron by construction. The observable part of the effect is determined by the second term in Eq. (6.233). Henceforth, m will denote the observable mass m_{exp}. We take the estimate $\Lambda \simeq m$ seriously; then Bethe estimate for the energy shift read

$$E^{(2)}_{\text{exp}} = m \frac{\alpha (Z\alpha)^4}{\pi n^3} F_n \, , \tag{6.237}$$

$$F_n = \frac{4}{3} \left[\frac{n^3}{2} \ln \frac{2}{(Z\alpha)^2} \langle \psi^0_{\text{at}} | (\hat{\mathsf{p}}_A)_i (\hat{\mathsf{h}}_0 - \varepsilon_0) (\hat{\mathsf{p}}_A)_i | \psi^0_{\text{at}} \rangle \right.$$
$$\left. - \frac{n^3}{2} \langle \psi^0_{\text{at}} | (\hat{\mathsf{p}}_A)_i (\hat{\mathsf{h}}_0 - \varepsilon_0) \ln \left[2(\hat{\mathsf{h}}_0 - \varepsilon_0) \right] (\hat{\mathsf{p}}_A)_i | \psi^0_{\text{at}} \rangle \right] .$$

Here, we made the transition from the natural to atomic units, Eqs. (6.83) and (6.204), where $E^{\text{at}}_I = E_0$.

The above performed procedure is commonly called the *mass renormalization* and is of key importance in the relativistic theory. Only a careful distinction between the observable and the bare electron provides a way to obtain numbers from the theory which are to be compared with the experiment. The observable electron is an electron surrounded by EM field which in turn acts back on its own source. The bare electron, on the other hand, is not surrounded by any EM field; it is a mere theoretical construction, though very useful.

In the relativistic theory, the problem is much more fundamental. We emphasized at the very beginning that the nonrelativistic approximation is applicable only for low frequencies of the virtual photon. Therefore, the unfavorable behavior of the expression (6.233) in the limit $\Lambda \to \infty$ is not bothersome at all since such a limit is evidently nonphysical. The problem lies in the self-energy of the electron being infinite in the relativistic theory as well. In such a case, we do not have an a priori reason to expect the limit $\Lambda \to \infty$ to be unphysical! We will return to this issue in the following chapter.

For now, we finish the evaluation of Eq. (6.237). For evaluation of the first term, we use identity (3.99) and definition of the reference state $(\hat{\mathsf{h}}_0 - \varepsilon_0)|\psi^0_{\text{at}}\rangle = 0$,

$$
\begin{aligned}
\langle\psi^0_{\text{at}}|(\hat{\mathsf{p}}_A)_i(\hat{\mathsf{h}}_0 - \varepsilon_0)(\hat{\mathsf{p}}_A)_i|\psi^0_{\text{at}}\rangle &= -\frac{1}{2}\langle\psi^0_{\text{at}}|\Big[(\hat{\mathsf{p}}_A)_i, \Big[(\hat{\mathsf{p}}_A)_i, \hat{\mathsf{h}}_0 - \varepsilon_0\Big]\Big]|\psi^0_{\text{at}}\rangle \\
&= -\frac{1}{2}\langle\psi^0_{\text{at}}|\nabla^2 r^{-1}|\psi^0_{\text{at}}\rangle = \frac{1}{2}\langle\psi^0_{\text{at}}|4\pi\delta(\mathbf{r})|\psi^0_{\text{at}}\rangle \\
&= \frac{1}{2}4\pi|\psi^0_{\text{at}}(0)|^2 = \frac{1}{2}\frac{4}{n^3}\delta_{l,0}\,, \qquad (6.238)
\end{aligned}
$$

where we used Eq. (4.101) in the last step. The second term in the expression for $E^{(2)}_{\text{exp}}$ is called the *Bethe logarithm* and its evaluation is more involved. Fortunately, we have already developed all that is needed for its calculation: it suffices to insert $f(x) = x \ln x$ for the function f into Eq. (6.205). The contribution of the continuous part of the spectrum is of crucial importance as it yields 96% of the effect. The shift of the $2p_{1/2}$ level is negligible in the first approximation, see the next section. For the difference between energy levels of $2s$ and $2p_{1/2}$ states, we obtain

$$(\Delta E)_{2s} - (\Delta E)_{2p_{1/2}} \doteq 1047.42\,\text{MHz}\,.$$

Bearing in mind all of the approximations we invoked, this result lies surprisingly close to the measured value 1058 MHz. However, there are several unclarities. For instance, it is not clear why Λ should equal exactly m and not $2m$ or generally Km, with K of the order of unity, instead. The logarithm is a slowly varying function, though, and the result is not overly sensitive to the precise value of K. Furthermore, it is not a priori clear, why the contribution of the short-wavelength photons yields only a relatively small correction to the contribution of the long-wavelength photons. Given the fact that the Bethe estimate and the experiment lie so close to each other, the meaning of the Bethe estimate is clear. The dominant contribution to the self-energy of the bound electron stems from the emission and absorption of long-wavelength photons. This will be an important clue for an accurate evaluation of the self-energy of the bound electron within the exact relativistic theory in the following chapter.

Finally, if the reference state ψ^0_{at} is the excited state, the energy has a nonvanishing imaginary part. With the use of the spectral decomposition of $\hat{\mathsf{h}}_0$, we find

$$(\hat{\mathsf{h}}_0 - \varepsilon_0)\ln(\hat{\mathsf{h}}_0 - \varepsilon_0) = \sum_n |n\rangle\left(\frac{1}{2N^2} - \frac{1}{2n^2}\right)\ln\left(\frac{1}{2N^2} - \frac{1}{2n^2}\right)\langle n| + \ldots\,,$$

where the contribution of the continuous part is not displayed explicitly. For $n < N$ we have $\ln\left(\frac{1}{2N^2} - \frac{1}{2n^2}\right) = \ln(-1) + \ln\left(\frac{1}{2n^2} - \frac{1}{2N^2}\right)$. If we take the branch of the logarithm corresponding to $\ln(-1) = -i\pi$, we obtain from Eq. (6.237)

$$\Im\left[E^{(2)}_{\text{exp}}\right] = \frac{2}{3} m\alpha(Z\alpha)^4 \sum_{n<N} \langle \psi^0_{\text{at}}|\hat{\mathsf{p}}_i|n\rangle \left(\frac{1}{2N^2} - \frac{1}{2n^2}\right) \langle n|\hat{\mathsf{p}}_i|\psi^0_{\text{at}}\rangle$$

in agreement with Eqs. (6.97), (6.99) and (6.197). The imaginary part of the energy shift corresponds to the possibility that the electron does not manage to absorb the emitted photon and falls down to a state of lower energy.

6.5.4 *Improved Bethe Estimate*

The relativistic theory we will deal with in the next chapter yields the Lamb shift as a sum of high- and low-energy parts, which are determined by contributions of short- and long-wavelength photons, respectively. For the low-energy part, we will derive in Sect. 7.5.5 the formula

$$F_{\text{low}} = \left\langle \psi^0_{\text{at}} \middle| (\hat{\mathsf{p}}_A)_i f(\hat{\mathsf{h}}_0 - \varepsilon_0)(\hat{\mathsf{p}}_A)_i \middle| \psi^0_{\text{at}} \right\rangle, \quad f(x) = n^3 x \int_0^1 \mathrm{d}y \int_0^1 \mathrm{d}w \frac{1-2w(1-w)}{y+2w(Z\alpha)^2 x}\,. \tag{6.239}$$

If the function $f(x)$ is expanded in series in x, one obtains for the leading order

$$f(\hat{\mathsf{h}}_0 - \varepsilon_0) \simeq n^3(\hat{\mathsf{h}}_0 - \varepsilon_0)\left\{\frac{13}{18} - \frac{2}{3}\ln(2) - \frac{2}{3}\ln\left[(Z\alpha)^2(\hat{\mathsf{h}}_0 - \varepsilon_0)\right] + \dots\right\}. \tag{6.240}$$

The last showed term is the Bethe nonrelativistic estimate (6.237). To derive expression (6.239), we need the relativistic theory. However, we do not need the relativistic theory at all to evaluate it, thus we will do so now. Recall that we have already learned how to calculate expressions of this type, see Eqs. (6.205) and (6.227).

To assess the accuracy of the numerical calculation, we use the summation rules for the ground state; namely Eq. (6.228) for $f = 1$ and

$$\langle(\hat{\mathsf{p}}_A)_i(\hat{\mathsf{h}}_0 - \varepsilon_0)(\hat{\mathsf{p}}_A)_i\rangle = 2 \tag{6.241}$$

for $f = \hat{\mathsf{h}}_0 - \varepsilon_0$, where the latter follows from Eq. (6.238). For $Z = 1$, the numerical calculation yields

$$F_{\text{low}}(1s) = 10.6360506\,. \tag{6.242}$$

In case of the excited states, one proceeds in the same way. The calculation of the overlap integrals between the reference and virtual states is slightly more complex, though. For $Z = 1$, the numerical calculation yields

$$F_{\text{low}}(2s) = 10.8658433\,, \qquad F_{\text{low}}(2p) = 0.0400340\,. \tag{6.243}$$

If we use the expansion (6.240) and Eq. (6.238), we find

$$F_{\text{low}} \simeq \delta_{l,0}\left(\frac{13}{9}-\frac{4}{3}\ln(Z\alpha)^2\right)-\frac{4}{3}\frac{n^3}{2}\langle\psi^0_{\text{at}}|(\hat{\mathsf{p}}_A)_i(\hat{\mathsf{h}}_0-\varepsilon_0)\ln\left[2(\hat{\mathsf{h}}_0-\varepsilon_0)\right](\hat{\mathsf{p}}_A)_i|\psi^0_{\text{at}}\rangle+\ldots, \tag{6.244}$$

where for three lowest hydrogen states, the Bethe logarithm equals

$$\frac{1}{2}\langle 1s|(\hat{\mathsf{p}}_A)_i(\hat{\mathsf{h}}_0-\varepsilon_0)\ln\left[2(\hat{\mathsf{h}}_0-\varepsilon_0)\right](\hat{\mathsf{p}}_A)_i|1s\rangle = 2.984128555\ldots, \tag{6.245}$$

$$\frac{2^3}{2}\langle 2s|(\hat{\mathsf{p}}_A)_i(\hat{\mathsf{h}}_0-\varepsilon_0)\ln\left[2(\hat{\mathsf{h}}_0-\varepsilon_0)\right](\hat{\mathsf{p}}_A)_i|2s\rangle = 2.811769893\ldots \tag{6.246}$$

and

$$\frac{2^3}{2}\langle 2p|(\hat{\mathsf{p}}_A)_i(\hat{\mathsf{h}}_0-\varepsilon_0)\ln\left[2(\hat{\mathsf{h}}_0-\varepsilon_0)\right](\hat{\mathsf{p}}_A)_i|2p\rangle = -0.030016709\ldots \tag{6.247}$$

It is more accurate and simpler[21] to calculate the improved Bethe estimate (6.239) than the Bethe logarithm (6.244). Nevertheless, we will use the Bethe logarithm later in Sect. 6.5.6.

Exercise 23: The Improved Bethe Estimate

In a programming language with a built-in procedure for numerical integration, write Eq. (6.227) for the function (6.239). Check the accuracy of the summation rules (6.228) and (6.241). Try to reproduce the result (6.242) as well as the results (6.243), (6.245), (6.246) and (6.247). The formulas (6.225) and (6.226) will be of great help to you.

Hint: in Eq. (6.239), integrate analytically over y and w prior to any numerical calculations.

[21] In case of the calculation of (6.239), we numerically integrate an expression which behaves for the s-states for large k_e as $k_e^{-4}\ln(k_e)$. On the other hand, in case of (6.244), we integrate an expression which behaves for the s-states for large k_e as $k_e^{-2}\ln(k_e)$, and thus is on the very border of convergence. These statement will become clear to the reader once he attempts the following exercise.

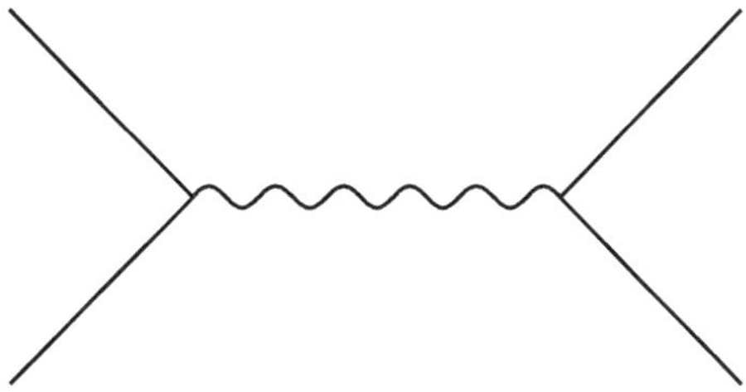

Fig. 6.12 One-photon exchange between two charged particles

6.5.5 One-Photon Exchange: Instantaneous Interaction

Let us now turn our attention to the last second-order process—one-photon exchange between two charged particles, see Fig. 6.12. We consider the effect of the interaction Hamiltonian (6.65) to the second order of the perturbation method

$$\Delta E_{\rm cc} = -\frac{e_1 e_2}{m_1 m_2}\langle\psi_{\rm at}^0|\langle 0|\Bigg\{\hat{\boldsymbol{A}}(\mathbf{r}_1)\cdot\hat{\boldsymbol{p}}_1\frac{1}{:\hat{\mathsf{H}}_{\rm EM}:+\hat{\mathsf{H}}_{\rm at}-E_0}\hat{\boldsymbol{A}}(\mathbf{r}_2)\cdot\hat{\boldsymbol{p}}_2$$
$$+\ \hat{\boldsymbol{A}}(\mathbf{r}_2)\cdot\hat{\boldsymbol{p}}_2\frac{1}{:\hat{\mathsf{H}}_{\rm EM}:+\hat{\mathsf{H}}_{\rm at}-E_0}\hat{\boldsymbol{A}}(\mathbf{r}_1)\cdot\hat{\boldsymbol{p}}_1\Bigg\}|0\rangle|\psi_{\rm at}^0\rangle\,.$$

Compare this equation with the first two equations in Sect. 6.5.3. The atomic Hamiltonian $\hat{\mathsf{H}}_{\rm at}$ is given by Eq. (6.64).

The reader will already manage to eliminate the field operators on his or her own and obtain

$$\Delta E_{\rm cc} = -\frac{e_1 e_2}{m_1 m_2 (2\pi)^3}\int\frac{{\rm d}^3\mathbf{k}}{2\omega}\left(\delta_{ij}-\eta_i\eta_j\right)$$
$$\times\ \langle\psi_{\rm at}^0|\Bigg\{(\hat{\mathsf{p}}_1)_i{\rm e}^{{\rm i}\mathbf{k}\cdot\hat{\boldsymbol{r}}_1}\frac{1}{\omega+\hat{\mathsf{H}}_{\rm at}-E_0}{\rm e}^{-{\rm i}\mathbf{k}\cdot\hat{\boldsymbol{r}}_2}(\hat{\mathsf{p}}_2)_j$$
$$+\ (\hat{\mathsf{p}}_2)_j{\rm e}^{-{\rm i}\mathbf{k}\cdot\hat{\boldsymbol{r}}_2}\frac{1}{\omega+\hat{\mathsf{H}}_{\rm at}-E_0}{\rm e}^{{\rm i}\mathbf{k}\cdot\hat{\boldsymbol{r}}_1}(\hat{\mathsf{p}}_1)_i\Bigg\}|\psi_{\rm at}^0\rangle\,.$$

In the second term, we substituted $\mathbf{k}\to-\mathbf{k}$ to convert the expression into a more symmetric form. We now rewrite the denominators as

$$\frac{1}{\omega+\hat{\mathsf{H}}_{\rm at}-E_0}=\frac{1}{\omega}-\frac{\hat{\mathsf{H}}_{\rm at}-E_0}{\omega(\omega+\hat{\mathsf{H}}_{\rm at}-E_0)}\,.$$

Consecutively, the one-photon exchange can be written as a sum of an instantaneous and a retarded interaction

$$\Delta E_{\rm cc}=\Delta E_{\rm cc,i}+\Delta E_{\rm cc,r}\,,$$

where

$$\Delta E_{\mathrm{cc,i}} = -\frac{e_1 e_2}{m_1 m_2 (2\pi)^3} \int \frac{\mathrm{d}^3\mathbf{k}}{\omega^2} \left(\delta_{ij} - \eta_i \eta_j\right) \left\langle \psi_{\mathrm{at}}^0 \middle| (\hat{\mathsf{p}}_1)_i \mathrm{e}^{\mathrm{i}\mathbf{k}\cdot(\hat{\boldsymbol{r}}_1 - \hat{\boldsymbol{r}}_2)} (\hat{\mathsf{p}}_2)_j \middle| \psi_{\mathrm{at}}^0 \right\rangle \tag{6.248}$$

and

$$\begin{aligned} \Delta E_{\mathrm{cc,r}} = {} & \frac{e_1 e_2}{m_1 m_2 (2\pi)^3} \int \frac{\mathrm{d}^3\mathbf{k}}{2\omega^2} \left(\delta_{ij} - \eta_i \eta_j\right) \\ & \times \left\langle \psi_{\mathrm{at}}^0 \right| \left\{ (\hat{\mathsf{p}}_1)_i \mathrm{e}^{\mathrm{i}\mathbf{k}\cdot\hat{\boldsymbol{r}}_1} \frac{\hat{\mathsf{H}}_{\mathrm{at}} - E_0}{\omega + \hat{\mathsf{H}}_{\mathrm{at}} - E_0} \mathrm{e}^{-\mathrm{i}\mathbf{k}\cdot\hat{\boldsymbol{r}}_2} (\hat{\mathsf{p}}_2)_j \right. \\ & + \left. (\hat{\mathsf{p}}_2)_j \mathrm{e}^{-\mathrm{i}\mathbf{k}\cdot\hat{\boldsymbol{r}}_2} \frac{\hat{\mathsf{H}}_{\mathrm{at}} - E_0}{\omega + \hat{\mathsf{H}}_{\mathrm{at}} - E_0} \mathrm{e}^{\mathrm{i}\mathbf{k}\cdot\hat{\boldsymbol{r}}_1} (\hat{\mathsf{p}}_1)_i \right\} \left| \psi_{\mathrm{at}}^0 \right\rangle . \end{aligned} \tag{6.249}$$

How do we know that this separation corresponds to the separation of the instantaneous and retarded interaction? If we perform the integration over **k**, see Sect. 3.6.1, we obtain expression (6.248) in the form

$$\Delta E_{\mathrm{cc,i}} = \left\langle \psi_{\mathrm{at}}^0 \middle| (\hat{\mathsf{H}}_1)_{\mathrm{cc}} \middle| \psi_{\mathrm{at}}^0 \right\rangle ,$$

where $(\hat{\mathsf{H}}_1)_{\mathrm{cc}}$ is the interaction Hamiltonian introduced in Eq. (3.127):

$$(\hat{\mathsf{H}}_1)_{\mathrm{cc}} = -\frac{e_1 e_2}{m_1 m_2 4\pi} \frac{1}{2 r_{12}} \left[\hat{\boldsymbol{p}}_1 \cdot \hat{\boldsymbol{p}}_2 + \mathbf{n}_{12} \cdot \hat{\boldsymbol{p}}_1 \, \mathbf{n}_{12} \cdot \hat{\boldsymbol{p}}_2 \right] . \tag{3.127}$$

This expression was derived in Sect. 3.6.1 from the classical equation (3.2) for vector potential where the retardation effect was neglected.

Exercise 24: One-Photon Exchange

Show that the effect of the spin-spin and the spin-other-orbit interactions on atomic energies can be obtained in two ways. Firstly,

$$\Delta E_{\mathrm{ss}} = \left\langle \psi_{\mathrm{at}}^0 \middle| \left((\hat{\mathsf{H}}_1)_{\mathrm{sst}} + (\hat{\mathsf{H}}_1)_{\mathrm{ssc}} \right) \middle| \psi_{\mathrm{at}}^0 \right\rangle,$$

$$\Delta E_{\mathrm{soo}} = \left\langle \psi_{\mathrm{at}}^0 \middle| (\hat{\mathsf{H}}_1)_{\mathrm{soo}} \middle| \psi_{\mathrm{at}}^0 \right\rangle,$$

where the interaction Hamiltonians $(\hat{\mathsf{H}}_1)_{\mathrm{sst}}$, $(\hat{\mathsf{H}}_1)_{\mathrm{ssc}}$ *and* $(\hat{\mathsf{H}}_1)_{\mathrm{soo}}$ *are given by Eqs. (3.125), (3.124) and (3.126), respectively. Secondly,*

$$\Delta E_{ss} = -\frac{e_1 e_2}{m_1 m_2}\langle\psi^0_{at}|\langle 0|\left\{\hat{\boldsymbol{S}}_1\cdot\hat{\boldsymbol{B}}(\mathbf{r}_1)\frac{1}{:\hat{\mathsf{H}}_{EM}:}\hat{\boldsymbol{S}}_2\cdot\hat{\boldsymbol{B}}(\mathbf{r}_2)\right.$$

$$\left. + \hat{\boldsymbol{S}}_2\cdot\hat{\boldsymbol{B}}(\mathbf{r}_2)\frac{1}{:\hat{\mathsf{H}}_{EM}:}\hat{\boldsymbol{S}}_1\cdot\hat{\boldsymbol{B}}(\mathbf{r}_1)\right\}|0\rangle|\psi^0_{at}\rangle,$$

$$\Delta E_{soo} = -\frac{e_1 e_2}{m_1 m_2}\langle\psi^0_{at}|\langle 0|\left\{\hat{\boldsymbol{A}}(\mathbf{r}_1)\cdot\hat{\boldsymbol{p}}_1\frac{1}{:\hat{\mathsf{H}}_{EM}:}\hat{\boldsymbol{S}}_2\cdot\hat{\boldsymbol{B}}(\mathbf{r}_2)\right.$$

$$+ \hat{\boldsymbol{S}}_2\cdot\hat{\boldsymbol{B}}(\mathbf{r}_2)\frac{1}{:\hat{\mathsf{H}}_{EM}:}\hat{\boldsymbol{A}}(\mathbf{r}_1)\cdot\hat{\boldsymbol{p}}_1$$

$$+ \hat{\boldsymbol{A}}(\mathbf{r}_2)\cdot\hat{\boldsymbol{p}}_2\frac{1}{:\hat{\mathsf{H}}_{EM}:}\hat{\boldsymbol{S}}_1\cdot\hat{\boldsymbol{B}}(\mathbf{r}_1)$$

$$\left. + \hat{\boldsymbol{S}}_1\cdot\hat{\boldsymbol{B}}(\mathbf{r}_1)\frac{1}{:\hat{\mathsf{H}}_{EM}:}\hat{\boldsymbol{A}}(\mathbf{r}_2)\cdot\hat{\boldsymbol{p}}_2\right\}|0\rangle|\psi^0_{at}\rangle,$$

where the effect of retardation has been neglected; in the denominators $(\hat{\mathsf{H}}_{at} - E_0)$ *was neglected in comparison with* $:\hat{\mathsf{H}}_{EM}:$.

6.5.6 One-Photon Exchange: Effect of Retardation

We now show how to further deal with the retarded part in Eq. (6.249). This part shifts the atomic energies at the order of α^5. Therefore, within accuracy to this order, we can express the term as a sum of two

$$\Delta E_{cc,r} \simeq \Delta E_{cc,r1} + \Delta E_{cc,r2}\,.$$

We now invoke the dipole approximation in the first part

$$\Delta E_{cc,r1} = \frac{e_1 e_2}{m_1 m_2 (2\pi)^3}\int\frac{d^3\mathbf{k}}{2\omega^2}\left(\delta_{ij} - \eta_i\eta_j\right)\langle\psi^0_{at}|\left\{(\hat{\mathsf{p}}_1)_i\frac{\hat{\mathsf{H}}_{at} - E_0}{\omega + \hat{\mathsf{H}}_{at} - E_0}(\hat{\mathsf{p}}_2)_j\right.$$

$$\left. + (\hat{\mathsf{p}}_2)_j\frac{\hat{\mathsf{H}}_{at} - E_0}{\omega + \hat{\mathsf{H}}_{at} - E_0}(\hat{\mathsf{p}}_1)_i\right\}|\psi^0_{at}\rangle\,.$$

In the second part, we again neglect the term $\hat{\mathsf{H}}_{\mathrm{at}} - E_0$ in the denominator with respect to ω:

$$\Delta E_{\mathrm{cc,r2}} = \frac{e_1 e_2}{m_1 m_2 (2\pi)^3} \int \frac{\mathrm{d}^3\mathbf{k}}{2\omega^3} \left(\delta_{ij} - \eta_i \eta_j\right) \times \left\langle \psi^0_{\mathrm{at}} \right| \left\{ (\hat{\mathsf{p}}_1)_i \mathrm{e}^{\mathrm{i}\mathbf{k}\cdot\hat{\boldsymbol{r}}_1} (\hat{\mathsf{H}}_{\mathrm{at}} - E_0) \mathrm{e}^{-\mathrm{i}\mathbf{k}\cdot\hat{\boldsymbol{r}}_2} (\hat{\mathsf{p}}_2)_j - (\hat{\mathsf{p}}_1)_i (\hat{\mathsf{H}}_{\mathrm{at}} - E_0)(\hat{\mathsf{p}}_2)_j + (\hat{\mathsf{p}}_2)_j \mathrm{e}^{-\mathrm{i}\mathbf{k}\cdot\hat{\boldsymbol{r}}_2} (\hat{\mathsf{H}}_{\mathrm{at}} - E_0) \mathrm{e}^{\mathrm{i}\mathbf{k}\cdot\hat{\boldsymbol{r}}_1} (\hat{\mathsf{p}}_1)_i - (\hat{\mathsf{p}}_2)_j (\hat{\mathsf{H}}_{\mathrm{at}} - E_0)(\hat{\mathsf{p}}_1)_i \right\} \left| \psi^0_{\mathrm{at}} \right\rangle .$$

Integration over the wave vectors in the first part yields the Bethe logarithm, see Eq. (6.237),

$$\Delta E_{\mathrm{cc,r1}} = \frac{e_1 e_2}{3 m_1 m_2 \pi^2} \left\langle \psi^0_{\mathrm{at}} \right| (\hat{\mathsf{p}}_1)_i (\hat{\mathsf{H}}_{\mathrm{at}} - E_0) \ln \left(\frac{\Lambda}{\hat{\mathsf{H}}_{\mathrm{at}} - E_0} \right) (\hat{\mathsf{p}}_2)_i \left| \psi^0_{\mathrm{at}} \right\rangle .$$

Here, the upper bound of the integration was stopped at Λ, and we neglected $(\hat{\mathsf{H}}_{\mathrm{at}} - E_0)$ in comparison with Λ. For hydrogen-like atoms in the center of mass frame, it holds that $e_2 = -Ze_1 = -Ze$, $\hat{\boldsymbol{p}}_2 = -\hat{\boldsymbol{p}}_1$ and $r_{12} = r$. The last equation then simplifies to, see Eq. (6.204), where m is replaced by m_r,

$$\Delta E_{\mathrm{cc,r1}} = \frac{4(Z\alpha)^5 m_r^3}{3 m_1 m_2 \pi} \left\langle \psi^0_{\mathrm{at}} \right| (\hat{\mathsf{p}}_A)_i (\hat{\mathsf{h}}_0 - \varepsilon_0) \ln \left(\frac{\Lambda}{m_r (Z\alpha)^2 (\hat{\mathsf{h}}_0 - \varepsilon_0)} \right) (\hat{\mathsf{p}}_A)_i \left| \psi^0_{\mathrm{at}} \right\rangle . \tag{6.250}$$

The second part, $\Delta E_{\mathrm{cc,r2}}$, is arranged by means of the operator identity[22]

$$\hat{\mathsf{A}}\hat{\mathsf{B}}\hat{\mathsf{C}} + \hat{\mathsf{C}}\hat{\mathsf{B}}\hat{\mathsf{A}} = \hat{\mathsf{A}}\hat{\mathsf{C}}\hat{\mathsf{B}} + \hat{\mathsf{B}}\hat{\mathsf{C}}\hat{\mathsf{A}} - [\hat{\mathsf{A}}, [\hat{\mathsf{C}}, \hat{\mathsf{B}}]]$$

and the definition of the reference state, $(\hat{\mathsf{H}}_{\mathrm{at}} - E_0)\left| \psi^0_{\mathrm{at}} \right\rangle = 0$,

$$\begin{aligned} \Delta E_{\mathrm{cc,r2}} &= -\frac{e_1 e_2}{m_1 m_2 (2\pi)^3} \int \frac{\mathrm{d}^3\mathbf{k}}{2\omega^3} \left(\delta_{ij} - \eta_i \eta_j\right) \\ &\quad \times \left\langle \psi^0_{\mathrm{at}} \right| \left\{ [(\hat{\mathsf{p}}_1)_i \mathrm{e}^{\mathrm{i}\mathbf{k}\cdot\hat{\boldsymbol{r}}_1}, [\mathrm{e}^{-\mathrm{i}\mathbf{k}\cdot\hat{\boldsymbol{r}}_2} (\hat{\mathsf{p}}_2)_j, \hat{\mathsf{H}}_{\mathrm{at}} - E_0]] - [(\hat{\mathsf{p}}_1)_i, [(\hat{\mathsf{p}}_2)_j, \hat{\mathsf{H}}_{\mathrm{at}} - E_0]] \right\} \left| \psi^0_{\mathrm{at}} \right\rangle \\ &= -\frac{e_1 e_2}{m_1 m_2 (2\pi)^3} \int \frac{\mathrm{d}^3\mathbf{k}}{2\omega^3} \left(\delta_{ij} - \eta_i \eta_j\right) \left\langle \psi^0_{\mathrm{at}} \right| (\mathrm{e}^{\mathrm{i}\mathbf{k}\cdot\mathbf{r}_{12}} - 1) \\ &\quad \times \left[(\hat{\mathsf{p}}_1)_i, \left[(\hat{\mathsf{p}}_2)_j, \frac{e_1 e_2}{4\pi} \frac{1}{r_{12}} \right] \right] \left| \psi^0_{\mathrm{at}} \right\rangle . \end{aligned}$$

---

[22]Clearly, $\hat{\mathsf{A}} = \hat{\mathsf{p}}_{1i} \mathrm{e}^{\mathrm{i}\mathbf{k}\cdot\mathbf{r}_1}$, $\hat{\mathsf{B}} = \hat{\mathsf{p}}_{2j} \mathrm{e}^{-\mathrm{i}\mathbf{k}\cdot\mathbf{r}_2}$ and $\hat{\mathsf{C}} = \hat{\mathsf{H}}_{\mathrm{at}} - E_0$. In addition, $\hat{\mathsf{p}}_i$ and $\mathrm{e}^{\mathrm{i}\mathbf{k}\cdot\mathbf{r}}$ can be freely commuted. Their commutator is $k_i \mathrm{e}^{\mathrm{i}\mathbf{k}\cdot\mathbf{r}}$, but the whole expression is multiplied by the projector P_{ij} and hence the commutator vanishes, see Eq. (6.33). This can be traced back to the Coulomb gauge $\nabla \cdot \hat{\boldsymbol{A}} = 0$.

We already calculated the commutator in Sect. 3.3.1

$$\left[(\hat{\mathrm{p}}_1)_i, \left[(\hat{\mathrm{p}}_2)_j, \frac{1}{r_{12}}\right]\right] = \frac{1}{3}4\pi\delta_{ij}\delta(\mathbf{r}_{12}) - \frac{\delta_{ij} - 3(n_{12})_i(n_{12})_j}{r_{12}^3}.$$

This expression is multiplied by $(\mathrm{e}^{\mathrm{i}\mathbf{k}\cdot\mathbf{r}_{12}} - 1)$, see the penultimate equation. Thus, the contribution of the first term on the rhs of the last equation vanishes. Multiplication of the second term by the projector P_{ij} yields

$$\left(\delta_{ij} - \eta_i\eta_j\right)\left(\delta_{ij} - 3n_in_j\right) = -1 + 3(\boldsymbol{\eta}\cdot\mathbf{n})^2.$$

Identifying the z-axis with the direction of the vector $\mathbf{n}$, the integration over the directions of the virtual photon is easily accomplished

$$\begin{aligned}\int \frac{\mathrm{d}^3\mathbf{k}}{(2\pi)^3\omega^3}&(-1 + 3(\boldsymbol{\eta}\cdot\mathbf{n})^2)(\mathrm{e}^{\mathrm{i}\mathbf{k}\cdot\mathbf{r}} - 1)\\ &= \frac{1}{(2\pi)^2}\int_0^\Lambda \frac{\mathrm{d}\omega}{\omega}\int_0^\pi \mathrm{d}\vartheta\,\sin\vartheta(-1 + 3\cos^2\vartheta)(\mathrm{e}^{\mathrm{i}\omega r\cos\vartheta} - 1)\\ &= -\frac{2}{(2\pi)^2}\int_0^\Lambda \frac{\mathrm{d}\omega}{\omega}\left(1 + 3\frac{\mathrm{d}^2}{\mathrm{d}z^2}\right)\frac{\sin(z)}{z}\bigg|_{z=\omega r}.\end{aligned}$$

For hydrogen-like atoms in the center of mass frame, we obtain from the last four equations and substitutions $r \to r/(m_rZ\alpha)$ and $\omega \to \omega(m_rZ\alpha)$,

$$\Delta E_{\mathrm{cc,r2}} = -\frac{(Z\alpha)^5 m_r^3}{m_1m_2\pi}\int_0^{\Lambda/(m_rZ\alpha)}\frac{\mathrm{d}\omega}{\omega}\int_0^\infty \frac{\mathrm{d}r}{r}|\psi_{\mathrm{at}}^0(r)|^2\left(1 + 3\frac{\mathrm{d}^2}{\mathrm{d}z^2}\right)\frac{\sin(z)}{z}\bigg|_{z=\omega r}. \tag{6.251}$$

It is advantageous to write the energy shift for a general s-state as a sum of state-dependent and state-independent parts

$$\Delta E_n = \frac{\Delta E_n - \Delta E_1}{n^3} + \frac{\Delta E_1}{n^3}.$$

The numerator of the first term, the "normalized" or "weighted" difference, behaves similar to the contribution of the p-states. For normalized difference of the s-states and for p-states, each of the expressions $E_{\mathrm{cc,r1}}$ and $E_{\mathrm{cc,\,r2}}$ is finite. The Bethe estimate (6.237) is in this case finite, too. As one can see from Eq. (6.237), and as we will see in the next section, the inadequacy of the nonrelativistic approximation appears at the order α^5 only for the state-independent part of the s-states. The advantage of the separation should be clear now. The state-dependent part of the energy shift of the s-states behaves much better than the complete energy shift. All the difficulties can be then reduced to the calculation of the ground state. Let us use this observation and split our further calculation into those for the ground state and those for the excited states.

- *Ground state*
  Integration over frequencies yields

$$\int_0^{\Lambda/(m_r Z\alpha)} \frac{d\omega}{\omega} \int_0^\infty \frac{dr}{r} e^{-\nu r} \left(1 + 3\frac{d^2}{dz^2}\right) \frac{\sin(z)}{z}\bigg|_{z=\omega r} = \frac{2}{3} \ln\left(\frac{\Lambda}{m_r Z\alpha \nu}\right) - \frac{8}{9}.$$

By means of this equation and of Eqs. (3.35), (4.101), and (6.238), we find for the total contribution of the retardation effect, i.e., the sum of Eqs. (6.250) and (6.251), for the state-independent part of the s-states that

$$\frac{\Delta E_{\text{cc,r}}(1s)}{n^3} = \frac{(Z\alpha)^5 m_r^3}{m_1 m_2 \pi n^3} \left[-\frac{4}{3} \langle 1s | (\hat{\mathsf{p}}_A)_i (\hat{\mathsf{h}}_0 - \varepsilon_0) \ln\left[2(\hat{\mathsf{h}}_0 - \varepsilon_0)\right] (\hat{\mathsf{p}}_A)_i | 1s \rangle \right.$$
$$\left. + \frac{32}{9} - \frac{8}{3} \ln(Z\alpha) + \frac{16}{3} \ln(2) \right]. \tag{6.252}$$

Note that the dependence on the upper bound of the integration Λ cancels out.

- *Excited states*
  Integration over the frequencies in Eq. (6.251) can now be extended to infinity:

$$\int_0^\infty \frac{d\omega}{\omega} \left(1 + 3\frac{d^2}{dz^2}\right) \frac{\sin(z)}{z}\bigg|_{z=\omega r} = \frac{2}{3}.$$

For the state-dependent part of the s-states $[n^3 \Delta E(ns) - \Delta E(1s)]/n^3$ and for the non-s-states, we obtain from Eqs. (6.250) and (6.251) the following expression for the energy shift

$$-\frac{(Z\alpha)^5 m_r^3}{m_1 m_2 \pi} \left\langle\!\!\left\langle \frac{4}{3} (\hat{\mathsf{p}}_A)_i (\hat{\mathsf{h}}_0 - \varepsilon_0) \ln\left[\hat{\mathsf{h}}_0 - \varepsilon_0\right] (\hat{\mathsf{p}}_A)_i + \frac{2}{3} \frac{1}{r_A^3} \right\rangle\!\!\right\rangle \tag{6.253}$$

where we introduced the notation

$$\left\langle\!\left\langle \hat{\mathsf{O}} \right\rangle\!\right\rangle = \delta_{l,0} \frac{n^3 \langle n,0,0|\hat{\mathsf{O}}|n,0,0\rangle - \langle 1,0,0|\hat{\mathsf{O}}|1,0,0\rangle}{n^3} + (1 - \delta_{l,0}) \langle n,l,m|\hat{\mathsf{O}}|n,l,m\rangle$$

for a general operator $\hat{\mathsf{O}}$. For the non-s-states, the meaning of the two brackets is clearly the same as merely one bracket: the average of the operator $\hat{\mathsf{O}}$ for a given atomic state. The Bethe logarithm of the three lowest hydrogen states was already given in Eqs. (6.245), (6.246), and (6.247). Furthermore, one can easily find, see Eqs. (3.95) and (3.96),

$$\left\langle\!\!\left\langle \frac{1}{r_A^3} \right\rangle\!\!\right\rangle_{2s} = \frac{4\ln(2) - 3}{8}, \qquad \left\langle\!\!\left\langle \frac{1}{r_A^3} \right\rangle\!\!\right\rangle_{2p} = \frac{1}{24}. \tag{6.254}$$

Comparison of Eqs. (6.252), (6.253), and (6.237) shows that the retardation effect shifts the atomic energy levels at the same order of α as the self-energy effect. In contrast to it, though, it depends on the ratio of particle masses m_1/m_2. This means that in case of the "ordinary" hydrogen-like atoms, this shift is very small. However, from helium on, the retardation effect of the one-photon exchange between two electrons is of the same order of magnitude as the self-energy effects of the individual electrons. In the next chapter, in Sect. 7.8.5, we use the formula (6.252) for an accurate determination of the spectral lines of positronium, in which case the mass ratio m_1/m_2 equals 1.

It clearly follows from Eqs. (6.237) and (6.252) that once the retardation is not neglected, it is not possible to write down the EM interaction between the charged particles in the form of the Schrödinger equation with a prescribed potential.

6.5.7 Two-Photon Exchange: Low Energies

Unlike the one-photon exchange, which we can describe using the nonrelativistic approximation with accuracy up to α^5, one needs to consider the relativistic theory when it comes to the two-photon exchange. The physical reason is similar to that in case of the self-energy problem: the contribution of the virtual states, where the relativistic description is necessary, is not negligible. Nevertheless, there is a part of the two-photon exchange which we can treat within the nonrelativistic approximation.

In case of the one-photon exchange, particles do not manage to cross the boundaries of the nonrelativistic regime. In the two-photon exchange, such a part is given by the second order correction due to the interaction $e^2\hat{\boldsymbol{A}}^2$, see Fig. 6.13,

$$\Delta E_{2\mathrm{pl}} = -\frac{e_1^2 e_2^2}{2m_1 2m_2}\langle\psi_{\mathrm{at}}^0|\langle 0|\Bigg[:\hat{\boldsymbol{A}}(\mathbf{r}_1)\cdot\hat{\boldsymbol{A}}(\mathbf{r}_1): \frac{1}{:\hat{\mathsf{H}}_{\mathrm{EM}}: + \hat{\mathsf{H}}_{\mathrm{at}} - E_0} :\hat{\boldsymbol{A}}(\mathbf{r}_2)\cdot\hat{\boldsymbol{A}}(\mathbf{r}_2):$$
$$+ :\hat{\boldsymbol{A}}(\mathbf{r}_2)\cdot\hat{\boldsymbol{A}}(\mathbf{r}_2): \frac{1}{:\hat{\mathsf{H}}_{\mathrm{EM}}: + \hat{\mathsf{H}}_{\mathrm{at}} - E_0} :\hat{\boldsymbol{A}}(\mathbf{r}_1)\cdot\hat{\boldsymbol{A}}(\mathbf{r}_1):\Bigg]|0\rangle|\psi_{\mathrm{at}}^0\rangle .$$

Provided the typical momenta of the particles which enter the process are small, how do we recognize which part of the effect can be treated nonrelativistically and which not? Roughly speaking, the nonrelativistic description is appropriate when the particles are on the outer lines on the corresponding diagram. If they are on the inner lines, as in the case of the self-energy effect, see Fig. 6.9, the relativistic approach is needed. The relativistic description is necessary if there is a mechanism creating, at least temporarily, particles with large momenta. In case of particles bound in light atoms, the only contributing mechanism is quantum mechanics: it enforces integration over all possible momenta on the inner lines of the corresponding diagrams, see discussion in Sect. 6.5.1.

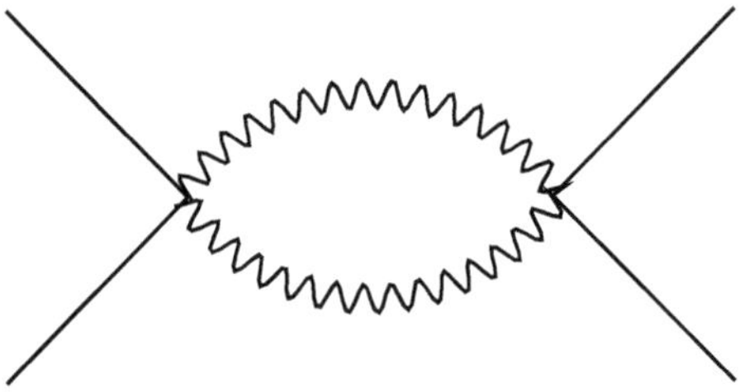

Fig. 6.13 Part of the two-photon exchange between charged particles amenable to nonrelativistic treatment

Obviously, from all possible processes contributing at the order $(e_1e_2)^2$, the one considered above is the only one with particles exclusively on outer lines, see Fig. 6.13. We consider the interaction $e^2\hat{\boldsymbol{A}}^2$ in the normally ordered form so that its vacuum expectation value vanishes. The reason is, we wish to describe the two-photon exchange between two particles, not two self-energy effects. The elimination of the field operators leads to the expression

$$\Delta E_{2\text{pl}} = -\frac{e_1^2e_2^2}{m_1m_2}\int\limits_{\omega_1<\Lambda}\frac{\mathrm{d}^3\mathbf{k}_1}{(2\pi)^3 2\omega_1}P_{ij}^{(1)}\int\limits_{\omega_2<\Lambda}\frac{\mathrm{d}^3\mathbf{k}_2}{(2\pi)^3 2\omega_2}P_{ij}^{(2)}\frac{\left\langle\psi_{\text{at}}^0\middle|\mathrm{e}^{\mathrm{i}(\mathbf{k}_1+\mathbf{k}_2)\cdot(\mathbf{r}_1-\mathbf{r}_2)}\middle|\psi_{\text{at}}^0\right\rangle}{\omega_1+\omega_2}, \tag{6.255}$$

where the projectors P_{ij} are given by Eq. (6.55). Their product reads

$$P_{ij}^{(1)}P_{ij}^{(2)} = 1 + [\boldsymbol{\eta}^{(1)}\cdot\boldsymbol{\eta}^{(2)}]^2\,.$$

For the sake of simplicity, we restrict our further considerations to the hydrogen-like atoms in the center of mass frame; that is, we set $r_{12} = r$ and $e_1e_2 = -Ze^2$. Furthermore, we change to the atomic units $\mathbf{r} = \mathbf{r}_A/(m_rZ\alpha)$ and substitute $\mathbf{k}_{1,2} \to \mathbf{k}_{1,2}(m_rZ\alpha)$.

Once again, we split the calculation into two parts: the ground and the excited states.

- *Ground state*

  For s-states, the angular integration is trivial

$$\left\langle\psi_{\text{at}}^0\middle|\mathrm{e}^{\mathrm{i}\mathbf{q}\cdot\mathbf{r}_A}\middle|\psi_{\text{at}}^0\right\rangle = \frac{1}{q}\int_0^\infty \mathrm{d}rr|\psi_{\text{at}}^{(0)}(r)|^2\sin(qr), \quad \mathbf{q} = \mathbf{k}_1 + \mathbf{k}_2\,.$$

  The integral over radial variable can be calculated by means of the formula

$$\frac{1}{q}\int_0^\infty \mathrm{d}rr\mathrm{e}^{-\nu r}\sin(qr) = \frac{2\nu}{(\nu^2+q^2)^2}.$$

  Obviously, the integrand (6.255) is symmetric with respect to the replacement of ω_1 and ω_2. Consecutively, the double integral over square $\omega_1 \in (0, \Lambda)$, $\omega_2 \in (0, \Lambda)$ can be expressed as a double of the integral over the triangle $\omega_1 \in (0, \Lambda)$,

$\omega_2 \in (0, \omega_1)$. Integral (6.255) then diverges in the upper bound of the integration over ω_1. Let us therefore subtract and add the divergent part. Substitution of the last three equations into Eq. (6.255) leads to

$$\Delta E_{2\mathrm{pl}} = -\frac{8(Z\alpha)^5 m_r^3}{\pi^2 m_1 m_2}\left[\Phi(2) + \frac{\pi}{4}\int_0^{\Lambda/(m_r Z\alpha)} \mathrm{d}\omega_1 \frac{\omega_1}{1+\omega_1^2}\right],$$

where

$$\Phi(\nu) = \int_0^\infty \mathrm{d}\omega_1 \left[\int_0^{\omega_1} \mathrm{d}\omega_2 \frac{\omega_1\omega_2}{\omega_1+\omega_2}\int_0^\pi \frac{\nu(1+\cos^2\vartheta)\sin\vartheta\,\mathrm{d}\vartheta}{(\nu^2+\omega_1^2+\omega_2^2+2\omega_1\omega_2\cos\vartheta)^2} - \frac{\pi}{4}\frac{\omega_1}{1+\omega_1^2}\right].$$

The first integral over ω_1, i.e., the integral from the expression inside the square brackets, is finite and can be evaluated numerically. The second integral diverges, however, it can be calculated analytically. The situation is in fact even better since the first integral can be calculated analytically as well. After simple integration over the angle ϑ, there are more involved integrations over ω_2 and ω_1 and an ambitious reader can test his abilities. The less ambitious one[23] can integrate over ω_1 numerically. For the state-independent part of the s-states, we find from the above equations and Eq. (3.35)

$$\frac{\Delta E_{2\mathrm{pl}}(1s)}{n^3} = \frac{(Z\alpha)^5 m_r^3}{m_1 m_2 \pi n^3}\left[\frac{14}{3} - \frac{8}{3}\ln 2 - 2\ln\left(\frac{\Lambda}{m_r Z\alpha}\right)\right]. \tag{6.256}$$

To eliminate the dependence on Λ, we have to use the relativistic theory, as we will do in Sect. 7.7.

- *Excited states*
  Integration over the frequencies of the virtual photons can be, in this case, extended to infinity; plane-wave decomposition into spherical waves, Eq. (5.30), knowledge of the spherical harmonics up to $l = 2$, the orthonormality of the spherical harmonics and integration over the frequencies ω_1 and ω_2 lead to

$$\int \frac{\mathrm{d}^3\mathbf{k}_1}{\omega_1}\int \frac{\mathrm{d}^3\mathbf{k}_2}{\omega_2}\{1+[\eta^{(1)}\cdot\eta^{(2)}]^2\}\frac{\mathrm{e}^{\mathrm{i}(\mathbf{k}_1+\mathbf{k}_2)\cdot\mathbf{r}_A}}{\omega_1+\omega_2} = \frac{(2\pi)^3}{r_A^3}.$$

The energy shift of the state-dependent part of the s-states $[n^3\Delta E(ns) - \Delta E(1s)]/n^3$ and non-s-states caused by the two-photon exchange is given by the last equation and Eq. (6.255)

$$-\frac{(Z\alpha)^5 m_r^3}{m_1 m_2 2\pi}\left\langle\left\langle\frac{1}{r_A^3}\right\rangle\right\rangle. \tag{6.257}$$

[23] Of course, an impatient reader will not worry at all.

The energy shift caused by one- and two-photon exchanges at the order α^5 for the state-dependent part of the s-states $[n^3 \Delta E(ns) - \Delta E(1s)]/n^3$ and non-s-states is given by Eqs. (6.253) and (6.257)

$$-\frac{(Z\alpha)^5 m_r^3}{m_1 m_2 \pi}\left\langle\left\langle\frac{4}{3}(\hat{\mathsf{p}}_A)_i(\hat{\mathsf{h}}_0-\varepsilon_0)\ln\left[\hat{\mathsf{h}}_0-\varepsilon_0\right](\hat{\mathsf{p}}_A)_i+\frac{7}{6}\frac{1}{r_A^3}\right\rangle\right\rangle . \tag{6.258}$$

The formula (6.256) will be used later in Sect. 7.7, and the formula (6.258) in Sect. 7.8.5 for a precise determination of the positronium spectral lines.

6.6 Formalism of the Second Quantization

Young man, in mathematics you don't understand things. You just get used to them.John von Neumann

So far, we regarded electrons as particles. However, we will see that unification of quantum mechanics and the special theory of relativity forces us to view electrons as quanta of an electron-positron field. Even within the nonrelativistic theory, one can perceive electrons as quanta of an electron field. Such a view has two advantages: one conceptual and one technical. We may ask why all electrons are indistinguishable, of the same mass, spin and charge? Because all of them are excitation of the same one electron field! In addition, while the wave function of two, three, and more electrons is a wave function in six-, nine-, and more dimensional spaces, the electron field is a field in an ordinary three-dimensional space, regardless of the number of quanta present. Later in the text, we shall see that already the analysis of the problem of three electrons which interact mutually and with the nucleus by the electrostatic interaction is much more easily penetrable via the field, rather than the particle, approach.

6.6.1 *Quantization of Free Fields*

Let us first summarize how we proceeded when quantizing the EM field. Quantization of the electron field will follow the same steps.

1. *Write an equation for the classical field.* In the Coulomb gauge, the classical equation for dynamical quantity, the vector potential, is of the form of Eq. (6.13), $\left(\frac{\partial^2}{\partial t^2} - \nabla^2\right)\mathbf{A} = 0$.
2. *Expand the classical field into suitable modes*; see Eq. (6.14), $\mathbf{A}(\mathbf{r}, t) = \sum_\tau q_\tau(t)\mathbf{T}_\tau(\mathbf{r})$, and Eq. (6.15), $-\nabla^2\mathbf{T}_\tau = \omega_\tau^2\mathbf{T}_\tau$.
3. *Determine the evolution equations for the coefficients of field decomposition into modes*; see Eq. (6.18), $\ddot{q}_\tau + \omega_\tau^2 q_\tau = 0$.
4. *Determine the canonical coordinates, the momenta, and the Hamiltonian for the classical theory.*

The classical equation (6.18) is equivalent to the Hamilton canonical equations (6.23), if the Hamiltonian is of the form (6.19) and (6.20), $H_{\text{EM}} = \sum_\tau \frac{p_\tau^2}{2} + \frac{1}{2}\omega_\tau^2 q_\tau^2$.

5. *Impose the canonical commutation relations on the classical coordinates and momenta. Insert the corresponding operators for the coordinates and momenta into the classical Hamiltonian to obtain the quantum Hamiltonian.* Canonical commutation relations are of the form (6.34), $[\hat{\mathsf{p}}_\tau, \hat{\mathsf{q}}_{\tau'}] = -\mathrm{i}\delta_{\tau\tau'}$, and the quantum Hamiltonian is of the form (6.38), $\hat{\mathsf{H}}_{\text{EM}} = \sum_\tau \frac{\hat{\mathsf{p}}_\tau^2}{2} + \frac{1}{2}\omega_\tau^2 \hat{\mathsf{q}}_\tau^2$.

However, the canonical coordinates and momenta are not determined uniquely, and the last two steps can be thus slightly modified. With this change, the similarity between quantization of the EM and electron fields will be more transparent.

4. Already in the classical theory, we can introduce suitable linear combinations of q_τ and p_τ

$$\begin{aligned} a_\tau &= \frac{1}{\sqrt{2\omega_\tau}} \left(\omega_\tau q_\tau + \mathrm{i} p_\tau\right) , \\ a_\tau^+ &= \frac{1}{\sqrt{2\omega_\tau}} \left(\omega_\tau q_\tau - \mathrm{i} p_\tau\right) . \end{aligned}$$

These relations can be inverted, i.e., we can express q_τ and p_τ as functions of a_τ and a_τ^+, and subsequently substituted into the Hamiltonian (6.20) to produce

$$H_{\text{EM}} = \sum_\tau \omega_\tau a_\tau^+ a_\tau \, . \tag{6.259}$$

The Hamilton equations (6.23) yield the time derivative of a_τ

$$\begin{aligned} \frac{\mathrm{d}}{\mathrm{d}t} a_\tau &= \frac{1}{\sqrt{2\omega_\tau}} \left(\omega_\tau \frac{\mathrm{d}}{\mathrm{d}t} q_\tau + \mathrm{i} \frac{\mathrm{d}}{\mathrm{d}t} p_\tau\right) = \frac{1}{\sqrt{2\omega_\tau}} \left(\omega_\tau \frac{\partial H}{\partial p_\tau} - \mathrm{i} \frac{\partial H}{\partial q_\tau}\right) \\ &= \frac{1}{\sqrt{2\omega_\tau}} (\omega_\tau p_\tau - \mathrm{i}\omega_\tau^2 q_\tau) = -\mathrm{i}\omega_\tau a_\tau \, ; \end{aligned}$$

likewise,

$$\frac{\mathrm{d}}{\mathrm{d}t} a_\tau^+ = \mathrm{i}\omega_\tau a_\tau^+ \, .$$

These equations can be obtained directly from the Hamiltonian (6.259), if a_τ is regarded as the canonical coordinate and $\mathrm{i}a_\tau^+$ as the canonically conjugated momentum:

$$\dot{a}_\tau = \frac{\partial H_{EM}}{\partial(\mathrm{i}a_\tau^+)} = -\mathrm{i}\omega_\tau a_\tau, \quad \mathrm{i}\dot{a}_\tau^+ = -\frac{\partial H_{EM}}{\partial a_\tau} = -\omega_\tau a_\tau^+ \, .$$

5. We change to the quantum theory by imposing canonical commutation relations

$$[\hat{a}_\tau, i\hat{a}^+_{\tau'}] = i\delta_{\tau,\tau'}\,. \tag{6.260}$$

The Hamiltonian of the quantum theory then reads

$$\hat{H}_{EM} = \sum_\tau \omega_\tau \hat{a}^+_\tau \hat{a}_\tau\,.$$

Let us add one last step:

6. *Define the vacuum state.* The ground state of the EM field is determined by the requirement that it is annihilated by all annihilation operators:

$$\hat{a}_\tau|0\rangle = 0\,.$$

Now we turn to the quantization of the electron field and follow the very same steps.

1. *Write an equation for the classical field.* Let us consider the Schrödinger equation for the wave function of one particle

$$\left(i\frac{\partial}{\partial t} - \hat{h}\right)\psi = 0\,, \tag{6.261}$$

where $\hat{h}$ is a one-particle Hamiltonian. For instance, we could think of

$$\hat{h}_0 = -\frac{\nabla^2}{2} - \frac{1}{r}\,, \tag{6.262}$$

however, it is better to regard $\hat{h}$ as an arbitrary effective one-particle operator. The wave function ψ is regarded as the classical field.
2. *Expand the field into suitable modes.* Suitable modes are the eigenstates of the Hamiltonian

$$\hat{h}U_\sigma(\mathbf{r}) = E_\sigma U_\sigma(\mathbf{r})\,. \tag{6.263}$$

Eigenstates of $\hat{h}$ form a complete one-particle basis. Thus, we can write

$$\psi(t,\mathbf{r}) = \sum_\sigma b_\sigma(t) U_\sigma(\mathbf{r})\,. \tag{6.264}$$

3. *Determine the evolution equations for the coefficients of the field decomposition into modes.* After inserting the decomposition (6.264) into the Schrödinger

equation (6.261), we find by means of orthonormality of the eigenstates of $\hat{\mathsf{h}}$ that

$$\left(\mathrm{i}\frac{\mathrm{d}}{\mathrm{d}t} - E_\sigma\right) b_\sigma(t) = 0\,. \tag{6.265}$$

Hermitian conjugation of the last equation yields

$$\left(-\mathrm{i}\frac{\mathrm{d}}{\mathrm{d}t} - E_\sigma\right) b_\sigma^+(t) = 0\,. \tag{6.266}$$

4. *Determine the canonical coordinates, the momenta and the Hamiltonian for the classical theory*. The Hamilton operator of the "classical" theory has the form

$$H_e = \sum E_\sigma b_\sigma^+ b_\sigma\,. \tag{6.267}$$

We derive the "classical" equations of motion, (6.265) and (6.266), from the Hamilton canonical equations if we regard b_σ as the canonical coordinate and $\mathrm{i}b_\sigma^+$ as the canonically conjugate momentum.

5. *Impose canonical anticommutation relations on the classical coordinates and momenta. Insert the corresponding operators for coordinates and momenta into the classical Hamiltonian to find the quantum Hamiltonian.* Here, we come to the only, though essential, difference between the quantization of the EM and electron fields. Quanta of the EM field behave as particles with spin 1, hence bosons. Owing to the *commutativity* of photon creation operators, the two-photon state is *symmetric* with respect to the interchange of the first and second photon, see discussion in Sect. 6.1.12. On the other hand, electrons, quanta of the electron field, are particles with spin $\frac{1}{2}$, hence fermions. The two-electron state must be *antisymmetric* with respect to the interchange of the first and second electron, see discussion in Sect. 5.1.1. Therefore, the electron creation operators must *anticommute*

$$\{\hat{\mathsf{b}}_\sigma^+, \hat{\mathsf{b}}_{\sigma'}^+\} = 0\,. \tag{6.268}$$

We will shortly clarify a connection between this requirement and the antisymmetry of the two-electron state. Hermitian conjugation of the last equation yields

$$\{\hat{\mathsf{b}}_\sigma, \hat{\mathsf{b}}_{\sigma'}\} = 0\,. \tag{6.269}$$

We will also shortly clarify that these relations enforce a replacement of the canonical commutation relation (6.260) by the anticommutation relation

$$\{\hat{\mathsf{b}}_\sigma, \mathrm{i}\hat{\mathsf{b}}_{\sigma'}^+\} = \mathrm{i}\delta_{\sigma\sigma'}\,. \tag{6.270}$$

The Hamilton operator of a free quantized electron field thus has the form

$$\hat{\mathsf{H}}_e = \sum E_\sigma \hat{\mathsf{b}}_\sigma^+ \hat{\mathsf{b}}_\sigma \,. \tag{6.271}$$

The quantized electron field is represented by the operator

$$\hat{\psi}(\mathbf{r}) = \sum_\sigma \hat{\mathsf{b}}_\sigma U_\sigma(\mathbf{r}) \,. \tag{6.272}$$

Obviously, this operator has been obtained from the decomposition (6.264) where the canonical coordinates $b_\sigma(t)$ were replaced by the corresponding operators $\hat{\mathsf{b}}_\sigma$.

6. *Define vacuum.* The vacuum state of the electron field is determined by equation

$$\hat{\mathsf{b}}_\sigma |0\rangle = 0 \tag{6.273}$$

valid for all modes σ.

6.6.2 *States of a Free Electron Field*

As follows from the last two equations, the lowest eigenvalue of $\hat{\mathsf{H}}_e$ is the vacuum state $|0\rangle$,

$$\hat{\mathsf{H}}_e |0\rangle = 0|0\rangle \,.$$

A state with one electron in the spin-orbital 1 is $\hat{\mathsf{b}}_1^+|0\rangle$ and has energy E_1

$$\hat{\mathsf{H}}_e \hat{\mathsf{b}}_1^+|0\rangle = E_1 \hat{\mathsf{b}}_1^+|0\rangle \,.$$

A state when two electrons are present, one in the spin-orbital 1, second in the spin-orbital 2, is $\hat{\mathsf{b}}_2^+\hat{\mathsf{b}}_1^+|0\rangle$ and has the energy $E_1 + E_2$

$$\hat{\mathsf{H}}_e \hat{\mathsf{b}}_2^+\hat{\mathsf{b}}_1^+|0\rangle = (E_1 + E_2)\hat{\mathsf{b}}_2^+\hat{\mathsf{b}}_1^+|0\rangle \,. \tag{6.274}$$

The last three equations were obtained by applying Eqs. (6.270)–(6.273). Owing to the anticommutation relation (6.268), equality

$$\hat{\mathsf{b}}_2^+\hat{\mathsf{b}}_1^+|0\rangle = -\hat{\mathsf{b}}_1^+\hat{\mathsf{b}}_2^+|0\rangle \tag{6.275}$$

holds. That is, the two-electron state is antisymmetric with respect to the interchange of the first and second electrons. After multiplying Eq. (6.274) from the left by $\langle 0|\hat{b}_1\hat{b}_2$, we obtain

$$\sum E_\sigma \langle 0|\hat{b}_1\hat{b}_2\hat{b}_\sigma^+\hat{b}_\sigma\hat{b}_2^+\hat{b}_1^+|0\rangle = E_1 + E_2.$$

The reader will easily convince himself that had we kept the anticommutation relations (6.268) and insisted on the commutation relations $[\hat{b}_\sigma, \mathrm{i}\hat{b}_{\sigma'}^+] = \mathrm{i}\delta_{\sigma\sigma'}$, the expression on the lhs of the last equation would automatically vanish.

If both electrons occupy the same spin-orbital 1, it follows from Eq. (6.275) that $\hat{b}_1^+\hat{b}_1^+|0\rangle = -\hat{b}_1^+\hat{b}_1^+|0\rangle$; hence

$$\hat{b}_1^+\hat{b}_1^+|0\rangle = 0\,.$$

As we already pointed out in Chap. 5, the Pauli exclusion principle follows from the antisymmetry of the wave function of two particles with spin 1/2, having taken this antisymmetry as an empirical fact. With the derivations above, we tried to point out that should this empirical fact hold, the operators creating electrons from the vacuum must obey anticommutation relations. In fact, the logic is quite the opposite. Pauli was the first who proved that—roughly speaking—should dynamics of quantized fields satisfy requirements of the special theory of relativity and should the state with the lowest energy exist, the operators creating particles with half-integer spin out of the vacuum must anticommute. Thus, the antisymmetry of the two-electron wave function does not need to be considered as an empirical fact, but rather something following from deeper principles. We shall touch this point later in Sect. 7.2.1. For history and review of proofs of the spin-statistics connection, the reader is referred to [6].

6.6.3 Self-interacting Electron Field

The Hamiltonian of an electron field evolving in the background Coulomb field of a nucleus is obtained as an average value of the one-particle Hamiltonian $\hat{h}_0$ in the state ψ, where ψ is replaced by the operator (6.272)

$$\hat{H}_0 = \int \hat{\psi}^+\hat{h}_0\hat{\psi}\,\mathrm{d}^3\mathbf{r} = \sum_{\sigma,\rho} \hat{b}_\sigma^+\hat{b}_\rho\varepsilon_{\sigma\rho}\,. \tag{6.276}$$

Here, $\varepsilon_{\sigma\rho}$ stands for

$$\varepsilon_{\sigma\rho} = \int U_\sigma^+(\mathbf{r})\hat{\mathsf{h}}_0 U_\rho(\mathbf{r})\,\mathrm{d}^3\mathbf{r}\,. \tag{6.277}$$

In case $\hat{\mathsf{h}}$ is taken to be equal to $\hat{\mathsf{h}}_0$, (6.262), then Eq. (6.263) and orthonormality of eigenfunctions U

$$\int U_\sigma^+(\mathbf{r})U_{\sigma'}(\mathbf{r})\,\mathrm{d}^3\mathbf{r} = \delta_{\sigma\sigma'} \tag{6.278}$$

yield

$$\varepsilon_{\sigma\rho} = E_\sigma \delta_{\sigma\rho}\,.$$

Equation (6.276) consequently reduces to Eq. (6.271).

The classical expression for energy of a charged cloud individual components of which are mutually interacting via electrostatic force reads

$$\frac{1}{2}\int \mathrm{d}^3\mathbf{r} \int \mathrm{d}^3\mathbf{r}' \frac{\rho(\mathbf{r})\rho(\mathbf{r}')}{|\mathbf{r}-\mathbf{r}'|}\,.$$

If we insert for charge density $\rho = \psi^+\psi$ and further replace ψ by the operator (6.272), we obtain the interaction Hamiltonian of the self-interacting electron field

$$\hat{\mathsf{H}}_1 = \frac{1}{2}\int \mathrm{d}^3\mathbf{r} \int \mathrm{d}^3\mathbf{r}' \frac{\hat{\psi}^+(\mathbf{r})\hat{\psi}(\mathbf{r})\hat{\psi}^+(\mathbf{r}')\hat{\psi}(\mathbf{r}')}{|\mathbf{r}-\mathbf{r}'|} = \frac{1}{2}\sum \hat{\mathsf{b}}_\sigma^+\hat{\mathsf{b}}_\rho\hat{\mathsf{b}}_{\sigma'}^+\hat{\mathsf{b}}_{\rho'} v_{\sigma\rho,\sigma'\rho'}\,, \tag{6.279}$$

where

$$v_{\sigma\rho,\sigma'\rho'} = \int \mathrm{d}^3\mathbf{r} \int \mathrm{d}^3\mathbf{r}' \frac{U_\sigma^+(\mathbf{r})U_\rho(\mathbf{r})U_{\sigma'}^+(\mathbf{r}')U_{\rho'}(\mathbf{r}')}{|\mathbf{r}-\mathbf{r}'|}\,. \tag{6.280}$$

Let us arrange the interaction Hamiltonian (6.279) into the normally ordered form, where the creation operators stand to the left of the annihilation operators, plus the "remainder"

$$\hat{\mathsf{H}}_1 = :\hat{\mathsf{H}}_1: + \hat{\mathsf{H}}_{11}\,, \tag{6.281}$$

where

$$:\hat{\mathsf{H}}_1: = -\frac{1}{2}\sum \hat{\mathsf{b}}_\sigma^+\hat{\mathsf{b}}_{\sigma'}^+\hat{\mathsf{b}}_\rho\hat{\mathsf{b}}_{\rho'} v_{\sigma\rho,\sigma'\rho'} \tag{6.282}$$

and the "remainder" reads

$$\hat{\mathsf{H}}_{11} = \frac{1}{2}\sum \hat{\mathsf{b}}_\sigma^+ \hat{\mathsf{b}}_{\rho'} v_{\sigma\rho,\rho\rho'} \,. \tag{6.283}$$

$\hat{\mathsf{H}}_{11}$ is obviously an one-particle operator. It describes the electrostatic interaction of an electron with itself and causes an infinite shift of the energy levels. However, the shift is state-independent:

$$\begin{aligned}\sum_\rho v_{\sigma\rho,\rho\rho'} &= \int \mathrm{d}^3\mathbf{r} \int \mathrm{d}^3\mathbf{r}' \frac{U_\sigma^+(\mathbf{r}) U_\rho(\mathbf{r}) U_\rho^+(\mathbf{r}') U_{\rho'}(\mathbf{r}')}{|\mathbf{r}-\mathbf{r}'|} \\ &= \sum_\rho \frac{4\pi}{(2\pi)^3} \int \frac{\mathrm{d}^3\mathbf{k}}{\omega^2} \int \mathrm{d}^3\mathbf{r} \int \mathrm{d}^3\mathbf{r}' U_\sigma^+(\mathbf{r}) \mathrm{e}^{\mathrm{i}\mathbf{k}\cdot\mathbf{r}} U_\rho(\mathbf{r}) U_\rho^+(\mathbf{r}') \mathrm{e}^{-\mathrm{i}\mathbf{k}\cdot\mathbf{r}'} U_{\rho'}(\mathbf{r}') \\ &= \frac{4\pi}{(2\pi)^3} \int \frac{\mathrm{d}^3\mathbf{k}}{\omega^2} \delta_{\sigma,\rho'} \,.\end{aligned}$$

Here, we used the Fourier transformation of the Coulomb potential, see Eq. (3.114), the completeness

$$\sum_\rho U_\rho(\mathbf{r}) U_\rho^+(\mathbf{r}') = \delta^{(3)}(\mathbf{r}-\mathbf{r}')$$

and the orthonormality (6.278) of the eigenfunctions U. Clearly, the term $\hat{\mathsf{H}}_{11}$ bears no observable consequences, since only the differences of the energy levels are observable. We will thus ignore the term henceforth.

Let us consider an electron in the spin-orbital 1 (for instance, the orbital 1s, with spin projection along the z-axis $+\frac{1}{2}$). The average value of the Hamilton operator

$$\frac{\hat{\mathsf{H}}}{Z^2} = \hat{\mathsf{H}}_0 + \frac{:\hat{\mathsf{H}}_1:}{Z}$$

in this state is

$$\frac{E}{Z^2} = \langle 0|\hat{\mathsf{b}}_1 \left(\hat{\mathsf{H}}_0 + \frac{:\hat{\mathsf{H}}_1:}{Z} \right) \hat{\mathsf{b}}_1^+ |0\rangle = \varepsilon_{11} \,. \tag{6.284}$$

Let us now consider two electrons; one in the spin-orbital 1 and the other in the spin-orbital 2. The average value of the Hamilton operator H in this state is

$$\frac{E}{Z^2} = \langle 0|\hat{\mathsf{b}}_1 \hat{\mathsf{b}}_2 \left(\hat{\mathsf{H}}_0 + \frac{:\hat{\mathsf{H}}_1:}{Z} \right) \hat{\mathsf{b}}_2^+ \hat{\mathsf{b}}_1^+ |0\rangle = \varepsilon_{11} + \varepsilon_{22} + \frac{1}{Z}(v_{11,22} - v_{12,21}) \,. \tag{6.285}$$

Clearly, the expression for the electrostatic self-interaction of the electron field automatically includes Coulomb and exchange interaction between the electrons. Recalling present notation, see Eqs. (6.277) and (6.280), the last expression is nearly identical with the expression (5.9), the only difference being that *orbitals a* and *b* appear in the expression (5.9) and that the spin part is an eigenstate of the total spin operator. On the other hand, *spin-orbitals* 1 and 2 appear in expression (6.285), and the state $\hat{\mathsf{b}}_2^+\hat{\mathsf{b}}_1^+|0\rangle$ is generally not an eigenstate of the total spin operator.

The reader can convince himself that the average value of the Hamilton operator $\hat{\mathsf{H}}$ in three-particle state is

$$\begin{aligned}\frac{E}{Z^2} &= \langle 0|\hat{\mathsf{b}}_1\hat{\mathsf{b}}_2\hat{\mathsf{b}}_3\left(\hat{\mathsf{H}}_0 + \frac{:\hat{\mathsf{H}}_1:}{Z}\right)\hat{\mathsf{b}}_3^+\hat{\mathsf{b}}_2^+\hat{\mathsf{b}}_1^+|0\rangle \\ &= \varepsilon_{11} + \varepsilon_{22} + \varepsilon_{33} + \frac{1}{Z}\left(v_{11,22} - v_{12,21} + v_{11,33} - v_{13,31} + v_{22,33} - v_{23,32}\right).\end{aligned}$$

Once we consider more than two particles, the advantage of the field approach becomes obvious. Try to merely imagine what it would mean to calculate the last expression within the particle viewpoint. We would have to calculate the average value of Hamilton operator (5.61) between the Slater determinants (5.62). This amounts to calculation of averages of 6 operators between 36 products of wave functions, altogether 216 terms. After some thinking, you would find that most of the terms vanish and finally arrive at the last equation. The field viewpoint bears the advantage that the need to think is substantially decreased. Briefly, this formalism is much more suitable for these and more complex calculations than the formalism based on the Slater determinants.

6.7 Final Notes

We did not get, with exception of Exercise 21, to the exposition of some important non-perturbative aspects of nonrelativistic quantum electrodynamics. Exercise 21 nicely illustrates the fact that the Fermi golden rule and its generalization, see Eqs. (6.79) and (6.177), are applicable only if there is an infinite amount of final states. Systems with a finite number of energy levels are oscillating and the perturbation method is unable to capture these oscillations. If the perturbation series is truncated after a finite number of terms, then in a finite time the perturbation solution will diverge from the exact solution. On the other hand, in real world, Rabi oscillations are often damped, as the finite lifetime of atomic states is often shorter than the duration of the EM pulses. For a qualitatively correct picture of what happens to an atom exposed to a strong EM field, it suffices, as in Exercise 21, to consider the atom as a two-level system and to consider only one mode of the external EM field. However, it is necessary, aside from the interaction of the atom with the one mode of the external EM field, to take into account also the interaction of the atom with an infinite number of other modes leading to spontaneous emission. Practical way how to do this is to solve the Bloch equations. For their exposition and their further application, e.g., for atomic cooling, we refer the reader to the excellent textbooks [4, 9]. If the reader managed to get this far in this textbook, he or she will encounter no difficulties when reading these textbooks.

We barely touched general features of the quantum-mechanical scattering theory. For an in-depth and still remarkably comprehensible treatment, we refer the reader to [14].

References

1. J. Benda, K. Houfek, Comput. Phys. Commun. **185**, 2903 (2014)
2. I. Bray, A.T. Stelbovics, Adv. At. Mol. Opt. Phys. **35**, 209 (1995)
3. M. Brune et al., Phys. Rev. Lett. **77**, 4887 (1996); S. Haroche, Rev. Mod. Phys. **85**, 1083 (2013); S. Haroche, M. Brune, J.M. Raimond, Phys. Today **66**, 27 (2013); S. Haroche, J.M. Raimond, *Exploring the Quantum, Atoms, Cavities, Photons* (Oxford University Press, Oxford, 2006)
4. C. Cohen-Tannoudji, J. Dupont-Roc, G. Grynberg, *Atom-Photon Inter Actions: Basic Processes and Applications*. Wiley Science Paperback Series (Wiley, London, 1992)
5. G. Drake, Phys. Rev. A **34**, 2871 (1986)
6. I. Duck, E.C.G. Sudarsahan, *Pauli and the Spin-Statistics Theorem* (World Scientific, Singapore, 1997)
7. R.P. Feynman, R.B. Leighton, M. Sands, *Feynman Lectures on Physics 3* (Addison-Wesley, Reading, 1971)
8. R.P. Feynman, R.B. Leighton, M. Sands, *Feynman Lectures on Physics 1, 2* (Addison-Wesley, Reading, 1977)
9. C.J. Foot, *Atomic Physics*. Oxford Master Series in Physics (Oxford University Press, Oxford, 2007)
10. S.S. Hodgman et al., Phys. Rev. Lett. **103**, 053002 (2009); G.W.F. Drake, Phys. Rev. A **3**, 908 (1971)
11. L.D. Landau, E.M. Lifshitz, *Quantum Mechanics, Non-relativistic Theory* (Elsevier/Butterworth-Heinemann, Amsterdam/London, 1976)
12. A. Messiah, *Quantum Mechanics* (North Holland, Amsterdam, 1961)
13. J. Schwinger et al., *Classical Electrodynamics* (Perseus Books, Cambridge, 1998)
14. J.R. Taylor, *Scattering Theory: The Quantum Theory of Nonrelativistic Collisions* (Dover Publications, New York, 2006)
15. S. Weinberg, *Cosmology* (Oxford University Press, Oxford, 2008)
16. A. Zee, *Einstein Gravity in a Nutshell* (Princeton University Press, Princeton, 2013)

Chapter 7
Dynamics: The Relativistic Theory

Relativistic quantum electrodynamics, *QED*, is a theory unifying quantum mechanics, special theory of relativity, and classical electrodynamics. In this chapter, we will focus on the following topics:

- *What essentially new is brought by the unification of quantum mechanics and special theory of relativity.* We show that although one can write the Schrödinger equation for an electron, i.e., the Dirac equation, to formally fulfill the requirements of the Lorentz invariance, difficulties associated with the existence of negative-energy solutions appear. These difficulties point to the existence of a positron, an antiparticle to the electron, and force us to regard electrons and positrons as quanta of an electron-positron field.
- *Motion of an electron bound in an atom is affected by fluctuations of the electromagnetic and electron-positron fields.* Despite the "normal" atoms, such as hydrogen, helium, and so on, being composed of merely nuclei and electrons, antiparticles still influence the positions of atomic spectral lines, as they appear in "virtual" states. We do not need to go to very high energies to see the relativistic quantum theory at work: its effects are observed for low energies if we are able to push the theory and experiment to sufficient accuracy. In particular, while the nonrelativistic Hamiltonian determines the differences of the energy levels at the order α^2 and the relativistic and magnetic effects make their appearance at the order α^4, the effects of relativistic quantum electrodynamics emerge at the order α^5. We will be concerned how to arrive at the most suitable mathematical formulation of these effects. While doing so, we will use a substantial amount of ingenious physical reasoning developed by Feynman.
- *The expressions for the fluctuations diverge when integrating over very large momenta of virtual particles.* Moreover, these expressions sometimes diverge also when integrating over very low momenta of virtual particles. However, the latter divergence is a mere consequence of extending the approximations beyond their realm of validity. The former divergence, on the other hand, is an inherent feature of the perturbation formulation of the theory. We will show how to tackle

J. Zamastil, J. Benda, *Quantum Mechanics and Electrodynamics*,
DOI 10.1007/978-3-319-65780-6_7

the infinities and how to obtain unambiguous numbers from the theory which we can in turn compare with the experiment.
- *QED predicts the existence and properties of positronium, an electron-positron bound state.*

Of all the QED processes, we will predominantly focus on the electron and photon self-energies. These two processes nearly completely determine the Lamb shift in "ordinary" hydrogen-like atoms mentioned already in the previous chapter. Calculation of these processes brings together everything that is conceptually new in QED; everything that distinguishes QED from the theories QED unifies.

Secondly, we will analyze positronium, the QED system *par excellence*, in considerable detail. For here, in contrast to the "ordinary" hydrogen-like atoms, additional QED effects are clearly visible; namely the nucleus recoil and virtual electron-positron annihilation.

As already mentioned in the Foreword, in contrast to the usual expositions of QED, we minimize discussion of purely formal aspects of QED. We rather try to convince the reader why some formal aspects of QED are of utmost practical importance. For instance, we will emphasize the role of the manifest Lorentz invariance for practical feasibility of the calculations, as well as for unambiguous predictions of the theory. We will discuss charge conjugation symmetry of QED as it leads to the correct form of the interaction Hamiltonian, to the positron wave function and so on. If a new formalism is introduced, such as the time-dependent Green functions, we will try to show the reader the necessity of such a step.

7.1 Relativistic Equation for an Electron

7.1.1 Relativistic Notation

For this chapter, it is necessary to introduce the relativistic notation. We assume that the reader is familiar with notions of rotational and Lorentz invariance. If not, the following is not likely to be comprehensible. In such a case, we recommend, e.g., [8, 43, 44] to the reader to acquire the requisite knowledge. A point in the space-time is represented by the four-vector

$$\boldsymbol{x} = (t, \mathbf{r})\,. \tag{7.1}$$

A proper (no reflections) homogeneous (no translations) Lorentz transformation consists of rotations and transitions to coordinate systems moving with respect to the original system by constant velocities

$$x'_\mu = \Lambda_{\mu\nu} x_\nu\,. \tag{7.2}$$

Rotations are special cases of Lorentz transformations involving only space components of the four-vector (7.1),

$$x'_j = R_{jk} x_k \,.$$

If coordinate axes transform in this way, components of a three-vector **A** transform according to

$$A'_j = R_{jk} A_k \,.$$

Likewise, in case of the Lorentz transformation, components of a four-vector transform according to

$$A'_\mu = \Lambda_{\mu\nu} A_\nu \,. \tag{7.3}$$

Useful examples of four-vectors are the four-momentum

$$\boldsymbol{p} = (E, \mathbf{p}) \,,$$

the four-potential

$$\boldsymbol{A} = (\varphi, \mathbf{A})$$

and the four-current

$$\boldsymbol{j} = (\rho, \mathbf{j}) \,.$$

The scalar product of three-vectors produces a number invariant with respect to the rotation of coordinate axes

$$\mathbf{A}' \cdot \mathbf{B}' = A'_j B'_j = R_{jk} R_{jl} A_k B_l = A_k B_k = \mathbf{A} \cdot \mathbf{B} \,. \tag{7.4}$$

Here, we used the fact that the rotation matrix R is orthogonal. That is, the inverse matrix equals the transpose matrix, $\mathsf{R}^{-1} = \mathsf{R}^\top$; in components

$$R_{jk} R_{jl} = R^\top_{kj} R_{jl} = \delta_{kl} \,. \tag{7.5}$$

Likewise, the scalar product of four-vectors produces a number invariant with respect to the Lorentz transformations

$$\boldsymbol{A}' \cdot \boldsymbol{B}' = A'_\mu B'_\mu = \Lambda_{\mu\sigma} \Lambda_{\mu\rho} A_\sigma B_\rho = A_\sigma B_\sigma = \boldsymbol{A} \cdot \boldsymbol{B} \,. \tag{7.6}$$

The Lorentz transformation (7.2) is a linear transformation of coordinates (one time and three space) conserving the infinitesimal interval

$$\mathrm{d}t^2 - \mathrm{d}\mathbf{r} \cdot \mathrm{d}\mathbf{r} = \mathrm{d}t'^2 - \mathrm{d}\mathbf{r}' \cdot \mathrm{d}\mathbf{r}' .$$

This is a mathematical expression of the requirement the speed of light be the same for all observers in inertial systems—if this interval vanishes in the original system, it vanishes in the primed system as well. The minus sign in the last equation enforces a definition of the scalar product of four-vectors as

$$\boldsymbol{A} \cdot \boldsymbol{B} = A_\mu B_\mu = A_0 B_0 - \mathbf{A} \cdot \mathbf{B} .$$

Should Eq. (7.6) hold, the summation over arbitrary two indices has to be understood in the same way, for instance

$$\Lambda^\top_{\sigma\mu} \Lambda_{\mu\rho} = \Lambda^\top_{\sigma 0} \Lambda_{0\rho} - \Lambda^\top_{\sigma j} \Lambda_{j\rho} .$$

Then the four-dimensional generalization of Eq. (7.5)

$$\Lambda_{\mu\sigma} \Lambda_{\mu\rho} = \Lambda^\top_{\sigma\mu} \Lambda_{\mu\rho} = \eta_{\sigma\rho} \Rightarrow \left(\Lambda^{-1}\right)_{\mu\sigma} = \Lambda_{\sigma\mu} , \tag{7.7}$$

used already in Eq. (7.6), indeed holds. In the last equation, we introduced the *metric tensor* η

$$\eta = \mathrm{diag}\{1, -1, -1, -1\} . \tag{7.8}$$

The four-gradient reads

$$\boldsymbol{\partial} = \left(\frac{\partial}{\partial t}, -\nabla \right) . \tag{7.9}$$

The gauge transformation (6.9) can then be compactly written as

$$A_\mu = A'_\mu - \partial_\mu \chi \tag{7.10}$$

and similarly the charge conservation law, Eq. (3.112), reads

$$\partial_\mu j_\mu = 0 . \tag{7.11}$$

Within this notation, the Maxwell equations (6.8) can be written as

$$\partial_\mu \partial_\mu A_\nu - \partial_\nu \partial_\mu A_\mu = j_\nu . \tag{7.12}$$

A word about the terminology. A vector transforming according to Eq. (7.3) is said to be *covariant* with respect to the Lorentz transformation or covariant for short. For instance the Maxwell equations (7.12) are covariant. On the other hand, as already mentioned, the magnitude of four-vectors, e.g.

$$p \cdot p = E^2 - \mathbf{p} \cdot \mathbf{p} = m^2 \tag{7.13}$$

or equations such as the charge conservation law, Eq. (7.11), are said to be Lorentz *invariant*.

7.1.2 Klein-Gordon Equation

Let us repeat the arguments that led us to the Klein-Gordon equation (3.97) and again, this time in depth, summarize the difficulties associated with this equation.

To obtain a relativistic equation for a free particle, one only needs to replace the four-vector $\boldsymbol{p}$ by the four-vector operator $\hat{\boldsymbol{p}}$ in Eq. (7.13)

$$[\hat{\boldsymbol{p}} \cdot \hat{\boldsymbol{p}} - m^2]|\psi\rangle = 0\,. \tag{7.14}$$

Changing to the momentum representation, we obtain Eq. (7.13). Going to the coordinate representation $\hat{\boldsymbol{p}} \to \mathrm{i}\boldsymbol{\partial}$, we find

$$\left(-\boldsymbol{\partial} \cdot \boldsymbol{\partial} - m^2\right) \psi(x) = 0\,. \tag{7.15}$$

As we already mentioned in Sect. 3.5.1, the Klein-Gordon equation was in fact discovered by Schrödinger. Moreover, he discovered this equation prior to its nonrelativistic approximation. Why did he abandon the relativistic theory and restricted himself merely to the nonrelativistic approximation of it?

- If we express the energy E from Eq. (7.13), the solutions are $E = \pm\sqrt{m^2 + |\mathbf{p}|^2}$. The energy spectrum of a free particle consists of two sheets $(-\infty, -m\rangle \cup \langle m, +\infty)$. If we consider the Klein-Gordon equation for stationary states in the Coulomb field

$$\left\{\left[E - e\varphi(\hat{\mathbf{r}})\right]^2 - \hat{\mathbf{p}} \cdot \hat{\mathbf{p}} - m^2\right\} |\psi\rangle = 0$$

  where $e\varphi(r) = -Z\alpha/r$, the continuous part of the energy spectrum is the same[1] as for the free particle and there will be bound states in the interval $(0, m)$. Obviously, such a spectrum is nonphysical since we already know that if an atom is in an excited state, spontaneous emission drives it to the ground state. However, there is no state of the lowest energy now. If we simply throw away the negative energy

[1] We have seen in the nonrelativistic theory that the presence of the Coulomb potential does not alter the eigenvalues of the continuous spectrum. The eigenfunctions are, of course, different. One can show this explicitly since the last equation can be solved exactly. However, such a solution is of minimal practical value, so we will not bother with it.

states as nonphysical, the positive energy states do not form a complete basis of states and we find sooner or later that the probability is not conserved.

- As we already mentioned in Sect. 3.5.1, the Klein-Gordon equation (7.14) does not contain anything connected with inner degrees of freedom such as spin. It is an equation for a spinless particle.
- Finally, Eq. (7.15) is of the second order in time. Therefore to know the time evolution, we need to know not only the wave function, but also its time derivative at a given time instant. However, this is in disagreement with the basic principle of quantum mechanics that the wave function, or more generally the state vector, fully describes the state of the system.

We will see that the first feature of the Klein-Gordon equation, in contrast to the other two, is shared by the Dirac equation as well. Moreover, this feature is inherent to all relativistic wave equations for massive particles. Today's view of the Klein-Gordon equation is very similar to that of the Maxwell equations: it is an equation for a classical massive scalar *field*. This field can be quantized much in the same way as the EM field. There are merely two differences: firstly, it has only one component, and secondly, it is massive. In the quantization procedure described in Chap. 6, this brings one single change: Eq. (6.15) is replaced by the equation

$$\left(-\nabla^2 + m^2\right) T_\tau = \omega_\tau^2 T_\tau \,.$$

Excitations of this field are massive scalar particles, the only candidate for such a particle among the known elementary particles being presently (2017) the Higgs boson. For an electron, we need to find something different.

7.1.3 Dirac Equation

The relation (7.13) has to hold for any massive free particle regardless of its spin and so has Eq. (7.14). However, as noted by Dirac, these equations can hold even if the fundamental equation is of the first order in time

$$(\boldsymbol{\gamma} \cdot \hat{\boldsymbol{p}} - m)|\psi\rangle = 0 \,. \tag{7.16}$$

This equation is commonly known as the *Dirac equation*.[2] Here, γs are matrices independent of space-time coordinates to be specified below. Multiplying this

[2] In literature, one encounters the so-called *Feynman slash notation*

$$\not{a} = \boldsymbol{\gamma} \cdot \boldsymbol{a} = \gamma_\mu a_\mu \,,$$

where $\boldsymbol{a}$ is an arbitrary four-vector. Since the advantage of this notation is minimal in the following calculations, we will not use it in this book.

equation by $(\boldsymbol{\gamma} \cdot \hat{\boldsymbol{p}} + m)$ from the left and using commutativity of the components of γ and $\hat{\mathsf{p}}$, we obtain

$$\left(\gamma_\nu \gamma_\mu \hat{\mathsf{p}}_\nu \hat{\mathsf{p}}_\mu - m^2\right) |\psi\rangle = 0\,.$$

However, this must be equivalent to the Klein-Gordon equation (7.14). Decomposing the product of momenta into symmetric and antisymmetric parts, $\hat{\mathsf{p}}_\nu \hat{\mathsf{p}}_\mu = \frac{1}{2}\left\{\hat{\mathsf{p}}_\nu, \hat{\mathsf{p}}_\mu\right\} + \frac{1}{2}\left[\hat{\mathsf{p}}_\nu, \hat{\mathsf{p}}_\mu\right]$, the latter part vanishes. Hence the product $\gamma_\mu \gamma_\nu$ can be replaced by its symmetric part

$$\left(\frac{1}{2}\left\{\gamma_\nu, \gamma_\mu\right\} \hat{\mathsf{p}}_\nu \hat{\mathsf{p}}_\mu - m^2\right) |\psi\rangle = 0\,.$$

This equation is equivalent to Eq. (7.14) if

$$\left\{\gamma_\mu, \gamma_\nu\right\} = 2\eta_{\mu\nu}\,. \tag{7.17}$$

This requirement states that the square of one particular component is one, $\gamma_\mu^2 = 1$, and different components of γ anticommute

$$\gamma_\mu \gamma_\nu = -\gamma_\nu \gamma_\mu \qquad (\mu \neq \nu)\,.$$

We easily find that if one particular set of γ-matrices obeys Eq. (7.17), so does another set $\gamma' = \mathsf{U}^+ \gamma \mathsf{U}$ where U is a unitary matrix. Hence, there is an infinite number of realizations of the relations (7.17), related mutually by unitary transformations. Which particular choice is the most suitable one depends on the situation. For transition to the nonrelativistic limit, the most convenient one is the *standard* or *Dirac realization*,

$$\gamma_0 = \begin{pmatrix} 1 & 0 \\ 0 & -1 \end{pmatrix}, \quad \boldsymbol{\gamma} = \begin{pmatrix} 0 & \boldsymbol{\sigma} \\ -\boldsymbol{\sigma} & 0 \end{pmatrix}, \tag{7.18}$$

where σ_i are Pauli matrices obeying the relations (1.25). All γ-matrices are square matrices 4×4 constituted by blocks 2×2. The wave function ψ has four components: it is a *bispinor*, i.e., composed of two two-component *spinors*. Why are γ-matrices of the order 4×4? There are no 4 matrices γ different from the unit matrix of the order 2×2 which obey (7.17). We will see later in Sect. 7.2.3 that there are other suitable choices of the γ-matrices.

7.1.4 External EM Field

In case of an electron moving under the influence of an external time-independent EM field described by the four-potential $\boldsymbol{A}(\mathbf{r})$, we perform the usual substitution

$\hat{p} \to \hat{\Pi} = \hat{p} - eA(\hat{r})$, see Eq. (3.1).[3] The Dirac equation thus has the form

$$(\gamma_\mu \hat{\Pi}_\mu - m)\psi_{\text{at}} = 0 \Rightarrow (\gamma_\mu (\mathrm{i}\partial - eA)_\mu - m)\psi_{\text{at}}(x) = 0 . \tag{7.19}$$

After multiplying this equation by γ_0 from the left and restricting ourselves to the stationary states $\psi_{\text{at}}(\mathbf{r}, t) = \mathrm{e}^{-\mathrm{i}Et}\psi_{\text{at}}(\mathbf{r})$, we obtain

$$\left[\gamma_0 \boldsymbol{\gamma} \cdot (\hat{\mathbf{p}} - e\mathbf{A}(\hat{\mathbf{r}})) + \gamma_0 m + e\varphi(\hat{\mathbf{r}})\right] \psi_{\text{at}}(\mathbf{r}) = E\psi_{\text{at}}(\mathbf{r}) . \tag{7.20}$$

In the matrix form, using Eq. (7.18) and writing $\psi_{\text{at}} = \begin{pmatrix} \psi_+ \\ \psi_- \end{pmatrix}$, the last equation reads

$$\begin{pmatrix} E - e\varphi(\hat{\mathbf{r}}) & 0 \\ 0 & E - e\varphi(\hat{\mathbf{r}}) \end{pmatrix} \begin{pmatrix} \psi_+ \\ \psi_- \end{pmatrix} = \begin{pmatrix} m & \boldsymbol{\sigma} \cdot (\hat{\mathbf{p}} - e\mathbf{A}(\hat{\mathbf{r}})) \\ \boldsymbol{\sigma} \cdot (\hat{\mathbf{p}} - e\mathbf{A}(\hat{\mathbf{r}})) & -m \end{pmatrix} \begin{pmatrix} \psi_+ \\ \psi_- \end{pmatrix} . \tag{7.21}$$

If we express the lower components of the wave function from the latter equation

$$\psi_- = \frac{1}{E + m - e\varphi} \boldsymbol{\sigma} \cdot (\hat{\mathbf{p}} - e\mathbf{A})\psi_+ ,$$

and substitute them into the former equation for the upper components of the wave function, ψ_+, we find

$$\left(E - m - e\varphi(\hat{\mathbf{r}})\right) \psi_+ = \boldsymbol{\sigma} \cdot (\hat{\mathbf{p}} - e\mathbf{A}(\hat{\mathbf{r}})) \frac{1}{E + m - e\varphi(\hat{\mathbf{r}})} \boldsymbol{\sigma} \cdot (\hat{\mathbf{p}} - e\mathbf{A}(\hat{\mathbf{r}}))\psi_+ .$$

This equation is exactly Eq. (3.98) which was the starting point of our discussion of the fine structure of atomic spectra back in Sect. 3.5.

It follows from this discussion and the penultimate equation that the nonrelativistic limit of the Dirac wave function reads[4]

$$\psi_{\text{at}} \simeq \begin{pmatrix} 1 \\ \frac{\boldsymbol{\sigma} \cdot (\hat{\mathbf{p}} - e\mathbf{A})}{2m} \end{pmatrix} \psi^0_{\text{at}} , \tag{7.22}$$

where ψ^0_{at} is the Pauli wave function obeying

---

[3]The vector and scalar potentials of the classical EM field will be denoted as $A(\hat{\mathrm{x}})$. A is a number-valued function of x and an operator, though only in the sense of the "first" quantization: $[\hat{\mathrm{p}}_i, A_j(\hat{\boldsymbol{r}})] = -\mathrm{i}\partial_i A_j$. The vector potential of the quantized EM field will be denoted $\hat{\mathrm{A}}(\hat{\mathrm{x}})$ or $\hat{\mathrm{A}}$ for short.

[4]The exact solution of the Dirac equation in the Coulomb field is of greater practical value than in the case of the Klein-Gordon equation. However, we will not need it here either. The reader can find one particularly elegant solution by doing Exercise 25.

$$\left\{\frac{[\sigma \cdot (\hat{\boldsymbol{p}} - e\mathbf{A})]^2}{2m} + e\varphi\right\} \psi^0_{\text{at}} = (E - m)\psi^0_{\text{at}} \tag{7.23}$$

and the operator on the lhs is the Pauli Hamiltonian (3.50).

Another way to find the nonrelativistic limit is to multiply the Dirac equation (7.19) by $(\boldsymbol{\gamma} \cdot \hat{\boldsymbol{\Pi}} + m)$ from the left

$$\hat{\mathscr{H}}\psi_{\text{at}} = 0\,, \qquad \hat{\mathscr{H}} = (\boldsymbol{\gamma}\cdot\hat{\boldsymbol{\Pi}} - m)(\boldsymbol{\gamma}\cdot\hat{\boldsymbol{\Pi}} + m) = \hat{\boldsymbol{\Pi}}\cdot\hat{\boldsymbol{\Pi}} - m^2 + \frac{1}{4}[\hat{\Pi}_\mu, \hat{\Pi}_\nu][\gamma_\mu, \gamma_\nu]\,, \tag{7.24}$$

where $\hat{\mathscr{H}}$ is usually referred to as the second order Dirac Hamiltonian and we will encounter it quite often in the following text. In case of the external Coulomb field

$$\hat{\boldsymbol{\Pi}} = \left(E + \frac{Z\alpha}{\hat{r}}, \hat{\boldsymbol{p}}\right)\,, \tag{7.25}$$

the nonrelativistic limit of this operator reads

$$\hat{\mathscr{H}} \simeq 2m\left(\hat{\Pi}_0 - m - \frac{\hat{p}^2}{2m}\right) = -2(mZ\alpha)^2(\hat{h}_0 - \varepsilon_0)\,, \tag{7.26}$$

where we used Eq. (6.204) in the second equality.

Exercise 25: Exact Relativistic Solution of Hydrogen

1. *Show that the second order Dirac Hamiltonian $\hat{\mathscr{H}}$, Eq. (7.24), takes for the potential (7.25) the form*

$$\begin{aligned}\hat{\mathscr{H}} &= \left(E + \frac{Z\alpha}{r}\right)^2 - \hat{\boldsymbol{p}}^2 - m^2 + \gamma_0\boldsymbol{\gamma}\cdot\left[\hat{\boldsymbol{p}}, \frac{Z\alpha}{r}\right] \\ &= E^2 - m^2 + 2\frac{EZ\alpha}{r} - \left(\hat{p}_r^2 + \frac{\hat{\Gamma}(\hat{\Gamma}-1)}{r^2}\right)\,,\end{aligned} \tag{7.27}$$

where the radial momentum operator $\hat{p}_r$ is given by Eq. (3.23) and the "relativistic orbital momentum" operator $\hat{\Gamma}$ reads

$$\hat{\Gamma} = \gamma_0\left(\hat{K} + \mathrm{i}(Z\alpha)\boldsymbol{\gamma}\cdot\mathbf{n}\right)\,,$$

where the "relativistic parity" operator $\hat{K}$ reads

$$\hat{K} = \gamma_0\left(\boldsymbol{\Sigma}\cdot\hat{\boldsymbol{L}} + 1\right)\,.$$

The symbol $\boldsymbol{\Sigma}$ stands for

$$\boldsymbol{\Sigma} = \begin{pmatrix} \boldsymbol{\sigma} & 0 \\ 0 & \boldsymbol{\sigma} \end{pmatrix} \tag{7.28}$$

and $\hat{\mathbf{L}}$ is the angular momentum operator.

2. *Show that the set of the operators $\{\hat{\mathscr{H}}, \hat{\mathsf{J}}^2, \hat{\mathsf{J}}_z, \hat{\mathsf{K}}, \hat{\Gamma}\}$ form a complete set of commuting operators. Here, $\hat{\mathbf{J}}$ is total angular momentum operator*

$$\hat{\mathbf{J}} = \hat{\mathbf{L}} + \frac{1}{2}\boldsymbol{\Sigma}\,.$$

3. *Show that*

$$\hat{\mathsf{K}}^2 = \hat{\mathsf{J}}^2 + \frac{1}{4}$$

and

$$\hat{\Gamma}^2 = \hat{\mathsf{K}}^2 - (Z\alpha)^2\,.$$

The eigenvalues of the operators $\hat{\mathsf{K}}$ and $\hat{\Gamma}$ are then

$$K = \pm(j + 1/2)$$

and

$$\Gamma = \pm\sqrt{(j + 1/2)^2 - (Z\alpha)^2}\,,$$

where as usually $j(j+1)$ denotes eigenvalue of $\hat{\mathsf{J}}^2$.

4. *Introduce the effective orbital quantum number l_Γ*

$$l_\Gamma(l_\Gamma + 1) = \Gamma(\Gamma - 1) \tag{7.29}$$

and make substitution

$$r = \frac{r_A}{EZ\alpha}\,.$$

This transforms the radial part of the Dirac second order Hamiltonian (7.27) to the form, that is, up to additive and multiplicative constants, formally identical to the radial part of the nonrelativistic radial Hamiltonian

$$\hat{\mathscr{H}} = E^2 - m^2 - 2(EZ\alpha)^2 \left[\frac{1}{2}\left(\left(\hat{\mathsf{p}}_r^2\right)_A + \frac{l_\Gamma(l_\Gamma+1)}{\hat{\mathsf{r}}_A^2}\right) - \frac{1}{\hat{\mathsf{r}}_A}\right]. \tag{7.30}$$

In order to have normalizable wave function, not diverging too strongly at the origin, the physical solution of Eq. (7.29) is

$$l_\Gamma = \Gamma - 1\,,(\Gamma > 0), \quad l_\Gamma = -\Gamma\,,(\Gamma < 0)\,.$$

Comparison of Eqs. (4.91), (7.24), and (7.30) finally yields equation for the energy E of the stationary state

$$E^2 - m^2 + \frac{(EZ\alpha)^2}{N^2} = 0\,,$$

where

$$N = n_r + l_\Gamma + 1\,, \quad n_r = 0, 1, 2, \ldots$$

7.1.5 Difficulties Associated with the Interpretation of the Dirac Equation and Their Resolution

The Dirac equation leads to the correct nonrelativistic limit, namely the Pauli equation. The first correction to this limit yields results which are in better agreement with experiment than the limit itself, see Sect. 3.5.2. On the other hand, the Dirac equation does not solve the problem of negative energies any better than the Klein-Gordon equation. The energy spectrum is the same for a free particle and qualitatively the same for a particle in the Coulomb field of a nucleus. There are bound states for $0 < E < m$, the energy spectrum is continuous for $E > m$ and $E < -m$, and there is no state of the lowest energy. Therefore, atoms and matter should be in general unstable according to relativistic quantum mechanics. However, this is clearly a nonsense as our very existence proves. For the final resolution of this problem, we are indebted to the physicist Paul Dirac, the same man who discovered Eq. (7.16) bearing his name. According to Dirac, matter is stable since all the states with negative energy are occupied. The Pauli exclusion principle forbidding two particles with half-integral spin exist in the same state guarantees that the electron in the state with the lowest *positive* energy cannot go into a state with negative energy. It can happen that two[5] photons with their total energy larger than $2m_e$ expel an electron from this *Dirac sea* to real particles with positive energy. In the sea, there will be now a *hole* with the same properties as an electron but for the opposite sign of charge.

[5]For kinematical reasons there must be two and not only one.

Fig. 7.1 Vacuum polarization

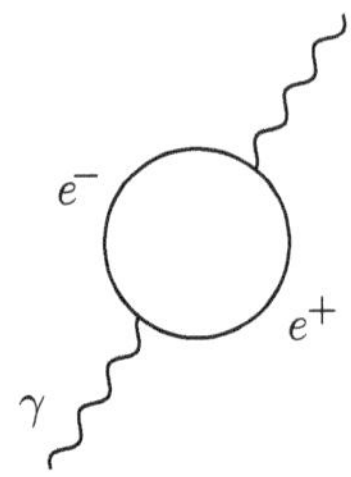

With this idea, Dirac not only reconciled the relativistic quantum theory with the stability of matter, he achieved much more: he predicted the existence of a *positron*, an antiparticle to the electron. He predicted processes such as a pair creation

$$\gamma + \gamma \to e^+ + e^- , \tag{7.31}$$

pair annihilation, "an electron falls into a hole,"

$$e^+ + e^- \to \gamma + \gamma , \tag{7.32}$$

vacuum polarization, where pairs virtually appear and disappear, see Fig. 7.1,

$$\gamma \to e^+ + e^- \to \gamma , \tag{7.33}$$

and so on. At this point, it is important to recognize that a correct interpretation of the one-particle relativistic wave equation requires an addition of infinitely many other particles, namely the Dirac sea of electrons occupying infinite number of negative energy states. It begs the question then: should we not regard electrons and positrons as quanta of an electron-positron field, much in the same way we regard photons as quanta of an EM field? If the particles do not appear or disappear, i.e., if they simply exist, it makes sense to perceive them as particles. That is, to assign a wave function to them. However, once we do so, we are not able to describe the creation or annihilation of the particles. Once we describe a particle with a wave function, there is a continuity equation for the probability density, see Eq. (6.115). To allow for the possibility of particle creation and annihilation, it is better to view particles as quanta of a field. We have never introduced a photon wave function since photon is too easily created or annihilated, emitted or absorbed. Likewise, in the presence of positrons or strong EM fields, an electron can be easily created or annihilated.[6] When energies of the involved particles are of small magnitude and there are no positrons present, reactions such as (7.31) and (7.32) do not occur. Thus

[6]Let us emphasize that the electric charge is strictly locally conserved, meaning the creation and annihilation of electrons is always (at least when neglecting the weak interactions) accompanied by creation and annihilation of positrons, see Eqs. (7.31) and (7.32). It holds even for virtual processes, see Eq. (7.33), and it led Feynman to another view of positrons, see Sect. 7.4.

the nonrelativistic description of matter is appropriate. However, in the relativistic theory, we deal with multi-particle states and transitions between them. The simplest way how to go from one-particle to many-particle theory represents the formalism of the second quantization introduced in Sect. 6.6. We will now turn to the application of this procedure to the Dirac field.

7.2 Hamiltonian of Relativistic Quantum Electrodynamics

7.2.1 *Quantization of the Electron-Positron Field*

Let us now extend the individual steps made in Sect. 6.6 to the relativistic case. In comparison with the procedure outlined in Sect. 6.6, there will be solely one, though essential, change concerning the definition of the vacuum state.

1. We multiply the Dirac equation (7.19) by the matrix γ_0 from the left and obtain Eq. (6.261), $(\mathrm{i}\frac{\partial}{\partial t} - \hat{\mathfrak{h}}_{\mathrm{D}})\psi = 0$, for the "classical" Dirac field ψ. The one-particle Hamiltonian $\hat{\mathfrak{h}}_{\mathrm{D}}$ reads, see Eq. (7.20),

$$\hat{\mathfrak{h}}_{\mathrm{D}} = \gamma_0 \boldsymbol{\gamma} \cdot [\hat{\boldsymbol{p}} - e\mathbf{A}(\hat{\boldsymbol{r}})] + \gamma_0 m + e\varphi(\hat{\boldsymbol{r}}) \,, \tag{7.34}$$

   where we recall that φ and $\mathbf{A}$ are potentials of a time-independent external field.
2. There are two kinds of eigenstates of the Dirac one-particle Hamiltonian $\hat{\mathfrak{h}}_{\mathrm{D}}$—with positive and negative energies. Thus, Eq. (7.20), $\hat{\mathfrak{h}}_{\mathrm{D}} X_\sigma = E_\sigma X_\sigma$, represents in fact two sets of equations

$$\hat{\mathfrak{h}}_{\mathrm{D}} X_{\sigma+} = E_{\sigma+} X_{\sigma+} \,, \qquad \hat{\mathfrak{h}}_{\mathrm{D}} X_{\sigma-} = E_{\sigma-} X_{\sigma-} \,,$$

   where clearly

$$E_{\sigma+} > 0, \qquad E_{\sigma-} < 0 \,.$$

   The expansion of the field into modes is then

$$\psi = \sum_{\sigma} c_\sigma X_\sigma = \sum_{\sigma+} c_{\sigma+} X_{\sigma+} + \sum_{\sigma-} c_{\sigma-} X_{\sigma-} \,. \tag{7.35}$$

3. By inserting Eq. (7.35) into Eq. (6.261) we obtain Eqs. (6.265) and (6.266), where we merely replace the coefficients b_σ by the coefficients c_σ,

$$\left(\mathrm{i}\frac{\mathrm{d}}{\mathrm{d}t} - E_\sigma\right) c_\sigma(t) = 0 \,, \qquad \left(-\mathrm{i}\frac{\mathrm{d}}{\mathrm{d}t} - E_\sigma\right) c_\sigma^+(t) = 0 \,.$$

4. The Hamiltonian of the "classical" theory is then given by Eq. (6.267), where b_σ is again replaced by c_σ, $H_\mathrm{D} = \sum_\sigma E_\sigma c_\sigma^+ c_\sigma$. Canonical coordinates and momenta are c_σ and $\mathrm{i}c_\sigma^+$, respectively.
5. Quantum anticommutation relations acquire the forms

$$\{\hat{\mathsf{c}}_\sigma, \hat{\mathsf{c}}_{\sigma'}^+\} = \delta_{\sigma,\sigma'} \tag{7.36}$$

and

$$\{\hat{\mathsf{c}}_\sigma, \hat{\mathsf{c}}_{\sigma'}\} = 0\,. \tag{7.37}$$

The Hamilton operator of a free quantized relativistic electron field reads

$$\hat{\mathsf{H}}_\mathrm{D} = \sum E_\sigma \hat{\mathsf{c}}_\sigma^+ \hat{\mathsf{c}}_\sigma\,. \tag{7.38}$$

6. Here, we come to the key step of the procedure. How do we define the vacuum? We could write

$$\hat{\mathsf{c}}_{\sigma+}|0\rangle = 0\,, \qquad \hat{\mathsf{c}}_{\sigma-}|0\rangle = 0\,, \tag{7.39}$$

however, we would consequently encounter the same troubles as in the one-particle case: there would be no state of the lowest energy of the field Hamiltonian (7.38). Here, it is necessary to emphasize that the difficulties in the interpretation of the Dirac equation are not solved by simply regarding electrons as quanta of the electron field; it is merely a different view of the same reality, a suitable formalism, nothing more. The difficulties with the interpretation of the Dirac equation are associated with the existence of negative energy states and are solved by the presence of the Dirac sea of occupied negative energy states. Thus there is an obvious requirement on the vacuum state, namely, "one cannot take out of the vacuum the particle with positive energy," as the first condition in Eq. (7.39) expresses. When all of the negative energy states are occupied, the second requirement imposed on the vacuum state must be: "one cannot create a particle with negative energy out of the vacuum" since all negative energy states are already occupied. However, the second condition in Eq. (7.39) does not correspond to this demand. The correct definition of the vacuum state reads

$$\hat{\mathsf{c}}_{\sigma+}|0\rangle = \hat{\mathsf{c}}_{\sigma-}^+|0\rangle = 0\,. \tag{7.40}$$

The operator $\hat{\mathsf{c}}_{\sigma-}$ annihilates an electron in a negative energy state, hence creates a hole in the Dirac sea, i.e., a positron. To comply with a general convention, we will modify our notation and terminology. Annihilation operators of the positive and negative energy states will be henceforth termed as electron annihilation operators and positron *creation* operators, respectively:

$$\hat{\mathsf{b}}_\sigma = \hat{\mathsf{c}}_{\sigma+}\,, \qquad \hat{\mathsf{d}}_\sigma^+ = \hat{\mathsf{c}}_{\sigma-}\,.$$

The vacuum definition (7.40) consequently reads in this new notation

$$\hat{\mathsf{b}}_\sigma|0\rangle = \hat{\mathsf{d}}_\sigma|0\rangle = 0\,.$$

The expansion of the quantized Dirac field $\hat{\psi}$ has the form, see Eq. (7.35),

$$\hat{\psi} = \sum_\sigma \hat{\mathsf{c}}_\sigma X_\sigma = \sum_\sigma \left(\hat{\mathsf{b}}_\sigma U_\sigma + \hat{\mathsf{d}}_\sigma^+ V_\sigma\right). \tag{7.41}$$

Once again, in agreement with the usual convention, we introduced the notation

$$X_{\sigma+} = U_\sigma\,, \qquad X_{\sigma-} = V_\sigma\,.$$

Using this notation and exploiting the anticommutation relations (7.36), the Hamiltonian (7.38) can be brought to the form

$$\hat{\mathsf{H}}_{\mathrm{D}} = \sum_\sigma \left(E_{\sigma+}\hat{\mathsf{b}}_\sigma^+\hat{\mathsf{b}}_\sigma - E_{\sigma-}\hat{\mathsf{d}}_\sigma^+\hat{\mathsf{d}}_\sigma\right) + \sum_\sigma E_{\sigma-}\,.$$

The last term is a constant independent of the state involved and corresponds to the infinite energy of the Dirac sea, that is, of the occupied negative energy states. But for the case of the gravity, one does not need to worry about additive constants in a Hamiltonian. We can thus write the Hamiltonian in a *normally ordered form* (creation operators to the left, annihilation to the right)

$$:\hat{\mathsf{H}}_{\mathrm{D}}: = \sum_\sigma \left(E_{\sigma+}\hat{\mathsf{b}}_\sigma^+\hat{\mathsf{b}}_\sigma - E_{\sigma-}\hat{\mathsf{d}}_\sigma^+\hat{\mathsf{d}}_\sigma\right). \tag{7.42}$$

Since by definition $E_{\sigma-} < 0$, the last Hamiltonian is a positive definite operator, its energy spectrum being bounded from below. Finally!

The reader can easily verify that had the creation and annihilation operators of the electron-positron field obeyed commutation relations, the operator $:\hat{\mathsf{H}}_{\mathrm{D}}:$ would not be positive definite. As already mentioned in Sect. 6.6, this shows that in the framework of the relativistic theory, the anticommutation relations (7.37) are a consequence of inner consistency of the theory.

7.2.2 *Interaction Hamiltonian*

In the nonrelativistic theory, we introduced interaction of a particle with a quantized EM field by the replacement

$$\hat{\boldsymbol{p}} \to \hat{\boldsymbol{p}} - e\hat{\boldsymbol{A}}\,,$$

where $\hat{\boldsymbol{A}}$ represents an operator of the vector potential of the quantized EM field. We will assume that the same prescription holds in the relativistic theory too, see the discussion in Sect. 7.5.3 regarding this point. Let us consider the interaction of the "classical" Dirac field, that is one Dirac particle, with an EM field first. Clearly, the Hamiltonian has the form[7]

$$\hat{\mathsf{H}} = :\hat{\mathsf{H}}_{\mathrm{EM}}: + \gamma_0 \boldsymbol{\gamma} \cdot (\hat{\boldsymbol{p}} - e\hat{\boldsymbol{A}} - e\mathbf{A}) + \gamma_0 m + e\varphi = :\hat{\mathsf{H}}_{\mathrm{EM}}: + \hat{\mathsf{h}}_{\mathrm{D}} + e\hat{\mathsf{h}}_1 \,. \tag{7.43}$$

Here, $\hat{\mathsf{h}}_{\mathrm{D}}$ is obviously the one-particle Dirac Hamiltonian (7.34) and already includes the interaction of an electron with an external time-independent EM field. The interaction Hamiltonian $\hat{\mathsf{h}}_1$ has the form

$$e\hat{\mathsf{h}}_1 = -\int \mathbf{j}(\mathbf{r}) \cdot \hat{\boldsymbol{A}}(\mathbf{r})\, \mathrm{d}^3\mathbf{r} \,, \tag{7.44}$$

where the charge current density reads

$$\mathbf{j} = e\delta(\mathbf{r})\gamma_0\boldsymbol{\gamma} \,.$$

Analogously, the Hamiltonian of an interacting EM and the electron-positron fields reads

$$\hat{\mathsf{H}} = :\hat{\mathsf{H}}_{\mathrm{EM}}: + :\hat{\mathsf{H}}_{\mathrm{D}}: + \hat{\mathsf{H}}_1 \,, \tag{7.45}$$

$$\hat{\mathsf{H}}_1 = \frac{1}{2}\int\int \frac{\hat{\rho}(\mathbf{r})\hat{\rho}(\mathbf{r}')}{4\pi|\mathbf{r}-\mathbf{r}'|}\, \mathrm{d}^3\mathbf{r}'\mathrm{d}^3\mathbf{r} - \int \hat{\boldsymbol{j}} \cdot \hat{\boldsymbol{A}}\, \mathrm{d}^3\mathbf{r} \,, \tag{7.46}$$

where the operators of charge density and charge current density for the Dirac field have the form

$$\hat{\rho} = e\hat{\psi}^+\hat{\psi} \,, \qquad \hat{\boldsymbol{j}} = e\hat{\psi}^+\gamma_0\boldsymbol{\gamma}\hat{\psi} \,.$$

Since $\gamma_0^2 = 1$, the last two relations can be written compactly using the four-dimensional notation

$$\hat{\mathsf{j}}_\mu = e\hat{\bar{\psi}}\gamma_\mu\hat{\psi} \,, \tag{7.47}$$

where

$$\hat{\bar{\psi}} = \hat{\psi}^+\gamma_0 \tag{7.48}$$

is usually referred to as *Dirac conjugation*.

---

[7]Recall that the term $(-e)\gamma_0\boldsymbol{\gamma} \cdot \mathbf{A}$ describes the interaction between an electron and an external magnetic field, such as the magnetic field created by nucleus. On the other hand, the term $(-e)\gamma_0\boldsymbol{\gamma}\cdot\hat{\boldsymbol{A}}$ captures the interaction between an electron and a quantized EM field.

The first term in Eq. (7.46) corresponds to the electrostatic interaction of the Dirac field with itself, the second term corresponds to the interaction of electric current density of the Dirac field with the transverse part of the EM field.

Calculation of the vacuum charge density should yield a zero as there are no charges in the vacuum. Does it? Let us calculate

$$\langle \hat{\mathrm{j}}_\mu \rangle = \langle 0|\hat{\mathrm{j}}_\mu|0\rangle = e\langle 0| \sum_{\sigma\rho} \left(U_\sigma^+ \hat{\mathrm{b}}_\sigma^+ + V_\sigma^+ \hat{\mathrm{d}}_\sigma \right) \gamma_0\gamma_\mu \left(U_\rho \hat{\mathrm{b}}_\rho + V_\rho \hat{\mathrm{d}}_\rho^+ \right) |0\rangle$$
$$= e \sum_{\sigma\rho} V_\sigma^+ \gamma_0\gamma_\mu V_\rho \delta_{\sigma\rho} .$$

Taking the zeroth component, we find

$$\langle \hat{\rho}(\mathbf{r}) \rangle = e \sum_\sigma |V_\sigma(\mathbf{r})|^2 \qquad \Rightarrow \qquad Q_0 = \int \langle \hat{\rho}(\mathbf{r}) \rangle \, \mathrm{d}^3\mathbf{r} = e \sum_{\sigma-} 1 .$$

This expression clearly does not equal zero. Is it surprising? We were guided by the picture of the Dirac sea of occupied negative energy states and the above result corresponds to its total (infinite) charge. While the energy of the Dirac sea is not of concern (unless one deals with problems on cosmological scales), the nonzero vacuum charge density is not to be readily dismissed. Nonzero vacuum charge density creates an electrostatic field which in turn acts on the "real" electron bound in an atom. This has observable consequences, as we will see later.

Heisenberg was the first one to notice that if an antisymmetrized expression is used instead of (7.47)

$$\hat{\mathrm{j}}_\mu = \frac{e}{2} \left[\hat{\psi}^+, \gamma_0\gamma_\mu \hat{\psi} \right] = \frac{e}{2} \sum_{\sigma\rho} X_\sigma^+ \gamma_0\gamma_\mu X_\rho [\hat{\mathrm{c}}_\sigma^+, \hat{\mathrm{c}}_\rho] , \tag{7.49}$$

one obtains (check it!)

$$\langle \hat{\mathrm{j}}_\mu \rangle = -\frac{e}{2} \sum_\sigma \left(U_\sigma^+ \gamma_0\gamma_\mu U_\sigma - V_\sigma^+ \gamma_0\gamma_\mu V_\sigma \right) . \tag{7.50}$$

This expression yields vacuum charge density

$$\langle \hat{\rho}(\mathbf{r}) \rangle = -\frac{e}{2} \sum_\sigma \left(|U_\sigma(\mathbf{r})|^2 - |V_\sigma(\mathbf{r})|^2 \right) .$$

For the vacuum charge, we find

$$Q_0 = \int \langle \hat{\rho}(\mathbf{r}) \rangle \, \mathrm{d}^3\mathbf{r} = -\frac{e}{2} \left(\sum_{\sigma+} 1 - \sum_{\sigma-} 1 \right) = 0 .$$

The last equality holds for an electron-positron field in a free space. In case of an electron-positron field existing on the background of an electrostatic field of a nucleus, charge asymmetry appears as the electrons are attracted to the nucleus while the positrons are repelled. Consequently, the vacuum charge density is nonzero. We will see later in the text that this effect bears observable consequences.

We can manipulate the charge density four-vector into the normally ordered form so that its vacuum expectation value vanishes. The original operator is a sum of the normally ordered operator and the vacuum expectation value:

$$\hat{\mathrm{j}}_\mu = :\hat{\mathrm{j}}_\mu: + \langle \hat{\mathrm{j}}_\mu \rangle \,.$$

We insert this decomposition into Eq. (7.46) to obtain the following form of the interaction Hamiltonian

$$\begin{aligned}\hat{\mathsf{H}}_1 = &-\int \langle \hat{\boldsymbol{j}} \rangle \cdot \hat{\boldsymbol{A}}\, \mathrm{d}^3\mathbf{r} - \int :\hat{\boldsymbol{j}}: \cdot \hat{\boldsymbol{A}}\, \mathrm{d}^3\mathbf{r} + \frac{1}{2}\frac{1}{4\pi}\int \frac{:\hat{\rho}(\mathbf{r})::\hat{\rho}(\mathbf{r}'):}{|\mathbf{r}-\mathbf{r}'|}\, \mathrm{d}^3\mathbf{r}\, \mathrm{d}^3\mathbf{r}' \\ &+ 2\frac{1}{2}\frac{1}{4\pi}\int \frac{\langle \hat{\rho}(\mathbf{r}) \rangle :\hat{\rho}(\mathbf{r}'):}{|\mathbf{r}-\mathbf{r}'|}\, \mathrm{d}^3\mathbf{r}\, \mathrm{d}^3\mathbf{r}' + \frac{1}{2}\frac{1}{4\pi}\int \frac{\langle \hat{\rho}(\mathbf{r}) \rangle \langle \hat{\rho}(\mathbf{r}') \rangle}{|\mathbf{r}-\mathbf{r}'|}\, \mathrm{d}^3\mathbf{r}\, \mathrm{d}^3\mathbf{r}' \,. \end{aligned} \tag{7.51}$$

The last term captures the electrostatic interaction of the vacuum charge density with itself. It merely shifts the atomic energy levels by the same infinite number and thus will not be considered henceforth.

7.2.3 *Note on Charge Symmetry*

There is another argument, apart from the one already given, why the expression (7.49) is better than (7.47). This argument is related to charge symmetry.[8] In classical electrodynamics, the equality $\boldsymbol{j}_c = -\boldsymbol{j}$ holds. Here, $\boldsymbol{j}$ and $\boldsymbol{j}_c$ are four-currents created by charge clouds differing merely by their signs. We will see that if we insist the same relation holds in quantum theory, we will find that the four-current density has to be of the form (7.49), instead of (7.47). To prove this assertion, we need to elucidate on the change of charge sign for the case of the Dirac field.

We begin with the change of charge sign for the Dirac wave function. Let us therefore consider the Dirac equation (7.19) for an electron in an external EM field

$$[\boldsymbol{\gamma} \cdot (\mathrm{i}\boldsymbol{\partial} - e\boldsymbol{A}) - m]\, \psi = 0 \,. \tag{7.52}$$

[8] In fact, it is the same argument as the first one, but in disguise.

Is there a way to obtain the positron wave function ψ_c from the electron wave function ψ

$$[\boldsymbol{\gamma} \cdot (\mathrm{i}\boldsymbol{\partial} + e\boldsymbol{A}) - m]\, \psi_c = 0\,?$$

For sure, there is. It suffices to take the complex conjugate of Eq. (7.52):

$$\left[-\boldsymbol{\gamma}^* \cdot (\mathrm{i}\boldsymbol{\partial} + e\boldsymbol{A}) - m\right] \psi^* = 0 \tag{7.53}$$

and exploit the advantage of the freedom in choosing a specific form of the γ-matrices. We perform a unitary transformation of the wave function

$$\psi = \mathsf{U}\psi^M \tag{7.54}$$

in the Dirac equation (7.52) and multiply Eq. (7.52) from the left by the inverse matrix U^+; we thus find

$$\left[\boldsymbol{\gamma}^M \cdot (\mathrm{i}\boldsymbol{\partial} - e\boldsymbol{A}) - m\right] \psi^M = 0\,,$$

where[9]

$$\gamma_\mu^M = \mathsf{U}^+\gamma_\mu \mathsf{U}.$$

If we opt for

$$\mathsf{U} = \frac{1}{\sqrt{2}} \begin{pmatrix} 1 & \sigma_2 \\ -\sigma_2 & 1 \end{pmatrix}, \tag{7.55}$$

we obtain *Majorana realization* of the Dirac γ-matrices

$$\gamma_0^M = \begin{pmatrix} 0 & \sigma_2 \\ \sigma_2 & 0 \end{pmatrix}, \quad \gamma_1^M = \begin{pmatrix} -\mathrm{i}\sigma_3 & 0 \\ 0 & -\mathrm{i}\sigma_3 \end{pmatrix}, \tag{7.56a}$$

$$\gamma_2^M = \begin{pmatrix} 0 & \sigma_2 \\ -\sigma_2 & 0 \end{pmatrix}, \quad \gamma_3^M = \begin{pmatrix} \mathrm{i}\sigma_1 & 0 \\ 0 & \mathrm{i}\sigma_1 \end{pmatrix}. \tag{7.56b}$$

The advantage of this particular choice is that the γ-matrices are purely imaginary

$$\boldsymbol{\gamma}^M = -(\boldsymbol{\gamma}^M)^* \tag{7.57}$$

[9]In more detail, $(\gamma_\mu^M)_{ab} = (U^+)_{ac}(\gamma_\mu)_{cd}U_{db}$. These are four matrix equations for μ running from 0 to 3, and a, b, c, d are spinor indices running from 1 to 4.

and the positron wave function is obtained from electron wave function by complex conjugation, see Eq. (7.53),

$$(\psi^M)_c = (\psi^M)^* . \tag{7.58}$$

It follows from Eq. (7.54) that the relation between the electron and positron wave functions in the standard Dirac realization reads

$$\psi_c = \mathsf{U}(\psi^M)_c = \mathsf{U}(\psi^M)^* = \mathsf{U}(\mathsf{U}^+)^*\psi^* = \gamma_2\psi^* , \tag{7.59}$$

where we used the explicit form of the U-matrix, Eq. (7.55), in the last equality. We will use this relation later in Sect. 7.3.2.

Now to the change of the charge sign for Dirac field. Rewriting Eq. (7.52) into the Hamiltonian form

$$\left(\mathrm{i}\frac{\partial}{\partial t} - \hat{\mathsf{h}}_{\mathrm{D}}(e)\right)\psi^M = 0 ,$$

where $\hat{\mathsf{h}}_{\mathrm{D}}(e)$ is given by Eq. (7.34), and taking the complex conjugate we obtain

$$\left(\mathrm{i}\frac{\partial}{\partial t} + \hat{\mathsf{h}}^*_{\mathrm{D}}(e)\right)(\psi^M)^* = 0 \Rightarrow \left(\mathrm{i}\frac{\partial}{\partial t} - \hat{\mathsf{h}}_{\mathrm{D}}(-e)\right)(\psi^M)^* = 0 .$$

Applying now the procedure of canonical quantization to the field ψ^M we find that the quantized field $\hat{\psi}^M$ annihilates electrons and creates positrons

$$\hat{\psi}^M = \sum_\sigma \hat{\mathsf{c}}_\sigma X^M_\sigma = \sum_\sigma \left(\hat{\mathsf{b}}_\sigma U^M_\sigma + \hat{\mathsf{d}}^+_\sigma V^M_\sigma\right) .$$

Likewise, applying the canonical quantization to the charge conjugated field $(\psi^M)^*$ we find that the quantized field $\hat{\psi}^M_c$ annihilates positrons and creates electrons

$$\hat{\psi}^M_c = \sum_\sigma \hat{\mathsf{c}}^+_\sigma (X^M_\sigma)^* = \sum_\sigma \left(\hat{\mathsf{b}}^+_\sigma (U^M_\sigma)^* + \hat{\mathsf{d}}_\sigma (V^M_\sigma)^*\right) .$$

Now we are sufficiently armed to tackle the question raised in the beginning of this section, namely the validity of the equality

$$(\hat{\mathsf{j}}_\mu)_c = -\hat{\mathsf{j}}_\mu . \tag{7.60}$$

Considering the four-current created by the charge conjugated field we obtain successively

$$(\hat{\mathsf{j}}_\mu)_c = \frac{e}{2}\left[(\hat{\psi}_c^M)^+, \gamma_0^M\gamma_\mu^M\hat{\psi}_c^M\right] = \frac{e}{2}\sum_{\sigma,\rho}(X_\sigma^M)^\top\gamma_0^M\gamma_\mu^M(X_\rho^M)^*[\hat{\mathsf{c}}_\sigma,\hat{\mathsf{c}}_\rho^+] =$$

$$= \frac{e}{2}\sum_{\sigma,\rho}(X_\rho^M)^+\gamma_0^M\gamma_\mu^M X_\sigma^M[\hat{\mathsf{c}}_\sigma,\hat{\mathsf{c}}_\rho^+] = -\frac{e}{2}\sum_{\sigma,\rho}(X_\rho^M)^+\gamma_0^M\gamma_\mu^M X_\sigma^M[\hat{\mathsf{c}}_\rho^+,\hat{\mathsf{c}}_\sigma] = -\hat{\mathsf{j}}_\mu\,,$$

where in the third equality we used

$$(X_\sigma^M)^\top\gamma_0^M\gamma_\mu^M(X_\rho^M)^* = (X_\rho^M)^+(\gamma_\mu^M)^+(\gamma_0^M)^+X_\sigma^M = (X_\rho^M)^+\gamma_0^M\gamma_\mu^M X_\sigma^M\,.$$

Here, in the first equality we used the rule for transposing the matrix product, $(\mathsf{AB})^\top = \mathsf{B}^\top\mathsf{A}^\top$, in Majorana representation $(\gamma_0^M\gamma_\mu^M)^\top = (\gamma_\mu^M)^+(\gamma_0^M)^+$, and the fact that the expression is a number, in the second equality we used the identity $\gamma_\mu^+\gamma_0^+ = \gamma_0\gamma_\mu$ which follows from the standard convention

$$\gamma_\mu^+ = \gamma_0\gamma_\mu\gamma_0\,, \tag{7.61}$$

confront with Eqs. (7.18) and (7.56).

It is not difficult to convince oneself about the following. Firstly, had we chosen the four-current operator in the form (7.47), the relation (7.60) would not follow thereof. Secondly, the expressions (7.47) and (7.49) differ only by their vacuum expectation values, their normally ordered form is the same.

7.2.4 Note on Gauge Invariance

We have already mentioned earlier that the four-potential $\boldsymbol{A}$ is determined but for the gauge transformation (7.10). However, this freedom is canceled by another freedom at our disposal: the freedom in the overall phase of the wave function. If the gauge transformation

$$A_\mu(\boldsymbol{x}) \to A_\mu(\boldsymbol{x}) + \partial_\mu\chi(\boldsymbol{x})$$

is accompanied by the phase transformation

$$\psi(\boldsymbol{x}) \to \mathrm{e}^{-ei\chi(\boldsymbol{x})}\psi(\boldsymbol{x})\,,$$

Eq. (7.19) remains unchanged. One crucial note follows from this observation: the form of the Hamilton operator (7.45) is not unique. When opting for a different gauge than the Coulomb one, the Hamiltonian acquires a different form. Nevertheless, physical predictions, such as the differences between eigenvalues of the Hamiltonian, must be independent of the choice of gauge. They indeed are, despite it being far from obvious now. Later on, we will return to this point, see Sects. 7.4.10, 7.6.2 and 7.7.4. For an in-depth treatment, we refer the reader to, e.g., [30, 42].

7.3 Ordinary Perturbation Method

Now that we have found the Hamiltonian for relativistic quantum electrodynamics, Eq. (7.45), we have all we need, don't we? Once we determine the Hamiltonian, it only remains to find its eigenvalues and eigenvectors. Recall that the relativistic Hamiltonian consists of three subHamiltonians: the Hamiltonian of a free EM field, given by Eq. (6.61), the Hamiltonian of a free electron-positron field, given by Eq. (7.42), and the interaction Hamiltonian, given by Eq. (7.51). We do not need to search for the exact solution of the eigenvalue problem, though. Nobody has found it, anyway. Perturbation treatment suffices. It will shortly become clear that the corrections to the energies of the "free" Hamiltonian

$$\hat{\mathsf{H}}_0 = :\hat{\mathsf{H}}_{\mathrm{EM}}: + :\hat{\mathsf{H}}_{\mathrm{D}}:$$

due to the interaction Hamiltonian are proportional to the square of the elementary charge, or—in the "correct," "natural" units—to the fine structure constant $\alpha \simeq 1/137.036$. We will henceforth denote the successive terms of the interaction Hamiltonian (7.51) as $(\hat{\mathsf{H}}_1)_{\perp}^{\mathrm{VA}}$, $(\hat{\mathsf{H}}_1)_{\perp}^{\mathrm{NO}}$, $(\hat{\mathsf{H}}_1)_{||}^{\mathrm{NO}}$, and $(\hat{\mathsf{H}}_1)_{||}^{\mathrm{VA}}$. The symbols VA and NO stand for the interaction of the vacuum expectation value and the normally ordered form of the four-current created by the Dirac field, respectively. From a physics point of view, this separation corresponds to splitting of the four-current into a part which is present even if there are no excitations of the Dirac field (labeled herein as VA) and into a part which is created by the excitations of the field (denoted herein as NO). The symbols $\perp$ and $||$ indicate whether the four-current interacts with the transverse (dynamical) or the longitudinal (static) parts of EM field, respectively.

Note that the individual terms are of different orders of magnitude. To obtain contribution of all of the terms at the order e^2, we need to take the transverse parts to the second order, while the longitudinal ones only to the first order.

Let us first introduce a few abbreviations, to save ink:

$$\begin{aligned}(\hat{\mathsf{H}}_1)_{\perp}^{\mathrm{NO}} &= -\int :\hat{\boldsymbol{j}}: \cdot \hat{\boldsymbol{A}}\,\mathrm{d}^3\mathbf{r} \\ &= -e \sum_{\sigma\sigma'\tau} \Big(w_{\tau}^{\sigma\sigma'} \hat{\mathsf{a}}_{\tau} (\hat{\mathsf{b}}_{\sigma}^{+} \hat{\mathsf{b}}_{\sigma'} - \hat{\mathsf{d}}_{\sigma'}^{+} \hat{\mathsf{d}}_{\sigma} + \hat{\mathsf{b}}_{\sigma}^{+} \hat{\mathsf{d}}_{\sigma'}^{+} + \hat{\mathsf{d}}_{\sigma} \hat{\mathsf{b}}_{\sigma'}) \\ &\quad + w_{-\tau}^{\sigma\sigma'} \hat{\mathsf{a}}_{\tau}^{+} (\hat{\mathsf{b}}_{\sigma}^{+} \hat{\mathsf{b}}_{\sigma'} - \hat{\mathsf{d}}_{\sigma'}^{+} \hat{\mathsf{d}}_{\sigma} + \hat{\mathsf{b}}_{\sigma}^{+} \hat{\mathsf{d}}_{\sigma'}^{+} + \hat{\mathsf{d}}_{\sigma} \hat{\mathsf{b}}_{\sigma'}) \Big), \end{aligned} \tag{7.62}$$

where we inserted for $\hat{\boldsymbol{j}}$ from Eq. (7.49) and for $\hat{\boldsymbol{A}}$ from Eq. (6.47) and denoted

$$w_{\tau}^{\sigma\sigma'} = \int (X_{\sigma}^{+})_a (\gamma_0 \boldsymbol{\gamma})_{ab} \cdot \mathbf{T}_{\tau} (X_{\sigma'})_b \frac{\mathrm{d}^3\mathbf{r}}{\sqrt{2\omega_{\tau}}}\,. \tag{7.63}$$

Obviously, the index τ runs over modes of the EM field, indices σ and σ' over modes of the Dirac field, and a and b are spinor indices. Modes of the Dirac field corresponding to positive or negative energies will be distinguished by the signs $+$ or $-$, respectively. For instance,[10]

$$w_\tau^{+\sigma,-\sigma'} = \int (U_\sigma^+)_a(\gamma_0\boldsymbol{\gamma})_{ab} \cdot \mathbf{T}_\tau (V_{\sigma'})_b \frac{\mathrm{d}^3\mathbf{r}}{\sqrt{2\omega_\tau}} \,. \tag{7.64}$$

Similarly, we rewrite the contribution of the longitudinal part of the interaction,

$$\begin{aligned}(\hat{\mathsf{H}}_1)_{||}^{\mathrm{NO}} &= \int \frac{1}{2}\frac{1}{4\pi}\frac{:\hat{\rho}(\mathbf{r})::\hat{\rho}(\mathbf{r}'):}{|\mathbf{r}-\mathbf{r}'|}\mathrm{d}^3\mathbf{r}\mathrm{d}^3\mathbf{r}' \\ &= \frac{e^2}{2}\sum_{\sigma\sigma'\rho\rho'} v_{\sigma\sigma',\rho\rho'}(\hat{\mathsf{b}}_\sigma^+\hat{\mathsf{b}}_{\sigma'} - \hat{\mathsf{d}}_{\sigma'}^+\hat{\mathsf{d}}_\sigma + \hat{\mathsf{b}}_\sigma^+\hat{\mathsf{d}}_{\sigma'}^+ + \hat{\mathsf{d}}_\sigma\hat{\mathsf{b}}_{\sigma'}) \\ &\quad (\hat{\mathsf{b}}_\rho^+\hat{\mathsf{b}}_{\rho'} - \hat{\mathsf{d}}_{\rho'}^+\hat{\mathsf{d}}_\rho + \hat{\mathsf{b}}_\rho^+\hat{\mathsf{d}}_{\rho'}^+ + \hat{\mathsf{d}}_\rho\hat{\mathsf{b}}_{\rho'}) \,,\end{aligned} \tag{7.65}$$

where we inserted for $\hat{\rho}$ from Eq. (7.49) and denoted

$$v_{\sigma\sigma',\rho\rho'} = \frac{1}{4\pi}\int \frac{X_\sigma^+(\mathbf{r})X_{\sigma'}(\mathbf{r})X_\rho^+(\mathbf{r}')X_{\rho'}(\mathbf{r}')}{|\mathbf{r}-\mathbf{r}'|}\mathrm{d}^3\mathbf{r}\mathrm{d}^3\mathbf{r}' \,. \tag{7.66}$$

If we further denote

$$ew_\tau^{00} = \int \langle\hat{\boldsymbol{j}}\rangle \cdot \mathbf{T}_\tau \frac{\mathrm{d}^3\mathbf{r}}{\sqrt{2\omega_\tau}} \tag{7.67}$$

and

$$ev_{\sigma\sigma',00} = \frac{1}{4\pi}\int \frac{X_\sigma^+(\mathbf{r})X_{\sigma'}(\mathbf{r})\,\langle\hat{\rho}(\mathbf{r}')\rangle}{|\mathbf{r}-\mathbf{r}'|}\mathrm{d}^3\mathbf{r}\mathrm{d}^3\mathbf{r}' \,, \tag{7.68}$$

we can write

$$\begin{aligned}(\hat{\mathsf{H}}_1)_{||}^{\mathrm{VA}} &= +2\frac{1}{2}\frac{1}{4\pi}\int \frac{\langle\hat{\rho}(\mathbf{r})\rangle:\hat{\rho}(\mathbf{r}'):}{|\mathbf{r}-\mathbf{r}'|}\mathrm{d}^3\mathbf{r}\mathrm{d}^3\mathbf{r}' \\ &= e^2\sum_{\sigma,\sigma'} v_{\sigma\sigma',00}(\hat{\mathsf{b}}_\sigma^+\hat{\mathsf{b}}_{\sigma'} - \hat{\mathsf{d}}_{\sigma'}^+\hat{\mathsf{d}}_\sigma + \hat{\mathsf{b}}_\sigma^+\hat{\mathsf{d}}_{\sigma'}^+ + \hat{\mathsf{d}}_\sigma\hat{\mathsf{b}}_{\sigma'})\end{aligned} \tag{7.69}$$

[10] To save ink, the notation is slightly inaccurate. We should also distinguish positive and negative modes in the indices w in Eq. (7.62) and other expressions; for instance, we should write $\sum_{\sigma\sigma'} w_\tau^{+\sigma,+\sigma'}\hat{\mathsf{a}}_\tau\hat{\mathsf{b}}_\sigma\hat{\mathsf{b}}_{\sigma'}$ instead of $\sum_{\sigma\sigma'} w_\tau^{\sigma,\sigma'}\hat{\mathsf{a}}_\tau\hat{\mathsf{b}}_\sigma\hat{\mathsf{b}}_{\sigma'}$, and $\sum_{\sigma\sigma'} w_\tau^{-\sigma,+\sigma'}\hat{\mathsf{a}}_\tau\hat{\mathsf{d}}_\sigma\hat{\mathsf{b}}_{\sigma'}$ instead of $\sum_{\sigma\sigma'} w_\tau^{\sigma,\sigma'}\hat{\mathsf{a}}_\tau\hat{\mathsf{d}}_\sigma\hat{\mathsf{b}}_{\sigma'}$.

and

$$(\hat{\mathsf{H}}_1)_\perp^{\mathrm{VA}} = -\int \langle \hat{\boldsymbol{j}} \rangle \cdot \mathbf{A}\, \mathrm{d}^3\mathbf{r} = -e \sum_\tau \left(w_\tau^{00} \hat{\mathsf{a}}_\tau + w_{-\tau}^{00} \hat{\mathsf{a}}_\tau^+ \right) . \tag{7.70}$$

7.3.1 *Interaction of a Bound Electron with Fluctuations of Fields*

In this section, we commence a seemingly modest project: calculation of the effect of the interaction Hamiltonian (7.51) on the positions of spectral lines of hydrogen-like atoms. Let us recall that the electrostatic interaction between a proton (or a nucleus in general) and an electron is already included in the Hamiltonian of a free Dirac field $:\hat{\mathsf{H}}_D:$. Energies of one-particle states of this field are solutions of the one-particle Dirac equation, and within the nonrelativistic limit, they become solutions of the Pauli equation. Let us consider a hydrogen-like atom in one of its stationary states, calling it 1, and the quantized electromagnetic field in the ground state. In the first, and already satisfying, approximation, such a state is an eigenstate of $\hat{\mathsf{H}}_0$

$$|\psi_0\rangle = \hat{\mathsf{b}}_1^+ |0\rangle \tag{7.71}$$

with energy

$$E_0 = E_{1+} \, .$$

The state $|0\rangle$ denotes the common vacuum state of both the electromagnetic and electron-positron fields. Clearly, the state 1 is one of the solutions of the Dirac equation (7.20) with positive energy. For instance, in the nonrelativistic limit, this could be the state where the electron is in the orbital $1s$ with its spin projection $+\frac{1}{2}$ along the z-axis. However, the electron interacts with fluctuations of the electromagnetic and electron-positron fields, and thus the energy E_0 is slightly shifted. Let us calculate to what extent. We will see that our seemingly modest project turns out to be a very ambitious one. We will have to master a number of new skills, at both technical and conceptual levels, to complete it successfully.

We calculate the effect of $\hat{\mathsf{H}}_1$ on the energy levels of $\hat{\mathsf{H}}_0$ to the second order of the perturbation method. At the first sight, the following calculations may appear slightly frightening. However, we would like to stress that they are in fact simple in terms of concepts. Besides the above discussed interaction Hamiltonian, we have already developed all of the requisite tools, namely the formulas for the first and second order energies, Eqs. (2.15), (2.16), and (6.186)

$$E_1 = \langle \psi_0 | \hat{\mathsf{H}}_1 | \psi_0 \rangle \, , \quad E_2 = -\langle \psi_0 | \hat{\mathsf{H}}_1 \frac{1}{\hat{\mathsf{H}}_0 - E_0} \hat{\mathsf{H}}_1 | \psi_0 \rangle \, ,$$

the (anti)commutation relations for creation and annihilation operators, Eqs. (6.37) and (7.36), and the definition of the vacuum, Eqs. (6.42) and (7.40).

We begin by evaluating the contribution of the longitudinal part of the EM field to the electron self-energy. In the first order of the perturbation method, we have

$$(E_1)^{\rm SE}_{||} = \langle\psi_0|(\hat{\mathsf{H}}_1)^{\rm NO}_{||}|\psi_0\rangle = \langle 0|\hat{\mathsf{b}}_1(\hat{\mathsf{H}}_1)^{\rm NO}_{||}\hat{\mathsf{b}}_1^+|0\rangle\,. \tag{7.72}$$

The method of calculating expressions of this type is universal and we have already used it in the last chapter. We know that the annihilation operators acting on the "right" vacuum, $|0\rangle$, and creation operators acting on the "left" vacuum, $\langle 0|$, yield zero. Hence all monomials of the ladder operators that either start with a creation operator or end with an annihilation operator vanish. In case of monomials comprising an annihilation operator on the left and a creation operator on the right, we use (anti)commutation relations to move the last annihilation operator to the very right and the first creation operator to the very left. In this way, strings of operators will be successively replaced by strings of (anti)commutators. However, those are numbers and can be taken out of the scalar product. Finally, we exploit the normalization of the vacuum state, $\langle 0|\,0\rangle = 1$. For instance, we insert Eq. (7.65) into Eq. (7.72) and we face the expression

$$\begin{aligned}(E_1)^{\rm SE}_{||} &= \frac{e^2}{2}\sum_{\sigma\sigma'\rho\rho'} v_{\sigma\sigma',\rho\rho'}\langle 0|\hat{\mathsf{b}}_1(\hat{\mathsf{b}}^+_\sigma\hat{\mathsf{b}}_{\sigma'} - \hat{\mathsf{d}}^+_{\sigma'}\hat{\mathsf{d}}_\sigma + \hat{\mathsf{b}}^+_\sigma\hat{\mathsf{d}}^+_{\sigma'} + \hat{\mathsf{d}}_\sigma\hat{\mathsf{b}}_{\sigma'})\\ &\quad\times(\hat{\mathsf{b}}^+_\rho\hat{\mathsf{b}}_{\rho'} - \hat{\mathsf{d}}^+_{\rho'}\hat{\mathsf{d}}_\rho + \hat{\mathsf{b}}^+_\rho\hat{\mathsf{d}}^+_{\rho'} + \hat{\mathsf{d}}_\rho\hat{\mathsf{b}}_{\rho'})\hat{\mathsf{b}}^+_1|0\rangle\,,\end{aligned}$$

which may appear frightening. However, we recognize that only monomials constituted by the same number of creation and annihilation operators, both electron and positron, can yield nonvanishing contribution

$$(E_1)^{\rm SE}_{||} = \frac{e^2}{2}\sum_{\sigma\sigma'\rho\rho'} v_{\sigma\sigma',\rho\rho'}\langle 0|\hat{\mathsf{b}}_1(\hat{\mathsf{b}}^+_\sigma\hat{\mathsf{b}}_{\sigma'}\hat{\mathsf{b}}^+_\rho\hat{\mathsf{b}}_{\rho'} + \hat{\mathsf{d}}_\sigma\hat{\mathsf{b}}_{\sigma'}\hat{\mathsf{b}}^+_\rho\hat{\mathsf{d}}^+_{\rho'})\hat{\mathsf{b}}^+_1|0\rangle\,.$$

Using now the anticommutation relations (7.36), we easily obtain

$$\begin{aligned}(E_1)^{\rm SE}_{||} &= \frac{e^2}{2}\sum_{\sigma\sigma'\rho\rho'} v_{\sigma\sigma',\rho\rho'}\left(\delta_{1,\sigma}\delta_{\sigma'+,\rho+}\delta_{\rho',1} - \delta_{1,\sigma'}\delta_{\sigma-,\rho'-}\delta_{\rho,1}\right)\\ &= \frac{e^2}{2}\sum_\rho (v_{1\rho+,\rho+1} - v_{\rho-1,1\rho-})\,,\end{aligned} \tag{7.73}$$

where we omitted the term $e^2\sum v_{\sigma\sigma',\rho\rho'}\delta_{\sigma'+,\rho+}\delta_{\sigma-,\rho'-}/2$, which emerges from the second term in the previous equation. It does not depend on the reference state 1; it is a constant which disappears when considering a difference of energy levels.

In a similar way, we calculate the contribution of the transverse part of the EM field to the electron self-energy. The only difference is that we now add the creation and annihilation of a virtual photon.

$$
\begin{aligned}
(E_2)_\perp^{\rm SE} &= -\langle 0|\hat{\mathsf b}_1(\hat{\mathsf H}_1)_\perp^{\rm NO}\frac{1}{\hat{\mathsf H}_0 - E_0}(\hat{\mathsf H}_1)_\perp^{\rm NO}\hat{\mathsf b}_1^+|0\rangle \\
&= (-e^2)\sum_{\tau\sigma\sigma'\rho\rho'} w_\tau^{\sigma\sigma'} w_{-\tau}^{\rho\rho'}\langle 0|\hat{\mathsf b}_1\left(\hat{\mathsf b}_\sigma^+\hat{\mathsf b}_{\sigma'}\frac{1}{:\hat{\mathsf H}_D: +\omega_\tau - E_0}\hat{\mathsf b}_\rho^+\hat{\mathsf b}_{\rho'}\right. \\
&\quad \left. +\hat{\mathsf d}_\sigma\hat{\mathsf b}_{\sigma'}\frac{1}{:\hat{\mathsf H}_D: +\omega_\tau - E_0}\hat{\mathsf b}_\rho^+\hat{\mathsf d}_{\rho'}^+\right)\hat{\mathsf b}_1^+|0\rangle \\
&= (-e^2)\sum_{\tau\sigma\sigma'\rho\rho'} w_\tau^{\sigma\sigma'} w_{-\tau}^{\rho\rho'}\left(\frac{\delta_{1,\sigma}\delta_{\sigma'+,\rho+}\delta_{\rho',1}}{E_{\rho+}+\omega_\tau - E_0} - \frac{\delta_{1,\sigma'}\delta_{\sigma-,\rho'-}\delta_{\rho,1}}{E_{\rho+}-E_{\rho'-}+E_0+\omega_\tau - E_0}\right) \\
&= (-e^2)\sum_{\tau\rho}\left(\frac{w_\tau^{1,\rho+}w_{-\tau}^{\rho+,1}}{\omega_\tau + E_{\rho+} - E_0} - \frac{w_\tau^{\rho-,1}w_{-\tau}^{1,\rho-}}{\omega_\tau + E_0 - E_{\rho-}}\right),
\end{aligned}
\tag{7.74}
$$

where we again omitted the constant term

$$
(-e^2)\sum_{\tau\sigma\sigma'\rho\rho'}\frac{w_\tau^{\sigma\sigma'}w_{-\tau}^{\rho\rho'}}{(\omega_\tau + E_{\rho+} - E_{\rho'-})}\delta_{\sigma'+,\rho+}\delta_{\sigma-,\rho'-}\,.
$$

The calculation of the vacuum polarization effect is even easier, as the vacuum expectation values of charge and current densities are merely numbers, unlike the operators in the previous expressions:

$$
(E_1)_{||}^{\rm VP} = \langle 0|\hat{\mathsf b}_1(\hat{\mathsf H}_1)_{||}^{\rm VA}\hat{\mathsf b}_1^+|0\rangle = e^2\sum_{\sigma,\sigma'} v_{\sigma\sigma'00}\langle 0|\hat{\mathsf b}_1\hat{\mathsf b}_\sigma^+\hat{\mathsf b}_{\sigma'}\hat{\mathsf b}_1^+|0\rangle = e^2 v_{1100}\,,
\tag{7.75}
$$

and

$$
\begin{aligned}
(E_2)_\perp^{\rm VP} &= -\langle 0|\hat{\mathsf b}_1\left((\hat{\mathsf H}_1)_\perp^{\rm VA}\frac{1}{:\hat{\mathsf H}_{\rm EM}: + :\hat{\mathsf H}_{\rm D}: - E_0}(\hat{\mathsf H}_1)_\perp^{\rm NO}\right. \\
&\quad \left. + (\hat{\mathsf H}_1)_\perp^{\rm NO}\frac{1}{:\hat{\mathsf H}_{\rm EM}: + :\hat{\mathsf H}_{\rm D}: - E_0}(\hat{\mathsf H}_1)_\perp^{\rm VA}\right)\hat{\mathsf b}_1^+|0\rangle \\
&= -e^2\sum w_{\tau'}^{00}w_\tau^{\sigma\sigma'}\langle 0|\hat{\mathsf b}_1\left(\hat{\mathsf a}_\tau\frac{1}{:\hat{\mathsf H}_{\rm EM}: + :\hat{\mathsf H}_{\rm D}: - E_0}\hat{\mathsf a}_{\tau'}^+\hat{\mathsf b}_\sigma^+\hat{\mathsf b}_{\sigma'}\right. \\
&\quad \left. + \hat{\mathsf a}_{\tau'}\hat{\mathsf b}_\sigma^+\hat{\mathsf b}_{\sigma'}\frac{1}{:\hat{\mathsf H}_{\rm EM}: + :\hat{\mathsf H}_{\rm D}: - E_0}\hat{\mathsf a}_\tau^+\right)\hat{\mathsf b}_1^+|0\rangle \\
&= -2e^2\sum_\tau\frac{w_\tau^{00}w_{-\tau}^{11}}{\omega_\tau}\,.
\end{aligned}
\tag{7.76}
$$

Again, we skipped the constant term $-\langle\psi_0|(\hat{\mathrm{H}}_1)^{\mathrm{VA}}_{\perp}\frac{1}{\hat{\mathrm{H}}_0-E_0}(\hat{\mathrm{H}}_1)^{\mathrm{VA}}_{\perp}|\psi_0\rangle$ which merely shifts all the atomic energy levels by the same value.

We have thus almost reached the end; it now remains only to insert into the last two equations from Eqs. (7.63), (7.67) and (7.68) for v and w, and to make transition from discrete to continuous modes of the EM field $\mathbf{T}_\tau(\mathbf{r}) \to (2\pi)^{-3/2}\varepsilon^{(\lambda)}\mathrm{e}^{\mathrm{i}\mathbf{k}\cdot\mathbf{r}}$, see Eq. (6.49). We thus obtain

$$(E_1)^{\mathrm{VP}}_{||} = \frac{e}{4\pi}\int\!\!\int U_1^+(\mathbf{r})U_1(\mathbf{r})\frac{1}{|\mathbf{r}-\mathbf{r}'|}\langle\hat{\rho}(\mathbf{r}')\rangle\,\mathrm{d}^3\mathbf{r}\,\mathrm{d}^3\mathbf{r}'$$

and

$$(E_2)^{\mathrm{VP}}_{\perp} = -2e\int \mathrm{d}^3\mathbf{r}\mathrm{d}^3\mathbf{r}'\int\frac{\mathrm{d}^3\mathbf{k}}{(2\pi)^3}\frac{1}{2\omega^2}\sum_\lambda U_1^+(\mathbf{r})\mathrm{e}^{\mathrm{i}\mathbf{k}\cdot\mathbf{r}}\gamma_0\boldsymbol{\gamma}\cdot\boldsymbol{\varepsilon}^{(\lambda)}U_1(\mathbf{r})\langle\hat{\boldsymbol{j}}(\mathbf{r}')\rangle\cdot\boldsymbol{\varepsilon}^{(\lambda)}\mathrm{e}^{-\mathrm{i}\mathbf{k}\cdot\mathbf{r}'}.$$

By inserting the Fourier transform of the Coulomb potential (3.114) and summing over polarizations (6.55), the last two equations can be united to one

$$(\Delta E)^{\mathrm{VP}} = -\frac{e}{(2\pi)^3}\int\frac{\mathrm{d}^3\mathbf{k}}{\omega^2}\int \mathrm{d}^3\mathbf{r}\,U_1^+(\mathbf{r})\gamma_0\gamma_\mu\mathrm{e}^{\mathrm{i}\mathbf{k}\cdot\mathbf{r}}U_1(\mathbf{r})P_{\mu\nu}\int \mathrm{d}^3\mathbf{r}'\langle\hat{\mathrm{j}}_\nu(\mathbf{r}')\rangle\mathrm{e}^{-\mathrm{i}\mathbf{k}\cdot\mathbf{r}'}, \tag{7.77}$$

where

$$P_{\mu\nu} = -\eta_{\mu\nu} + \frac{k_\mu k_\nu}{k^2}, \tag{7.78}$$

$$\boldsymbol{k} = (0,\mathbf{k}), \qquad \boldsymbol{k}^2 = -\omega^2. \tag{7.79}$$

We can further manipulate Eq. (7.50) for vacuum four-current density into the form

$$\begin{aligned}\langle\hat{\mathrm{j}}_\nu(\mathbf{r})\rangle &= \frac{e}{2}\sum_\sigma\left[\left(V_a^\sigma(\mathbf{r})\right)^+(\gamma_0\gamma_\nu)_{ab}V_b^\sigma(\mathbf{r}) - \left(U_a^\sigma(\mathbf{r})\right)^+(\gamma_0\gamma_\nu)_{ab}U_b^\sigma(\mathbf{r})\right] \qquad (7.80)\\ &= \frac{e}{2}(\gamma_0\gamma_\nu)_{ab}\sum_\sigma\langle\mathbf{r}|\left[\hat{\mathrm{P}}^{\sigma-}-\hat{\mathrm{P}}^{\sigma+}\right]_{ba}|\mathbf{r}\rangle,\end{aligned}$$

where the projection operators onto the subspaces of positive and negative energy states were introduced

$$P^{\rho+}_{ba}(\mathbf{r},\mathbf{r}') = U_b^\rho(\mathbf{r})(U_a^\rho)^+(\mathbf{r}'), \qquad P^{\rho-}_{ba}(\mathbf{r},\mathbf{r}') = V_b^\rho(\mathbf{r})(V_a^\rho)^+(\mathbf{r}'). \tag{7.81}$$

Likewise, by inserting for v from Eq. (7.66), and the Fourier transform of the Coulomb potential, Eq. (3.114), Eq. (7.73) can be brought into the form

$$(E_1)^{\mathrm{SE}}_{||} = e^2\int\frac{\mathrm{d}^3\mathbf{k}}{(2\pi)^3}\int\!\!\int \mathrm{d}^3\mathbf{r}\,\mathrm{d}^3\mathbf{r}'M_{||}(\mathbf{k},\mathbf{r},\mathbf{r}'), \tag{7.82a}$$

$$M_{||}(\mathbf{k},\mathbf{r},\mathbf{r}') = \frac{1}{2\omega^2}\sum_\rho U_1^+(\mathbf{r})\mathrm{e}^{\mathrm{i}\mathbf{k}\cdot\mathbf{r}}\left(P^{\rho+}(\mathbf{r},\mathbf{r}') - P^{\rho-}(\mathbf{r},\mathbf{r}')\right)\mathrm{e}^{-\mathrm{i}\mathbf{k}\cdot\mathbf{r}}U_1(\mathbf{r}')\,. \tag{7.82b}$$

Finally, by inserting for w Eq. (7.63), changing from discrete to continuous number of modes of the EM field, Eq. (6.49), and summing over polarizations, Eq. (6.55), we can rewrite Eq. (7.74) as

$$(E_2)^{\mathrm{SE}}_{\perp} = (-e^2)\int\frac{\mathrm{d}^3\mathbf{k}}{(2\pi)^3}\int\int \mathrm{d}^3\mathbf{r}\,\mathrm{d}^3\mathbf{r}' M_\perp(\mathbf{k},\mathbf{r},\mathbf{r}')\,, \tag{7.83a}$$

$$M_\perp(\mathbf{k},\mathbf{r},\mathbf{r}') = \frac{P_{ij}}{2\omega}\sum_\rho U_1^+(\mathbf{r})\gamma_0\gamma_i\mathrm{e}^{\mathrm{i}\mathbf{k}\cdot\mathbf{r}}\left(\frac{P^{\rho+}(\mathbf{r},\mathbf{r}')}{\omega + E_{\rho+} - E_0} - \frac{P^{\rho-}(\mathbf{r},\mathbf{r}')}{\omega - E_{\rho-} + E_0}\right)\mathrm{e}^{-\mathrm{i}\mathbf{k}\cdot\mathbf{r}'}\gamma_0\gamma_j U_1(\mathbf{r}')\,. \tag{7.83b}$$

We thus see that by means of the "good old" perturbation method, we were able to derive all of the leading QED corrections to the atomic energy levels. From this point of view, the only conceptually new step involved in transition from the nonrelativistic to relativistic quantum theory is to tackle the problem of negative energy states of one-particle Dirac equation by a convenient definition of the vacuum state, see Eq. (7.40). All other aspects of the formalism have been already discussed earlier.

Although the above viewpoint is possible, it is rather unsuitable for the following reasons:

- The summations in Eqs. (7.80), (7.82), and (7.83) over the modes of the Dirac field, that is over the solutions of the one-particle Dirac equation (7.20), have to be understood symbolically. In fact, it is a summation and integration over the discrete and continuous parts of the hydrogen spectrum, respectively. The presence of the positive and negative energy projectors in Eqs. (7.80), (7.82), and (7.83) makes the calculation of the integrals exceedingly difficult, as becomes clear almost immediately once one tries to evaluate these expression any further. The negative energy states remain to be troublemakers.
- The Lorentz invariance of the theory, hidden already in the starting Hamiltonian (7.45), is further obscured by the necessity to distinguish between the positive- and negative-energy virtual states together with the necessity to distinguish between the transverse and longitudinal parts of the EM field. The notion of the positive and negative energy states, and for that matter, the notion of the longitudinal and transverse parts of EM field as well, are not Lorentz-invariant notions. A positive energy state in one inertial frame is a mixture of positive and negative energy states in a different inertial frame. Likewise, a transverse part of the EM field in one inertial frame is a mixture of transverse and longitudinal parts of the EM field in a different inertial frame. We emphasize that the lack of the manifest Lorentz invariance is not a mere formal insufficiency. It is closely related to the previous point, namely the difficulty of performing actual calculations.

- Last and worst: the corrections to the energy levels expected to be small because of the small value of the fine-structure constant α turn out to be infinite! We already encountered this in Sect. 6.5 and showed there that a careful distinction between a theoretical construction and actual measured value removes the worst divergences. One could therefore expect a similar approach could solve the problem within the relativistic theory too. In fact, the situation is even better in this case: the so-called renormalization makes the relativistic quantum electrodynamics finite. However, to tackle the infinities unambiguously, the manifest Lorentz invariance is again of key importance.

In summary, one could build up the relativistic quantum theory as closely as possible to the nonrelativistic theory. However, that is certainly not the best choice. Such approach lacks an essential insight how to derive expressions for the mutual interaction of the electromagnetic and electron-positron fields in a manifestly Lorentz-invariant form. The therein absent insight was provided by an American physicist Richard P. Feynman.

7.3.2 *Positronium I*

Prior to the exposition of the Feynman approach, we make a small digression. Despite having emphasized in length that the usual formalism of quantum mechanics, with its separation of time and space variables, is generally not suitable for calculations of relativistic processes, there are a few simple problems where the application of this formalism is not so unfavorable. It is always worth realizing what one can and cannot do with the so far acquired knowledge. For instance, were we to calculate the Lamb shift of atomic energy levels, we would need to master additional skills. However, our present knowledge suffices to explain the most important features of positronium, a bound state of an electron and a positron.

The ground state of the positronium lives about 10^{-10} s, then vanishes with a flash of two photons, see the first two diagrams in Fig. 7.2. The gross structure of positronium is within a reasonable approximation given by Eq. (3.36), where we insert $Z = 1, m_j = m_e$ to obtain $\nu_{12} = R_\infty c \times \frac{3}{8} \simeq 1{,}233{,}690\,\text{GHz}$ which is in

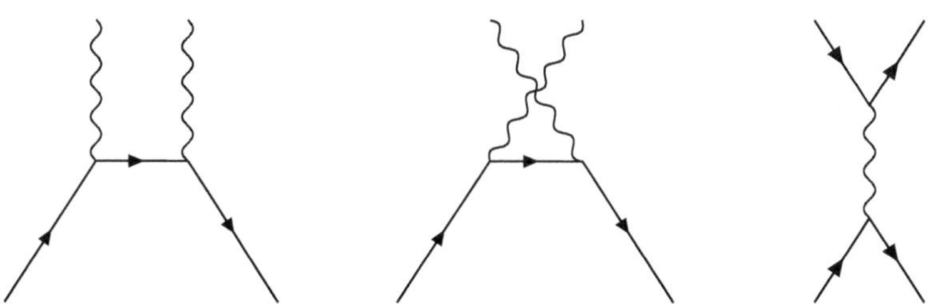

Fig. 7.2 Real and virtual annihilation of an electron-positron pair

a good agreement with the experiment. However, the fine splitting of the ground state is not given by Eq. (3.58). There is another effect responsible for it, namely the virtual positronium annihilation, see the third diagram in Fig. 7.2. We now show how to calculate the lifetime and fine-splitting of the positronium ground state.

1. *Relation of bound- and free-state wave functions*
   When we were considering interaction of a bound electron with fluctuations of the fields, we expanded the electron-positron field into the eigenstates of the Dirac Hamiltonian (7.34) and neglected the proton motion. Since a proton is about 2000 times more massive than an electron, it is clearly not a bad first approximation for which we shall correct in due time. However, in case of positronium, the situation is completely different as one must treat the positron in the same way as an electron. For our current purposes, it suffices to express the bound state as a linear combination of free states

$$\psi^0_{\text{at}}(\mathbf{r}) = \int \mathrm{d}^3\mathbf{p}\,\psi^0_{\text{at}}(\mathbf{p})\frac{e^{\mathrm{i}\mathbf{p}\cdot\mathbf{r}}}{(2\pi)^{3/2}} = \int \mathrm{d}^3\mathbf{p}\,\psi^0_{\text{at}}(\mathbf{p})\frac{e^{\mathrm{i}\mathbf{p}\cdot(\mathbf{r}_1-\mathbf{r}_2)}}{(2\pi)^{3/2}}\,. \tag{7.84}$$

   Here, we used the expansion of the wave function in the coordinate representation into the momentum representation, a three-dimensional generalization of Eqs. (1.94) and (1.95), and we subsequently returned to the original coordinates of the electron and positron, see Eq. (3.4). We took the bound-state wave function in the nonrelativistic approximation which well suffices for the accuracy we aim for. The last equation can be rewritten in the Dirac notation as

$$\langle \mathbf{r}_1,\mathbf{r}_2|\,\psi^0_{\text{at}}\rangle = (2\pi)^{3/2}\int \mathrm{d}^3\mathbf{p}\,\psi^0_{\text{at}}(\mathbf{p})\langle \mathbf{r}_1|\,\mathbf{p}\rangle\langle \mathbf{r}_2|-\mathbf{p}\rangle\,. \tag{7.85}$$

   We make one more step in abstraction and leave out the projection onto the coordinate eigenstates:

$$|\psi^0_{\text{at}}\rangle = (2\pi)^{3/2}\int \mathrm{d}^3\mathbf{p}\,\psi^0_{\text{at}}(\mathbf{p})|\mathbf{p}\rangle_1|-\mathbf{p}\rangle_2 \tag{7.86}$$

   We express the one-particle electron and positron states in terms of action of the corresponding operators on the vacuum

$$|\psi^0_{\text{at}}\rangle = \sum_{\mathbf{p}} \psi^0_{\text{at}}(\mathbf{p})\hat{\mathsf{b}}^+_{\mathbf{p}}\hat{\mathsf{d}}^+_{-\mathbf{p}}|0\rangle\,. \tag{7.87}$$

   To avoid complications stemming from continuum of the states, we assumed, as in the case of EM field, that the electron-positron field is periodic with period L and we will take the continuum limit $L \to \infty$ at the end of our derivations. The factor $(2\pi)^{3/2}$ disappeared upon transition from continuum to discrete basis. The state (7.87) is now normalized to unity, in the same way as the state (7.84) we started with[11]:

$$\langle\psi^0_{\rm at}|\,\psi^0_{\rm at}\rangle = \sum_{\mathbf{p}}\sum_{\mathbf{p}'}[\psi^0_{\rm at}(\mathbf{p}')]^*\psi^0_{\rm at}(\mathbf{p})\langle 0|\hat{\mathsf{d}}_{-\mathbf{p}'}\hat{\mathsf{b}}_{\mathbf{p}'}\hat{\mathsf{b}}^+_{\mathbf{p}}\hat{\mathsf{d}}^+_{-\mathbf{p}}|0\rangle = \sum_{\mathbf{p}}|\psi^0_{\rm at}(\mathbf{p})|^2$$

$$= \sum_{\mathbf{p}}\langle\psi^0_{\rm at}|\,\mathbf{p}\rangle\langle\mathbf{p}|\,\psi^0_{\rm at}\rangle = \langle\psi^0_{\rm at}|\,\psi^0_{\rm at}\rangle = \int \mathrm{d}^3\mathbf{r}\langle\psi^0_{\rm at}|\,\mathbf{r}\rangle\langle\mathbf{r}|\,\psi^0_{\rm at}\rangle = \int \mathrm{d}^3\mathbf{r}|\psi^0_{\rm at}(\mathbf{r})|^2 = 1\,.$$

In the second equality, we used the anticommutation relations (7.36), and in the fourth equality, we used the discrete three-dimensional generalization of the completeness relation (1.92)

$$\sum_{\mathbf{p}}|\mathbf{p}\rangle\langle\mathbf{p}| = 1\,. \tag{7.88}$$

The most important contribution to the processes of the real and virtual annihilation comes from the states of the expansion (7.87) which do not move at all

$$|\psi^0_{\rm at}\rangle \simeq \sum_{\mathbf{p}}\psi^0_{\rm at}(\mathbf{p})\hat{\mathsf{b}}^+_0\hat{\mathsf{d}}^+_0|0\rangle\,. \tag{7.89}$$

Again, this approximation suffices for our purposes.

We now decompose the electron-positron field into the eigenstates of the free Dirac Hamiltonian. Namely, we follow the procedure outlined in Sect. 7.2.1 with the only exception of setting $\mathbf{A} = \varphi = 0$ in Eq. (7.34).

2. *Interaction Hamiltonian*

Only the part (7.62) of the interaction Hamiltonian (7.51) is of interest now; the other terms do not contribute to the process in question. Let us examine it:

$$(\hat{\mathsf{H}}_1)^{\rm NO}_\perp = (\hat{\mathsf{H}}_1)_{ee} + (\hat{\mathsf{H}}_1)_{pp} + (\hat{\mathsf{H}}_1)_{ep,0} + (\hat{\mathsf{H}}_1)_{0,ep},$$

where

$$\begin{aligned}(\hat{\mathsf{H}}_1)_{ee} &= -e\sum_{\sigma\sigma'}u^{\sigma,\sigma'}\hat{\mathsf{b}}^+_\sigma\hat{\mathsf{b}}_{\sigma'}, \qquad &(\hat{\mathsf{H}}_1)_{pp} &= -e\sum_{\sigma\sigma'}u^{\sigma,\sigma'}(-1)\hat{\mathsf{d}}^+_{\sigma'}\hat{\mathsf{d}}_\sigma\,,\\ (\hat{\mathsf{H}}_1)_{ep,0} &= -e\sum_{\sigma\sigma'}u^{\sigma,\sigma'}\hat{\mathsf{b}}^+_\sigma\hat{\mathsf{d}}^+_{\sigma'}, \qquad &(\hat{\mathsf{H}}_1)_{0,ep} &= -e\sum_{\sigma\sigma'}u^{\sigma,\sigma'}\hat{\mathsf{d}}_\sigma\hat{\mathsf{b}}_{\sigma'}\end{aligned} \tag{7.90}$$

[11] On the other hand, the norm of the states (7.86) and (7.85) is infinite. How is it possible? Upon transition from Eqs. (7.84) to (7.85), we tacitly moved from a function of *one* variable to a function of *two* variables. As a function of one variable $\mathbf{r} = \mathbf{r}_1 - \mathbf{r}_2$, the bound state function is normalized to unity, while as a function of two variables, $\mathbf{r}_1, \mathbf{r}_2$, it is normalized to the δ-function.

and where we introduced the abbreviations

$$u^{\sigma,\sigma'} = \sum_{\tau} \left(w_{\tau}^{\sigma\sigma'} \hat{\mathsf{a}}_{\tau} + w_{-\tau}^{\sigma\sigma'} \hat{\mathsf{a}}_{\tau}^{+} \right) .$$

The term $(\hat{\mathsf{H}}_1)_{ee}$ annihilates an electron and creates another one, the term $(\hat{\mathsf{H}}_1)_{pp}$ annihilates a positron and creates another one, the term $(\hat{\mathsf{H}}_1)_{ep,0}$ creates an clectron-positron pair out of vacuum, and the term $(\hat{\mathsf{H}}_1)_{0,ep}$ annihilates an electron-positron pair.

3. *Second order of the perturbation method—virtual annihilation*
The energy shift of positronium caused by its virtual annihilation reads, see Eqs. (2.16) and (6.186),

$$\begin{aligned} E_2 = -\langle \psi_{\rm at}^0 | \Big[(\hat{\mathsf{H}}_1)_{0,ep} \frac{1}{:\hat{\mathsf{H}}_D: + :\hat{\mathsf{H}}_{EM}: - E_0} (\hat{\mathsf{H}}_1)_{ep,0} \\ + (\hat{\mathsf{H}}_1)_{ep,0} \frac{1}{:\hat{\mathsf{H}}_D: + :\hat{\mathsf{H}}_{EM}: - E_0} (\hat{\mathsf{H}}_1)_{0,ep} \Big] | \psi_{\rm at}^0 \rangle . \end{aligned} \tag{7.91}$$

At the second order of perturbation method, there is also contribution from

$$\begin{aligned} \langle \psi_{\rm at}^0 | \Big[(\hat{\mathsf{H}}_1)_{ee} \frac{1}{:\hat{\mathsf{H}}_D: + :\hat{\mathsf{H}}_{EM}: - E_0} (\hat{\mathsf{H}}_1)_{pp} \\ + (\hat{\mathsf{H}}_1)_{pp} \frac{1}{:\hat{\mathsf{H}}_D: + :\hat{\mathsf{H}}_{EM}: - E_0} (\hat{\mathsf{H}}_1)_{ee} \Big] | \psi_{\rm at}^0 \rangle . \end{aligned}$$

These terms represent one-photon exchange between the electron and positron and a part of the electron and positron self-energies. The one-photon exchange was already analyzed in Sects. 6.5.5 and 6.5.6, the self-energy effect was discussed in the previous Sect. 7.3.1. For now, we ignore these terms and focus solely on the calculation of Eq. (7.91); by means of Eqs. (7.89) and (7.90) one obtains

$$\begin{aligned} E_2 &= -e^2 \left| \sum_{\mathbf{p}} \psi_{\rm at}^0(\mathbf{p}) \right|^2 \sum_{\tau} \left[\frac{w_{\tau}^{-0,+0} w_{-\tau}^{+0,-0}}{\omega + E_0} + \frac{w_{\tau}^{+0,-0} w_{-\tau}^{-0,+0}}{\omega - E_0} \right] \\ &= -e^2 \left| \sum_{\mathbf{p}} \psi_{\rm at}^0(\mathbf{p}) \right|^2 L^3 U_0^{+} \gamma_0 \gamma_i V_0 V_0^{+} \gamma_0 \gamma_i U_0 \lim_{\omega \to 0} \frac{1}{2\omega} \left(\frac{1}{\omega + E_0} + \frac{1}{\omega - E_0} \right) \\ &= \frac{e^2}{4m^2} \left| \sum_{\mathbf{p}} \psi_{\rm at}^0(\mathbf{p}) \right|^2 L^3 U_0^{+} \gamma_0 \gamma_i V_0 V_0^{+} \gamma_0 \gamma_i U_0 . \end{aligned} \tag{7.92}$$

In the second equality, we substituted, see Eqs. (6.43), (6.44), and (7.64),

$$w_\tau^{-0,+0} = \frac{1}{\sqrt{2\omega_\tau}} \int \mathrm{d}^3\mathbf{r} V_0^+ \gamma_0 \boldsymbol{\gamma} \cdot \mathbf{T}_\tau(\mathbf{r}) U_0 = \delta_{\mathbf{k}_\tau,0} \frac{1}{\sqrt{2\omega_\tau}} V_0^+ \gamma_0 \boldsymbol{\gamma} \cdot \boldsymbol{\varepsilon}^{(\tau)} U_0 L^{3/2},$$

and in the third equality we also substituted

$$E_0 = E_{+0} - E_{-0} = m - (-m) = 2m. \tag{7.93}$$

Clearly, only the zeroth mode of the EM field $\mathbf{k}_\tau = 0$ contributes! Therefore, there is a limit $\omega \to 0$ in the second row of Eq. (7.92). Since the zeroth mode possesses no direction, only the first term δ_{ij} in the projector (6.32) contributes.

Let us recall that U_0 and V_0 describe the electron and positron at rest, respectively. These states depend only on the spin, they are space-independent, therefore they are to be regarded as constants when integrating over $\mathbf{r}$. Note that we omitted additional terms originating from Eq. (7.91). These terms correspond either to (1) virtual creation and annihilation of electron-positron pair and photon, or (2) another part of electron and positron self-energies. The former is state-independent and bears no observable consequences, and we omit the latter now.

4. *Second order of the perturbation method—real annihilation*
For the probability amplitude of the positronium real annihilation, we find from Eqs. (6.181), (7.89), and (7.90)

$$\begin{aligned} b_2 &= -e^2 \langle 0|\hat{\mathsf{a}}_1 \hat{\mathsf{a}}_2 (\hat{\mathsf{H}}_1)_{0,ep} \frac{1}{:\hat{\mathsf{H}}_D : + : \hat{\mathsf{H}}_{EM} : -E_0} \left[(\hat{\mathsf{H}}_1)_{ee} + (\hat{\mathsf{H}}_1)_{pp} \right] |\psi_{\mathrm{at}}^0\rangle \\ &= -e^2 \sum_{\mathbf{p}} \psi_{\mathrm{at}}^0(\mathbf{p}) \sum_\sigma \left[\frac{w_1^{-0,+\sigma} w_2^{+\sigma,+0}}{-E_{-0} + E_{+\sigma} + \omega_2 - (E_{+0} - E_{-0})} \right. \\ &\quad - \frac{w_2^{-\sigma,+0} w_1^{-0,-\sigma}}{E_{+0} - E_{-\sigma} + \omega_1 - (E_{+0} - E_{-0})} + \frac{w_2^{-0,+\sigma} w_1^{+\sigma,+0}}{-E_{-0} + E_{+\sigma} + \omega_1 - (E_{+0} - E_{-0})} \\ &\quad \left. - \frac{w_1^{-\sigma,+0} w_2^{-0,-\sigma}}{E_{+0} - E_{-\sigma} + \omega_2 - (E_{+0} - E_{-0})} \right]. \end{aligned} \tag{7.94}$$

The four terms present multiplication of two possibilities, the "first" photon is radiated by either electron or positron, by two other possibilities, either the "first" photon has frequency ω_1 and the "second" ω_2 or vice versa.

We can manipulate this expression into more a suitable form. Each of the two pairs of fractions in the square brackets can be rearranged by means of the conservation of energy, see Eq. (6.177),

$$E_{+0} - E_{-0} = \omega_1 + \omega_2$$

in such a way that only one of the frequencies ω_1 or ω_2 will be present. For instance, we substitute for w from the definition (7.64) and find for the first pair of fractions in the square brackets in Eq. (7.94)

$$\sum_\sigma \left(\frac{w_1^{-0,+\sigma} w_2^{+\sigma,+0}}{-E_{-0} + E_{+\sigma} + \omega_2 - (E_{+0} - E_{-0})} - \frac{w_2^{-\sigma,+0} w_1^{-0,-\sigma}}{E_{+0} - E_{-\sigma} + \omega_1 - (E_{+0} - E_{-0})} \right)$$

$$= \frac{1}{\sqrt{(2\omega_1)(2\omega_2)}} \int \mathrm{d}^3\mathbf{r}\mathrm{d}^3\mathbf{r}' V_0^+ \gamma_0 \boldsymbol{\gamma} \cdot \mathbf{T}_1(\mathbf{r}) \sum_\sigma \left(\frac{U_\sigma(\mathbf{r}) U_\sigma^+(\mathbf{r}')}{E_{+\sigma} + \omega_2 - E_{+0}} + \frac{V_\sigma(\mathbf{r}) V_\sigma^+(\mathbf{r}')}{E_{-\sigma} + \omega_2 - E_{+0}} \right) \gamma_0 \boldsymbol{\gamma} \cdot \mathbf{T}_2(\mathbf{r}') U_0 \,.$$

The reader now surely recognizes the spectral decomposition of the Dirac Hamiltonian (7.34) with $\mathbf{A} = \varphi = 0$ in the bracketed expression in the last equation

$$\begin{aligned}
\sum_\sigma \left(\frac{U_\sigma(\mathbf{r}) U_\sigma^+(\mathbf{r}')}{E_{+\sigma} + \omega_2 - E_{+0}} + \frac{V_\sigma(\mathbf{r}) V_\sigma^+(\mathbf{r}')}{E_{-\sigma} + \omega_2 - E_{+0}} \right) &= \langle \mathbf{r} | \frac{1}{\hat{\mathsf{h}}_\mathrm{D} + \omega_2 - E_{+0}} | \mathbf{r}' \rangle \\
&= \langle \mathbf{r} | \frac{1}{\omega_2 - E_{+0} + \gamma_0 \boldsymbol{\gamma} \cdot \hat{\boldsymbol{p}} + \gamma_0 m} | \mathbf{r}' \rangle \\
&= \sum_\mathbf{p} \langle \mathbf{r} | \mathbf{p} \rangle \frac{1}{\omega_2 - E_{+0} + \gamma_0 \boldsymbol{\gamma} \cdot \mathbf{p} + \gamma_0 m} \langle \mathbf{p} | \mathbf{r}' \rangle \\
&= \sum_\mathbf{p} \langle \mathbf{r} | \mathbf{p} \rangle \frac{\omega_2 - E_{+0} - \gamma_0 \boldsymbol{\gamma} \cdot \mathbf{p} - \gamma_0 m}{(\omega_2 - E_{+0})^2 - \mathbf{p} \cdot \mathbf{p} - m^2} \langle \mathbf{p} | \mathbf{r}' \rangle \,.
\end{aligned} \tag{7.95}$$

In the equality between the second and the third row, we used the completeness relation (7.88). When going from the third to the fourth row, both the numerator and denominator were multiplied by $\omega_2 - E_{+0} - \gamma_0 \boldsymbol{\gamma} \cdot \mathbf{p} - \gamma_0 m$. After substituting the last equation into the penultimate one, the integration over the space variables yields momentum conservation when going from "real" to "virtual" states and back, see Eqs. (6.43) and (6.44):

$$\int \mathrm{d}^3\mathbf{r} \mathbf{T}_1(\mathbf{r}) \langle \mathbf{r} | \mathbf{p} \rangle = \boldsymbol{\varepsilon}^{(1)} \int \mathrm{d}^3\mathbf{r} \frac{e^{\mathrm{i}\mathbf{k}_1 \cdot \mathbf{r}}}{L^{3/2}} \frac{e^{\mathrm{i}\mathbf{p} \cdot \mathbf{r}}}{L^{3/2}} = \boldsymbol{\varepsilon}^{(1)} \delta_{-\mathbf{p},\mathbf{k}_1} \,,$$

$$\int \mathrm{d}^3\mathbf{r} \mathbf{T}_2(\mathbf{r}) \langle \mathbf{p} | \mathbf{r} \rangle = \boldsymbol{\varepsilon}^{(2)} \delta_{\mathbf{p},\mathbf{k}_2} \,.$$

It follows from the last two equations that $\mathbf{p} \cdot \mathbf{p} = \omega_1^2 = \omega_2^2$. If we further recall Eq. (7.93), the denominator in Eq. (7.95) equals

$$(\omega_2 - E_{+0})^2 - \mathbf{p} \cdot \mathbf{p} - m^2 = -2\omega_2 m \,.$$

By substituting these partial results into Eq. (7.94), we obtain

$$b_2 = -e^2 \sum_{\mathbf{p}} \frac{\psi^0_{\mathrm{at}}(\mathbf{p})}{\sqrt{(2\omega_1)(2\omega_2)}} V_0^+ \left[\gamma_0 \boldsymbol{\gamma} \cdot \boldsymbol{\varepsilon}^{(1)} \frac{\omega_2 - m - \gamma_0 \boldsymbol{\gamma} \cdot \mathbf{k}_2 - \gamma_0 m}{-2\omega_2 m} \gamma_0 \boldsymbol{\gamma} \cdot \boldsymbol{\varepsilon}^{(2)} \right.$$
$$\left. + \gamma_0 \boldsymbol{\gamma} \cdot \boldsymbol{\varepsilon}^{(2)} \frac{\omega_1 - m - \gamma_0 \boldsymbol{\gamma} \cdot \mathbf{k}_1 - \gamma_0 m}{-2\omega_1 m} \gamma_0 \boldsymbol{\gamma} \cdot \boldsymbol{\varepsilon}^{(1)} \right] U_0 \delta_{\mathbf{k}_1, -\mathbf{k}_2} . \tag{7.96}$$

5. *Relation between the positron and electron states*
Before we proceed further, we need to ascertain the form of the positron state V_0. The state of an electron at rest is formally a solution to the Dirac equation (7.16)

$$|\psi(t)\rangle = \mathrm{e}^{-imt}|U_0\rangle\,, \qquad (\gamma_0 - 1)|U_0\rangle = 0\,,$$

and depends only on the electron spin state

$$|U_0\rangle = \frac{1}{L^{3/2}} \begin{pmatrix} |S\rangle_e \\ 0 \end{pmatrix}, \tag{7.97}$$

where $|S\rangle = |\pm\rangle$ denotes the state vector describing the electron states with positive or negative spin projections along the z-axis. It follows from Eq. (7.59)

$$|V_0\rangle = \gamma_2 \frac{1}{L^{3/2}} \begin{pmatrix} |S\rangle_p^* \\ 0 \end{pmatrix} = \frac{1}{L^{3/2}} \begin{pmatrix} 0 \\ -\sigma_2 |S\rangle_p^* \end{pmatrix}. \tag{7.98}$$

In the following text, we will consider only real spinors $|S\rangle_p$, thus we will omit the star. Why not to take for V_0 an arbitrary solution to Eq. (7.16) corresponding to the rest state of a positron,

$$|\psi(t)\rangle = \mathrm{e}^{+imt}|V_0\rangle\,, \qquad (\gamma_0 + 1)|V_0\rangle = 0\,?$$

In other words, why do we need to consider

$$\begin{pmatrix} 0 \\ -\sigma_2 |S\rangle_p^* \end{pmatrix} \quad \text{instead of} \quad \begin{pmatrix} 0 \\ |S\rangle_p \end{pmatrix} ?$$

The reason is we need to compare electron and positron spins in the same coordinate system. The solution for the positron rest state differs from the solution for the electron rest state by the replacement $t \to -t$. In the positron rest frame, the time flows in the opposite direction than in the electron rest frame. In a frame where the time flows in the opposite direction, spin will be reversed. But for the phase factor, this corresponds to the replacement $|\pm\rangle \to \mp\mathrm{i}|\mp\rangle = -\sigma_2|\pm\rangle$.

6. *Calculation of matrix elements*
After substituting Eqs. (7.97) and (7.98) into Eq. (7.92) and using the identities[12]

$$(\sigma_i)_{ab}(\sigma_i)_{dc} = -\frac{1}{2}(\sigma_i)_{ac}(\sigma_i)_{db} + \frac{3}{2}\delta_{ac}\delta_{db}$$

and

$$(\sigma_2\sigma_i\sigma_2)^T = -\sigma_i \, , \qquad \sigma_2^2 = 1 \, ,$$

we rearrange Eq. (7.92) into the form of the matrix element of spin operators between the initial and final positronium spin states

$$E_2 = \frac{e^2}{2m^2}\left|\sum_{\mathbf{p}} \frac{\psi^0_{\text{at}}(\mathbf{p})}{L^{3/2}}\right|^2 \langle S|_e \langle S|_p \left(\hat{\boldsymbol{S}}_e \cdot \hat{\boldsymbol{S}}_p + \frac{3}{4}\right) |S\rangle_e |S\rangle_p \, . \tag{7.99}$$

After inserting Eqs. (7.97) and (7.98) into Eq. (7.96) and performing several straightforward algebraic manipulations involving the property (1.25) of Pauli spin matrices, we obtain

$$|b_2| = \frac{e^2}{m}\sum_{\mathbf{p}} \frac{|\psi^0_{\text{at}}(\mathbf{p})|}{L^3} \frac{\delta_{\mathbf{k}_1,-\mathbf{k}_2}}{\sqrt{(2\omega_1)(2\omega_2)}} |\boldsymbol{\eta}_1 \cdot (\boldsymbol{\varepsilon}^{(1)} \times \boldsymbol{\varepsilon}^{(2)})| |\langle S|_p \sigma_2 |S\rangle_e| \, . \tag{7.100}$$

As usual, $\boldsymbol{\eta}_1$ denotes the direction of the $\mathbf{k}_1$-vector, $\mathbf{k}_1 = \omega_1 \boldsymbol{\eta}_1$.
Positronium annihilation rate is given by the Fermi golden rule (6.177)

$$w = \frac{2\pi}{2} \sum_{\mathbf{k}_1,\mathbf{k}_2,\boldsymbol{\varepsilon}^{(1)},\boldsymbol{\varepsilon}^{(2)}} \delta(E_0 - \omega_1 - \omega_2)|b_2|^2 \, , \tag{7.101}$$

where the factor one half comes from indistinguishability of photons and the summation runs over all possible photon wave and polarization vectors.

7. *Transition to continuous basis*
We now take the limit $L \to \infty$, which produces the following changes:

$$\sum_{\mathbf{p}} \frac{\psi^0_{\text{at}}(\mathbf{p})}{L^{3/2}} \to \int d^3\mathbf{p} \frac{\psi^0_{\text{at}}(\mathbf{p})}{(2\pi)^{3/2}} = \psi^0_{\text{at}}(\mathbf{r} = 0) \, , \tag{7.102}$$

[12]The first identity is most easily proved by means of Eq. (1.35) and the explicit form of the spin matrices, Eqs. (1.21), (1.22) and (1.23). For instance, the lhs equals $\frac{1}{2}\left[(\sigma_+)_{ab}(\sigma_-)_{dc} + (\sigma_-)_{ab}(\sigma_+)_{dc}\right] + (\sigma_3)_{ab}(\sigma_3)_{dc}$. Further, $(\sigma_+)_{ab} = 2\delta_{a1}\delta_{b2}$, $(\sigma_-)_{ab} = 2\delta_{a2}\delta_{b1}$ and $(\sigma_3)_{ab} = \delta_{a1}\delta_{b1} - \delta_{a2}\delta_{b2}$. Indices a, b, c, d equal only 1 or 2, therefore, for instance, $\delta_{ac} = \delta_{a1}\delta_{c1} + \delta_{a2}\delta_{c2}$.

where we used Eq. (7.84) in the last step, and

$$\sum_{\mathbf{k}_1,\mathbf{k}_2} \frac{\delta_{\mathbf{k}_1,-\mathbf{k}_2}}{L^3} \to \int \mathrm{d}^3\mathbf{k}_1 \mathrm{d}^3\mathbf{k}_2 \delta(\mathbf{k}_1+\mathbf{k}_2) \frac{1}{(2\pi)^3} . \tag{7.103}$$

8. *Integration over the wave vectors*
After inserting the replacements (7.103) and Eq. (7.100) into Eq. (7.101), we integrate over the "phase space of the photon final states"[13]

$$\frac{1}{2}\int \mathrm{d}^3\mathbf{k}_1 \mathrm{d}^3\mathbf{k}_2 \delta(\mathbf{k}_1+\mathbf{k}_2) \frac{1}{(2\pi)^3} \frac{1}{2\omega_1 2\omega_2} 2\pi\delta(E_0-\omega_1-\omega_2) f(\boldsymbol{\eta}_1) \tag{7.104}$$

$$= \frac{1}{2}\int \mathrm{d}^3\mathbf{k}_1 \frac{1}{2^2(2\pi)^2\omega_1^2} \delta(E_0-2\omega_1) f(\boldsymbol{\eta}_1) = \int \frac{\mathrm{d}\Omega_1}{64\pi^2} f(\boldsymbol{\eta}_1) ,$$

where in the last equation we introduced spherical coordinates in the wave vector space.
9. *Virtual annihilation—the final result*
After substitution of Eq. (7.102) into Eq. (7.99), we find

$$E_2 = \frac{\pi\alpha}{m^2} |\psi^0_{\mathrm{at}}(\mathbf{r}=0)|^2 \left\langle \hat{\mathrm{S}}^2 \right\rangle , \tag{7.105}$$

where we introduced total positronium spin operator $\hat{\boldsymbol{S}} = \hat{\boldsymbol{S}}_e + \hat{\boldsymbol{S}}_p$. Transition to the atomic units $\mathbf{r} = \mathbf{r}_A/m_r\alpha$, $m_r = m/2$, and value of the positronium wave function at the origin, see Eq. (4.101),

$$|\psi^0_{\mathrm{at}}(\mathbf{r}=0)|^2 = (m_r\alpha)^3 |\psi^0_{\mathrm{at}}(\mathbf{r}_A=0)|^2 = \left(\frac{m\alpha}{2}\right)^3 \frac{1}{4\pi} \frac{4}{n^3} \delta_{l,0} \tag{7.106}$$

lead to the fine-splitting between the singlet and triplet positronium ground states

$$\nu_{\mathrm{theo}}(1^3s - 1^1s) = \frac{2R_\infty c\alpha^2}{8}\left(\frac{8}{3}+2\right) = R_\infty c\alpha^2 \frac{7}{6} = 204.4\ \mathrm{GHz} , \tag{7.107}$$

where $8/3$ in the bracket stems from the spin-spin interaction, Eq. (3.58), and the number 2 from the virtual annihilation, Eq. (7.105). In the last equality, we used Eq. (3.12). Our theoretical prediction is in a satisfying agreement with the experimental value [25]

$$\nu_{\mathrm{exp}}(1^3s - 1^1s) = 203.3875(16)\ \mathrm{GHz} . \tag{7.108}$$

[13] Note that $\int \mathrm{d}\omega_1 \delta(E_0 - 2\omega_1) = 1/2$.

It is worth noting the difference between this splitting in positronium and the one in hydrogen, Eq. (3.59), namely notice the substantially greater value in case of the former. Why is it so?

10. *Real annihilation—final result*
By substituting Eqs. (7.100), (7.102), (7.103), and (7.104) into Eq. (7.101), we obtain for the positronium annihilation rate

$$w = \frac{e^4}{16\pi m^2}|\psi^0_{\rm at}(\mathbf{r}=0)|^2 \int \frac{\mathrm{d}\Omega_1}{4\pi} \sum_{\boldsymbol{\varepsilon}^{(1)},\boldsymbol{\varepsilon}^{(2)}} |\boldsymbol{\eta}_1\cdot(\boldsymbol{\varepsilon}^{(1)}\times\boldsymbol{\varepsilon}^{(2)})|^2 |\langle S|_p\sigma_2|S\rangle_e|^2 \,. \tag{7.109}$$

To perform the summation over polarization, we note that the expression $\boldsymbol{\eta}_1 \cdot (\boldsymbol{\varepsilon}^{(1)}\times\boldsymbol{\varepsilon}^{(2)})$ is a number independent of the coordinate system. Therefore, we evaluate it in the frame (6.31), $\boldsymbol{\eta}_1 = (0,0,1)$. In case the first photon is polarized along the x-axis, $\boldsymbol{\varepsilon}^{(1)} = (1,0,0)$, the second photon must be polarized along the y-axis, $\boldsymbol{\varepsilon}^{(2)} = (0,1,0)$, and vice versa. We sum over both possibilities to find

$$\sum_{\boldsymbol{\varepsilon}^{(1)},\boldsymbol{\varepsilon}^{(2)}} |\boldsymbol{\eta}_1\cdot(\boldsymbol{\varepsilon}^{(1)}\times\boldsymbol{\varepsilon}^{(2)})|^2 = 2\,. \tag{7.110}$$

We should be more specific regarding the spin dependence of the positronium state. Instead of the state (7.87), where the spin part was completely unspecified, we should write the singlet positronium state in the form

$$|\psi^0_{\rm at}\rangle = \sum_{\mathbf{p}} \psi^0_{\rm at}(\mathbf{p}) \frac{1}{\sqrt{2}} \left(\hat{\mathsf{b}}^+_{\mathbf{p},+}\hat{\mathsf{d}}^+_{-\mathbf{p},-} - \hat{\mathsf{b}}^+_{\mathbf{p},-}\hat{\mathsf{d}}^+_{-\mathbf{p},+} \right) |0\rangle\,,$$

where the additional signs specify the spin orientations with respect to the z-axis. Calculation is then identical to the one outlined above but for the replacement

$$\left|\langle S|_p\sigma_2|S\rangle_e\right|^2 \longrightarrow \left|\frac{1}{\sqrt{2}}\left(\langle -|\sigma_2|+\rangle - \langle +|\sigma_2|-\rangle\right)\right|^2 = 2\,. \tag{7.111}$$

The reader can easily convince himself that for an arbitrary triplet state, the probability of two-photon annihilation vanishes and this statement holds regardless of approximations made. For a general proof, see, e.g., [15].[14] By

[14] The general proof is based on the recognition that the electromagnetic interactions are invariant with respect to the replacement of particles by antiparticles. Hence, if the weak force is omitted, the determination of positive and negative charge is purely a matter of convention, like the notion of the left and right-hand side. Analogously to the conservation of "normal" parity, the conservation of "charge" parity holds. The two-photon state is of even charge parity, while the triplet positronium s-state is of odd charge parity. With an even charge parity Hamiltonian, the latter cannot make transition to the former and vice versa. In more detail, a general singlet and triplet positronium states can be written in a good approximation in the form

substituting Eqs. (7.106), (7.110), and (7.111) into Eq. (7.109), we find the lifetime of the positronium ground state

$$\tau_{\text{theo}}(1^1s) = \frac{2}{m\alpha^5} \to \frac{1}{2R_\infty c\alpha^3\pi} = 1.245 \times 10^{-10}\ \text{s},$$

where the replacement (6.101) was made. This value is in a good agreement with the experimental value [1]

$$\tau_{\text{exp}}(1^1s) = 1.251(2) \times 10^{-10}\ \text{s}.$$

The triplet state 1^3s annihilates into three photons; to describe this process, one has to reach for the third order of the perturbation method, see e.g. [2, 15]. Lifetime of the lowest triplet state is approximately 10^{-7} s.

$$\sum_{\mathbf{p}} \psi^0_{\text{at}}(\mathbf{p}) \frac{1}{\sqrt{2}} \left(\hat{\mathsf{b}}^+_{\mathbf{p},+} \hat{\mathsf{d}}^+_{-\mathbf{p},-} - \hat{\mathsf{b}}^+_{\mathbf{p},-} \hat{\mathsf{d}}^+_{-\mathbf{p},+} \right) |0\rangle$$

and

$$\sum_{\mathbf{p}} \psi^0_{\text{at}}(\mathbf{p}) \hat{\mathsf{b}}^+_{\mathbf{p}} \hat{\mathsf{d}}^+_{-\mathbf{p}} |0\rangle,$$

respectively. Charge conjugation swaps electron states to positronium states and vice versa

$$\sum_{\mathbf{p}} \psi^0_{\text{at}}(\mathbf{p}) \frac{1}{\sqrt{2}} \left(\hat{\mathsf{d}}^+_{\mathbf{p},+} \hat{\mathsf{b}}^+_{-\mathbf{p},-} - \hat{\mathsf{d}}^+_{\mathbf{p},-} \hat{\mathsf{b}}^+_{-\mathbf{p},+} \right) |0\rangle$$
$$= \sum_{\mathbf{p}} \psi^0_{\text{at}}(-\mathbf{p}) \frac{1}{\sqrt{2}} \left(\hat{\mathsf{b}}^+_{\mathbf{p},+} \hat{\mathsf{d}}^+_{-\mathbf{p},-} - \hat{\mathsf{b}}^+_{\mathbf{p},-} \hat{\mathsf{d}}^+_{-\mathbf{p},+} \right) |0\rangle$$

and

$$\sum_{\mathbf{p}} \psi^0_{\text{at}}(\mathbf{p}) \hat{\mathsf{d}}^+_{\mathbf{p}} \hat{\mathsf{b}}^+_{-\mathbf{p}} |0\rangle = -\sum_{\mathbf{p}} \psi^0_{\text{at}}(-\mathbf{p}) \hat{\mathsf{b}}^+_{\mathbf{p}} \hat{\mathsf{d}}^+_{-\mathbf{p}} |0\rangle,$$

where we used the anticommutativity of $\hat{\mathsf{b}}^+$ and $\hat{\mathsf{d}}^+$ operators and made the replacement $\mathbf{p} \to -\mathbf{p}$. The reader will easily convince himself that if the wave function is even/odd in the coordinate representation, it is even/odd in momentum representation as well. Hence it should be clear, see Eq. (5.60), that the charge parity of positronium is

$$(-1)^{\Pi_C} = (-1)^{l+s},$$

where l and s are the total orbital and spin angular momenta. The charge parity of the vector potential and consequently of the one-photon state is odd; a change of the charge sign, $e \to -e$, leads to a change of the potential sign, $\hat{\boldsymbol{A}} \to -\hat{\boldsymbol{A}}$, see the Maxwell equations (6.11). The charge parity of the n-photon state is

$$(-1)^{\Pi_C} = (-1)^n.$$

Since $(\hat{\boldsymbol{j}})_c = -\hat{\boldsymbol{j}}$, see Eq. (7.60), the interaction Hamiltonian (7.62) is of even charge parity.

7.4 Feynman Space-Time Approach

7.4.1 Electron in an External EM Field

Feynman's starting point was to turn away from the notion of a quantized Dirac field.[15] Instead, he created his own approach based on one-particle interpretation of the Dirac equation.[16] We have already seen that this interpretation encounters difficulties when it comes to negative energies, see Sect. 7.1.5. However, note that this problem arises merely from the separation of the space and time variables. Therefore, if we follow the electron evolution in the whole space-time, we might better understand what negative energies correspond to and how to interpret them properly. Let us thus consider electron motion in an external, generally time-dependent, EM field, see Eq. (7.44),

$$\mathrm{i}\frac{\mathrm{d}}{\mathrm{d}t}|\psi\rangle = (\hat{\mathsf{H}}_0 + e\hat{\mathsf{H}}_1)|\psi\rangle\,, \tag{7.112}$$

$$\hat{\mathsf{H}}_0 = \gamma_0\boldsymbol{\gamma}\cdot\hat{\boldsymbol{p}} + \gamma_0 m\,, \quad \hat{\mathsf{H}}_1 = -\gamma_0\boldsymbol{\gamma}\cdot\mathbf{A}(t,\hat{\boldsymbol{r}})\,,$$

where $\hat{\mathsf{H}}_0$ is the free-electron Hamiltonian and $\hat{\mathsf{H}}_1$ describes interaction of the electron with the external classical EM field.[17]

A particle state at time t is determined by the action of an evolution operator, *propagator* $\hat{\mathsf{U}}$, on the particle state at time t_0, see Eq. (1.14),

$$|\psi(t)\rangle = \hat{\mathsf{U}}(t,t_0)|\psi(t_0)\rangle\,. \tag{7.113}$$

After substituting Eq. (7.113) into Eq. (7.112), we obtain a differential equation for the propagator[18]

$$\left(\mathrm{i}\frac{\mathrm{d}}{\mathrm{d}t} - \hat{\mathsf{H}}_0\right)\hat{\mathsf{U}}(t,t_0) = e\hat{\mathsf{H}}_1(t)\hat{\mathsf{U}}(t,t_0) \tag{7.114}$$

---

[15] See [35, p. 456].

[16] Other expositions of the Feynman approach we are aware of can be found either from the "master himself" [7] or in [3, 11, 33].

[17] Let us emphasize once again that the components of **A** are operators, but solely in the sense of the "first" quantization. The vector potential is a function of particle coordinates $\mathbf{A} = \mathbf{A}(t,\hat{\boldsymbol{r}})$, which are represented by operators and do not generally commute with the components of the momentum operator. Hence the components of $\hat{\boldsymbol{p}}$ and **A** generally do not commute with each other. In the next Sect. 7.4.2 we show that the interaction of the electron with the quantized EM field can be brought to the form very similar to Eq. (7.112).

[18] We should write $\hat{\mathsf{H}}_1 = \hat{\mathsf{H}}_1(t,\hat{\boldsymbol{r}})$, but let us omit the space dependence for now.

and solve it by the method of *Green functions*.[19] We attempt to find a function obeying the equation

$$\left(\mathrm{i}\frac{\mathrm{d}}{\mathrm{d}t} - \hat{\mathsf{H}}_0\right)\hat{\mathsf{G}}_0(t, t_0) = \delta(t - t_0)\,. \tag{7.115}$$

Once we have $\hat{\mathsf{G}}_0$, the solution to Eq. (7.114) is obtained by integration

$$\hat{\mathsf{U}}(t, t_0) = \hat{\mathsf{U}}_0(t, t_0) + e\int_{t_0}^{\infty}\hat{\mathsf{G}}_0(t, t')\hat{\mathsf{H}}_1(t')\hat{\mathsf{U}}(t', t_0)\mathrm{d}t'\,, \tag{7.116}$$

where $\hat{\mathsf{U}}_0$ is an evolution operator of a free particle

$$\left(\mathrm{i}\frac{\mathrm{d}}{\mathrm{d}t} - \hat{\mathsf{H}}_0\right)\hat{\mathsf{U}}_0(t, t_0) = 0\,, \quad \hat{\mathsf{U}}_0(t_0, t_0) = 1\,. \tag{7.117}$$

To verify that the expression (7.116) indeed solves Eq. (7.114), we act with the operator $\left(\mathrm{i}\frac{\mathrm{d}}{\mathrm{d}t} - \hat{\mathsf{H}}_0\right)$ on both sides of Eq. (7.116)

$$\left(\mathrm{i}\frac{\mathrm{d}}{\mathrm{d}t} - \hat{\mathsf{H}}_0\right)\hat{\mathsf{U}}(t, t_0) = \underbrace{\left(\mathrm{i}\frac{\mathrm{d}}{\mathrm{d}t} - \hat{\mathsf{H}}_0\right)\hat{\mathsf{U}}_0(t, t_0)}_{0}$$
$$+e\int_{t_0}^{\infty}\underbrace{\left(\mathrm{i}\frac{\mathrm{d}}{\mathrm{d}t} - \hat{\mathsf{H}}_0\right)\hat{\mathsf{G}}_0(t, t')}_{\delta(t-t')}\hat{\mathsf{H}}_1(t')\hat{\mathsf{U}}(t', t_0)\mathrm{d}t' = e\hat{\mathsf{H}}_1(t)\hat{\mathsf{U}}(t, t_0).$$

How are $\hat{\mathsf{G}}_0(t, t_0)$ and $\hat{\mathsf{U}}(t, t_0)$ related to each other? There are two basic possibilities[20]

$$\hat{\mathsf{G}}_0^{(\pm)}(t, t_0) = \mp\mathrm{i}\theta(\pm t \mp t_0)\hat{\mathsf{U}}_0(t, t_0)\,, \tag{7.118}$$

as can be easily verified by substituting directly into Eq. (7.115) and using Eq. (7.117). These two basic cases are referred to as *retarded* and *advanced* Green functions.

Equation (7.116) can be solved iteratively: we substitute $\hat{\mathsf{U}}_0$ for $\hat{\mathsf{U}}$ on the rhs and denote the result as $\hat{\mathsf{U}}_1$,

$$\hat{\mathsf{U}}_1(t, t_0) = e\int_{t_0}^{\infty}\hat{\mathsf{G}}_0(t, t')\hat{\mathsf{H}}_1(t')\hat{\mathsf{U}}_0(t', t_0)\mathrm{d}t'\,,$$

[19] In the following text, we will not distinguish between the Green *function* and Green *operator*. We will deal mainly with the Green operator, however, we will often refer to it as the Green function.

[20] The Heaviside step-function is defined as $\theta(x) = 1$ for $x > 0$ and $\theta(x) = 0$ for $x < 0$.

and again substitute on the rhs (7.116) and denote the result as $\hat{\mathsf{U}}_2$,

$$\hat{\mathsf{U}}_2(t,t_0) = e\int_{t_0}^{\infty} \hat{\mathsf{G}}_0(t,t'')\hat{\mathsf{H}}_1(t'')\hat{\mathsf{U}}_1(t'',t_0)\mathrm{d}t''$$
$$= e^2\int_{t_0}^{\infty}\int_{t_0}^{\infty} \hat{\mathsf{G}}_0(t,t'')\hat{\mathsf{H}}_1(t'')\hat{\mathsf{G}}_0(t'',t')\hat{\mathsf{H}}_1(t')\hat{\mathsf{U}}_0(t',t_0)\mathrm{d}t'\mathrm{d}t'' \tag{7.119}$$

and so on. Subsequently, we use spectral decomposition of the interaction Hamiltonian $\hat{\mathsf{H}}_1$,

$$\hat{\mathsf{H}}_1(t') = \int H_1(t',\mathbf{r})|\mathbf{r}\rangle\langle\mathbf{r}|\,\mathrm{d}^3\mathbf{r}\,,$$

and insert it into Eq. (7.119). We project this equation onto coordinate eigenstates:

$$U_2(\boldsymbol{x},\boldsymbol{x}_0) = -e^2\int\int \mathrm{d}^4\boldsymbol{x}'\mathrm{d}^4\boldsymbol{x}''G_0(\boldsymbol{x},\boldsymbol{x}'')H_1(\boldsymbol{x}'')G_0(\boldsymbol{x}'',\boldsymbol{x}')H_1(\boldsymbol{x}')U_0(\boldsymbol{x}',\boldsymbol{x}_0)\,, \tag{7.120}$$

where

$$U_2(\boldsymbol{x},\boldsymbol{x}_0) = \langle\mathbf{r}|\hat{\mathsf{U}}_2(t,t_0)|\mathbf{r}_0\rangle$$

and

$$G_0(\boldsymbol{x},\boldsymbol{x}_0) = \langle\mathbf{r}|\hat{\mathsf{G}}_0(t,t_0)|\mathbf{r}_0\rangle\,, \qquad U_0(\boldsymbol{x},\boldsymbol{x}_0) = \langle\mathbf{r}|\hat{\mathsf{U}}_0(t,t_0)|\mathbf{r}_0\rangle\,.$$

If we use the retarded Green operator, insertion of Eq. (7.118) into Eq. (7.120) yields

$$U_2(\boldsymbol{x},\boldsymbol{x}_0) = -e^2\int_{t_0}^{t}\mathrm{d}t'\int_{t_0}^{t'}\mathrm{d}t''\int\mathrm{d}^3\mathbf{r}'\int\mathrm{d}^3\mathbf{r}''U_0(\boldsymbol{x},\boldsymbol{x}'')H_1(\boldsymbol{x}'')U_0(\boldsymbol{x}'',\boldsymbol{x}')H_1(\boldsymbol{x}')U_0(\boldsymbol{x}',\boldsymbol{x}_0)\,.$$

The last equation allows for a simple physical interpretation, see the left diagram of Fig. 7.3. It is the probability amplitude that a particle originally at time t_0 at the point $\mathbf{r}_0$ will be found at later time t at the point $\mathbf{r}$. Between these two events, the particle interacts twice with the external EM field: firstly at $(t',\mathbf{r}')$, secondly at $(t'',\mathbf{r}'')$, $t''>t'$. These two events can occur anytime between t_0 and t and anywhere in space. Thus, one has to integrate over all possibilities, which brings surprising consequences.

The inequality $t'<t''$ stemming from the choice of the retarded Green function enforces definite ordering of events, which in turn restricts the integration over t. The integration itself does not give rise to any problems, but quantum mechanics forces us to integrate over all possible space coordinates $\mathbf{r}'$ and $\mathbf{r}''$. Consequently, a possibility arises that the event at $(t'',\mathbf{r}'')$ will be located out of the light cone outgoing from $(t',\mathbf{r}')$. In the special-relativistic jargon, the events will be space-like

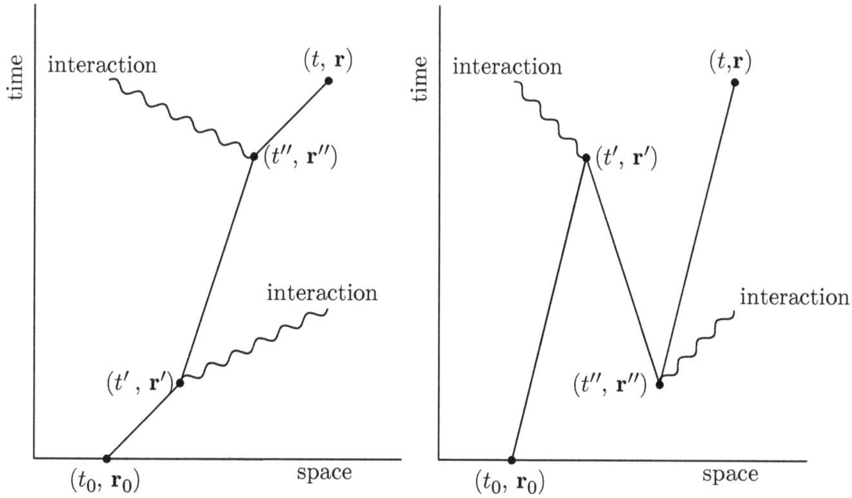

Fig. 7.3 Second order propagator

separated. One can then find an inertial frame where the time ordering of the events will be opposite, $t' > t''$![21] However, this means that an observer in such an inertial frame will see the particle moving backwards in time! How does such a particle look like? The answer is very simple: the only difference with respect to a particle moving forward in time is the sign of the charge current. There are two ways how one can perform the change of the current sign: we change either the charge sign or the time flow:

$$I = \frac{\mathrm{d}Q}{\mathrm{d}t} \Rightarrow \frac{\mathrm{d}Q}{\mathrm{d}(-t)} = -I = \frac{\mathrm{d}(-Q)}{\mathrm{d}t} \, .$$

Hence a particle moving backward in time with charge Q can be interpreted as a particle moving forward in time with charge $-Q$, that is, as an antiparticle! Note that this new route, due to Feynman, leads to the existence of antiparticles as well and yet it follows solely from the basic principles of quantum mechanics and the special theory of relativity. According to the latter, a massive particle can move only inside its light cone. However, the very notion of the light cone assumes that one is able to specify exactly the particle's coordinate and velocity at a given time instant. Quantum mechanics forbids that, though: the light cone is always slightly "blurred." Hence there is a small but nonvanishing probability that a particle in a virtual state leaves the light cone. According to special theory of relativity then, there is an inertial frame where the particle moves backward in time. Thus to every massive particle, there is an antiparticle—a particle with the same mass, spin, etc. and opposite charge.

[21] See literature on the special theory of relativity, for instance [39].

Now, roughly speaking, if observers in different inertial frames should agree on what they see,[22] one needs to choose the Green function in such a way that it leads to the following physical picture. The transition amplitude for a particle evolving from the initial incidence $\boldsymbol{x}_0$ to the final incidence $\boldsymbol{x}$ and interacting twice with the external EM field is a sum of two amplitudes. The first amplitude corresponds to the process where the particle interacts with the EM field firstly at time t', then moves forward in time and in later time t'' interacts with EM field for the second time. The second amplitude corresponds to a process with the opposite time ordering: the particle moves between events $\boldsymbol{x}'$ and $\boldsymbol{x}''$ backward in time, see the right diagram of Fig. 7.3. One can interpret this second process that a particle-antiparticle pair is created at time t'', the antiparticle then annihilates the "original" particle at time t', while the particle born at time t'' replaces the "original" particle. However, the choice of the retarded Green function does not lead to this picture; it enforces the particular time ordering $t'' > t'$. Note that if we do not choose in Eq. (7.120) the time direction by the specific choice of the retarded Green function, Eq. (7.120) allows for both time orderings, $t'' > t'$ as well as $t'' < t'$. What Green function should we then opt for?

Feynman suggested the following solution. The positive energy states correspond to particle states and particles move forward in time. Now, we recall Sect. 7.2.3: in the Majorana representation of the γ-matrices, the interchange of a charge sign for free particle corresponds to the interchange of the energy sign. To see this it suffices to take complex conjugate to

$$E\psi^M = [-\mathrm{i}\gamma_0^M \boldsymbol{\gamma}^M \cdot \nabla + \gamma_0^M m]\psi^M ;$$

in Majorana representation $\gamma_0^M \boldsymbol{\gamma}^M$ is real; thus

$$-E(\psi^M)^* = [-\mathrm{i}\gamma_0^M \boldsymbol{\gamma}^M \cdot \nabla + \gamma_0^M m](\psi^M)^* .$$

Hence, the negative energy states correspond to antiparticle states and antiparticles move backward in time. In summary, we are to consider the retarded Green function for positive energy solutions and the advanced Green function for negative energy solutions. We will now show how to do so in detail.

We start with Eq. (7.115) and insert the Fourier transforms

$$\hat{\mathsf{G}}_0(t, t_0) = \frac{1}{2\pi} \int \hat{\mathsf{G}}_0(E) \mathrm{e}^{-\mathrm{i}E(t-t_0)} \mathrm{d}E , \qquad \delta(t - t_0) = \frac{1}{2\pi} \int \mathrm{e}^{-\mathrm{i}E(t-t_0)} \mathrm{d}E$$

therein. For the Fourier transform of the Green operator, we find

$$\hat{\mathsf{G}}_0(E) = \frac{1}{E - \hat{\mathsf{H}}_0} .$$

[22] More precisely, if one shall arrive at a Lorentz-invariant transition amplitude between two events.

Using spectral decomposition of the Hamiltonian $\hat{\mathsf{H}}_0$, the Green operator reads

$$\hat{\mathsf{G}}_0(t, t_0) = \frac{1}{2\pi} \int \mathrm{d}E \sum_n |n\rangle \frac{\mathrm{e}^{-\mathrm{i}E(t-t_0)}}{E - E_n} \langle n| \,, \tag{7.121}$$

where $\hat{\mathsf{H}}_0|n\rangle = E_n|n\rangle$. For the moment, we assume that appropriate boundary conditions are adjusted in such a way that the spectrum of $\hat{\mathsf{H}}_0$ is countable. The integral over the energies hits a large number of poles and its result depends on the way we encircle the poles: the Green operator is not determined uniquely by Eq. (7.121), as we already know, see Eq. (7.118) and the discussion concerning it. In order to determine uniquely how to encircle the poles, we add positive or negative infinitesimal imaginary parts to the poles, $E_n \to E_n \mp \mathrm{i}\varepsilon$. As we will shortly show, the choice of infinitesimal negative or positive imaginary parts leads to the choice of retarded or advanced Green operator! The uniquely determined Green operator has the form

$$\hat{\mathsf{G}}_0^{(\pm)}(t, t_0) = \frac{1}{2\pi} \lim_{\varepsilon \to 0+} \int \mathrm{d}E \sum_n |n\rangle \frac{\mathrm{e}^{-\mathrm{i}E(t-t_0)}}{E - (E_n \mp \mathrm{i}\varepsilon)} \langle n| \,. \tag{7.122}$$

To be specific, we now consider the case of the upper sign. If we integrate along the real axis and close the contour by a semi-circle in the upper half of the complex plane $E \in \mathbb{C}$, $\Im E > 0$, the integral is convergent for $t < t_0$ and its value vanishes owing to the residue theorem as there are no singularities inside the integration contour. On the other hand, if we calculate the integral for $t > t_0$, we have to close the integration contour in the lower half of the complex plane where $\Im E < 0$, see the upper diagram of Fig. 7.4. There are now many poles inside the integration contour, their positions being $E_n - \mathrm{i}\varepsilon$, and the residue of the integrated function in each of the poles equals $\mathrm{e}^{-\mathrm{i}E_n(t-t_0)}$. Thus by virtue of the residue theorem

$$\oint f(z)\mathrm{d}z = 2\pi\mathrm{i} \sum_i \mathrm{Res}_{z_i} f \,, \tag{7.123}$$

we find from Eq. (7.122) the following expression for the Green functions

$$\hat{\mathsf{G}}_0^{(+)}(t, t_0) = \begin{cases} 0 & t < t_0, \\ \sum_n \mathrm{e}^{-\mathrm{i}E_n(t-t_0)}(-\mathrm{i})|n\rangle\langle n| & t > t_0 \end{cases} \tag{7.124}$$

and

$$\hat{\mathsf{G}}_0^{(-)}(t, t_0) = \begin{cases} \sum_n \mathrm{e}^{-\mathrm{i}E_n(t-t_0)}(+\mathrm{i})|n\rangle\langle n| & t < t_0, \\ 0 & t > t_0. \end{cases} \tag{7.125}$$

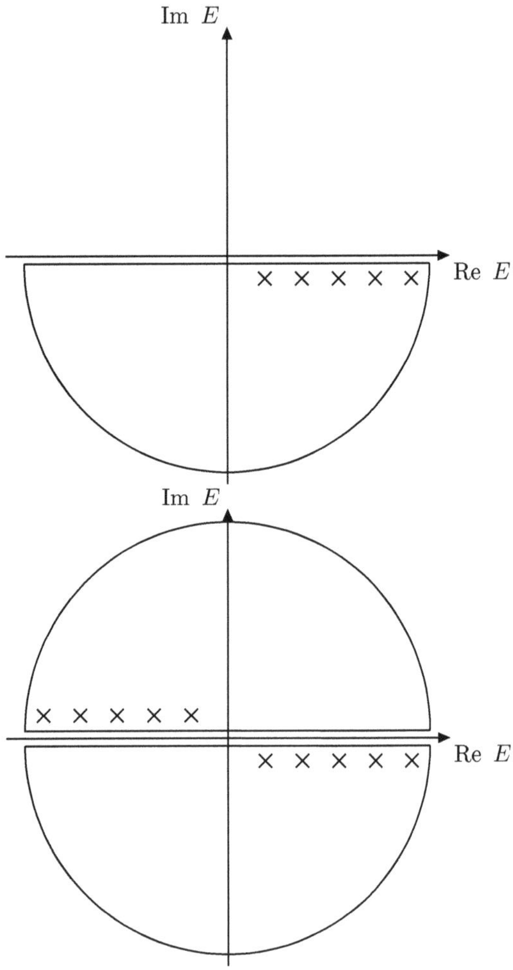

Fig. 7.4 Integration contours for the Green functions (retarded and Feynman, respectively)

The negative sign of $\hat{\mathsf{G}}_0^{(+)}$ originates from the direction of the integration contour. When applying the residue theorem, Eq. (7.123), we set the counterclockwise direction as positive. Hence for $\hat{\mathsf{G}}_0^{(+)}$ we close the contour along the real axis from the right to the left. The advanced Green operator is tackled similarly, the only difference is the poles are moved into the upper half of the complex plane now, which enforces the opposite direction of the time flow. Notice that Eqs. (7.124) and (7.125) are the same as Eq. (7.118).

We have not discussed the sign of the energies of the Hamiltonian $\hat{\mathsf{H}}_0$ yet. If the energy spectrum is positive, nothing needs to be added. If $\hat{\mathsf{H}}_0$ is the one-particle Dirac Hamiltonian (7.34), states with positive energies correspond to a particle moving forward in time and we add infinitesimal negative imaginary part to the poles lying on the positive real axis. States with negative energies correspond to a particle moving backward in time and we add infinitesimal positive imaginary part to the poles lying on the negative real axis. We thus arrive at the *Feynman form of the electron Green function*

$$\hat{\mathsf{G}}_0^{(F)}(t,t_0) = \frac{1}{2\pi}\int \frac{\mathrm{e}^{-\mathrm{i}E(t-t_0)}}{E-\hat{\mathsf{H}}_0}\mathrm{d}E \tag{7.126}$$

$$= \frac{1}{2\pi}\lim_{\varepsilon\to 0+}\int \mathrm{d}E \sum_n \left(\frac{\hat{\mathsf{P}}^{n+}}{E-(E_{n+}-\mathrm{i}\varepsilon)} + \frac{\hat{\mathsf{P}}^{n-}}{E-(E_{n-}+\mathrm{i}\varepsilon)}\right)\mathrm{e}^{-\mathrm{i}E(t-t_0)}.$$

Recall that $\hat{\mathsf{P}}^{n\pm}$ are projectors onto the n-th eigenstate of $\hat{\mathsf{H}}_0$ with positive or negative energy, see Eq. (7.81). According to the sign of $t-t_0$, we integrate over upper or lower semicircle on the lower diagram of Fig. 7.4. We integrate over E to obtain from Eq. (7.126)

$$\hat{\mathsf{G}}_0^{(F)}(t,t_0) = -\mathrm{i}\left[\theta(t-t_0)\sum_{n+}\mathrm{e}^{-\mathrm{i}E_{n+}(t-t_0)}\hat{\mathsf{P}}^{n+} - \theta(t_0-t)\sum_{n-}\mathrm{e}^{-\mathrm{i}E_{n-}(t-t_0)}\hat{\mathsf{P}}^{n-}\right]. \tag{7.127}$$

We will show that this choice of the one-particle Green operator produces the same results as the theory based on the quantized Dirac field with the ground state (7.40) determined in accordance with the hole picture of positrons.

To sum up the foregoing considerations, the Feynman solution of the negative energy problem is as follows. The key idea is not to separate space and time in our considerations. If we follow the particle track in space-time, the positive-energy solutions of the Dirac equation are moving forward in time, while the negative-energy solutions backward in time. In terms of mathematics, this corresponds to the introduction of a new integration variable E and shifting the energies to the complex plane appropriately, see Eq. (7.126). Until integration over E is performed, the time flow between two virtual processes remains undetermined and we can treat both processes depicted in Fig. 7.3 in a unified manner.

In the next sections, we show how this point of view leads to a manifestly Lorentz-invariant expression for the process of emission and absorption of a virtual photon.

7.4.2 Electron Interacting with Its Own EM Field

With Feynman's interpretation of positrons as electrons moving backward in time, one can keep the one-particle interpretation of the Dirac equation. Let us now consider the Hamiltonian describing the electron interaction with a quantized EM field, Eq. (7.43). To apply the above introduced formalism of the Green functions to it, we need to eliminate the term corresponding to the energy of the EM field $:\hat{\mathsf{H}}_{\mathrm{EM}}:$

from the free Hamiltonian. This can be achieved by a unitary transformation of the state vector

$$|\psi\rangle = \mathrm{e}^{\mathrm{i}\hat{\mathsf{S}}}|\psi'\rangle,$$

where $\hat{\mathsf{S}}$ is a hermitian operator to be determined in a moment. Differentiating $|\psi\rangle$ with respect to time yields

$$\mathrm{i}\frac{\mathrm{d}}{\mathrm{d}t}|\psi\rangle = \mathrm{i}\left[\mathrm{i}\frac{\mathrm{d}\hat{\mathsf{S}}}{\mathrm{d}t}\mathrm{e}^{\mathrm{i}\hat{\mathsf{S}}}|\psi'\rangle + \mathrm{e}^{\mathrm{i}\hat{\mathsf{S}}}\frac{\mathrm{d}}{\mathrm{d}t}|\psi'\rangle\right].$$

At the same time, the state vector describing the electron and EM field evolves according to the Schrödinger equation

$$\mathrm{i}\frac{\mathrm{d}}{\mathrm{d}t}|\psi\rangle = (\hat{\mathsf{h}}_{\mathrm{D}} + :\hat{\mathsf{H}}_{\mathrm{EM}}: + e\hat{\mathsf{H}}_1)|\psi\rangle.$$

If we now choose $\mathrm{i}\hat{\mathsf{S}} = -\mathrm{i}:\hat{\mathsf{H}}_{\mathrm{EM}}:(t-t_0)$, we find

$$\mathrm{i}\frac{\mathrm{d}}{\mathrm{d}t}|\psi'\rangle = (\hat{\mathsf{h}}_{\mathrm{D}} + e\hat{\mathsf{H}}_1(t))|\psi'\rangle, \tag{7.128}$$

where

$$\hat{\mathsf{H}}_1(t) = \mathrm{e}^{-\mathrm{i}\hat{\mathsf{S}}}\hat{\mathsf{H}}_1\mathrm{e}^{\mathrm{i}\hat{\mathsf{S}}} = \exp(+\mathrm{i}:\hat{\mathsf{H}}_{\mathrm{EM}}:(t-t_0))\hat{\mathsf{H}}_1\exp(-\mathrm{i}:\hat{\mathsf{H}}_{\mathrm{EM}}:(t-t_0)). \tag{7.129}$$

Equation (7.128) now formally matches Eq. (7.112) and the introduced Green function formalism can be readily applied to it.

We consider the amplitude of the following process. The initial state $|\psi_{\mathrm{at}}\rangle|0\rangle$, describing an electron in a spin-orbital $|\psi_{\mathrm{at}}\rangle$ and the EM field in the ground state $|0\rangle$, evolves in time as the electron interacts twice with the transverse EM field and ends up in the original state. This amplitude is given by projection of Eq. (7.119) onto the initial and final states,

$$\mathscr{A}_\perp \simeq e^2\int_{t_0}^{\infty}\int_{t_0}^{\infty}\langle 0|\langle\psi_{\mathrm{at}}(t)|\hat{\mathsf{G}}_0^{(F)}(t,t'')\hat{\mathsf{H}}_1(t'')\hat{\mathsf{G}}_0^{(F)}(t'',t')\hat{\mathsf{H}}_1(t')\hat{\mathsf{U}}_0(t',t_0)|\psi_{\mathrm{at}}(t_0)\rangle|0\rangle\,\mathrm{d}t'\,\mathrm{d}t'', \tag{7.130}$$

where $\hat{\mathsf{H}}_1(t)$ is given by Eqs. (7.129) and (7.44):

$$e\hat{\mathsf{H}}_1(t) = -e\gamma_0\boldsymbol{\gamma}\cdot\int \mathrm{d}^3\mathbf{r}|\mathbf{r}\rangle\hat{\boldsymbol{A}}(t,\mathbf{r})\langle\mathbf{r}|, \qquad \hat{\boldsymbol{A}}(t,\mathbf{r}) = \mathrm{e}^{+\mathrm{i}:\hat{\mathsf{H}}_{\mathrm{EM}}:(t-t_0)}\hat{\boldsymbol{A}}(\mathbf{r})\mathrm{e}^{-\mathrm{i}:\hat{\mathsf{H}}_{\mathrm{EM}}:(t-t_0)}. \tag{7.131}$$

For $\hat{\mathsf{G}}_0$ we take the Feynman form of the Green operator (7.126) where for $\hat{\mathsf{H}}_0$ we consider the one-particle Dirac Hamiltonian $\hat{\mathsf{h}}_{\mathrm{D}}$, Eq. (7.34).

Exercise 26: Pair Annihilation à la Feynman

Try to rederive Eq. (7.96) using the Feynman space-time approach.

7.4.3 Photon Propagator and Time Ordered Operator Product

Now we have to recognize that we integrate over all possible times t' and t'' in Eq. (7.130). Therefore, when eliminating operators of the EM field, we have to distinguish between two situations: if $t'' > t'$ the operator $\hat{\mathsf{H}}_1(t')$ acts first and then the operator $\hat{\mathsf{H}}_1(t'')$, otherwise, i.e., if $t'' < t'$, the action of the operators is reversed. This means that after we substitute into Eq. (7.130) from Eq. (7.131), it is necessary to consider a *time ordered product* of two operators

$$\begin{aligned} D_{ij}(x'' - x') &= \langle 0|\mathscr{T}\left(\hat{\mathsf{A}}_i(t'', \mathbf{r}'')\hat{\mathsf{A}}_j(t', \mathbf{r}')\right)|0\rangle \\ &= \theta(t'' - t')\langle 0|\hat{\mathsf{A}}_i(t'', \mathbf{r}'')\hat{\mathsf{A}}_j(t', \mathbf{r}')|0\rangle \\ &\quad + \theta(t' - t'')\langle 0|\hat{\mathsf{A}}_j(t', \mathbf{r}')\hat{\mathsf{A}}_i(t'', \mathbf{r}'')|0\rangle \,, \end{aligned} \tag{7.132}$$

instead of a plain product of the two operators, $\langle 0|\left(\hat{\mathsf{A}}_i(t'', \mathbf{r}'')\hat{\mathsf{A}}_j(t', \mathbf{r}')\right)|0\rangle$. The physical reasoning behind this step is as follows. In an analogy to the amplitude of the process where an electron interacts twice with an external EM field, the amplitude of the process where an electron interacts twice with its own EM field is in fact a sum of two amplitudes. The corresponding processes are depicted in Fig. 7.5. The first amplitude corresponds to the process when an electron at time t' emits a photon, both the electron and the photon move forward in time, and at later time t'' the electron absorbs the photon back. The second amplitude corresponds to the process when the time ordering of the events is opposite: an electron emits a photon at time t', both the electron and photon move backward in time and at earlier time t'' the electron absorbs the photon back. This implies that once we consider the advanced solution for the electron, we have to consider the advanced solution for the photon as well! The second process may be "conventionally" interpreted that at time t'' the system of electromagnetic and electron-positron fields "borrows" energy out of nothing: an electron-positron pair and a photon are created from vacuum. The system then "returns" the energy back at later time t' as the photon is absorbed back and the positron annihilates the "original" electron. In this view, the backward propagation of the electron between the times t' and t'' is realized by a forward propagation of a positron between the times t'' and t'. The electron born at the time t'' then plays the role of the "original" electron. Looking now at the time ordered product (7.132), we see that in the first case the creation operator acts on the vacuum at time t' and the annihilation operator at time t''. In the second case, the creation operator acts at the

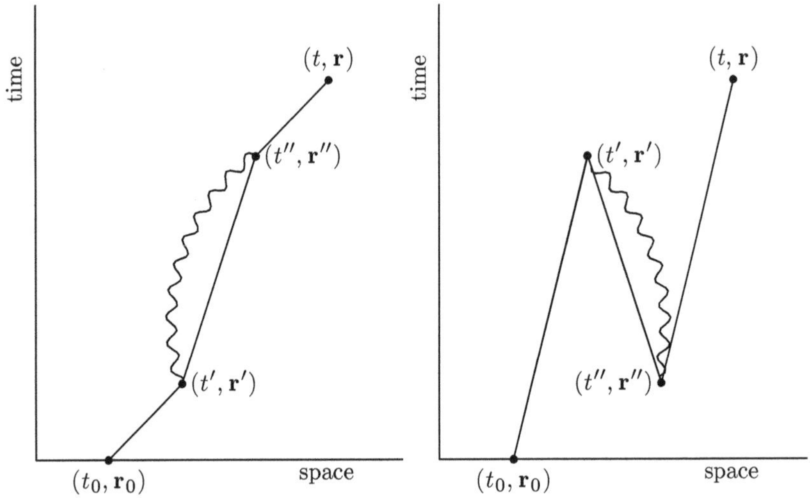

Fig. 7.5 Space-time ordering of emission and absorption of a photon by an electron

time t'' and the annihilation operator at time t'. Clearly, the expression (7.132) does correspond to the above-described physical picture.

Since $:\hat{\mathsf{H}}_{\mathrm{EM}}:|0\rangle = 0$, then also $\mathrm{e}^{-\mathrm{i}:\hat{\mathsf{H}}_{\mathrm{EM}}:(t-t_0)}|0\rangle = |0\rangle$. Hence Eq. (7.132) can be simplified to

$$D_{ij}(\boldsymbol{x}'' - \boldsymbol{x}') = \theta(t'' - t')\langle 0|\hat{\mathsf{A}}_i(\mathbf{r}'')\mathrm{e}^{-\mathrm{i}:\hat{\mathsf{H}}_{\mathrm{EM}}:(t''-t')}\hat{\mathsf{A}}_j(\mathbf{r}')|0\rangle$$
$$+ \theta(t' - t'')\langle 0|\hat{\mathsf{A}}_j(\mathbf{r}')\mathrm{e}^{-\mathrm{i}:\hat{\mathsf{H}}_{\mathrm{EM}}:(t'-t'')}\hat{\mathsf{A}}_i(\mathbf{r}'')|0\rangle .$$

If we now use the expansion (6.47), $\hat{\boldsymbol{A}}(\mathbf{r}) = \sum_\tau \frac{1}{\sqrt{2\omega_\tau}}\left[\hat{\mathsf{a}}_\tau \mathbf{T}_\tau(\mathbf{r}) + \hat{\mathsf{a}}_\tau^+ \mathbf{T}_{-\tau}(\mathbf{r})\right]$, we finally obtain

$$D_{ij}(\boldsymbol{x}'' - \boldsymbol{x}') = \theta(t'' - t')\sum_\tau \frac{1}{2\omega_\tau} T_{\tau_i}(\mathbf{r}'')T_{-\tau_j}(\mathbf{r}')\mathrm{e}^{-\mathrm{i}\omega_\tau(t''-t')}$$
$$+ \theta(t' - t'')\sum_\tau \frac{1}{2\omega_\tau} T_{\tau_j}(\mathbf{r}')T_{-\tau_i}(\mathbf{r}'')\mathrm{e}^{+\mathrm{i}\omega_\tau(t''-t')} .$$

Now we take the continuous limit: we replace the summation over the mode index τ by an integral over the momenta $\mathbf{k}$ and the summation over polarizations, Eq. (6.55); since we integrate over all directions we can replace $\mathbf{k} \to -\mathbf{k}$ in the second term,

$$D_{ij}(\boldsymbol{x}'' - \boldsymbol{x}') = \int \frac{1}{2\omega} P_{ij}\left(\theta(t'' - t')\mathrm{e}^{-\mathrm{i}\omega(t''-t')}\right.$$
$$\left. + \theta(t' - t'')\mathrm{e}^{\mathrm{i}\omega(t''-t')}\right)\mathrm{e}^{-\mathrm{i}\mathbf{k}\cdot(\mathbf{r}'-\mathbf{r}'')}\frac{\mathrm{d}^3\mathbf{k}}{(2\pi)^3} . \qquad (7.133)$$

Introducing now an additional integration variable, namely the time component k_0 of a four-vector $\boldsymbol{k}$, we can achieve, as in the case of electron, that both retarded and advanced waves are contained in only one expression:

$$\lim_{\varepsilon\to 0+} \mathrm{i}\int \frac{\mathrm{e}^{-\mathrm{i}k_0(t''-t')}}{(k_0-\omega+\mathrm{i}\varepsilon)(k_0+\omega-\mathrm{i}\varepsilon)}\frac{\mathrm{d}k_0}{2\pi} = \frac{1}{2\omega}\left(\theta(t''-t')\mathrm{e}^{-\mathrm{i}\omega(t''-t')}+\theta(t'-t'')\mathrm{e}^{\mathrm{i}\omega(t''-t')}\right).$$

Analogously to the electron case, we added infinitesimal negative and positive imaginary parts to positive and negative energies, respectively. For the vacuum expectation value of a time ordered product of vector potential components, commonly known as the photon propagator in the Coulomb gauge, we find

$$D_{ij}(\boldsymbol{x}''-\boldsymbol{x}') = \langle 0|\mathscr{T}\left(A_i(t'',\mathbf{r}'')A_j(t',\mathbf{r}')\right)|0\rangle = \mathrm{i}\int P_{ij}\frac{\mathrm{e}^{-\mathrm{i}\boldsymbol{k}\cdot(\boldsymbol{x}''-\boldsymbol{x}')}}{\boldsymbol{k}^2}\frac{\mathrm{d}^4\boldsymbol{k}}{(2\pi)^4}. \tag{7.134}$$

Here, $\boldsymbol{k}=(k_0,\mathbf{k})$ stands for a wave four-vector, the magnitude of its space component being $|\mathbf{k}|=\omega$. Notice that the last expression is identical with the Green function of the vector Maxwell equation (6.11).

7.4.4 Electron Self-energy via Green Functions

After substituting the interaction Hamiltonian, Eq. (7.131), and the photon propagator, Eq. (7.134), into Eq. (7.130), we arrive at

$$\mathscr{A}_\perp \simeq e^2\int\!\!\int \mathrm{d}^4\boldsymbol{x}''\mathrm{d}^4\boldsymbol{x}' D_{ij}(\boldsymbol{x}''-\boldsymbol{x}')\langle\psi_{\mathrm{at}}(t)|\hat{\mathsf{G}}_0^{(F)}(t,t'')|\mathbf{r}''\rangle\gamma_0\gamma_j \times \langle\mathbf{r}''|\hat{\mathsf{G}}_0^{(F)}|\mathbf{r}'\rangle\gamma_0\gamma_i\langle\mathbf{r}'|\hat{\mathsf{U}}_0(t',t_0)|\psi_{\mathrm{at}}(t_0)\rangle. \tag{7.135}$$

The action of the propagators $\hat{\mathsf{G}}_0^{(F)}$ and $\hat{\mathsf{U}}_0$ on ψ_{at} is found easily, see Eq. (7.126),

$$\langle\psi_{\mathrm{at}}(t)|\hat{\mathsf{G}}_0^{(F)}(t,t'') = \langle\psi_{\mathrm{at}}(t)|\int\frac{\mathrm{e}^{-\mathrm{i}E(t-t'')}}{E-\hat{\mathsf{h}}_{\mathrm{D}}}\frac{\mathrm{d}E}{2\pi} = \langle\psi_{\mathrm{at}}(t)|\int\frac{\mathrm{e}^{-\mathrm{i}E(t-t'')}}{E-E_0+\mathrm{i}\varepsilon}\frac{\mathrm{d}E}{2\pi} = -\mathrm{i}\theta(t-t'')\langle\psi_{\mathrm{at}}(t'')| \tag{7.136}$$

and Eq. (7.113)

$$\hat{\mathsf{U}}_0(t',t_0)|\psi_{\mathrm{at}}(t_0)\rangle = |\psi_{\mathrm{at}}(t')\rangle, \tag{7.137}$$

respectively. After substituting the last two equations into Eq. (7.135), we obtain

$$\mathscr{A}_\perp \simeq -\mathrm{i}e^2 \int\int \mathrm{d}^4x''\mathrm{d}^4x'\theta(t-t'')D_{ij}(x''-x')\psi^+_{\mathrm{at}}(x'')\gamma_0\gamma_i G_0^{(F)}(x'',x')\gamma_0\gamma_j\psi_{\mathrm{at}}(x')\,. \tag{7.138}$$

Now we come to the decisive point. We could distinguish the temporal ordering of the events at x' and x''. In such a case, we would insert the expressions (7.133) and (7.127) for the photon and electron propagators, respectively, into the last equation. It is not difficult to see then that, after integrating over the time variables t' and t'', we would arrive at an expression that matches the expression (7.83) but for an inessential multiplicative constant. We would thus reach the same dead end as earlier.

We do not have to distinguish the temporal ordering of the events at x' *and* x'', *though.* We can insert the expressions (7.134) and (7.126) into the last equation for the photon and electron propagators, respectively, where the temporal ordering of the virtual events is hidden in the way we integrate over k_0 and E, respectively. Thus we see that the crucial point is merely not to hurry with these integrations.

We insert $\psi_{\mathrm{at}}(x) = \psi(\mathbf{r})\mathrm{e}^{-\mathrm{i}E_0t}$ into Eq. (7.138) and factor out the time dependences coming from the initial and final state vectors and from the Green functions. At the limit $t_0 \to -\infty$, we obtain

$$\int_{-\infty}^{\infty} \mathrm{e}^{\mathrm{i}(k_0+E-E_0)t'}\mathrm{d}t' = 2\pi\delta(k_0+E-E_0)\,.$$

The integration over E is now trivial; the presence of the δ-function leads to the condition $k_0 + E - E_0 = 0$, which is the law of conservation of "energy" in virtual process. The electron "energy" E in virtual state is given by the difference between the initial electron energy E_0 and the virtual photon "energy" k_0. The integration over the second time variable t'' yields by the virtue of the previous equation[23]

$$\int_{t_0}^{t} \mathrm{e}^{-\mathrm{i}(k_0+E-E_0)t''}\mathrm{d}t'' = t - t_0 = T\,.$$

Hence the trivial integrations over t', t'', and E in Eq. (7.138) result in

$$\lim_{t_0\to-\infty}\mathscr{A}_\perp = Te^2\int \frac{\mathrm{d}^4k}{(2\pi)^4}\frac{1}{k^2}P_{ij}\langle\psi_{\mathrm{at}}|\gamma_0\gamma_i\mathrm{e}^{\mathrm{i}\mathbf{k}\cdot\hat{\mathbf{r}}}\frac{1}{E_0-\hat{\mathrm{h}}_{\mathrm{D}}-k_0}\mathrm{e}^{-\mathrm{i}\mathbf{k}\cdot\hat{\mathbf{r}}}\gamma_0\gamma_j|\psi_{\mathrm{at}}\rangle\,. \tag{7.139}$$

We have thus made a huge step in our striving for a manifestly Lorentz-invariant formulation of the perturbation method. In the last equation, we managed to find a unified expression for the both positive and negative energy modes, confront with Eq. (7.83). However, the way we reached the desired result goes against all our previous experience. We already knew that the use of symmetry always

[23]The reader may ask why we do not take the limit $t_0 \to -\infty$ in the integration over t'' as well. The reason is we want to avoid the infinity at this stage of calculation. This is justifiable as we will see later on, see Eq. (7.142), since this number drops out from the expression for the energy shift that is of physical relevance.

leads to a simplification of the problem at hand. However, until now, this use of symmetry resulted in a *reduction* of the dimensionality. In this case, by contrast, the simplification appears by an *increase* in the dimensionality. We avoided the separation of positive and negative energy modes by adding time into our considerations. By introducing additional integration variables E and k_0, we managed to keep the electron and photon propagators in the past-future symmetric forms.

7.4.5 Integration over k_0

Note that Eq. (7.139) contains no longer any time-dependent variables. However, we emphasized above that the way of encircling the poles in the complex plane is determined by the temporal ordering of the events. How do we proceed now when all of the time variables were integrated out? Everything is now hidden in the integration over k_0! Let us thus show now how it works. In Eq. (7.139), we encounter the integral

$$\frac{\mathrm{i}}{2\pi}\int\frac{\mathrm{d}k_0}{(k_0^2-\omega^2)(k_0-E_0+\hat{\mathsf{h}}_\mathrm{D})}\,. \tag{7.140}$$

This integral is determined uniquely by an infinitesimal shift of the poles from the real axis to the complex plane. We argued before that the poles on the positive and negative real axis correspond to states moving forward and backward in time, respectively. Therefore, we add an infinitesimally small negative and positive imaginary parts, respectively, to them. For the moment, we assume that the energy E_0 is the lowest state with positive energy. Consequently, we have $E_0-E_{n+}<0$ for the *positive* energy eigenstates of the Hamiltonian $\hat{\mathsf{h}}_\mathrm{D}$. Hence these states lie on the *negative* real axis of k_0. Similarly, the *negative* energy eigenstates of $\hat{\mathsf{h}}_\mathrm{D}$ lie on the *positive* real axis of k_0:

$$\frac{1}{k_0-E_0+\hat{\mathsf{h}}_\mathrm{D}}=\sum_n\left(\frac{\hat{\mathsf{P}}_{n+}}{k_0-(E_0-E_{n+}+\mathrm{i}\varepsilon)}+\frac{\hat{\mathsf{P}}_{n-}}{k_0-(E_0-E_{n-}-\mathrm{i}\varepsilon)}\right)$$

and

$$\frac{1}{k_0^2-\omega^2}=\frac{1}{(k_0-(\omega-\mathrm{i}\varepsilon))(k_0-(-\omega+\mathrm{i}\varepsilon))}\,.$$

To isolate the contributions of the individual poles, we decompose the integrand into partial fractions:

$$\frac{1}{k_0^2-\omega^2}\frac{1}{k_0-\Delta}=-\frac{1}{2\omega(\Delta-\omega)}\frac{1}{k_0-\omega}+\frac{1}{2\omega(\Delta+\omega)}\frac{1}{k_0+\omega}$$
$$+\frac{1}{(\Delta-\omega)(\Delta+\omega)}\frac{1}{k_0-\Delta}\,.$$

We close the integration path in the lower half of the complex plane. Hence the poles shifted into the upper half of the complex plane do not contribute. Contribution of poles from the lower half of the complex plane equals $-2\pi\mathrm{i}$ times the residue

$$-\frac{\mathrm{i}}{2\pi}\int\frac{\mathrm{d}k_0}{(k_0^2-\omega^2)(E_0-\hat{\mathsf{h}}_\mathrm{D}-k_0)}=\frac{1}{2\omega}\sum_n\left(\frac{\hat{\mathsf{p}}_{n+}}{E_{n+}-E_0+\omega}+\frac{\hat{\mathsf{p}}_{n-}}{E_{n-}-E_0-\omega}\right).$$

Comparison of Eq. (7.139) and the last equation with Eq. (7.83) shows that the latter can be rewritten into the form ($\langle\mathbf{r}|\,\psi_\mathrm{at}\rangle=U_1(\mathbf{r})$)

$$(E_2)_\perp^\mathrm{SE}=\mathrm{i}e^2\int\frac{\mathrm{d}^4\boldsymbol{k}}{(2\pi)^4}\frac{1}{\boldsymbol{k}^2}\left\{\langle\psi_\mathrm{at}|\gamma_0\gamma_i\mathrm{e}^{\mathrm{i}\mathbf{k}\cdot\hat{\boldsymbol{r}}}\frac{1}{E_0-\hat{\mathsf{h}}_\mathrm{D}-k_0}\gamma_0\gamma_i\mathrm{e}^{-\mathrm{i}\mathbf{k}\cdot\hat{\boldsymbol{r}}}|\psi_\mathrm{at}\rangle\right.$$
$$\left.-\frac{1}{\omega^2}\langle\psi_\mathrm{at}|\gamma_0\boldsymbol{\gamma}\cdot\mathbf{k}\,\mathrm{e}^{\mathrm{i}\mathbf{k}\cdot\hat{\boldsymbol{r}}}\frac{1}{E_0-\hat{\mathsf{h}}_\mathrm{D}-k_0}\gamma_0\boldsymbol{\gamma}\cdot\mathbf{k}\,\mathrm{e}^{-\mathrm{i}\mathbf{k}\cdot\hat{\boldsymbol{r}}}|\psi_\mathrm{at}\rangle\right\}. \quad (7.141)$$

In case the reference state ψ_at is an excited state, the part of the positive energy states lying below the reference state has to be shifted into the lower half of the complex plane. That is, we need to treat them as negative-energy states.

By comparing now Eq. (7.139) with the last equation, we find an important relation between the probability amplitude $\mathscr{A}$ that a system evolves under the influence of a perturbation for a time interval T back into the initial state and the energy shift of the system ΔE caused by the perturbation:

$$\mathscr{A}=-\mathrm{i}T\Delta E, \quad (7.142)$$

which is a general relation and will be used in Sects. 7.7 and 7.8. This relation should be not at all surprising: the probability amplitude that a stationary state evolves in time $T=t-t_0$ again to the stationary state equals, see Eq. (1.18),

$$\langle\psi_\mathrm{at}(t)|\mathrm{e}^{-\mathrm{i}E_0(t-t_0)}|\psi_\mathrm{at}(t_0)\rangle=1,$$

where clearly E_0 represents the energy of the stationary state. If this stationary state is under the influence of a perturbation, but remains approximately a stationary state, the action of the perturbation merely shifts the state phase:

$$\langle\psi_\mathrm{at}(t)|\mathrm{e}^{-\mathrm{i}E(t-t_0)}|\psi_\mathrm{at}(t_0)\rangle\simeq1-\mathrm{i}(E-E_0)(t-t_0),$$

where the second term on the rhs corresponds to the rhs of Eq. (7.142).

7.4.6 Electron Self-energy: Cancellation of the Non-covariant Terms

I was very surprised to discover that it was not known at that time that every one of the formulas that had been worked out so patiently by separating longitudinal and transverse waves could be obtained from the formula for the transverse waves alone if instead of summing over only the two perpendicular polarization directions you would sum over all four possible directions of polarization.R.P. Feynman [35]

We are nearly at the end of our way to the relativistic perturbation method. The last piece to show is that the non-covariant parts of electron interaction with transverse and longitudinal parts of EM field cancel out. For an arbitrary "reasonable" function f, the relation

$$e^{i\mathbf{k}\cdot\hat{\boldsymbol{r}}} f(\hat{\boldsymbol{p}}) e^{-i\mathbf{k}\cdot\hat{\boldsymbol{r}}} = f(\hat{\boldsymbol{p}} - \mathbf{k}), \tag{7.143}$$

holds. We use this equality in the expression (7.141) with the Hamiltonian (7.34) and obtain

$$e^{i\mathbf{k}\cdot\hat{\boldsymbol{r}}} (\hat{\mathsf{h}}_D - E_0 + k_0)^{-1} e^{-i\mathbf{k}\cdot\hat{\boldsymbol{r}}} = (\hat{\mathsf{h}}_D - E_0 + k_0 - \gamma_0 \boldsymbol{\gamma}\cdot\mathbf{k})^{-1}.$$

The second expression in the curly brackets in Eq. (7.141) now takes the form

$$\frac{1}{\omega^2} \langle \psi_{at} | \gamma_0 \boldsymbol{\gamma}\cdot\mathbf{k} \frac{1}{E_0 - \hat{\mathsf{h}}_D - k_0 + \gamma_0 \boldsymbol{\gamma}\cdot\mathbf{k}} \gamma_0 \boldsymbol{\gamma}\cdot\mathbf{k} | \psi_{at} \rangle. \tag{7.144}$$

Action of the Hamiltonian on the reference state ψ_{at} is from its very definition $(E_0 - \hat{\mathsf{h}}_D)\psi_{at} = 0$; hence

$$\begin{aligned}
&\langle \psi_{at} | \gamma_0 \boldsymbol{\gamma}\cdot\mathbf{k} \frac{1}{E_0 - \hat{\mathsf{h}}_D - k_0 + \gamma_0 \boldsymbol{\gamma}\cdot\mathbf{k}} \gamma_0 \boldsymbol{\gamma}\cdot\mathbf{k} | \psi_{at} \rangle \\
&= \langle \psi_{at} | \left(\gamma_0 \boldsymbol{\gamma}\cdot\mathbf{k} + E_0 - \hat{\mathsf{h}}_D - k_0 + k_0 \right) \frac{1}{E_0 - \hat{\mathsf{h}}_D - k_0 + \gamma_0 \boldsymbol{\gamma}\cdot\mathbf{k}} \gamma_0 \boldsymbol{\gamma}\cdot\mathbf{k} | \psi_{at} \rangle \\
&= \langle \psi_{at} | \gamma_0 \boldsymbol{\gamma}\cdot\mathbf{k} | \psi_{at} \rangle + k_0 \langle \psi_{at} | \frac{1}{E_0 - \hat{\mathsf{h}}_D - k_0 + \gamma_0 \boldsymbol{\gamma}\cdot\mathbf{k}} \gamma_0 \boldsymbol{\gamma}\cdot\mathbf{k} | \psi_{at} \rangle.
\end{aligned}$$

After inserting the last equation into Eq. (7.141), we easily find that the first term vanishes as we integrate an odd function over a symmetric interval. In case of the second term, we use the same trick:

$$
\begin{aligned}
k_0 \langle \psi_{\rm at} | \frac{1}{E_0 - \hat{\mathsf{h}}_{\rm D} - k_0 + \gamma_0 \boldsymbol{\gamma} \cdot \mathbf{k}} \gamma_0 \boldsymbol{\gamma} \cdot \mathbf{k} | \psi_{\rm at} \rangle \\
&= k_0 \langle \psi_{\rm at} | \frac{1}{E_0 - \hat{\mathsf{h}}_{\rm D} - k_0 + \gamma_0 \boldsymbol{\gamma} \cdot \mathbf{k}} (\gamma_0 \boldsymbol{\gamma} \cdot \mathbf{k} + E_0 - \hat{\mathsf{h}}_{\rm D} - k_0 + k_0) | \psi_{\rm at} \rangle \\
&= k_0 + k_0^2 \langle \psi_{\rm at} | \frac{1}{E_0 - \hat{\mathsf{h}}_{\rm D} - k_0 + \gamma_0 \boldsymbol{\gamma} \cdot \mathbf{k}} | \psi_{\rm at} \rangle ;
\end{aligned}
$$

application of these rearrangements results in

$$
(7.144) = \frac{k_0^2}{\omega^2} \langle \psi_{\rm at} | \frac{1}{E_0 - \hat{\mathsf{h}}_{\rm D} - k_0 + \gamma_0 \boldsymbol{\gamma} \cdot \mathbf{k}} | \psi_{\rm at} \rangle - \frac{k_0}{\omega^2} .
$$

After substituting the last equation into Eq. (7.141) and using the identity

$$
\frac{k_0^2}{\boldsymbol{k}^2 \omega^2} = \frac{1}{\boldsymbol{k}^2} + \frac{1}{\omega^2} ,
$$

where we used the relation $\boldsymbol{k}^2 = k_0^2 - \omega^2$, we arrive at

$$
(E_2)_\perp^{\rm SE} = (\Delta E)^{\rm SE} + (E_2)_\perp^{\rm SE} - (\Delta E)^{\rm SE} . \tag{7.145}
$$

In the first part, a Lorentz-invariant combination of time and space components of γ-matrices appears

$$
(\Delta E)^{\rm SE} = -{\rm i} e^2 \int \frac{{\rm d}^4 \boldsymbol{k}}{(2\pi)^4 \boldsymbol{k}^2} \langle \psi_{\rm at} | \gamma_0 \gamma_\mu \frac{1}{E_0 - \hat{\mathsf{h}}_{\rm D} - k_0 + \gamma_0 \boldsymbol{\gamma} \cdot \mathbf{k}} \gamma_0 \gamma_\mu | \psi_{\rm at} \rangle . \tag{7.146}
$$

The second part contains non-covariant terms

$$
(E_2)_\perp^{\rm SE} - (\Delta E)^{\rm SE} = -{\rm i} e^2 \int \frac{{\rm d}^4 \boldsymbol{k}}{(2\pi)^4} \frac{1}{\omega^2} \left\{ \frac{k_0}{\boldsymbol{k}^2} - \langle \psi_{\rm at} | {\rm e}^{{\rm i}\mathbf{k}\cdot\hat{\boldsymbol{r}}} \frac{1}{-E_0 + \hat{\mathsf{h}}_{\rm D} + k_0} {\rm e}^{-{\rm i}\mathbf{k}\cdot\hat{\boldsymbol{r}}} | \psi_{\rm at} \rangle \right\} . \tag{7.147}
$$

Notice, though, that this non-covariant piece exactly cancels out with the contribution of the longitudinal part $(E_1)_{||}^{\rm SE}$, Eq. (7.82). To show that this is true, we integrate in Eq. (7.147) over k_0. We split the first integral into two and use the residue theorem:

$$
\begin{aligned}
\frac{{\rm i} e^2}{2\pi} \int \frac{{\rm d}k_0 (k_0 - \omega + \omega)}{k_0^2 - \omega^2} &= \frac{{\rm i} e^2}{2\pi} \left[\int \frac{{\rm d}k_0}{k_0 + \omega - {\rm i}\varepsilon} + \omega \int \frac{{\rm d}k_0}{(k_0 - \omega + {\rm i}\varepsilon)(k_0 + \omega - {\rm i}\varepsilon)} \right] \\
&= \frac{{\rm i} e^2}{2\pi} \left[-\omega 2\pi {\rm i} \cdot \frac{1}{2\omega} \right] = \frac{e^2}{2} ,
\end{aligned}
$$

where the integration path leads, as usual, along the real axis and is closed in the lower half of the complex plane. The first integral in this expression does not decrease rapidly enough to zero for large k_0, though. However, the contribution of the lower semicircle of this expression cancels out with the contribution of lower semicircle of the second term in Eq. (7.147). Now, using the decomposition of unity

$$1 = \langle\psi_{\mathrm{at}}|\mathrm{e}^{\mathrm{i}\mathbf{k}\cdot\hat{\boldsymbol{r}}}\sum_n\left(\hat{\mathsf{P}}^{n+} + \hat{\mathsf{P}}^{n-}\right)\mathrm{e}^{-\mathrm{i}\mathbf{k}\cdot\hat{\boldsymbol{r}}}|\psi_{\mathrm{at}}\rangle\,,$$

we obtain the first integral in Eq. (7.147) in the desired form

$$-\mathrm{i}e^2\int\frac{k_0\mathrm{d}^4\boldsymbol{k}}{\boldsymbol{k}^2\omega^2(2\pi)^4} = -e^2\int\frac{\mathrm{d}^3\mathbf{k}}{2\omega^2(2\pi)^3}\langle\psi_{\mathrm{at}}|\mathrm{e}^{\mathrm{i}\mathbf{k}\cdot\hat{\boldsymbol{r}}}\sum_n\left(\hat{\mathsf{P}}^{n+} + \hat{\mathsf{P}}^{n-}\right)\mathrm{e}^{-\mathrm{i}\mathbf{k}\cdot\hat{\boldsymbol{r}}}|\psi_{\mathrm{at}}\rangle\,.$$

The second integral in Eq. (7.147) is calculated again by the residue theorem:

$$\frac{\mathrm{i}e^2}{2\pi}\int\frac{\mathrm{d}k_0}{k_0+\hat{\mathsf{h}}_{\mathrm{D}}-E_0} = \frac{\mathrm{i}e^2}{2\pi}\int\mathrm{d}k_0\sum_n\left(\frac{\hat{\mathsf{P}}^{n+}}{k_0-(-E_{n+}+E_0+\mathrm{i}\varepsilon)}\right.$$
$$\left.+\frac{\hat{\mathsf{P}}^{n-}}{k_0-(-E_{n-}+E_0-\mathrm{i}\varepsilon)}\right) = e^2\sum_n\hat{\mathsf{P}}^{n-}.$$

Equation (7.147) then becomes

$$(E_2)_\perp^{\mathrm{SE}} - (\Delta E)^{\mathrm{SE}} = -e^2\int\frac{\mathrm{d}^3\mathbf{k}}{2\omega^2(2\pi)^3}\langle\psi_{\mathrm{at}}|\mathrm{e}^{\mathrm{i}\mathbf{k}\cdot\hat{\boldsymbol{r}}}\sum_n\left(\hat{\mathsf{P}}^{n+} - \hat{\mathsf{P}}^{n-}\right)\mathrm{e}^{-\mathrm{i}\mathbf{k}\cdot\hat{\boldsymbol{r}}}|\psi_{\mathrm{at}}\rangle\,,$$

which matches but for the opposite sign Eq. (7.82) for the contribution of longitudinal part of EM field:

$$(E_2)_\perp^{\mathrm{SE}} - (\Delta E)^{\mathrm{SE}} + (E_1)_{||}^{\mathrm{SE}} = 0\,.$$

Finally, we rearrange the integral (7.146). By means of the identity $\mathsf{A}^{-1}\mathsf{B}^{-1} = (\mathsf{BA})^{-1}$, valid for two arbitrary matrices A and B, for which it makes sense, the second γ_0-matrix can be moved to the denominator. The denominator then takes the form

$$\gamma_0\left[E_0-\hat{\mathsf{h}}_{\mathrm{D}}-k_0+\gamma_0\boldsymbol{\gamma}\cdot\mathbf{k}\right] = \gamma_0(E_0-e\varphi)-\boldsymbol{\gamma}\cdot(\hat{\boldsymbol{p}}-e\mathbf{A})-m-\gamma_0k_0+\boldsymbol{\gamma}\cdot\mathbf{k}$$
$$= \gamma_\mu(\hat{\Pi}-k)_\mu - m = \boldsymbol{\gamma}\cdot(\hat{\boldsymbol{\Pi}}-\boldsymbol{k})-m\,,$$

where in the first equality we recalled Eq. (7.34). We arrive at the final form of the shift of the atomic energy levels due to the self-energy effect

$$(\Delta E)^{\mathrm{SE}} = -\mathrm{i}e^2 \int \frac{\mathrm{d}^4\boldsymbol{k}}{(2\pi)^4 k^2} \langle \overline{\psi}_{\mathrm{at}} | \gamma_\mu \frac{1}{\boldsymbol{\gamma} \cdot (\hat{\boldsymbol{\Pi}} - \boldsymbol{k}) - m} \gamma_\mu | \psi_{\mathrm{at}} \rangle \,, \tag{7.148}$$

where we used the Dirac conjugation introduced in Eq. (7.48).

7.4.7 Vacuum Polarization: Covariant Formulation

Much less effort is needed to manipulate Eq. (7.80),

$$\langle \hat{\mathsf{j}}_\nu(\mathbf{r}) \rangle = \frac{e}{2} (\gamma_0 \gamma_\nu)_{ab} \sum_\sigma \langle \mathbf{r} | \left[\hat{\mathsf{P}}^{\sigma-} - \hat{\mathsf{P}}^{\sigma+} \right]_{ba} | \mathbf{r} \rangle \,, \tag{7.80}$$

for the vacuum expectation value of the four-current created by the Dirac field, into a manifestly Lorentz-covariant form. By comparing Eq. (7.126),

$$\begin{aligned} \hat{\mathsf{G}}_0^{(F)}(t, t_0) &= \frac{1}{2\pi} \int \frac{\mathrm{e}^{-\mathrm{i}E(t-t_0)}}{E - \hat{\mathsf{H}}_0} \mathrm{d}E \\ &= \frac{1}{2\pi} \lim_{\varepsilon \to 0+} \int \mathrm{d}E \sum_n \left(\frac{\hat{\mathsf{P}}^{n+}}{E - (E_{n+} - \mathrm{i}\varepsilon)} + \frac{\hat{\mathsf{P}}^{n-}}{E - (E_{n-} + \mathrm{i}\varepsilon)} \right) \mathrm{e}^{-\mathrm{i}E(t-t_0)} \,, \end{aligned} \tag{7.126}$$

and (7.127),

$$\hat{\mathsf{G}}_0^{(F)}(t, t_0) = -\mathrm{i} \left[\theta(t - t_0) \sum_{n+} \mathrm{e}^{-\mathrm{i}E_{n+}(t-t_0)} \hat{\mathsf{P}}^{n+} - \theta(t_0 - t) \sum_{n-} \mathrm{e}^{-\mathrm{i}E_{n-}(t-t_0)} \hat{\mathsf{P}}^{n-} \right] , \tag{7.127}$$

for the Green function of the Dirac equation with Eq. (7.80), we see that

$$\begin{aligned} \langle \hat{\mathsf{j}}_\mu(\mathbf{r}) \rangle &= -\frac{\mathrm{i}e}{2} (\gamma_0 \gamma_\mu)_{ab} \langle \mathbf{r} | \left(\lim_{t \to t_0-} \hat{\mathsf{G}}_0^F(t, t_0) + \lim_{t \to t_0+} \hat{\mathsf{G}}_0^F(t, t_0) \right)_{ba} | \mathbf{r} \rangle \\ &= -\mathrm{i}e \int_\Gamma \frac{\mathrm{d}E}{2\pi} \mathrm{Tr} \, \langle \mathbf{r} | \gamma_0 \gamma_\mu \frac{1}{E - \hat{\mathsf{h}}_{\mathrm{D}}} | \mathbf{r} \rangle \,. \end{aligned}$$

The integration along the path Γ denotes the average between closing the integration along the real axis over the upper and lower semicircles, see the lower part of Fig. 7.4. Owing to the cyclicity of the trace $\mathrm{Tr}\,\mathsf{AB} = \mathrm{Tr}\,\mathsf{BA}$, γ_0-matrix can be again moved to the denominator

$$\langle \hat{\mathsf{j}}_\mu(\mathbf{r}) \rangle = -\mathrm{i}e \int_\Gamma \frac{\mathrm{d}E}{2\pi} \mathrm{Tr} \, \langle \mathbf{r} | \gamma_\mu \frac{1}{\boldsymbol{\gamma} \cdot \hat{\boldsymbol{\Pi}} - m} | \mathbf{r} \rangle \,. \tag{7.149}$$

7.4.8 Discussion of the Lorentz Invariance

Are the expressions (7.77), (7.148), and (7.149) invariant with respect to Lorentz transformations? If not, what part thereof causes it? By comparing the expressions (7.77), (7.148) and (7.149) with the Dirac equations (7.16) or (7.19), we readily see that at least part of the expressions (7.77), (7.148), and (7.149) is Lorentz invariant provided the Dirac equations (7.16) or (7.19) are invariant as well. Are they? The rest mass m is Lorentz invariant, hence obviously the product $\boldsymbol{\gamma} \cdot \hat{\boldsymbol{p}}$ must be invariant as well. When considering the Lorentz transformation (7.2), $x'_\mu = \Lambda_{\mu\nu} x_\nu$, the four-momentum $\hat{\boldsymbol{p}}$ clearly transforms as a four-vector

$$\hat{\mathsf{p}}'_\mu = -\mathrm{i}\frac{\partial}{\partial x'_\mu} = -\mathrm{i}\frac{\partial x_\rho}{\partial x'_\mu}\frac{\partial}{\partial x_\rho} = (\Lambda^{-1})_{\rho\mu}(-\mathrm{i})\frac{\partial}{\partial x_\rho} = \Lambda_{\mu\rho}\hat{\mathsf{p}}_\rho \,, \tag{7.150}$$

where we used Eq. (7.7) in the last equality. Thus the matrices $\boldsymbol{\gamma}$ have to transform as components of a four-vector as well. Do they?

Now, we have to recognize that when going from one inertial system to another, one needs to transform both the coordinates and the wave function. In a primed inertial system, the Dirac equation (7.16) takes the form

$$(\gamma'_\mu \hat{\mathsf{p}}'_\mu - m)\psi'(\boldsymbol{x}') = 0 \,. \tag{7.151}$$

Hence, should the Dirac equation be invariant with respect to Lorentz transformations, it has to be possible to find an invertible matrix S transforming the wave function

$$\psi'(\boldsymbol{x}') = \mathsf{S}\psi(\boldsymbol{x}) \,. \tag{7.152}$$

By inserting the last equation into Eq. (7.151) and multiplying the obtained equation by the inverse matrix S^{-1} from the left, we find

$$(\mathsf{S}^{-1}\gamma'_\mu \mathsf{S}\hat{\mathsf{p}}'_\mu - m)\psi(\boldsymbol{x}) = 0$$

and see that this expression matches Eq. (7.16) if

$$\mathsf{S}^{-1}\gamma'_\mu \mathsf{S}\hat{\mathsf{p}}'_\mu = \gamma_\rho \hat{\mathsf{p}}_\rho \,. \tag{7.153}$$

Using Eq. (7.150)

$$\mathsf{S}^{-1}\gamma'_\mu \mathsf{S}\hat{\mathsf{p}}'_\mu = \Lambda_{\mu\rho}\mathsf{S}^{-1}\gamma'_\mu \mathsf{S}\hat{\mathsf{p}}_\rho \,,$$

and comparing the rhs of the last two equations, we arrive at the requirement

$$\gamma_\rho = \Lambda_{\mu\rho}\mathsf{S}^{-1}\gamma'_\mu \mathsf{S} \Rightarrow \gamma'_\mu = \Lambda_{\mu\rho}\mathsf{S}\gamma_\rho \mathsf{S}^{-1} \,. \tag{7.154}$$

What is the relation between the matrices $\boldsymbol{\gamma}'$ and $\boldsymbol{\gamma}$? The relation between the four-momentum and the rest mass must hold in the primed system

$$(\hat{\mathsf{p}}'_\mu \hat{\mathsf{p}}'_\mu - m^2)\psi'(\boldsymbol{x}') = 0$$

as well as in original system, Eq. (7.14). That is, from comparison of Eq. (7.151) with the last equation, we find that in the primed system the anticommutation relation (7.17) for γ-matrices has to hold as well

$$\left\{\gamma'_\mu, \gamma'_\nu\right\} = 2\eta_{\mu\nu}\,.$$

However, up to a unitary transformation $\gamma'_\mu = \mathsf{U}^+\gamma_\mu\mathsf{U}$, the γ-matrices are determined by the anticommutation relation (7.17) uniquely. Hence, should the Dirac equation be invariant with respect to Lorentz transformation, it must be possible to find a matrix S obeying

$$\mathsf{U}^+\gamma_\mu\mathsf{U} = \Lambda_{\mu\rho}\mathsf{S}\gamma_\rho\mathsf{S}^{-1}\,.$$

Such a matrix S indeed exists.[24] The knowledge of the existence of the matrix S suffices for our purposes, we will not need its explicit form. There is only one property to remember: in the sense of Eq. (7.154), the Dirac γ-matrices behave with respect to Lorentz transformations as components of a four-vector.

Clearly, see Eq. (7.18), the γ-matrices are not hermitian. The hermitian conjugated γ-matrices are related to the original matrices via Eq. (7.61), $\gamma_\mu^+ = \gamma_0\gamma_\mu\gamma_0$. An interesting question arises: what is the Lorentz invariant form of the Dirac equation for Hermite conjugated wave function ψ^+?

Hermitian conjugation of Eq. (7.16) together with Eq. (7.61) yield

$$\psi^+(\gamma_\mu^+\hat{\mathsf{p}}_\mu - m) = \psi^+(\gamma_0\gamma_\mu\gamma_0\hat{\mathsf{p}}_\mu - m) = \overline{\psi}(\gamma_\mu\hat{\mathsf{p}}_\mu - m)\gamma_0 = 0\,, \tag{7.155}$$

where $\overline{\psi} = \psi^+\gamma_0$ is the Dirac conjugated function, see Eq. (7.48). Its meaning should be now clear: the Lorentz invariant equation

$$\overline{\psi}(\gamma_\mu\hat{\mathsf{p}}_\mu - m) = 0 \tag{7.156}$$

is obtained for the Dirac conjugated function $\bar{\psi}$, not for the Hermite conjugated function ψ^+.

There is one more point worth attention prior to leaving this formal analysis of the Dirac equation. In a primed inertial system, Eq. (7.156) takes the form

$$\psi'^+(x')\gamma'_0(\gamma'_\mu\hat{\mathsf{p}}'_\mu - m) = \psi^+(x)\mathsf{S}^+\gamma'_0\mathsf{S}(\gamma_\mu\hat{\mathsf{p}}_\mu - m)\mathsf{S}^{-1} = 0\,,$$

[24] We do not even need to exploit the freedom in the unitary transformation; that is, we can set U to a unit matrix. For the precise form of matrix S see, e.g., [30, 33, 38].

where we used Eqs. (7.152) and (7.153). Comparison of the last two equations yields

$$\mathsf{S}^{+}\gamma_0'\mathsf{S} = \gamma_0 \qquad \Rightarrow \qquad \mathsf{S}^{-1} = \gamma_0\mathsf{S}^{+}\gamma_0', \tag{7.157}$$

hence the matrix S is not unitary. That is, when going from one inertial system to another it is not the scalar product $\psi^{+}\varphi$ of the bispinors ψ and φ that is preserved, but

$$\overline{\psi'}\varphi' = \psi'^{+}\gamma_0'\varphi' = \psi^{+}\mathsf{S}^{+}\gamma_0'\mathsf{S}\varphi = \psi^{+}\gamma_0\gamma_0\mathsf{S}^{+}\gamma_0'\mathsf{S}\varphi = \psi^{+}\gamma_0\varphi = \overline{\psi}\varphi .$$

Now, it should be clear that the only non-covariant parts of Eq. (7.148) are the three-dimensional integrations $d^3\mathbf{r}$ and $d^3\mathbf{r}'$: recall that $\langle\psi|\hat{\mathsf{O}}|\psi\rangle$ stands for $\int\int d^3\mathbf{r}d^3\mathbf{r}'\psi^{+}(\mathbf{r})\langle\mathbf{r}|\hat{\mathsf{O}}|\mathbf{r}'\rangle\psi(\mathbf{r}')$. Lorentz covariant is a four-dimensional integration d^4x, not the three-dimensional one. Similarly, in Eq. (7.77) there are non-covariant three dimensional integrations $d^3\mathbf{r}$, $d^3\mathbf{r}'$ and $d^3\mathbf{k}$, while the rest is covariant. However, this is not a problem: we calculate the energy shift which is clearly not a Lorentz-invariant quantity. We will exploit the knowledge of Lorentz-invariant parts of the expressions (7.148) and (7.77) later in Sect. 7.8.

7.4.9 *What View of Positrons Is the Correct One?*

> Did I speak of an electron going backward in time? Did I mumble something about a sea of negative energy electrons? This metaphorical language (...) confused generations of physics students and physicists. The presentation given here is in the modern spirit, which seeks to avoid these potentially confusing metaphors. A. Zee [42]

> I may be old-fashioned, but I still regard the "hole theory" as a good way of describing positron theory. It seems to contain much of the essence of the situation (...). Of course one realizes that the picture is symmetric and one could just as easily take the electron as a hole in the sea of positrons with negative energy. (...) The consistency of the alternative way of looking at the situation always seemed to me to come out by a miracle, unless one knew it was equivalent to the "hole" description. But of course it is a matter of taste, since the two ways of looking at the problem are equivalent. R. Peierls [35]

> On looking back over the work, I can only feel a kind of regret for the enormous amount of physical reasoning and mathematical re-expression which ends by merely re-expressing what was previously known, although in a form which is much more efficient for the calculation of specific problems. R.P. Feynman [35]

Dirac's picture of an occupied sea of negative energy levels naturally leads to the determination of the correct ground state (7.40) and to the expressions for the quantized Dirac field (7.41) and the Hamiltonian of the Dirac field (7.42). However, one can avoid this insight and arrive at these equations by imposing a formal requirement of positive definiteness of the Hamilton operator of the quantized Dirac field.

Feynman's picture of positrons moving backward in time naturally leads to the view that the process A when an electron radiates a photon at $\boldsymbol{x}'$ and absorbs it back at $\boldsymbol{x}''$ and the process B when a photon and an electron-positron pair are created at $\boldsymbol{x}'$ and annihilated at $\boldsymbol{x}''$ differ merely in the temporal ordering of the events at $\boldsymbol{x}'$ and at $\boldsymbol{x}''$. This in turn leads to the use of Green functions where the temporal ordering of the events is hidden in the way the poles are encircled in the complex plane. This further leads to a manifest Lorentz invariance of the pertinent expressions. However, one can avoid this insight as well. Freeman Dyson, inspired by Feynman's work, showed how to maintain the manifest Lorentz invariance within the framework of the ordinary quantum field theory when electrons and positrons are regarded as quanta of an electron-positron field. To do so, the perturbation method is reformulated in time-dependent and time-symmetric ways, see e.g., [2, 15, 24, 30, 33].[25]

It is worth recalling that the Dirac equation was published in 1928, but Dirac's prediction of positron came out only in 1931. Similarly, quantum electrodynamics was formulated by Heisenberg and Pauli in 1929, but the expressions derived in this book appeared for the first time in Feynman's papers published in 1949. In both cases, the formal derivation appeared only after an intuitive one, although in both cases there was plenty of time for anyone to do so. This clearly shows that it was impossible to arrive at the heart of the matter without Dirac's and Feynman's insights. Therefore, we decided not to hide them as is the case of all modern expositions of this subject. We are led to this by the recognition that, at least in our case, an intuitive approach leads to significantly less confusion than a purely formal one. The pictures, not necessarily regarded as poetic, enable us to understand "where all the formalism comes from," or "why the things are done in the way they are."

The usual argument against the hole picture, namely that it does not hold for bosons, is somewhat misleading. The reason is bosons differ from fermions at the very fundamental level. When quantizing EM field, or for that matter any other boson field, one does not encounter the question of positive definiteness of the quantum Hamiltonian at all. It merely follows from the canonical quantization procedure and masslessness of the EM field does not play any role therein. On the other hand, when quantizing a fermion field, positive definiteness of the quantum Hamiltonian has to be *imposed*. The Dirac sea is a physical illustration of what a formal requirement of a positive definite Hamiltonian means. Once we accustom to it, we can move away from the illustrative picture.

To conclude, positrons are best regarded as quanta of an electron-positron field. However, it is worth keeping other views in mind as well. They are not likely, at least in our view, to lead to confusion any more than any other concept one is to accept.

---

[25]There are also other ways of arriving at the manifest Lorentz invariance of amplitudes for relativistic processes. The most popular one reformulates quantum theory via path integrals first and only then uses the perturbation method, see, e.g., [4, 30, 42]. This was, in principle, also Feynman's original approach; however, details of the original Feynman's and modern derivations differ.

7.4.10 Note on the Feynman Diagrams and Feynman Rules

> Here there is a most curious situation: the resulting machinery is far better than originating theory.[38]

It follows from discussions in Sects. 7.4.5 and 7.4.6 that if we add a similar amplitude for the electron interaction with the longitudinal part of EM field to the amplitude (7.138) for the electron interaction with the transverse part of EM field, we arrive at a Lorentz invariant amplitude

$$\lim_{\substack{t\to+\infty \\ t_0\to-\infty}} \mathscr{A} \simeq -\mathrm{i}e^2 \int\int \mathrm{d}^4\boldsymbol{x}''\mathrm{d}^4\boldsymbol{x}' D_{\mu\nu}(\boldsymbol{x}''-\boldsymbol{x}')\overline{\psi}_{\mathrm{at}}(\boldsymbol{x}'')\gamma_\mu G_0^{(F)}(\boldsymbol{x}'',\boldsymbol{x}')\gamma_0\gamma_\nu\psi_{\mathrm{at}}(\boldsymbol{x}')\,, \tag{7.158}$$

where the *Feynman form of the photon Green function* reads

$$D_{\mu\nu}(\boldsymbol{x}''-\boldsymbol{x}') = -\mathrm{i}\eta_{\mu\nu}\int \frac{\mathrm{e}^{-\mathrm{i}\boldsymbol{k}\cdot(\boldsymbol{x}''-\boldsymbol{x}')}}{\boldsymbol{k}^2}\frac{\mathrm{d}^4\boldsymbol{k}}{(2\pi)^4}\,. \tag{7.159}$$

If we look at Fig. 7.6, we see that Eq. (7.158) could have been guessed from Fig. 7.6 had we made the following correspondence between mathematical symbols and elements of Fig. 7.6.

1. Assign the state vectors of initial and final states, $|\psi_{\mathrm{at}}\rangle$ and $\langle\overline{\psi}_{\mathrm{at}}|$, respectively, to the external electron lines.
2. Assign $\mathrm{i}e\boldsymbol{\gamma}$ to the interaction vertices.
3. Assign the propagator $\mathrm{i}G_0^{(F)}(\boldsymbol{x}'',\boldsymbol{x}')\gamma_0$ to the internal electron line.
4. Assign the propagator $D_{\mu\nu}(\boldsymbol{x}''-\boldsymbol{x}')$ to the internal photon line.
5. Integrate over all space-time coordinates of the interaction events.

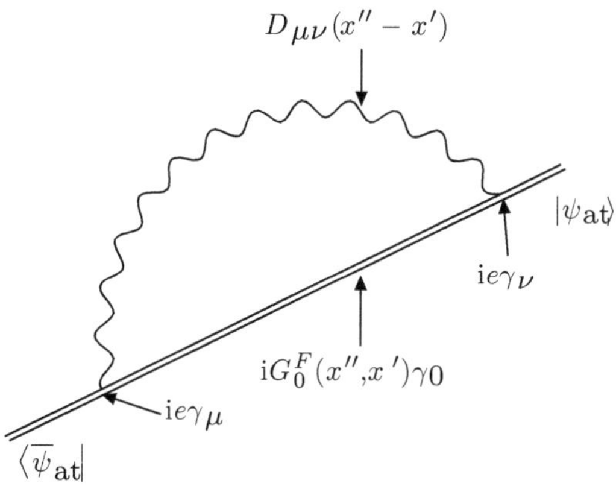

Fig. 7.6 Correspondence between mathematical symbols and elements of Feynman diagram in coordinate space

This correspondence between the "pictures" illustrating the processes and the mathematical expressions providing the transition amplitudes for these processes is called *Feynman rules* in coordinate space and the corresponding pictures are referred to as *Feynman diagrams*. We did not try at all to systematically derive this correspondence and neither will we formulate it in a systematic and exhaustive manner since we will not need it in this book. (It is definitely worth knowing about it nevertheless.) For a systematic treatment of the Feynman rules in both coordinate and momentum space for scattering problems, we refer the reader to any modern quantum field theory textbook, such as [2, 15, 24, 30].

Feynman diagrams have several important features; we will mention the two of them which we consider the most important. Firstly, they elucidate the physical meaning of the expressions to be calculated. Secondly, they form an indispensable tool for organizing calculations in higher orders of the perturbation method. Especially the latter property makes them a ubiquitous tool in various areas of physics, well beyond the borders of relativistic quantum electrodynamics they were originally invented for, see, e.g., [22].

Diagrams containing closed loops, such as those in Fig. 6.9 or in Fig. 7.1, are referred to as *loop diagrams*. If the diagrams do not comprise closed loops, such as those in Fig. 7.2, they are referred to as *tree diagrams*.

Note that the photon propagator (7.159) corresponds to the Green function of the Maxwell equations in the *Lorenz* gauge $\partial_\mu A_\mu = 0$. Although we have been working in the Coulomb gauge the whole time, we do not recognize it from expressions (7.148) nor (7.158)! This thus gives us a hint on the gauge invariance of the physical predictions of quantum electrodynamics mentioned in Sect. 7.2.4.

Our starting points, the Maxwell equations (6.3) and (6.4) and the Dirac equation (7.16), are manifestly invariant with respect to the Lorentz and gauge transformations. The Feynman diagrams and rules yield expressions for transition amplitudes manifestly invariant with respect to the Lorentz and gauge transformations as well. It should be clear, though, that our starting and ending points feature properties superior to those of any of the in between steps.

To quantize the EM field, we had to eliminate the gauge freedom by choosing one particular gauge; we opted for the Coulomb gauge. We emphasized in Chap. 6 that this gauge is suitable especially for the study of atom-photon interactions, and at the same time we mentioned in Sect. 7.3.1 that this gauge is not Lorentz invariant. However, although the Lorenz gauge does not feature this particular drawback, it does possess other deficiencies, see, e.g., [15, 24]. In addition, the choice of the Lorenz gauge leads to essential complications in the most of the calculations performed so far as well as in some of those to be performed, see Sect. 7.7.

Furthermore, the whole procedure of the canonical quantization is, in its orthodox form as outlined herein, alien to the "relativistic spirit." The classical Hamiltonian corresponds to the total energy of a system which is not a Lorentz invariant quantity. Canonical coordinates and momenta are determined for a fixed time instant, hence canonical commutation relations are clearly not Lorentz invariant either. When quantizing the electron-positron field, we had to distinguish between stationary solutions of the Dirac equation with positive and negative energies, yet the notion of positive or negative energy state is, as already mentioned in Sect. 7.3.1, not Lorentz invariant.

Despite all the listed shortcomings, we have seen that in fact none of them matter. The loss of the manifest[26] Lorentz and gauge invariance, enforced by the canonical quantization procedure, is reversible! To conclude this section, we wish to note that the common statement "the Coulomb law (6.12) conflicts with the special theory of relativity since it assumes action at a distance" is somewhat misleading. We derived this law back in Sect. 6.1.3 from the Maxwell equations which are manifestly invariant with respect to the Lorentz transformation, see Eq. (7.12). As we stressed in Sects. 6.1.5 and 6.5.5, a part of the EM interaction between the charged particles *is* instantaneous and only the retarded part of the interaction is mediated by the EM field carrying momentum and thus being experimentally detectable. The gauge invariance is the very reason responsible for the fact that the separation of the EM interaction into instantaneous and retarded parts does not conflict with special theory of relativity. It is the mismatch between the number of mathematical quantities needed for the correct description of the EM field, namely four, and the measurable internal degrees of freedom of the EM field, namely two, as mentioned already in Sects. 6.1.4 and 6.1.5, that at the end of the day reconciles the relativity with the action-at-distance.

7.5 Electron Self-energy: Calculation

To evaluate the expression (7.148) it is advantageous, for reason we will give in a moment, to multiply the electron propagator from the right by $(\boldsymbol{\gamma} \cdot (\hat{\boldsymbol{\Pi}} - \boldsymbol{k}) + m)^{-1}(\boldsymbol{\gamma} \cdot (\hat{\boldsymbol{\Pi}} - \boldsymbol{k}) + m)$,

$$\frac{1}{\boldsymbol{\gamma} \cdot (\hat{\boldsymbol{\Pi}} - \boldsymbol{k}) - m} = \frac{1}{\boldsymbol{k}^2 - 2\boldsymbol{k} \cdot \hat{\boldsymbol{\Pi}} + \hat{\mathscr{H}}} (\boldsymbol{\gamma} \cdot (\hat{\boldsymbol{\Pi}} - \boldsymbol{k}) + m) , \tag{7.160}$$

where the operator $\hat{\mathscr{H}}$ is given by Eq. (7.24).

By substituting Eq. (7.160) into Eq. (7.148) we find the operator $(\boldsymbol{\gamma} \cdot (\hat{\boldsymbol{\Pi}} - \boldsymbol{k}) + m)$ acting on $\gamma_\mu |\psi_{\mathrm{at}}\rangle$. By means of the anticommutation relations for the Dirac matrices (7.17) and the Dirac equation (7.19), we are able to simplify this action to

$$(\boldsymbol{\gamma} \cdot (\hat{\boldsymbol{\Pi}} - \boldsymbol{k}) + m)\gamma_\mu |\psi_{\mathrm{at}}\rangle = (2\hat{\Pi}_\mu + \gamma_\mu(-\boldsymbol{\gamma} \cdot \hat{\boldsymbol{\Pi}} + m) - \boldsymbol{\gamma} \cdot \boldsymbol{k}\gamma_\mu)|\psi_{\mathrm{at}}\rangle = (2\hat{\Pi}_\mu - \boldsymbol{\gamma} \cdot \boldsymbol{k}\gamma_\mu)|\psi_{\mathrm{at}}\rangle . \tag{7.161}$$

[26]Note that this word is crucial. We have shown that the expressions derived in Sect. 7.3.1 are equivalent to the manifestly Lorentz and gauge invariant expressions. Hence the expressions derived in Sect. 7.3.1 have to be Lorentz and gauge invariant as well. However, this invariance is completely obscured, it is not manifest.

7.5.1 Regularization

The integral over the photon four-momenta in Eq. (7.148) is divergent (in common, not especially appealing, jargon called ultraviolet divergent) as we will shortly show, and the treatment thereof will be a subject to further considerations. Usually, one begins by *regularizing* the expression (7.148) to transform it into a well-defined mathematical object. There are, of course, various ways how one can obtain such a form. We will opt for the one suggested by Feynman: the photon propagator in momentum space $\boldsymbol{k}^{-2}$ is replaced by

$$\frac{1}{\boldsymbol{k}^2} \rightarrow \frac{1}{\boldsymbol{k}^2} - \frac{1}{\boldsymbol{k}^2 - \Lambda^2} = -\int_0^{\Lambda^2} \frac{d\lambda}{(\boldsymbol{k}^2 - \lambda)^2} \,. \tag{7.162}$$

Using this replacement and the above Eqs. (7.160) and (7.161), in Eq. (7.148), we find

$$(\Delta E)^{\mathrm{SE}} = \frac{\alpha}{\pi} \int_0^{\Lambda^2} \mathrm{d}\lambda \int \frac{\mathrm{d}^4\boldsymbol{k}_F}{(\boldsymbol{k}^2 - \lambda)^2} \langle \overline{\psi}_{\mathrm{at}} | \gamma_\mu \frac{1}{\boldsymbol{k}^2 - 2\hat{\boldsymbol{\Pi}} \cdot \boldsymbol{k} + \hat{\mathscr{H}}} (2\hat{\Pi}_\mu - \boldsymbol{\gamma} \cdot \boldsymbol{k} \gamma_\mu) | \psi_{\mathrm{at}} \rangle \,, \tag{7.163}$$

where we introduced the notation

$$\mathrm{d}^4\boldsymbol{k}_F = \mathrm{i}\frac{\mathrm{d}^4\boldsymbol{k}}{(2\pi)^2} \,.$$

Obviously, the original expression (7.148) can be obtained from the regularized one, Eq. (7.163), in the limit $\Lambda \rightarrow \infty$.

From a physical point of view, the regularization (7.162) represents an electron interacting with—apart from the massless photons—also "massive photons" of mass Λ, the electron couplings to the massless and massive photons featuring opposite signs. Since Λ acquires large values, the interaction with massive photons is significant only at very small distances. The origin of this idea could be traced back to Poincare's attempts to solve the self-energy problem in the framework of classical electrodynamics.[27] It should be pointed out, however, that the idea to "repair" electrodynamics on the classical level and only then proceed to the quantum theory has never been successful and this result is not likely to change. For, even if the electron self-energy is solved in this way, we are still left with the photon self-energy, Eq. (7.149), which is a purely quantum effect with no classical analog.[28] A modern

[27]See, e.g., [8, 32] where other attempts to solve the self-energy problem within classical electrodynamics are summarized as well.

[28]Furthermore, a photon always interacts virtually with bosons $W^{\pm}$, mediating thus weak interactions, see, e.g., [13], and with quark-antiquark pairs which interact with each other also via strong force. When calculating shifts of the atomic energy levels, these virtual interactions are utterly negligible. However, they are present and they do contribute to the electron inertial mass. Therefore, any attempt to solve the origin of the electron mass is necessarily incomplete within the framework of the electromagnetic theory.

view of the regularization (7.162) is of a purely formal nature. Its only purpose is to transform the integral over the four-momenta of the virtual photon into a convergent one in a way that preserves the Lorentz invariance. A large, though finite, value of Λ suppresses contributions from the region of very large $\boldsymbol{k}^2$. Unfortunately, we possess no experimental information about this region and hence do not know what happens there. Apparently, we cannot assume that everything remains the same. The theory itself points out to its own limitations.

Clearly, the regularization itself does not solve anything even if we knew the magnitude of Λ, which we do not. The divergence is unambiguously removed by renormalization which we will focus on in a moment. Prior to coming to it, we need to turn our attention to the evaluation of the expression (7.163).

7.5.2 Integration over the Four-Momenta of the Virtual Photon

The compact notation hides the actual complexity of the expression (7.163). What we are to calculate is an 11-dimensional integral: 4 integrations over the photon momenta, 3+3 integrations over the electron variables when calculating the matrix elements of γ_μ and $(2\hat{\Pi}_\mu - \boldsymbol{k}\cdot\boldsymbol{\gamma}\gamma_\mu)$ between the reference and virtual atomic states, and one integration plus one summation over the continuous and discrete parts of the atomic virtual states, respectively. In addition, as mentioned above, the whole expression diverges in the limit $\Lambda \to \infty$ and we need to find a way how to arrive at a finite expression. The last and worst, for the most interesting case of an external Coulomb field, Eq. (7.25), the expression (7.163) cannot be calculated exactly and we need to rely on a perturbation expansion. However, the whole expression contains not only one, but many significant regions. On one hand, we need to describe well the region of large frequencies ω of the virtual photon as this region causes the aforementioned ultraviolet divergences. On the other hand, Bethe showed with his simple estimate, see Sect. 6.5.3, that the dominant contribution to the whole effect comes from the region of low frequencies ω of the virtual photon. Moreover, the electron momentum in virtual states can acquire both large and small values and we need to find out which one will be of greater importance. All of the listed problems can be solved [41]. In this book, we will restrict ourselves to the leading order estimate of the effect, when a number of simplifications appears.

To find the perturbation expansion of the electron propagator $(\boldsymbol{k}^2 - 2\boldsymbol{k}\cdot\hat{\boldsymbol{\Pi}} + \hat{\mathscr{H}})^{-1}$, that yields a sufficiently accurate estimate of the effect after considering the first few terms, it suffices to make a single assumption. Namely, we assume that the electron four-momentum $\hat{\boldsymbol{\Pi}}$ is dominated by the four-momentum $\boldsymbol{\varepsilon}$ of the electron at rest,

$$\boldsymbol{\varepsilon} = (m, 0)\,.$$

For instance, after changing to atomic units, we find for the external Coulomb field that

$$\hat{\boldsymbol{\Pi}} - \boldsymbol{\varepsilon} = \left(E - m + \frac{m(Z\alpha)^2}{\hat{\mathsf{r}}_A}, (mZ\alpha)\hat{\mathbf{p}}_A \right).$$

Hence time and space components of $\hat{\boldsymbol{\Pi}} - \boldsymbol{\varepsilon}$ are suppressed by the factors $(Z\alpha)^2$ and $(Z\alpha)$ with respect to the rest mass m. However, we are to integrate over all possible wave numbers k_e of the electron in virtual states. The suppression of the momentum components $\hat{\boldsymbol{p}}$ with respect to the rest mass m clearly fails for electron virtual states with the wave numbers k_e larger than $(Z\alpha)^{-1}$. Nevertheless, we have seen in the case of the photoelectric effect, see Sect. 6.3.5, that within the dipole approximation, the overlap between the ground and continuum states is the largest for the states lying closely above the ionization threshold and rapidly decreases with increasing wave numbers k_e. This means that of all the continuum states, those lying closely above the ionization threshold, i.e., the ones with $k_e \sim 1$, $|\mathbf{p}| \sim mZ\alpha$, dominate. The dipole approximation corresponds here to the negligence of the term $2\mathbf{k} \cdot \hat{\boldsymbol{p}}$ in the electron propagator, see Eqs. (7.143) and (7.163), which is the space part of $2\boldsymbol{k} \cdot (\hat{\boldsymbol{\Pi}} - \boldsymbol{\varepsilon})$. We can expect the dipole approximation to apply as the nonrelativistic Bethe estimate discussed in Sect. 6.5.3 lies close to the experimental value. In summary, the contribution of the states $k_e > (Z\alpha)^{-1}$ is indeed suppressed with respect to the contribution of the states $k_e \simeq 1$ and the expansion of electron propagator in powers of $2\boldsymbol{k} \cdot (\hat{\boldsymbol{\Pi}} - \boldsymbol{\varepsilon})$ will be appropriate.[29]

As follows from the foregoing considerations, we expand the electron propagator as:

$$\begin{aligned}\frac{1}{\boldsymbol{k}^2 - 2\hat{\boldsymbol{\Pi}} \cdot \boldsymbol{k} + \hat{\mathscr{H}}} = &\frac{1}{\boldsymbol{k}^2 - 2\boldsymbol{\varepsilon} \cdot \boldsymbol{k} + \hat{\mathscr{H}}} \\ &+ \frac{1}{\hat{\mathscr{L}}_{01}} 2\boldsymbol{k} \cdot (\hat{\boldsymbol{\Pi}} - \boldsymbol{\varepsilon})_1 \frac{1}{\hat{\mathscr{L}}_{12}} \\ &+ \frac{1}{\hat{\mathscr{L}}_{01}} 2\boldsymbol{k} \cdot (\hat{\boldsymbol{\Pi}} - \boldsymbol{\varepsilon})_1 \frac{1}{\hat{\mathscr{L}}_{12}} 2\boldsymbol{k} \cdot (\hat{\boldsymbol{\Pi}} - \boldsymbol{\varepsilon})_2 \frac{1}{\hat{\mathscr{L}}_{23}} + \dots,\end{aligned} \tag{7.164}$$

where

$$\hat{\mathscr{L}}_{mn} = \boldsymbol{k}^2 - 2\boldsymbol{\varepsilon} \cdot \boldsymbol{k} + \hat{\mathscr{H}}_{mn}$$

and we used Eq. (6.198). The components of the operator $\hat{\Pi}$ commute neither mutually nor with the operator $\hat{\mathscr{H}}$; therefore we attach subscripts to the operators $\hat{\Pi}$ and $\hat{\mathscr{H}}$ to keep the order of their action. Once we do so, we can treat the operators $\hat{\Pi}$ and $\hat{\mathscr{H}}$ as numbers! Clearly, the double subscripts of $\hat{\mathscr{H}}$ are given by the subscripts of two adjacent operators $\hat{\Pi}$. It is convenient to introduce the notation

[29] Had we aimed at great precision, we would have to pay special attention to the states of the continuum part of the spectrum with very large wave numbers. The contribution of these states to the self-energy is not described well by the proposed expansion. If the reader wishes to learn how to do so, we refer him to [41].

$$D_1\left[f(\hat{\mathscr{H}})\right] = \frac{1}{\hat{\mathscr{H}}_{12}-\hat{\mathscr{H}}_{01}}f(\hat{\mathscr{H}}_{12}) + \frac{1}{\hat{\mathscr{H}}_{01}-\hat{\mathscr{H}}_{12}}f(\hat{\mathscr{H}}_{01}), \tag{7.165}$$

$$D_2\left[f(\hat{\mathscr{H}})\right] = \frac{1}{(\hat{\mathscr{H}}_{01}-\hat{\mathscr{H}}_{12})(\hat{\mathscr{H}}_{01}-\hat{\mathscr{H}}_{23})}f(\hat{\mathscr{H}}_{01}) + \frac{1}{(\hat{\mathscr{H}}_{12}-\hat{\mathscr{H}}_{01})(\hat{\mathscr{H}}_{12}-\hat{\mathscr{H}}_{23})}f(\hat{\mathscr{H}}_{12}) + \frac{1}{(\hat{\mathscr{H}}_{23}-\hat{\mathscr{H}}_{01})(\hat{\mathscr{H}}_{23}-\hat{\mathscr{H}}_{12})}f(\hat{\mathscr{H}}_{23}). \tag{7.166}$$

and so on. We apply it to Eq. (7.164) to obtain

$$\frac{1}{\boldsymbol{k}^2-2\hat{\boldsymbol{\Pi}}\cdot\boldsymbol{k}+\hat{\mathscr{H}}} = \frac{1}{\boldsymbol{k}^2-2\boldsymbol{k}\cdot\boldsymbol{\varepsilon}+\hat{\mathscr{H}}} + 2\boldsymbol{k}\cdot(\hat{\boldsymbol{\Pi}}-\boldsymbol{\varepsilon})_1 D_1\left[-\frac{1}{\boldsymbol{k}^2-2\boldsymbol{k}\cdot\boldsymbol{\varepsilon}+\hat{\mathscr{H}}}\right] + 2\boldsymbol{k}\cdot(\hat{\boldsymbol{\Pi}}-\boldsymbol{\varepsilon})_1 2\boldsymbol{k}\cdot(\hat{\boldsymbol{\Pi}}-\boldsymbol{\varepsilon})_2 D_2\left[\frac{1}{\boldsymbol{k}^2-2\boldsymbol{k}\cdot\boldsymbol{\varepsilon}+\hat{\mathscr{H}}}\right] + \ldots$$

Notice that we succeeded in writing the original fractions as a sum of simpler fractions. However, a problem seems to appear as the number of components of $\boldsymbol{k}$ in the numerator increases with the increasing order of the expansion, while the powers of $\boldsymbol{k}$ in the denominators do not change! Thus the integrals over $\boldsymbol{k}$ seem to be becoming more rapidly divergent as the order of the expansion grows. Recall, though, that if we return to the original form of the expansion, Eq. (7.164), it becomes obvious that if the first term of the expansion leads to a convergent integral, so do the other terms. Thus the above mentioned difficulty can be avoided by differentiating the individual terms on the rhs of the last equation with respect to $\hat{\mathscr{H}}$ and integrating back[30]

[30]If the reader is scared to differentiate and integrate with respect to the operator $\hat{\mathscr{H}}$, the use of its spectral decomposition could return his courage:

$$\frac{1}{\boldsymbol{k}^2-2\boldsymbol{k}\cdot\boldsymbol{\varepsilon}+\hat{\mathscr{H}}} = \sum_n \frac{|\psi_n\rangle\langle\psi_n|}{\boldsymbol{k}^2-2\boldsymbol{k}\cdot\boldsymbol{\varepsilon}+E_n}\,.$$

Here, E_n and ψ_n are eigenvalues and eigenstates of $\hat{\mathscr{H}}$. Differentiation and integration with respect to $\hat{\mathscr{H}}$ then clearly means differentiation and integration of the n-th term on the rhs with respect to E_n.

$$
\begin{aligned}
\frac{1}{\boldsymbol{k}^2 - 2\boldsymbol{k}\cdot\boldsymbol{\varepsilon} + \hat{\mathscr{H}}} &= \int_0^{\hat{\mathscr{H}}} \frac{\mathrm{d}}{\mathrm{d}w_1}\frac{1}{\boldsymbol{k}^2 - 2\boldsymbol{k}\cdot\boldsymbol{\varepsilon} + w_1}\mathrm{d}w_1 \\
&= \int_0^{\hat{\mathscr{H}}}\int_0^{w_2} \frac{\mathrm{d}^2}{\mathrm{d}w_1^2}\frac{1}{\boldsymbol{k}^2 - 2\boldsymbol{k}\cdot\boldsymbol{\varepsilon} + w_1}\mathrm{d}w_1\mathrm{d}w_2 \\
&= \int_0^{\hat{\mathscr{H}}} (\hat{\mathscr{H}} - w)\frac{\mathrm{d}^2}{\mathrm{d}w^2}\frac{1}{\boldsymbol{k}^2 - 2\boldsymbol{k}\cdot\boldsymbol{\varepsilon} + w}\mathrm{d}w \\
&= \frac{1}{(n-1)!}\int_0^{\hat{\mathscr{H}}} (\hat{\mathscr{H}} - w)^{n-1}\frac{\mathrm{d}^n}{\mathrm{d}w^n}\frac{1}{\boldsymbol{k}^2 - 2\boldsymbol{k}\cdot\boldsymbol{\varepsilon} + w}\mathrm{d}w
\end{aligned}
$$

and so on, where we used the formula for iterated integration

$$
\int_0^a \mathrm{d}w_k \dots \int_0^{w_3} \mathrm{d}w_2 \int_0^{w_2} \mathrm{d}w_1 f(w_1) = \frac{1}{(k-1)!}\int_0^a \mathrm{d}w (a-w)^{(k-1)} f(w) .
$$

The components of k in the numerators, k_α, $k_\alpha k_\beta$, etc., can now be eliminated via parametric differentiation:

$$
\begin{aligned}
\frac{\partial}{\partial \varepsilon_\lambda}\frac{1}{\boldsymbol{k}^2 - 2\boldsymbol{k}\cdot\boldsymbol{\varepsilon} + w} &= -\frac{\mathrm{d}}{\mathrm{d}w}\frac{2k_\lambda}{\boldsymbol{k}^2 - 2\boldsymbol{k}\cdot\boldsymbol{\varepsilon} + w} , \\
\frac{\partial^2}{\partial \varepsilon_\alpha \partial \varepsilon_\beta}\frac{1}{\boldsymbol{k}^2 - 2\boldsymbol{k}\cdot\boldsymbol{\varepsilon} + w} &= \frac{\mathrm{d}^2}{\mathrm{d}w^2}\frac{2k_\alpha 2k_\beta}{\boldsymbol{k}^2 - 2\boldsymbol{k}\cdot\boldsymbol{\varepsilon} + w}
\end{aligned}
$$

and so on. Here we used the identity following from the definition of the four-dimensional gradient (7.9)

$$
\frac{\partial \varepsilon_\mu}{\partial \varepsilon_\nu} = \eta_{\mu\nu} .
$$

Collecting all the parts together, we have

$$
\begin{aligned}
-\frac{2k_\lambda}{\boldsymbol{k}^2 - 2\boldsymbol{k}\cdot\boldsymbol{\varepsilon} + \hat{\mathscr{H}}} &= \frac{\partial}{\partial \varepsilon_\lambda}\int_0^{\hat{\mathscr{H}}} \frac{\mathrm{d}w}{\boldsymbol{k}^2 - 2\boldsymbol{k}\cdot\boldsymbol{\varepsilon} + w} , \\
\frac{2k_\lambda 2k_\rho}{\boldsymbol{k}^2 - 2\boldsymbol{k}\cdot\boldsymbol{\varepsilon} + \hat{\mathscr{H}}} &= \frac{\partial^2}{\partial \varepsilon_\lambda \partial \varepsilon_\rho}\int_0^{\hat{\mathscr{H}}} \frac{\mathrm{d}w(\hat{\mathscr{H}} - w)}{\boldsymbol{k}^2 - 2\boldsymbol{k}\cdot\boldsymbol{\varepsilon} + w}
\end{aligned}
$$

and so on. The expansion (7.164) of the electron propagator can be thus written in a concise form

$$\frac{1}{\boldsymbol{k}^2 - 2\boldsymbol{k}\cdot\hat{\boldsymbol{\Pi}} + \hat{\mathscr{H}}} = \frac{1}{\boldsymbol{k}^2 - 2\boldsymbol{k}\cdot\boldsymbol{\varepsilon} + \hat{\mathscr{H}}} + (\hat{\boldsymbol{\Pi}} - \boldsymbol{\varepsilon})_{1\lambda}\frac{\partial}{\partial\varepsilon_\lambda}D_1\left[\int_0^{\hat{\mathscr{H}}}\frac{\mathrm{d}w}{\boldsymbol{k}^2 - 2\boldsymbol{k}\cdot\boldsymbol{\varepsilon} + w}\right]$$

$$+(\hat{\boldsymbol{\Pi}} - \boldsymbol{\varepsilon})_{1\lambda}(\hat{\boldsymbol{\Pi}} - \boldsymbol{\varepsilon})_{2\rho}\frac{\partial^2}{\partial\varepsilon_\lambda\partial\varepsilon_\rho}D_2\left[\int_0^{\hat{\mathscr{H}}}\frac{\mathrm{d}w(\hat{\mathscr{H}} - w)}{\boldsymbol{k}^2 - 2\boldsymbol{k}\cdot\boldsymbol{\varepsilon} + w}\right] + \ldots$$

We substitute this expression into Eq. (7.163) and see that the only integrals over the four-momentum $\boldsymbol{k}$ to be explicitly performed are only the following two: (we write them compactly as one, using vector notation)

$$\int_0^{\Lambda^2}\mathrm{d}\lambda\int\frac{\mathrm{d}^4\boldsymbol{k}_F(1,k_\nu)}{(\boldsymbol{k}^2-\lambda)^2}\frac{1}{(n-1)!}\int_0^{\hat{\mathscr{H}}}\mathrm{d}w\frac{(\hat{\mathscr{H}}-w)^{n-1}}{\boldsymbol{k}^2 - 2\boldsymbol{k}\cdot\boldsymbol{\varepsilon} + w}. \tag{7.167}$$

These integrals are most easily calculated by the method of *Feynman parameters.* When integrating over the four-momentum of the virtual photon $\boldsymbol{k}$, it is more advantageous to deal with only one fraction rather than a product of two. Why? Recall that when integrating over k_0 we need to calculate the residues of poles, see Sect. 7.4.5. Feynman, inspired by earlier Schwinger's work, realized that by the introduction of an additional integration variable, the product of two fractions can be reduced to one fraction

$$\frac{1}{ab} = \int_0^1\frac{\mathrm{d}y}{(ay + b(1-y))^2}$$

and similarly for higher powers, for instance

$$\frac{1}{ab^2} = -\frac{\partial}{\partial b}\frac{1}{ab} = \int_0^1\frac{2(1-y)\mathrm{d}y}{(ay+b(1-y))^3}.$$

The integration variable y is usually called the Feynman parameter.

The integrals (7.167) can be rearranged by means of the last equation and substitution $k \to k + \varepsilon y$

$$\int\frac{\mathrm{d}^4\boldsymbol{k}_F(1,k_\nu)}{(\boldsymbol{k}^2-\lambda)^2(\boldsymbol{k}^2 - 2\boldsymbol{k}\cdot\boldsymbol{\varepsilon} + w)} = \int_0^1\mathrm{d}y2(1-y)\int\mathrm{d}^4\boldsymbol{k}_F\frac{(1,k_\nu+\varepsilon_\nu y)}{(\boldsymbol{k}^2 - \boldsymbol{\varepsilon}^2y^2 + wy - \lambda(1-y))^3}.$$

The only integrals over the four-momenta of the virtual photon to be explicitly performed are the integrals

$$\int\frac{\mathrm{d}^4\boldsymbol{k}_F(1,k_\nu)}{(\boldsymbol{k}^2-L)^3} = \frac{(1,0)}{8L}. \tag{7.168}$$

The first integral is calculated as follows: we add infinitesimal negative and positive imaginary parts to the poles lying on the positive and negative real axis, respectively. We integrate over a closed curve composed of the real axis and a semicircle in the lower half of the complex plane, see Sect. 7.4.5. The result of the integration over k_0 reads $-2\pi \mathrm{i}$ times the residues in the poles lying in the lower half of the complex plane

$$\int \frac{\mathrm{d}^4 k_F}{(k^2 - L)^3} = \frac{\mathrm{i}}{2\pi} \int \frac{\mathrm{d}^3 \mathbf{k}}{2\pi} \int_{-\infty}^{\infty} \frac{\mathrm{d}k_0}{(k_0^2 - \omega^2 - L)^3} ,$$

$$\mathrm{Res}_{k_0 = \sqrt{\omega^2 + L}} \left\{ \frac{1}{(k_0^2 - \omega^2 - L)^3} \right\} = \frac{6}{(2\sqrt{\omega^2 + L})^5} .$$

For the integration over $\mathrm{d}^3\mathbf{k}$, we introduce spherical coordinates and obtain

$$\int \frac{\mathrm{d}^4 k_F}{(k^2 - L)^3} = \frac{3}{8} \int_0^{\infty} \frac{\omega^2 \mathrm{d}\omega}{(\omega^2 + L)^{5/2}} = \frac{1}{8L} .$$

The reader can easily verify that the result is the same if we close the integration path in the upper half of the complex plane. The second integral in (7.168) is odd in the components of k_ν, hence vanishes.

Note that we managed to perform the integration over the four-momenta of the virtual photon without separating the Dirac Hamiltonian into positive and negative energy modes! Owing to the rearrangement (7.160), we obtain the Dirac second order Hamiltonian $\hat{\mathscr{H}}$ in the denominator. This Hamiltonian is negative definite if the reference state is the ground state, as one can see from the following reasoning. The Dirac second order Hamiltonian in the nonrelativistic limit (7.26)[31] has the eigenvalues $-(Z\alpha m)^2 \left(k_e^2 + \frac{1}{N^2}\right)$ and $-(Z\alpha m)^2 \left(-\frac{1}{n^2} + \frac{1}{N^2}\right)$ for continuous and discrete parts of the spectrum, respectively. If the reference state is the ground state, then necessarily $n \geq N$. If the reference state is an excited state, a part of the discrete spectrum eigenvalues, $n < N$, is positive. This breakdown of the negative definiteness leads to nonzero imaginary parts of the excited state energies discussed in Sect. 6.5.3.

Owing to the negative definiteness of the operator $\hat{\mathscr{H}}$, we will not encounter the separation of the positive and negative energies any further. From a computational point of view, this point is of key importance, although usually not emphasized at all. The introduction of a fourth integration variable, for instance k_0, proves practically indeed advantageous *only* as one eliminates the first order Dirac Hamiltonian $\hat{\mathfrak{h}}_\mathrm{D}$ in favor of the second order Dirac Hamiltonian $\hat{\mathscr{H}}$ in the denominators of the integrands of the integrals over the four-momenta of the virtual particles.

[31] We will need the spectral decomposition of $\hat{\mathscr{H}}$ only at this limit. For a more complete treatment, see [41].

7.5.3 Mass Renormalization

After substituting Eq. (7.168) into Eq. (7.167) we find

$$\int_0^{\Lambda^2} d\lambda \int \frac{d^4 k_F(1, k_\nu)}{(k^2-\lambda)^2(k^2-2k\cdot\varepsilon+w)}$$
$$= \int_0^{\Lambda^2} d\lambda \int_0^1 dy \frac{1}{4}(1-y)\frac{(1,\varepsilon_\nu y)}{\varepsilon^2 y^2 - wy + \lambda(1-y)}$$
$$= \frac{1}{4}\int_0^1 dy(1,\varepsilon_\nu y)\ln\frac{\Lambda^2(1-y)}{\varepsilon^2 y^2 - wy} + O(1/\Lambda^2)\,. \tag{7.169}$$

These integrals clearly diverge (logarithmically) in the limit $\Lambda \to \infty$, which we need to tackle somehow.[32] Let us start by considering Eq. (7.163) for a free electron. What is the effect of the interaction of the free electron with its own EM field? There is only one possibility: this interaction adds the "electromagnetic mass" to the mass m of the "bare" electron, i.e., a hypothetical electron which is not surrounded by any EM field. The electromagnetic mass is the part of the electron mass which arises from the interaction of a real electron with other charged particles via the EM field. This EM field carries energy and hence, according to the special relativity, mass. The observable electron mass m_{exp} is a sum of the "bare" and "electromagnetic" masses

$$m_{\text{exp}} = m + \Delta m\,. \tag{7.170}$$

The energy of free particle equals $E = \sqrt{\mathbf{p}^2 + m^2}$. Differentiation of this equality with respect to m for a fixed $\mathbf{p}$ yields

$$(\Delta E)_0^{\text{SE}} = \frac{m}{E}\Delta m\,. \tag{7.171}$$

We clearly have $\hat{\boldsymbol{\Pi}} = \hat{\boldsymbol{p}}$ for a free particle. Consequently, the second order Hamiltonian $\hat{\mathscr{H}} = \hat{\boldsymbol{p}}^2 - m^2$ commutes with the γ-matrices and the electron propagator $(\boldsymbol{k}^2 - 2\boldsymbol{k}\cdot\hat{\boldsymbol{p}} + \hat{\mathscr{H}})^{-1}$ in Eq. (7.163) acts directly on the free-particle reference state which we will denote as $|p\rangle$. We find from the definition of the reference state that $\hat{\mathscr{H}}|p\rangle = \left(\hat{\boldsymbol{p}}^2 - m^2\right)|p\rangle = (\boldsymbol{p}^2 - m^2)\,|p\rangle = 0$, hence

$$\frac{1}{\boldsymbol{k}^2 - 2\boldsymbol{k}\cdot\hat{\boldsymbol{p}} + \hat{\mathscr{H}}}|p\rangle = \frac{1}{\boldsymbol{k}^2 - 2\boldsymbol{k}\cdot\boldsymbol{p}}|p\rangle\,.$$

---

[32] The reader surely notices that the following consideration resembles the one made in Sect. 6.5.

Here, $\boldsymbol{p}$ is obviously the four-momentum of the reference state $\boldsymbol{p}^2 = m^2$. It follows from the last two relations and Eq. (7.169) that in case of a free particle the integration over the four-momentum $\boldsymbol{k}$ of the virtual photon yields

$$\int_0^{\Lambda} \mathrm{d}\lambda \int \frac{\mathrm{d}^4\boldsymbol{k}_F}{(\boldsymbol{k}^2-\lambda)^2} \frac{(1, k_\nu)}{\boldsymbol{k}^2 - 2\boldsymbol{k}\cdot\boldsymbol{p}} = \frac{1}{4}\int_0^1 \mathrm{d}y(1, p_\nu y) \ln \frac{\Lambda^2(1-y)}{m^2y^2} .$$

The expression (7.163) in the case of a free particle leads to

$$(\Delta E)_0^{\mathrm{SE}} = \frac{\alpha}{2\pi}\int_0^1 \mathrm{d}y \ln \frac{\Lambda^2(1-y)}{m^2y^2} \langle \overline{p}|\gamma_\mu \left(\hat{\mathsf{p}}_\mu - \frac{y}{2}\hat{\mathsf{p}}_\nu\gamma_\nu\gamma_\mu\right)|p\rangle \tag{7.172}$$

$$= \frac{\alpha}{2\pi} m \int_0^1 \mathrm{d}y(1+y) \ln \frac{\Lambda^2(1-y)}{m^2y^2} \langle \overline{p}|p\rangle ,$$

where we used the Dirac equation for a free particle, Eq. (7.16), $\boldsymbol{\gamma}\cdot\hat{\boldsymbol{p}}|p\rangle = m|p\rangle$, and the identity for the Dirac matrices

$$\gamma_\mu\gamma_\nu\gamma_\mu = -2\gamma_\nu \tag{7.173}$$

which follows from the anticommutation relations (7.17). The reader can easily show from the solution of the Dirac equation (7.16) that the relation

$$\langle \overline{p}|p\rangle = \frac{m}{E} \tag{7.174}$$

holds for a free particle. By comparing Eqs. (7.172) and (7.174) with Eq. (7.171), we obtain for the electromagnetic electron mass

$$\Delta m = \frac{\alpha}{2\pi} m \int_0^1 (1+y) \ln \frac{\Lambda^2(1-y)}{m^2y^2} \mathrm{d}y , \tag{7.175}$$

which is clearly invariant with respect to Lorentz transformations.

In Eq. (7.169), we now separate the part present also for a free electron and the part present only for a bound electron. Since a logarithm of a product is a sum of logarithms, we have

$$(7.169) = \frac{1}{4}\int_0^1 \mathrm{d}y(1, \varepsilon_\nu y) \ln \frac{\Lambda^2(1-y)}{m^2y^2} - \frac{1}{4}\int_0^1 \mathrm{d}y(1, \varepsilon_\nu y) \ln \frac{\boldsymbol{\varepsilon}^2y^2 - wy}{m^2y^2} .$$

We write the result of the integration in Eq. (7.163) in the form

$$(\Delta E)^{\mathrm{SE}} = (\Delta E)^{\mathrm{SE}}_{\mathrm{EM}} + (\Delta E)^{\mathrm{SE}}_{\mathrm{exp}} , \tag{7.176}$$

where

$$
\begin{aligned}
(\Delta E)^{\rm SE}_{\rm EM} &= \frac{\alpha}{2\pi}\int_0^1 {\rm d}y \ln\frac{\Lambda^2(1-y)}{m^2y^2} \\
&\times \langle\overline{\psi}_{\rm at}|\gamma_\mu\left(\hat{\sqcap}_\mu - \frac{1}{2}y\left(\varepsilon_\nu + (\hat{\boldsymbol{\Pi}} - \boldsymbol{\varepsilon})_{1\lambda}\frac{\partial}{\partial\varepsilon_\lambda}D_1\left[\varepsilon_\nu\int_0^{\hat{\mathscr{H}}} {\rm d}w\right]\right)\gamma_\nu\gamma_\mu\right)|\psi_{\rm at}\rangle \\
&= \frac{\alpha}{2\pi}\int_0^1 {\rm d}y \ln\frac{\Lambda^2(1-y)}{m^2y^2}\langle\overline{\psi}_{\rm at}|\gamma_\mu\left(\hat{\sqcap}_\mu - \frac{y}{2}\hat{\sqcap}_\nu\gamma_\nu\gamma_\mu\right)|\psi_{\rm at}\rangle \\
&= \frac{\alpha}{2\pi}\int_0^1 {\rm d}y(1+y)\ln\frac{\Lambda^2(1-y)}{m^2y^2}\langle\overline{\psi}_{\rm at}|\psi_{\rm at}\rangle = \Delta m\langle\overline{\psi}_{\rm at}|\psi_{\rm at}\rangle\,.
\end{aligned}
\tag{7.177}
$$

In the second equality, we used the identity

$$
D_1[\mathscr{H}] = \frac{\hat{\mathscr{H}}_{01}}{\hat{\mathscr{H}}_{01}-\hat{\mathscr{H}}_{12}} + \frac{\hat{\mathscr{H}}_{12}}{\hat{\mathscr{H}}_{12}-\hat{\mathscr{H}}_{01}} = 1
$$

and in the third equality we used the Dirac equation (7.19), $\boldsymbol{\gamma}\cdot\hat{\boldsymbol{\Pi}}\,|\psi_{\rm at}\rangle = m|\psi_{\rm at}\rangle$.

Now we insert into the Hamiltonian (7.34) of a Dirac particle in an external EM field from Eq. (7.170),

$$
\hat{\sqcap}_{\rm D}(m) = \gamma_0\boldsymbol{\gamma}\cdot(\hat{\mathbf{p}} - e\mathbf{A}(\hat{\mathbf{r}})) + e\varphi(\hat{\mathbf{r}}) + \gamma_0 m = \hat{\sqcap}_{\rm D}(m_{\rm exp}) - \gamma_0\Delta m\,,
$$

where the last term is of the first order in α, see Eq. (7.175), and should be treated perturbatively. Its effect on the bound state energies evaluated in the first order of the perturbation method is clearly

$$
-\langle\overline{\psi}_{\rm at}|\psi_{\rm at}\rangle\Delta m\,,
$$

which matches but for the opposite sign Eq. (7.177).

The observable part of the effect of the electron interaction with its own EM field is given by the second term in Eq. (7.176), where the replacement $m \to m_{\rm exp}$ has to be made. *This term is independent of the cutoff Λ and clearly finite.* Notice that an unambiguous determination of which part of the self-energy effect can be ascribed to the electron electromagnetic mass and which one is the observable part is substantially simplified by the Lorentz invariance of the self-energy operator and the Lorentz invariant regularization (7.162). Early efforts aiming at this distinction within the framework of the ordinary perturbation method and using a Lorentz non-invariant regularization led to considerable confusion, see [35]. We emphasize that it is not true that the "renormalization consists of throwing away the infinite terms."

The reader can perform the integration in Eq. (7.172) to easily find that the electron electromagnetic mass comprises also terms which are finite in the limit $\Lambda \to \infty$. We will see in Sect. 7.6 that the preservation of the manifest gauge invariance of the regularized expression is by no means a lesser issue than the preservation of the Lorentz invariance.

The removal of the infinite "small correction" Δm to the mass of a particle, as we made above, is a procedure ubiquitous in all field theories. It is termed the *renormalization*, here in particular the *mass renormalization*. Usually, there is more than one parameter to be renormalized; there is usually more than one loop diagram yielding an infinite result. For instance, in quantum electrodynamics there is also a charge renormalization which stems from the need to make the contributions from the diagrams with a closed fermion loop finite, see the next Sect. 7.6. In general, quantum field theories involving a *finite* number of parameters to be renormalized are called *renormalizable*. Such theories form only a minor subset of all conceivable relativistic quantum field theories as the requirement of renormalizability imposes severe restrictions on the form of the interaction Hamiltonian. For instance, the basic principles of quantum mechanics and special relativity do not provide any reason why the interaction Hamiltonian of QED should not, besides the term $\int \hat{\mathsf{j}}_\mu \hat{\mathsf{A}}_\mu \mathrm{d}^3\mathbf{r}$, contain also the term $\int \hat{\mathsf{m}}_{\mu\nu}(\partial_\nu \hat{\mathsf{A}}_\mu - \partial_\mu \hat{\mathsf{A}}_\nu)\mathrm{d}^3\mathbf{r}$, where $\hat{\mathsf{m}}_{\mu\nu}$ is a magnetic moment created by the Dirac field, or other interaction terms containing even more space-time derivatives of EM potentials for that matter. The only theoretical reason to exclude these terms is the loss of renormalizability of the theory.

It is worth emphasizing that the requirement of renormalizability, together with the obvious requirement the theory reproduce known experimental facts, was up to now *the only reliable* clue[33] for the construction of quantum field theories describing weak and strong interactions that are in agreement with new experimental facts. To be more precise, the requirement of renormalizability applies only when one deals with elementary fields. An effort to create a renormalizable theory of the strong interactions where the basic fields are proton and neutron fields exchanging π-mesons proved to be a dead end. In the modern renormalizable theory of strong interactions, namely in quantum chromodynamics, the basic fields are quark fields exchanging gluons.

The requirement of renormalizability leads to theories which are in an excellent agreement with the experiment within the energy range going from 10^{-9} to 10^{-12} eV (details of the fine and hyperfine structures of atomic spectral lines) up to $10^{12} - 10^{13}$ eV (collider experiments on LHC in CERN). All of this is more remarkable given that the renormalization procedure has been regarded as "necessary evil" from its very inception. American physicist Kenneth Wilson was the first one to show that given the interaction Hamiltonian containing both renormalizable and non-renormalizable terms and integrating successively high energy degrees of freedom, the renormalizable terms dominate over non-renormalizable ones at low energies, see, e.g., [30]. Since then, renormalizable quantum field theories started to

[33]Freeman Dyson was the first one to propose that we take this clue seriously.

be regarded not as fundamental theories but rather as an effective description valid at low energies.[34] The LHC discovery of Higgs boson (and only of Higgs boson) points out, however, that our successful renormalizable theories of electromagnetic, weak and strong interactions are somewhat more fundamental and less effective than we had thought.

7.5.4 Calculation of the Observable Part of the Effect

As already pointed out above, the observable part of the energy shift is given by the second term in Eq. (7.176). The treatment of this term aiming at obtaining a number involves lengthy algebraic manipulations. Thus this and the following two sections were written for an indeed devoted reader. If desired, one may skip this derivation.

We show in this section that the observable part of $(\Delta E)^{\mathrm{SE}}$ can be expressed as a sum of low- and high-energy parts of the effect and discuss each of the two parts in Sect. 7.5.5 (the end result being Eq. (7.198)) and 7.5.6 (the end result being Eq. (7.202)), respectively. The two sections following these sections focus on the evaluation of this effect in the most interesting cases of purely magnetic (Sect. 7.5.7) and Coulomb (Sect. 7.5.8) external fields. In case of the former, the self-energy effect leads to the anomalous magnetic moment of the electron and in the latter case to the Lamb shift of spectral lines discussed already in Sect. 6.5. Vacuum polarization contributes to the Lamb shift in hydrogen-like atoms as well and we will evaluate its effect in Sect. 7.6. At the end of this section, we compare the theory and experiment. An impatient reader can now proceed directly to Sect. 7.5.7. A devoted reader may continue reading further!

As follows from previous two sections, the observable part of Eq. (7.163) is given by

$$(\Delta E)^{\mathrm{SE}}_{\mathrm{exp}} = -\frac{\alpha}{2\pi}\left\langle \gamma_\mu \left(\hat{\mathsf{G}}_4 \hat{\Pi}_\mu - \frac{1}{2}\hat{\boldsymbol{G}}\cdot\boldsymbol{\gamma}\gamma_\mu \right)\right\rangle , \tag{7.178}$$

where for an arbitrary operator $\hat{\mathsf{O}}$ we introduced the notation

$$\left\langle \hat{\mathsf{O}} \right\rangle = \langle \overline{\psi}_{\mathrm{at}} | \hat{\mathsf{O}} | \psi_{\mathrm{at}} \rangle \tag{7.179}$$

and

$$\hat{\mathsf{G}}_4 = \int_0^1 \left(L(m^2, \hat{\mathscr{H}}) + (\hat{\boldsymbol{\Pi}} - \boldsymbol{\varepsilon})_{1\lambda} \frac{\partial}{\partial \varepsilon_\lambda} D_1 \left[\int_0^{\hat{\mathscr{H}}} L(\boldsymbol{\varepsilon}^2, w)\mathrm{d}w \right] + \ldots \right) \mathrm{d}y , \tag{7.180}$$

[34]What is regarded as "low energy" depends of course on the cultural and historical situation, see also [42]. For a brief yet incisive exposition of the effective field theories see [21].

$$\hat{\mathsf{G}}_\nu = \int_0^1 \left\{ \varepsilon_\nu L(m^2, \hat{\mathscr{H}}) + (\hat{\boldsymbol{\Pi}} - \boldsymbol{\varepsilon})_{1\lambda} \frac{\partial}{\partial \varepsilon_\lambda} D_1 \left[\varepsilon_\nu \int_0^{\hat{\mathscr{H}}} L(\boldsymbol{\varepsilon}^2, w) \mathrm{d}w \right] + \ldots \right\} y \,\mathrm{d}y \,. \tag{7.181}$$

Here, symbol $L(\boldsymbol{\varepsilon}^2, w)$ represents

$$L(\boldsymbol{\varepsilon}^2, w) = \ln \frac{\boldsymbol{\varepsilon}^2 y - w}{m^2 y}$$

and by m stands for $m_{\exp}$.

Let us proceed with the calculation. We need to perform differentiations of $L(\boldsymbol{\varepsilon}^2, w)$ with respect to the components of $\boldsymbol{\varepsilon}$:

$$\frac{\partial}{\partial \varepsilon_\lambda} L = \frac{\partial \boldsymbol{\varepsilon}^2}{\partial \varepsilon_\lambda} \frac{\partial}{\partial (\boldsymbol{\varepsilon}^2)} L = 2\varepsilon_\lambda L' \,,$$

$$\frac{\partial^2}{\partial \varepsilon_\lambda \partial \varepsilon_\rho} L = \frac{\partial}{\partial \varepsilon_\rho} \left(2\varepsilon_\lambda L' \right) = 2\eta_{\lambda\rho} L' + 4\varepsilon_\lambda \varepsilon_\rho L'' \,,$$

$$\frac{\partial^3}{\partial \varepsilon_\lambda \partial \varepsilon_\rho \partial \varepsilon_\sigma} L = 4(\eta_{\lambda\rho}\varepsilon_\sigma + \eta_{\lambda\sigma}\varepsilon_\rho + \eta_{\sigma\rho}\varepsilon_\lambda) L'' + 8\varepsilon_\lambda \varepsilon_\rho \varepsilon_\sigma L''' \,,$$

$$\frac{\partial^4}{\partial \varepsilon_\lambda \partial \varepsilon_\rho \partial \varepsilon_\sigma \partial \varepsilon_\kappa} L = 4(\eta_{\lambda\rho}\eta_{\sigma\kappa} + \eta_{\lambda\sigma}\eta_{\rho\kappa} + \eta_{\sigma\rho}\eta_{\lambda\kappa}) L'' + \ \ldots$$

and so on where the primes denote differentiation with respect to $\boldsymbol{\varepsilon}^2$. We also need to differentiate the product $\varepsilon_\nu L$:

$$\frac{\partial}{\partial \varepsilon_\lambda} (\varepsilon_\nu L) = \eta_{\nu\lambda} L + 2\varepsilon_\nu \varepsilon_\lambda L' \,,$$

$$\frac{\partial^2}{\partial \varepsilon_\lambda \partial \varepsilon_\rho} (\varepsilon_\nu L) = 2(\eta_{\nu\lambda}\varepsilon_\rho + \eta_{\nu\rho}\varepsilon_\lambda + \eta_{\lambda\rho}\varepsilon_\nu) L' + 4\varepsilon_\nu \varepsilon_\lambda \varepsilon_\rho L'' \,,$$

$$\frac{\partial^3}{\partial \varepsilon_\lambda \partial \varepsilon_\rho \partial \varepsilon_\sigma} (\varepsilon_\nu L) = 2(\eta_{\nu\lambda}\eta_{\rho\sigma} + \eta_{\nu\rho}\eta_{\lambda\sigma} + \eta_{\lambda\rho}\eta_{\nu\sigma}) L' + \ \ldots$$

Inserting these relations into G_4 and G_μ we arrive at formidably looking expressions

$$\hat{\mathsf{G}}_4 = \int_0^1 \mathrm{d}y \left\{ L(m^2, \hat{\mathscr{H}}) + (\hat{\boldsymbol{\Pi}} - \boldsymbol{\varepsilon})_{1\lambda} \varepsilon_\lambda D_1 \left[2 \int_0^{\hat{\mathscr{H}}} L' \mathrm{d}w \right] \right.$$

$$+ (\hat{\boldsymbol{\Pi}} - \boldsymbol{\varepsilon})_{1\lambda} (\hat{\boldsymbol{\Pi}} - \boldsymbol{\varepsilon})_{2\lambda} D_2 \left[2 \int_0^{\hat{\mathscr{H}}} (\hat{\mathscr{H}} - w) L' \mathrm{d}w \right]$$

$$+ (\hat{\boldsymbol{\Pi}} - \boldsymbol{\varepsilon})_{1\lambda}(\hat{\boldsymbol{\Pi}} - \boldsymbol{\varepsilon})_{2\rho}\varepsilon_\lambda\varepsilon_\rho D_2\left[4\int_0^{\mathscr{H}}(\mathscr{H} - w)L''\mathrm{d}w\right]$$

$$+ (\hat{\boldsymbol{\Pi}} - \boldsymbol{\varepsilon})_{1\lambda}(\hat{\boldsymbol{\Pi}} - \boldsymbol{\varepsilon})_{2\rho}(\hat{\boldsymbol{\Pi}} - \boldsymbol{\varepsilon})_{3\sigma}(\varepsilon_\lambda\eta_{\rho\sigma} + \varepsilon_\sigma\eta_{\lambda\rho} + \varepsilon_\rho\eta_{\lambda\sigma})$$

$$\times D_3\left[4\frac{1}{2!}\int_0^{\hat{\mathscr{H}}}(\hat{\mathscr{H}} - w)^2L''\mathrm{d}w\right]$$

$$+ (\hat{\boldsymbol{\Pi}} - \boldsymbol{\varepsilon})_{1\lambda}(\hat{\boldsymbol{\Pi}} - \boldsymbol{\varepsilon})_{2\rho}(\hat{\boldsymbol{\Pi}} - \boldsymbol{\varepsilon})_{3\sigma}(\hat{\boldsymbol{\Pi}} - \boldsymbol{\varepsilon})_{4\kappa}(\eta_{\lambda\rho}\eta_{\sigma\kappa} + \eta_{\lambda\sigma}\eta_{\rho\kappa} + \eta_{\sigma\rho}\eta_{\lambda\kappa})$$

$$\times D_4\left[4\frac{1}{3!}\int_0^{\hat{\mathscr{H}}}(\hat{\mathscr{H}} - w)^3L''\mathrm{d}w\right]$$

$$+ \ldots\Bigg\},$$

$$\hat{\mathsf{G}}_\nu = \int_0^1 y\,\mathrm{d}y\Bigg\{\varepsilon_\nu L(m^2, \hat{\mathscr{H}}) + (\hat{\boldsymbol{\Pi}} - \boldsymbol{\varepsilon})_{1\nu}D_1\left[\int_0^{\hat{\mathscr{H}}}L\mathrm{d}w\right]$$

$$+ (\hat{\boldsymbol{\Pi}} - \boldsymbol{\varepsilon})_{1\lambda}\varepsilon_\lambda\varepsilon_\nu D_1\left[2\int_0^{\hat{\mathscr{H}}}L'\mathrm{d}w\right]$$

$$+ (\hat{\boldsymbol{\Pi}} - \boldsymbol{\varepsilon})_{1\lambda}(\hat{\boldsymbol{\Pi}} - \boldsymbol{\varepsilon})_{2\rho}(\eta_{\nu\lambda}\varepsilon_\rho + \eta_{\nu\rho}\varepsilon_\lambda + \eta_{\lambda\rho}\varepsilon_\nu)D_2\left[2\int_0^{\hat{\mathscr{H}}}(\hat{\mathscr{H}} - w)L'\right]$$

$$+ (\hat{\boldsymbol{\Pi}} - \boldsymbol{\varepsilon})_{1\lambda}(\hat{\boldsymbol{\Pi}} - \boldsymbol{\varepsilon})_{2\rho}(\hat{\boldsymbol{\Pi}} - \boldsymbol{\varepsilon})_{3\sigma}(\eta_{\nu\lambda}\eta_{\rho\sigma} + \eta_{\nu\rho}\eta_{\lambda\sigma} + \eta_{\lambda\rho}\eta_{\nu\sigma})$$

$$\times D_3\left[2\frac{1}{2!}\int_0^{\mathscr{H}}(\mathscr{H} - w)^2L'\mathrm{d}w\right] + \ldots\Bigg\}.$$

We regroup the terms into the form

$$\begin{aligned}\hat{\mathsf{G}}_4 = \phi_0 &+ (\hat{\boldsymbol{\Pi}} - \boldsymbol{\varepsilon})_{1\lambda}\varepsilon_\lambda D_1[\phi_2'] + (\hat{\boldsymbol{\Pi}} - \boldsymbol{\varepsilon})_{1\lambda}(\hat{\boldsymbol{\Pi}} - \boldsymbol{\varepsilon})_{2\lambda}D_2[\phi_2]\\ &+ (\hat{\boldsymbol{\Pi}} - \boldsymbol{\varepsilon})_{1\lambda}(\hat{\boldsymbol{\Pi}} - \boldsymbol{\varepsilon})_{2\rho}D_2[\phi_4'']\\ &+ (\hat{\boldsymbol{\Pi}} - \boldsymbol{\varepsilon})_{1\lambda}(\hat{\boldsymbol{\Pi}} - \boldsymbol{\varepsilon})_{2\rho}(\hat{\boldsymbol{\Pi}} - \boldsymbol{\varepsilon})_{3\sigma}(\eta_{\lambda\rho}\varepsilon_\sigma + \eta_{\lambda\sigma}\varepsilon_\rho + \eta_{\sigma\rho}\varepsilon_\lambda)D_3[\phi_4']\\ &+ (\hat{\boldsymbol{\Pi}} - \boldsymbol{\varepsilon})_{1\lambda}(\hat{\boldsymbol{\Pi}} - \boldsymbol{\varepsilon})_{2\rho}(\hat{\boldsymbol{\Pi}} - \boldsymbol{\varepsilon})_{3\sigma}(\hat{\boldsymbol{\Pi}} - \boldsymbol{\varepsilon})_{4\kappa}\\ &\quad\times (\eta_{\lambda\rho}\eta_{\sigma\kappa} + \eta_{\lambda\sigma}\eta_{\rho\kappa} + \eta_{\sigma\rho}\eta_{\lambda\kappa})D_4[\phi_4] + \ldots,\end{aligned} \tag{7.182}$$

$$
\begin{aligned}
\hat{\mathsf{G}}_\nu = \varepsilon_\nu \phi_1' &+ (\hat{\boldsymbol{\Pi}} - \boldsymbol{\varepsilon})_{1\lambda} D_1[\phi_1] + (\hat{\boldsymbol{\Pi}} - \boldsymbol{\varepsilon})_{1\lambda} \varepsilon_\lambda \varepsilon_\nu D_1[\phi_3''] \\
&+ (\hat{\boldsymbol{\Pi}} - \boldsymbol{\varepsilon})_{1\lambda} (\hat{\boldsymbol{\Pi}} - \boldsymbol{\varepsilon})_{2\rho} (\eta_{\nu\lambda}\varepsilon_\rho + \eta_{\nu\rho}\varepsilon_\lambda + \eta_{\lambda\rho}\varepsilon_\nu) D_2[\phi_3'] \\
&+ (\hat{\boldsymbol{\Pi}} - \boldsymbol{\varepsilon})_{1\lambda} (\hat{\boldsymbol{\Pi}} - \boldsymbol{\varepsilon})_{2\rho} (\hat{\boldsymbol{\Pi}} - \boldsymbol{\varepsilon})_{3\sigma} \\
&\quad \times (\eta_{\nu\lambda}\eta_{\rho\sigma} + \eta_{\nu\rho}\eta_{\lambda\sigma} + \eta_{\lambda\rho}\eta_{\nu\sigma}) D_3[\phi_3] + \ldots ,
\end{aligned} \tag{7.183}
$$

where

$$
\phi_0 = \int_0^1 \ln \frac{m^2 y - \hat{\mathscr{H}}}{m^2 y} \, \mathrm{d}y , \tag{7.184}
$$

$$
\phi_{2n} = \int_0^1 \int_0^{\hat{\mathscr{H}}} \frac{(\hat{\mathscr{H}} - w)^{2n-1}}{(2n-1)!} 2^n \frac{\partial^n}{\partial(\boldsymbol{\varepsilon}^2)^n} \ln(\boldsymbol{\varepsilon}^2 y - w) \bigg|_{\boldsymbol{\varepsilon}^2 = m^2} \mathrm{d}w \, \mathrm{d}y \tag{7.185}
$$

and

$$
\begin{aligned}
\phi_1 &= \int_0^1 y \int_0^{\hat{\mathscr{H}}} \ln \frac{m^2 y - w}{m^2 y} \, \mathrm{d}w \, \mathrm{d}y , \\
\phi_{2n+1} &= \int_0^1 y \int_0^{\hat{\mathscr{H}}} \frac{(\hat{\mathscr{H}} - w)^{2n}}{(2n)!} 2^n \frac{\partial^n}{\partial(\boldsymbol{\varepsilon}^2)^n} \ln(\boldsymbol{\varepsilon}^2 y - w) \bigg|_{\boldsymbol{\varepsilon}^2 = m^2} \mathrm{d}w \, \mathrm{d}y .
\end{aligned}
$$

The integrals ϕ_n are calculated easily and neither is it difficult to calculate, at least in the case of the Coulomb external field, the expressions of the form $(\hat{\boldsymbol{\Pi}} - \boldsymbol{\varepsilon})_1 \ldots (\hat{\boldsymbol{\Pi}} - \boldsymbol{\varepsilon})_i D_i[\phi]$ appearing in Eqs. (7.182) and (7.183). The whole calculation thus can be performed without invoking any further approximation. Recall that the only approximation made so far is the expansion of the electron propagator in the series (7.164). However, as we noted before, we will restrict ourselves to the calculation of the leading part of the effect. This allows for additional approximations which in turn lead to substantial simplifications. The calculation remains difficult enough nevertheless. Shall the reader be interested in a more accurate calculation, we refer him to [41].

After calculation of the functions ϕ we discover that they can be expanded in a series in $\hat{\mathscr{H}}$. In case of the Coulomb external field (7.25), the operator $\hat{\mathscr{H}}$ is proportional to $(Z\alpha)^2$, see Eq. (7.26). The series constitutes, apart from the increasing powers of $\hat{\mathscr{H}}$, also a term proportional to $\ln(\hat{\mathscr{H}})$ multiplied by a power of $\hat{\mathscr{H}}$. As we will see shortly, calculations involving powers of $\hat{\mathscr{H}}$ can be performed analytically to the very end, in contrast to the calculation of the terms involving logarithms of $\hat{\mathscr{H}}$. We expand the functions ϕ in the powers of $\hat{\mathscr{H}}$ up to the logarithmic term. It turns out that the calculation is both simpler and more accurate if the remaining part of the function ϕ is not expanded any further:

$$
\phi_2 = \frac{\hat{\mathscr{H}}^2}{m^2} + \tilde{\phi}_2 , \tag{7.186}
$$

$$\phi_1 \simeq -\frac{\hat{\mathscr{H}}^2}{2m^2}, \tag{7.187}$$

$$\phi_3 \simeq \frac{\hat{\mathscr{H}}^3}{6m^2} + \frac{\hat{\mathscr{H}}^4}{12m^4}, \qquad \phi_4 \simeq -\frac{\hat{\mathscr{H}}^4}{6m^4}, \qquad \phi_5 \simeq -\frac{\hat{\mathscr{H}}^5}{60m^4}.$$

To calculate the terms with powers of $\hat{\mathscr{H}}$, we exploit the identities

$$D_1[a] = \frac{a}{\hat{\mathscr{H}}_{01} - \hat{\mathscr{H}}_{12}} + \frac{a}{\hat{\mathscr{H}}_{12} - \hat{\mathscr{H}}_{01}} = 0, \qquad D_1[\hat{\mathscr{H}}] = 1,$$

$$D_1[\hat{\mathscr{H}}^2] = \frac{\hat{\mathscr{H}}_{01}^2}{\hat{\mathscr{H}}_{01} - \hat{\mathscr{H}}_{12}} + \frac{\hat{\mathscr{H}}_{12}^2}{\hat{\mathscr{H}}_{12} - \hat{\mathscr{H}}_{01}} = \hat{\mathscr{H}}_{01} + \hat{\mathscr{H}}_{12},$$

$$D_2[a] = D_2[\hat{\mathscr{H}}] = 0,$$

$$D_2[\hat{\mathscr{H}}^2] = \frac{\hat{\mathscr{H}}_{01}^2}{(\hat{\mathscr{H}}_{01} - \hat{\mathscr{H}}_{12})(\hat{\mathscr{H}}_{01} - \hat{\mathscr{H}}_{23})} + \frac{\hat{\mathscr{H}}_{12}^2}{(\hat{\mathscr{H}}_{12} - \hat{\mathscr{H}}_{01})(\hat{\mathscr{H}}_{12} - \hat{\mathscr{H}}_{23})}$$
$$+ \frac{\hat{\mathscr{H}}_{23}^2}{(\hat{\mathscr{H}}_{23} - \hat{\mathscr{H}}_{01})(\hat{\mathscr{H}}_{23} - \hat{\mathscr{H}}_{12})} = 1,$$

$$D_2[\hat{\mathscr{H}}^3] = \frac{\hat{\mathscr{H}}_{01}^3}{(\hat{\mathscr{H}}_{01} - \hat{\mathscr{H}}_{12})(\hat{\mathscr{H}}_{01} - \hat{\mathscr{H}}_{23})} + \frac{\hat{\mathscr{H}}_{12}^3}{(\hat{\mathscr{H}}_{12} - \hat{\mathscr{H}}_{01})(\hat{\mathscr{H}}_{12} - \hat{\mathscr{H}}_{23})}$$
$$+ \frac{\hat{\mathscr{H}}_{23}^3}{(\hat{\mathscr{H}}_{23} - \hat{\mathscr{H}}_{01})(\hat{\mathscr{H}}_{23} - \hat{\mathscr{H}}_{12})}$$
$$= \hat{\mathscr{H}}_{01} + \hat{\mathscr{H}}_{12} + \hat{\mathscr{H}}_{23},$$

and so on. Consequently, the expressions in Eqs. (7.182) and (7.183) simplify substantially. For instance, if we insert the first term on the rhs of Eq. (7.186) into the second and third terms on the rhs of Eq. (7.182), we find

$$(\hat{\boldsymbol{\Pi}} - \boldsymbol{\varepsilon})_{1\lambda}\varepsilon_\lambda D_1[2\hat{\mathscr{H}}] + (\hat{\boldsymbol{\Pi}} - \boldsymbol{\varepsilon})_{1\lambda}(\hat{\boldsymbol{\Pi}} - \boldsymbol{\varepsilon})_{2\lambda} D_2[\hat{\mathscr{H}}^2] = \hat{\boldsymbol{\Pi}} \cdot \hat{\boldsymbol{\Pi}} - m^2 = \tilde{\Pi}^2 .$$

Likewise, insertion of Eq. (7.187) into the first and second terms on the rhs of Eq. (7.183) leads to

$$\varepsilon_\nu 2\hat{\mathscr{H}} + (\hat{\boldsymbol{\Pi}} - \boldsymbol{\varepsilon})_{1\lambda} D_1[\hat{\mathscr{H}}^2] = \{\hat{\Pi}_\nu, \hat{\mathscr{H}}\}$$

and so on.

With this approximation, Eqs. (7.182) and (7.183) simplify to

$$\hat{\mathsf{G}}_4 \simeq \frac{\tilde{\Pi}^2}{m^2} - \frac{\overline{\tilde{\Pi}^4}}{6m^4} + \phi_0 + (\hat{\boldsymbol{\Pi}} - \boldsymbol{\varepsilon})_{1\lambda}\varepsilon_\lambda D_1[\tilde{\phi}_2'] + (\hat{\boldsymbol{\Pi}} - \boldsymbol{\varepsilon})_{1\lambda}(\hat{\boldsymbol{\Pi}} - \boldsymbol{\varepsilon})_{2\lambda} D_2[\tilde{\phi}_2]\,, \tag{7.188}$$

$$\hat{\mathsf{G}}_\nu \simeq -\frac{1}{2m^2}\{\hat{\Pi}_\nu, \hat{\mathscr{H}}\} + \frac{1}{6m^2}\overline{\tilde{\Pi}^2\hat{\Pi}_\nu} + \frac{1}{12m^2}\overline{\tilde{\Pi}^2\hat{\Pi}_\nu\hat{\mathscr{H}}} - \frac{1}{60m^4}\overline{\tilde{\Pi}^4\hat{\Pi}_\nu}\,, \tag{7.189}$$

where the bar over the product of the operators denotes a sum of all possible orderings; for instance

$$\overline{\tilde{\Pi}^2\hat{\Pi}_\nu} = \tilde{\Pi}_\mu\tilde{\Pi}_\mu\hat{\Pi}_\nu + \tilde{\Pi}_\mu\hat{\Pi}_\nu\tilde{\Pi}_\mu + \hat{\Pi}_\nu\tilde{\Pi}_\mu\tilde{\Pi}_\mu$$

or

$$\overline{\tilde{\Pi}^4} = \tilde{\Pi}_\mu\tilde{\Pi}_\mu\tilde{\Pi}_\nu\tilde{\Pi}_\nu + \tilde{\Pi}_\mu\tilde{\Pi}_\nu\tilde{\Pi}_\mu\tilde{\Pi}_\nu + \tilde{\Pi}_\mu\tilde{\Pi}_\nu\tilde{\Pi}_\nu\tilde{\Pi}_\mu\,.$$

Thus, the symbol $\overline{\tilde{\Pi}^2\hat{\Pi}_\nu\hat{\mathscr{H}}}$ represents a sum of 12 terms. Substituting for $\hat{\mathsf{G}}_4$ and $\hat{\mathsf{G}}_\mu$ from Eqs. (7.188) and (7.189), respectively, into Eq. (7.178) we obtain a manageable expression. We can further rearrange it by means of the Dirac equation (7.19) to

$$\left\langle \gamma_\mu \hat{\mathsf{G}}_4 \hat{\Pi}_\mu \right\rangle = \frac{1}{2}\left\langle \gamma_\mu \hat{\mathsf{G}}_4 \hat{\Pi}_\mu + \hat{\Pi}_\mu \hat{\mathsf{G}}_4 \gamma_\mu \right\rangle = m\left\langle \hat{\mathsf{G}}_4 \right\rangle - \frac{1}{2}\left\langle [\hat{\Pi}_\mu, [\gamma_\mu, \hat{\mathsf{G}}_4]] \right\rangle \tag{7.190}$$

and

$$-\frac{1}{2}\left\langle \gamma_\mu \hat{\boldsymbol{G}}\cdot\boldsymbol{\gamma}\gamma_\mu \right\rangle = -\frac{1}{4}\left\langle \gamma_\mu \{\hat{\mathsf{G}}_\nu, \gamma_\nu\}\gamma_\mu \right\rangle = \frac{1}{2}\left\langle \{\hat{\mathsf{G}}_\nu, \gamma_\nu\} \right\rangle - \frac{1}{8}\left\langle \{\gamma_\nu, [\gamma_\mu, [\gamma_\mu, \hat{\mathsf{G}}_\nu]]\} \right\rangle. \tag{7.191}$$

For the Coulomb external field, the second terms on the rhs of the last two equation are smaller than the first terms by the factor $Z\alpha$.

We split the expression for the energy shift into two parts: the low-energy one containing the logarithmic terms in $\hat{\mathscr{H}}$ coming from ϕ_0 and $\tilde{\phi}_2$, and the high-energy one coming from additional power terms,

$$(\Delta E)^{\mathrm{SE}}_{\mathrm{exp}} = (\Delta E)^{\mathrm{SE}}_{\mathrm{low}} + (\Delta E)^{\mathrm{SE}}_{\mathrm{high}}\,.$$

The nomenclature high- and low-energy parts stems from the integration over the high and low energies of virtual photons, respectively. This is hidden in the integration over the parameter y where the regions $y \to 1$ and $y \to 0$ correspond to high- and low-energy regions, respectively.

7.5.5 *Low-Energy Part of the Effect*

After inserting the last three terms in Eq. (7.188) into Eq. (7.178), we obtain

$$(\Delta E)^{\rm SE}_{\rm low} = -\frac{\alpha}{2\pi}\Big\langle \gamma_\mu\Big(\phi_0 + (\hat{\boldsymbol{\Pi}} - \boldsymbol{\varepsilon})_{1\lambda}\varepsilon_\lambda D_1\big[\tilde{\phi}_2'\big] + (\hat{\boldsymbol{\Pi}} - \boldsymbol{\varepsilon})_{1\lambda}(\hat{\boldsymbol{\Pi}} - \boldsymbol{\varepsilon})_{2\lambda} D_2\big[\tilde{\phi}_2\big]\Big)\hat{\Pi}_\mu\Big\rangle. \tag{7.192}$$

We further simplify this expression. By means of the identities

$$[\gamma_\mu, \hat{\mathscr{H}}] = 2[\hat{\Pi}_\mu, \hat{\Pi}_\lambda]\gamma_\lambda\,, \tag{7.193}$$

$$\phi(\hat{\mathscr{H}}) = (\boldsymbol{\gamma}\cdot\hat{\boldsymbol{\Pi}} - m)\frac{\phi(\hat{\mathscr{H}})}{\hat{\mathscr{H}}}(\boldsymbol{\gamma}\cdot\hat{\boldsymbol{\Pi}} + m)\,,$$

following from Eqs. (7.17) and (7.24), we can write

$$\Big\langle \gamma_\mu\phi_0\hat{\Pi}_\mu\Big\rangle = \Big\langle [\gamma_\mu, \hat{\mathscr{H}}]\frac{\phi_0}{\hat{\mathscr{H}}}\hat{\Pi}_\mu\Big\rangle = \Big\langle 2[\hat{\Pi}_\mu, \hat{\Pi}_\lambda]\gamma_\lambda\frac{\phi_0}{\hat{\mathscr{H}}}\hat{\Pi}_\mu\Big\rangle \tag{7.194}$$

and

$$\begin{aligned}\Big\langle \hat{\Pi}_\mu\phi_0\hat{\Pi}_\mu\Big\rangle &= \Big\langle \hat{\Pi}_\mu(\boldsymbol{\gamma}\cdot\hat{\boldsymbol{\Pi}} - m)\frac{\phi_0(\hat{\mathscr{H}})}{\hat{\mathscr{H}}}(\boldsymbol{\gamma}\cdot\hat{\boldsymbol{\Pi}} + m)\hat{\Pi}_\mu\Big\rangle \\ &= 2m\Big\langle [\hat{\Pi}_\mu, \hat{\Pi}_\lambda]\gamma_\lambda\frac{\phi_0}{\hat{\mathscr{H}}}\hat{\Pi}_\mu\Big\rangle + \Big\langle \gamma_\rho[\hat{\Pi}_\mu, \hat{\Pi}_\rho]\frac{\phi_0}{\hat{\mathscr{H}}}\gamma_\nu[\hat{\Pi}_\mu, \hat{\Pi}_\nu]\Big\rangle.\end{aligned} \tag{7.195}$$

We used the second order Dirac equation (7.24) in the first equality in Eq. (7.194) and the first order Dirac equation (7.19) in the second equality in Eq. (7.195). The second term on the rhs of the last equation is by a factor of $Z\alpha$ smaller than the first one. When we neglect it and combine Eqs. (7.194) and (7.195), we obtain

$$\Big\langle \gamma_\mu\phi_0\hat{\Pi}_\mu\Big\rangle \simeq \frac{1}{m}\Big\langle \hat{\Pi}_\mu\phi_0\hat{\Pi}_\mu\Big\rangle. \tag{7.196}$$

By means of Eq. (7.190) we can write, neglecting again the terms suppressed by the additional factor of $Z\alpha$,

$$\Big\langle \gamma_\mu\Big((\hat{\boldsymbol{\Pi}} - \boldsymbol{\varepsilon})_{1\lambda}\varepsilon_\lambda D_1\big[\tilde{\phi}_2'\big] + (\hat{\boldsymbol{\Pi}} - \boldsymbol{\varepsilon})_{1\lambda}(\hat{\boldsymbol{\Pi}} - \boldsymbol{\varepsilon})_{2\lambda} D_2\big[\tilde{\phi}_2\big]\Big)\hat{\Pi}_\mu\Big\rangle \simeq m\Big\langle \hat{\Pi}_\mu\frac{\tilde{\phi}_2}{\hat{\mathscr{H}}^2}\hat{\Pi}_\mu\Big\rangle, \tag{7.197}$$

where we used the identities

$$D_2[\tilde{\phi}_2]_{\hat{\mathscr{H}}_{01}=\hat{\mathscr{H}}_{23}=0} = \frac{\tilde{\phi}_2(\hat{\mathscr{H}}_{12})}{\hat{\mathscr{H}}_{12}^2}\,, \qquad D_1[\tilde{\phi}_2']_{\hat{\mathscr{H}}_{01}=\hat{\mathscr{H}}_{12}=0} = 0\,,$$

which follow from the definitions (7.165), (7.166) and $\tilde{\phi}_2(\mathscr{H} \to 0) \to \mathscr{H}^3$, see Eqs. (7.185) and (7.186). After inserting Eqs. (7.196) and (7.197) into Eq. (7.192), we have

$$(\Delta E)^{\rm SE}_{\rm low} \simeq -\frac{\alpha}{2\pi}\left\langle \hat{\Pi}_\mu \left(\frac{\phi_0}{m} + m\frac{\tilde{\phi}_2}{\hat{\mathscr{H}}^2}\right)\hat{\Pi}_\mu\right\rangle \simeq \frac{\alpha}{2\pi}\left\langle \hat{\mathsf{p}}_i \left(\frac{\phi_0}{m} + m\frac{\tilde{\phi}_2}{\hat{\mathscr{H}}^2}\right)\hat{\mathsf{p}}_i\right\rangle \tag{7.198}$$

$$= \frac{m\alpha}{\pi n^3}(Z\alpha)^4 F_{\rm low}\,,$$

where in the second equality we neglected the contribution of the time components of $\hat{\boldsymbol{\Pi}}$ with respect to the contribution of space components as it is again suppressed by an additional factor of $Z\alpha$. In the last equality, we made transition to the atomic units, (6.83),

$$F_{\rm low} = \left\langle (\hat{\mathsf{p}}_A)_i f(\hat{\mathscr{H}})(\hat{\mathsf{p}}_A)_i \right\rangle, \qquad f(\hat{\mathscr{H}}) = \frac{n^3}{2(Z\alpha)^2}\left(\phi_0 + m^2\frac{\tilde{\phi}_2}{\hat{\mathscr{H}}^2}\right). \tag{7.199}$$

From the definition of the functions ϕ_0 and $\tilde{\phi}_2$, see Eqs. (7.184), (7.185), and (7.186), we can write

$$\phi_0 = \int_0^1 \mathrm{d}y \ln\frac{m^2 y - \hat{\mathscr{H}}}{m^2 y} = \frac{-\hat{\mathscr{H}}}{m^2}\int_0^1 \mathrm{d}w \int_0^1 \mathrm{d}y \frac{1}{y - w\frac{\hat{\mathscr{H}}}{m^2}}$$

and

$$m^2\frac{\tilde{\phi}_2}{\hat{\mathscr{H}}^2} = \frac{2}{\hat{\mathscr{H}}^2}\int_0^{\hat{\mathscr{H}}} \mathrm{d}w\, w(\hat{\mathscr{H}} - w)\int_0^1 \frac{\mathrm{d}y}{m^2 y - w} = \frac{\hat{\mathscr{H}}}{m^2}\int_0^1 \mathrm{d}w\, w(1-w)\int_0^1 \mathrm{d}y \frac{2}{y - w\frac{\hat{\mathscr{H}}}{m^2}}\,.$$

After substituting the last two equations into Eq. (7.199) we have

$$f(\hat{\mathscr{H}}) = \frac{n^3}{2(Z\alpha)^2}\frac{-\hat{\mathscr{H}}}{m^2}\int_0^1 \mathrm{d}y \int_0^1 \mathrm{d}w \frac{1 - 2w(1-w)}{y - w\frac{\hat{\mathscr{H}}}{m^2}}\,. \tag{7.200}$$

The expression (7.199) could be calculated as it is. However, it suffices for our purposes to make a nonrelativistic approximation. In this approximation $\psi_{\rm at} \simeq \psi^0_{\rm at}$, see Eqs. (7.22)–(7.23), and $\frac{-\hat{\mathscr{H}}}{m^2} \simeq 2(Z\alpha)^2(\hat{\mathsf{h}}_0 - \varepsilon_0)$, see Eq. (7.26). We thus finally obtain Eq. (6.239) from Eqs. (7.199) and (7.200)

$$F_{\rm low} = \langle \psi^0_{\rm at}|(\hat{\mathsf{p}}_A)_i f(\hat{\mathsf{h}}_0 - \varepsilon_0)(\hat{\mathsf{p}}_A)_i|\psi^0_{\rm at}\rangle, \quad f(x) = n^3 x\int_0^1 \mathrm{d}y \int_0^1 \mathrm{d}w \frac{1 - 2w(1-w)}{y + 2w(Z\alpha)^2 x}\,. \tag{6.239}$$

7.5.6 *High-Energy Part of the Effect*

After inserting the first two terms in Eq. (7.188) and all of the terms in Eq. (7.189) into Eq. (7.178), using Eqs. (7.190) and (7.191), and neglecting again the terms suppressed by an additional power of $Z\alpha$, we obtain

$$\Delta E^{\rm SE}_{\rm high} = -\frac{\alpha}{2\pi}\left\langle \frac{\overline{\tilde{\Pi}^2}}{m} - \frac{\overline{\tilde{\Pi}^4}}{6m^3} - \frac{1}{4m^2}\{\gamma_\nu, \{\hat{\Pi}_\nu, \hat{\mathscr{H}}\}\} \right.$$
$$+ \frac{1}{16m^2}\left\{\gamma_\nu, \left[\gamma_\mu, \left[\gamma_\mu, \{\hat{\Pi}_\nu, \hat{\mathscr{H}}\}\right]\right]\right\} + \gamma_\nu \frac{\overline{\tilde{\Pi}^2\hat{\Pi}_\nu}}{6m^2}$$
$$\left. + \frac{1}{24m^4}\left\{\gamma_\nu, \overline{\hat{\Pi}_\nu \tilde{\Pi}^2 \hat{\mathscr{H}}}\right\} - \gamma_\nu \frac{\overline{\tilde{\Pi}^4\hat{\Pi}_\nu}}{60m^4} + \dots \right\rangle. \tag{7.201}$$

We calculate the contribution of the individual terms in Eq. (7.201) by means of the operator identity (3.99) and the Dirac equations (7.19) and (7.24). For instance,

$$\frac{1}{m^2}\left\langle \tilde{\Pi}^2 \right\rangle = -\frac{1}{4m^2}\left\langle [\gamma_\mu, \gamma_\nu][\hat{\Pi}_\mu, \hat{\Pi}_\nu]\right\rangle,$$

where we used the Dirac equation (7.24),

$$\frac{1}{6m^3}\left\langle \overline{\tilde{\Pi}^2\hat{\Pi}_\nu}\gamma_\nu \right\rangle = \frac{1}{6m^3}\left\langle \left(\frac{3}{2}\{\tilde{\Pi}^2, \hat{\Pi}_\nu\} - \frac{1}{2}[\hat{\Pi}_\mu, [\hat{\Pi}_\mu, \hat{\Pi}_\nu]]\right)\gamma_\nu \right\rangle$$
$$= -\frac{1}{8m^2}\left\langle [\gamma_\mu, \gamma_\nu][\hat{\Pi}_\mu, \hat{\Pi}_\nu]\right\rangle - \frac{1}{12m^3}\left\langle [\hat{\Pi}_\mu, [\hat{\Pi}_\mu, \hat{\Pi}_\nu]]\gamma_\nu \right\rangle,$$

where we successively used the identity (3.99), the Dirac equations (7.19) and (7.24). We further have

$$-\frac{1}{4m^3}\left\langle \left\{\gamma_\mu, \left\{\hat{\Pi}_\mu, \hat{\mathscr{H}}\right\}\right\}\right\rangle = \frac{1}{4m^3}\left\langle \left[\hat{\Pi}_\mu, [\gamma_\mu, \hat{\mathscr{H}}]\right]\right\rangle = \frac{1}{2m^3}\left\langle [\hat{\Pi}_\mu, [\hat{\Pi}_\mu, \hat{\Pi}_\lambda]]\gamma_\lambda \right\rangle,$$

$$\frac{1}{16m^3}\left\langle \left\{\gamma_\nu, \left[\gamma_\mu, \left\{[\gamma_\mu, \hat{\mathscr{H}}], \hat{\Pi}_\nu\right\}\right]\right\}\right\rangle = \frac{1}{8m^3}\left\langle \{\gamma_\nu, [\gamma_\mu, \gamma_\lambda]\}\left\{[\hat{\Pi}_\mu, \hat{\Pi}_\lambda], \hat{\Pi}_\nu\right\}\right\rangle$$

$$= \frac{1}{8m^3}\left\langle [\gamma_\nu, [\gamma_\mu, \gamma_\lambda]][\hat{\Pi}_\mu, \hat{\Pi}_\lambda]\hat{\Pi}_\nu + [[\gamma_\mu, \gamma_\lambda], \gamma_\nu]\hat{\Pi}_\nu[\hat{\Pi}_\mu, \hat{\Pi}_\lambda] + 4m[\gamma_\mu, \gamma_\lambda][\hat{\Pi}_\mu, \hat{\Pi}_\lambda]\right\rangle$$

$$= -\frac{1}{m^3}\left\langle [\hat{\Pi}_\nu, [\hat{\Pi}_\nu, \hat{\Pi}_\lambda]]\gamma_\lambda \right\rangle + \frac{1}{2m^2}\left\langle [\gamma_\mu, \gamma_\nu][\hat{\Pi}_\mu, \hat{\Pi}_\nu]\right\rangle,$$

$$-\frac{1}{6m^4}\left\langle \overline{\tilde{\Pi}^4} \right\rangle \simeq -\frac{1}{6m^4}\left\langle [\hat{\Pi}_\mu, [\hat{\Pi}_\mu, \tilde{\Pi}_\nu\tilde{\Pi}_\nu]]\right\rangle \simeq -\frac{1}{6m^4}\left\langle \left[\hat{\Pi}_\mu, \left[\hat{\Pi}_\mu, \mathscr{H}\right]\right]\right\rangle,$$

$$\frac{1}{24m^5}\left\langle \left\{\gamma_\nu, \overline{\tilde{\Pi}^2\hat{\Pi}_\nu\hat{\mathscr{H}}}\right\}\right\rangle \simeq \frac{1}{3m^4}\left\langle \overline{\tilde{\Pi}^2\mathscr{H}}\right\rangle = -\frac{1}{6m^4}\left\langle [\hat{\Pi}_\mu, [\hat{\Pi}_\mu, \mathscr{H}]]\right\rangle$$

and

$$-\frac{1}{60m^5}\left\langle\overline{\tilde{\Pi}^4\hat{\Pi}_\nu\gamma_\nu}\right\rangle \simeq -\frac{1}{12m^4}\left\langle\overline{\tilde{\Pi}^4}\right\rangle,$$

where we used Eq. (7.193) in the first two equations and neglected again the terms suppressed by an additional factor of $Z\alpha$ in the last three equations. Considering Eqs. (7.24) and (7.19) and neglecting the terms suppressed by an additional factor of $Z\alpha$, we have

$$\left\langle[\hat{\Pi}_\nu,[\hat{\Pi}_\nu,\hat{\mathscr{H}}]]\right\rangle \simeq 2m\left\langle[\hat{\Pi}_\nu,[\hat{\Pi}_\nu,\hat{\Pi}_\lambda]]\gamma_\lambda\right\rangle.$$

Finally, we collect all the parts to obtain

$$\Delta E^{\mathrm{SE}}_{\mathrm{high}} \simeq -\frac{m\alpha}{2\pi}\left\langle\frac{1}{8m^2}[\gamma_\mu,\gamma_\nu][\hat{\Pi}_\mu,\hat{\Pi}_\nu] - \frac{5}{12m^3}[\hat{\Pi}_\mu,[\hat{\Pi}_\mu,\hat{\Pi}_\lambda]]\gamma_\lambda\right\rangle. \tag{7.202}$$

7.5.7 Electron Anomalous Magnetic Moment

In this section, we derive the most cited consequence of the quantum-electrodynamic corrections from the last equation (7.202). Consider the case of an external magnetic field

$$\hat{\boldsymbol{\Pi}} = \left(E,\hat{\mathbf{p}} - e\mathbf{A}(\hat{\boldsymbol{r}})\right)\;;$$

then

$$\left\langle[\gamma_\mu,\gamma_\nu][\hat{\Pi}_\mu,\hat{\Pi}_\nu]\right\rangle = \left\langle[\gamma_i,\gamma_j][\hat{\Pi}_i,\hat{\Pi}_j]\right\rangle = \left\langle 4e\,\Sigma_k\varepsilon_{ijk}\frac{\partial A_j}{\partial x_i}\right\rangle = \langle 4e\boldsymbol{\Sigma}\cdot\mathbf{B}\rangle\;,$$

where we used the formula for the Dirac γ-matrices, see Eqs. (7.18) and (7.28),

$$[\gamma_i,\gamma_j] = -2\mathrm{i}\varepsilon_{ijk}\Sigma_k\,, \tag{7.203}$$

and, see Sect. 3.3.2,

$$[\hat{\Pi}_i,\hat{\Pi}_j] = \mathrm{i}e\left(\frac{\partial A_j}{\partial x_i} - \frac{\partial A_i}{\partial x_j}\right).$$

In the nonrelativistic limit $\psi_{\mathrm{at}} \simeq \psi^0_{\mathrm{at}}$, we obtain from the first term of Eq. (7.202)

$$\Delta E = -\frac{e}{m}\left\langle\hat{\mathbf{S}}\cdot\mathbf{B}\right\rangle\frac{\alpha}{2\pi}\,,$$

which is the celebrated[35] "anomalous" correction to the electron gyromagnetic ratio. This correction was for the first time calculated by American physicist Julian S. Schwinger in 1947 within the framework of the old-fashioned formalism [35] (with the separation of the contributions from the positive and negative energies in virtual states, transverse and longitudinal parts of the EM field) nearly immediately after it was measured by Isidor Rabi and his coworkers using the method explained in Sect. 1.2.7.

It follows from the Dirac equation that the ratio of the electron magnetic and mechanical moments equals e/m_e, see Sects. 3.3.2 and 3.5.1. By taking into account the electron interaction with its own EM field, we find

$$(g_e)_{\text{theo}} = \frac{\frac{\mu}{S}}{\frac{e}{m_e}} = 1 + \frac{\alpha}{2\pi} + \ldots \simeq 1.001161. \tag{7.204}$$

The most accurate experimental value at present is [27]

$$(g_e)_{\text{exp}} = 1.00115965218085(76), \tag{7.205}$$

with the relative error of the measurement of 6.6×10^{-10}! It is worth noting that the theoretical value has been calculated up to the order of α^4! Presently, the comparison of the theory and experiment in this case yields the most accurate determination of the fine structure constant α. For our "modest" accuracy, i.e., for the comparison of Eqs. (7.204) and (7.205), it suffices to take the fine-structure constant α from the measurement of the quantum Hall effect, see, e.g., [19].

In Sect. 3.3.3, we derived an expression for the hyperfine splitting of the hydrogen ground state, see Eq. (3.58),

$$\nu_{\text{theo}}(1^3s - 1^1s) = \frac{8}{3}\frac{m_e}{m_p}\frac{2R_\infty c\alpha^2}{\left(1 + \frac{m_e}{m_p}\right)^3} g_p g_e \,. \tag{7.206}$$

If we insert 1 for g_e, the last equation yields 1418.4 MHz. If we insert for g_e from Eq. (7.204), the last equation produces the result 1420.1 MHz which agrees better with the experimental value 1420.4 MHz, see Eq. (3.59). We achieve an even more impressive improvement of the theoretical prediction if we replace the proton by a muon. If we insert $g_p = g_e = 1$ and the appropriate mass ratio, Eq. (7.206) yields 4453.8 MHz. If we insert for both g_p and g_e from Eq. (7.204), Eq. (7.206) yields 4464.2 MHz in much better agreement with the experimental value 4463.3 MHz, see Eq. (3.60).

[35] More precisely, celebrated among those interested physics; even more precisely, celebrated among those captivated by the quantum field theory.

7.5.8 Lamb Shift

Considering the nonrelativistic limit of the Dirac wave function, Eq. (7.22), and the definitions (7.48) and (7.179), we find for the Coulomb potential (7.25)

$$\left\langle [\hat{\sqcap}_\mu, [\hat{\sqcap}_\mu, \hat{\sqcap}_\lambda]]\gamma_\lambda \right\rangle = \left\langle -[\hat{\mathsf{p}}_i, [\hat{\mathsf{p}}_i, \hat{\sqcap}_0]]\gamma_0 \right\rangle \tag{7.207}$$

and

$$\left\langle [\gamma_\mu, \gamma_\nu][\hat{\sqcap}_\mu, \hat{\sqcap}_\nu] \right\rangle = \left\langle 4\gamma_0\gamma_i[\hat{\mathsf{p}}_i, \hat{\sqcap}_0] \right\rangle \simeq -\frac{2}{m}\langle \psi^0_{\text{at}}| \left[\boldsymbol{\sigma}\cdot\hat{\boldsymbol{p}}, \left[\boldsymbol{\sigma}\cdot\hat{\boldsymbol{p}}, \hat{\sqcap}_0\right]\right] |\psi^0_{\text{at}}\rangle \tag{7.208}$$

$$= -\frac{2}{m}\langle \psi^0_{\text{at}}| [\hat{\mathsf{p}}_i, [\hat{\mathsf{p}}_i, \hat{\sqcap}_0]] + 4Z\alpha\frac{\hat{\boldsymbol{S}}\cdot\hat{\boldsymbol{L}}}{\hat{\mathsf{r}}^3} |\psi^0_{\text{at}}\rangle,$$

where we used Eq. (3.106) in the last equality. After inserting Eqs. (7.207) and (7.208) into Eq. (7.202), we obtain

$$\Delta E^{\text{SE}}_{\text{high}} \simeq -\frac{m\alpha}{2\pi}\langle \psi^0_{\text{at}}| \frac{1}{6m^3}[\hat{\mathsf{p}}_i, [\hat{\mathsf{p}}_i, \hat{\sqcap}_0]] - \frac{Z\alpha}{m^3}\frac{\hat{\boldsymbol{S}}\cdot\hat{\boldsymbol{L}}}{\hat{\mathsf{r}}^3} |\psi^0_{\text{at}}\rangle,$$

where, see Eq. (6.238),

$$\frac{1}{m^3}\langle \psi^0_{\text{at}}| [\hat{\mathsf{p}}_i, [\hat{\mathsf{p}}_i, \hat{\sqcap}_0]] |\psi^0_{\text{at}}\rangle = (Z\alpha)^4\frac{4}{n^3}\delta_{l,0}\,.$$

The final result for the shift of the hydrogen-like energy levels caused by the emission and absorption of a virtual photon reads[36]

$$(\Delta E)^{\text{SE}}_{\text{exp}} = \frac{m\alpha}{\pi n^3}(Z\alpha)^4 F(Z\alpha, n, l, j)\,, \qquad F = F_{\text{high}} + F_{\text{low}}\,, \tag{7.209}$$

where the expression for F_{low} was already given in Eq. (6.239),

$$F_{\text{low}} = \langle \psi^0_{\text{at}}| (\hat{\mathsf{p}}_A)_i f(\hat{\mathsf{h}}_0 - \varepsilon_0)(\hat{\mathsf{p}}_A)_i |\psi^0_{\text{at}}\rangle, \quad f(x) = n^3 x\int_0^1 \mathrm{d}y\int_0^1 \mathrm{d}w\frac{1-2w(1-w)}{y+2w(Z\alpha)^2 x}\,. \tag{6.239}$$

From the above equations we have for F_{high}

$$F_{\text{high}} = -\frac{1}{3}\delta_{l,0} + n^3\langle \psi^0_{\text{at}}| \frac{\hat{\boldsymbol{S}}\cdot\hat{\boldsymbol{L}}}{2\hat{\mathsf{r}}^3_A} |\psi^0_{\text{at}}\rangle. \tag{7.210}$$

[36]The first correct expression for the energy shift of hydrogen-like atoms with accuracy up to $\alpha(Z\alpha)^4$ was for the first time published [20] in the framework of the old-fashioned formalism of Sect. 7.3.1.

Comparing Eqs. (6.242), (6.243), (7.210), (3.108) and (3.111), we see that the total energy shift is for the s-states by two orders of magnitude greater than for the p-states. For s-states, the dominant contribution is given by the low-energy part, while the high-energy part contributes the most in case of p-states. We postpone the comparison with the experiment until we have calculated the second contribution of quantum electrodynamics to the shift of the hydrogen-like energy levels, namely vacuum polarization.

7.5.9 Nuclear Motion Effect

So far, we considered an electron moving in an external EM field and neglected the nuclear motion. To take it into account it suffices, see Sect. 3.2.1, to replace m by m_r in the kinetic term $\frac{\hat{\mathrm{p}}^2}{2m}$ in Eq. (7.26) and to do the same when changing to the atomic units, Eq. (6.83); then

$$\frac{-\hat{\mathscr{H}}}{m^2} \simeq 2(Z\alpha)^2\mu(\hat{\mathrm{h}}_0 - \varepsilon_0)\,,$$

where

$$\mu = \frac{m_r}{m}\,.$$

The function $f(x)$ in Eq. (6.239) consequently changes to

$$f(x,\mu) = n^3 x \int_0^1 \mathrm{d}y \int_0^1 \mathrm{d}w \frac{1-2w(1-w)}{y+2wx\mu(Z\alpha)^2}\,.$$

By means of the expansion (6.240) we can write with sufficient accuracy

$$f(x,\mu) \simeq f(x) - \frac{2}{3}n^3 x \ln\mu\,.$$

Equation (7.209) then changes to

$$(\Delta E)^{\mathrm{SE}}_{\mathrm{exp}} = \frac{m\alpha}{\pi n^3}(Z\alpha)^4\left(\frac{m_r}{m}\right)^3\left[F_{\mathrm{high}}\left(\frac{m_r}{m}\right) + F_{\mathrm{low}}\left(\frac{m_r}{m}\right)\right], \tag{7.211}$$

where the multiplicative factor $(m_r/m)^3$ stems from the transition to the atomic units with reduced mass, $r = r_A/(m_r Z\alpha)$, and

$$F_{\mathrm{low}}\left(\frac{m_r}{m}\right) = (6.239) - \delta_{l,0}\frac{4}{3}\ln\frac{m_r}{m}\,. \tag{7.212}$$

If we take the nuclear motion into account, we should also in Eq. (3.126), in the term describing electron spin interaction with orbital motion of the nucleus, consider the Schwinger correction to the gyromagnetic ratio (7.204). The high-energy part F_{high} is then modified to

$$F_{\text{high}}\left(\frac{m_r}{m}\right) = -\frac{1}{3}\delta_{l,0} + \frac{m}{m_r}n^3\langle\psi_{\text{at}}^0|\frac{\hat{\boldsymbol{S}}\cdot\hat{\boldsymbol{L}}}{2\hat{r}_A^3}|\psi_{\text{at}}^0\rangle . \tag{7.213}$$

Exercise 27: Hydrogen Fine Structure

Consider the interaction of the electron spin with the proton orbit and the effect of the electron self-energy on the spin-orbital interaction, Eq. (7.213). How does it change the theoretical prediction of the interval $2p_{3/2} - 2p_{1/2}$? We estimated this interval back in Sect. 3.5.2. Compare the result with the experimental value (3.109). Do the same for muonium where the "official" value [28] for this interval reads 10,922 MHz.

7.6 Vacuum Polarization: Calculation

7.6.1 Propagator Expansion

We now turn our attention to the second quantum electrodynamic process that contributes to the Lamb shift in hydrogen-like atoms—vacuum polarization. We begin by recalling the relations we have already derived. The shift of the atomic energy levels caused by vacuum polarization is

$$(\Delta E)^{\text{VP}} = -\frac{e}{(2\pi)^3}\int\frac{\mathrm{d}^3\mathbf{k}}{\omega^2}\langle\psi_{\text{at}}|\gamma_0\gamma_\mu \mathrm{e}^{\mathrm{i}\mathbf{k}\cdot\hat{\mathbf{r}}}|\psi_{\text{at}}\rangle P_{\mu\nu}\int \mathrm{d}^3\mathbf{r}'\,\langle j_\nu(\mathbf{r}')\rangle\,\mathrm{e}^{-\mathrm{i}\mathbf{k}\cdot\mathbf{r}'}\,, \tag{7.77}$$

where the vacuum expectation value of the four-current reads

$$\langle j_\mu(\mathbf{r})\rangle = -\mathrm{i}e\int_\Gamma \frac{\mathrm{d}E}{2\pi}\mathrm{Tr}\,\langle\mathbf{r}|\gamma_\mu\frac{1}{\boldsymbol{\gamma}\cdot\hat{\boldsymbol{\Pi}} - m}|\mathbf{r}\rangle\,. \tag{7.149}$$

It follows from Eq. (7.77) that we need to calculate the Fourier transform of Eq. (7.149). For the sake of simplicity, we will try to calculate it approximately: we expand the electron propagator in the powers of the potential of the external EM field. The physical momentum is $\hat{\boldsymbol{\Pi}} = \hat{\boldsymbol{p}} - e\boldsymbol{A}(\hat{\mathrm{r}})$, hence in the spirit of Eq. (6.198) we write

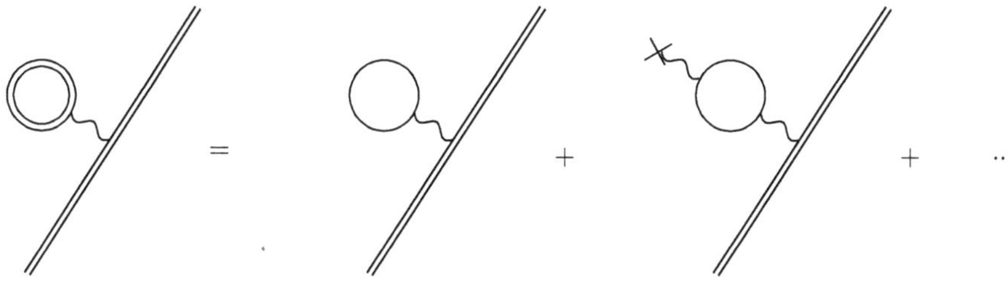

Fig. 7.7 Potential expansion of current density

$$\frac{1}{\boldsymbol{\gamma}\cdot\hat{\boldsymbol{\Pi}}-m}=\frac{1}{\boldsymbol{\gamma}\cdot\hat{\boldsymbol{p}}-m}+\frac{1}{\boldsymbol{\gamma}\cdot\hat{\boldsymbol{p}}-m}e\boldsymbol{\gamma}\cdot\boldsymbol{A}(\hat{\boldsymbol{r}})\frac{1}{\boldsymbol{\gamma}\cdot\hat{\boldsymbol{p}}-m}+\dots .$$

Inserting this expansion into Eq. (7.149) leads to

$$\begin{aligned}\langle j_\mu(\mathbf{r})\rangle &= -\mathrm{i}e\int_\Gamma \frac{\mathrm{d}E}{2\pi}\mathrm{Tr}\,\langle\mathbf{r}|\gamma_\mu\left(\frac{1}{\boldsymbol{\gamma}\cdot\hat{\boldsymbol{p}}-m}+\frac{1}{\boldsymbol{\gamma}\cdot\hat{\boldsymbol{p}}-m}\boldsymbol{\gamma}\cdot\boldsymbol{V}(\hat{\boldsymbol{r}})\frac{1}{\boldsymbol{\gamma}\cdot\hat{\boldsymbol{p}}-m}+\dots\right)|\mathbf{r}\rangle \\ &= \langle j_\mu(\mathbf{r})\rangle_0+\langle j_\mu(\mathbf{r})\rangle_1+\dots , \end{aligned} \tag{7.214}$$

where obviously

$$\boldsymbol{V}=e\boldsymbol{A}\,. \tag{7.215}$$

The expansion is schematically depicted in Fig. 7.7. The first term, called the "tadpole," vanishes identically and so do all the other terms with an odd number of adjacent photon lines. Let us show where this rule, commonly termed as the *Furry theorem*, stems from. We extend the fraction, analogously to Eq. (7.160),

$$\frac{1}{\boldsymbol{\gamma}\cdot\hat{\boldsymbol{p}}-m}=\frac{\boldsymbol{\gamma}\cdot\hat{\boldsymbol{p}}+m}{\hat{\boldsymbol{p}}^2-m^2}\,. \tag{7.216}$$

The charge current created by the virtual creation and annihilation of a free electron-positron pair is given as

$$\langle j_\mu(\mathbf{r})\rangle_0=-\frac{\mathrm{i}e}{2\pi}\int_\Gamma \mathrm{Tr}\,\langle\mathbf{r}|\gamma_\mu\frac{\boldsymbol{\gamma}\cdot\hat{\boldsymbol{p}}+m}{\hat{\boldsymbol{p}}^2-m^2}|\mathbf{r}\rangle\,\mathrm{d}E\,. \tag{7.217}$$

The mass term is proportional to $\mathrm{Tr}\,\gamma_\mu$, hence vanishes. To prove this in such a way that it can be easily generalized to an arbitrary odd number of γ-matrices, we introduce a fifth Dirac matrix γ_5, defined by the requirements $\gamma_5^2=1$ and $\{\gamma_\mu,\gamma_5\}=0$. The reader can easily verify that in the Dirac realization one can opt for instance for

$$\gamma_5=\begin{pmatrix}0&1\\1&0\end{pmatrix}. \tag{7.218}$$

Consequently we find

$$\operatorname{Tr}\gamma_\mu = \operatorname{Tr}\gamma_\mu\gamma_5^2 = \begin{cases} \operatorname{Tr}\gamma_5\gamma_\mu\gamma_5 & \text{(cyclicity of trace)}, \\ -\operatorname{Tr}\gamma_5\gamma_\mu\gamma_5 & \text{(anticommutator)}, \end{cases}$$

hence indeed $\operatorname{Tr}\gamma_\mu = 0$ and by the same token $\operatorname{Tr}\gamma_\mu\gamma_\nu\gamma_\rho = 0$ and so on. To calculate the trace of an even number of γ-matrices, we again use the cyclicity of the trace and the anticommutation relations for the γ-matrices (7.17). Recalling also that γ-matrices are matrices 4×4, we easily find

$$\begin{aligned} \operatorname{Tr} 1 &= 4, \\ \operatorname{Tr}\gamma_\mu &= 0, \\ \operatorname{Tr}\gamma_\mu\gamma_\nu &= 4\eta_{\mu\nu}, \\ \operatorname{Tr}\gamma_\mu\gamma_\nu\gamma_\rho &= 0, \\ \operatorname{Tr}\gamma_\mu\gamma_\nu\gamma_\rho\gamma_\sigma &= 4(\eta_{\mu\nu}\eta_{\rho\sigma} - \eta_{\mu\rho}\eta_{\nu\sigma} + \eta_{\mu\sigma}\eta_{\nu\rho}). \end{aligned}$$

In Eq. (7.217), we use these traces to obtain

$$\langle j_\mu(\mathbf{r})\rangle_0 = -4\frac{\mathrm{i}e}{2\pi}\int_\Gamma \langle\mathbf{r}|\frac{\hat{\mathrm{p}}_\mu}{\hat{\boldsymbol{p}}^2 - m^2}|\mathbf{r}\rangle\,\mathrm{d}E.$$

Now, using the spectral decomposition of the momentum operator, the last equation reads

$$\langle j_\mu(\mathbf{r})\rangle_0 = -4\frac{\mathrm{i}e}{2\pi}\int\int_\Gamma \langle\mathbf{r}|\,\mathbf{p}\rangle\frac{p_\mu}{\boldsymbol{p}^2 - m^2}\langle\mathbf{p}|\,\mathbf{r}\rangle\,\mathrm{d}E\,\mathrm{d}^3\mathbf{p} = -4\mathrm{i}e\int\frac{\mathrm{d}^4\boldsymbol{p}}{(2\pi)^4}\frac{p_\mu}{\boldsymbol{p}^2 - m^2} = 0,$$

where the last integral is odd in $\boldsymbol{p}$, hence vanishes, see Eq. (7.168). The first nonvanishing contribution to the Fourier transform of $\langle j_\mu(\mathbf{r})\rangle$ comes from the second term in Eq. (7.214). By inserting twice the completeness relation in the basis of the momentum eigenstates, we find

$$\begin{aligned} &\int\frac{\mathrm{d}^3\mathbf{r}}{(2\pi)^3}\langle j_\mu(\mathbf{r})\rangle_1\,\mathrm{e}^{-\mathrm{i}\mathbf{k}\cdot\mathbf{r}} \\ &= -\frac{\mathrm{i}e}{(2\pi)^4}\int_\Gamma \mathrm{d}E\int\mathrm{d}^3\mathbf{p}_1\mathrm{d}^3\mathbf{p}_2\operatorname{Tr}\left[\gamma_\mu\frac{1}{\boldsymbol{\gamma}\cdot\boldsymbol{p}_1 - m}\langle\mathbf{p}_1|\boldsymbol{\gamma}\cdot\boldsymbol{V}|\mathbf{p}_2\rangle\frac{1}{\boldsymbol{\gamma}\cdot\boldsymbol{p}_2 - m}\right. \\ &\quad\left.\times\int\mathrm{d}^3\mathbf{r}\langle\mathbf{r}|\,\mathbf{p}_1\rangle\langle\mathbf{p}_2|\,\mathbf{r}\rangle\mathrm{e}^{-\mathrm{i}\mathbf{k}\cdot\mathbf{r}}\right]. \end{aligned} \tag{7.219}$$

The integral over the coordinate $\mathbf{r}$ yields a δ-function $\delta^{(3)}(\mathbf{p}_1 - \mathbf{k} - \mathbf{p}_2)$, which represents the momentum conservation at the interaction vertex, see Fig. 7.8. By means of the spectral decomposition of the potential operator

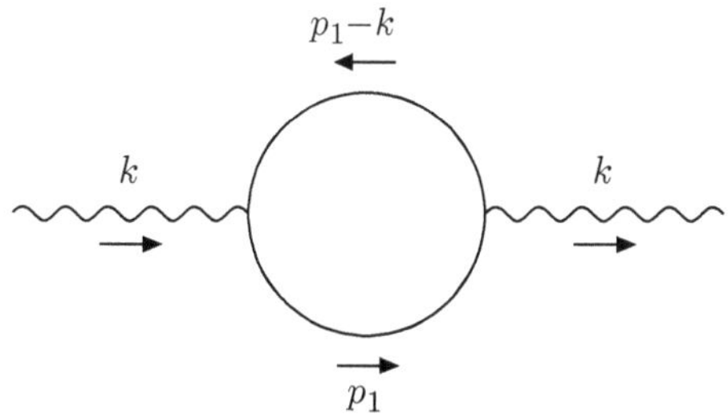

Fig. 7.8 Feynman diagram for the vacuum polarization in the momentum space. Conservation of the momentum holds in the interaction vertices, thus what "flows" inside equals what "flows" outside. Note that the previous Feynman diagram for the self-energy effect, Fig. 7.6, is in the coordinate space

$$V_\nu(\hat{\boldsymbol{r}}) = \int \mathrm{d}^3\mathbf{r}|\mathbf{r}\rangle\langle\mathbf{r}|V_\nu(\mathbf{r}),$$

we easily find that the matrix elements of the potential components between the momentum eigenstates equal the Fourier transform of the potential components

$$\begin{aligned}\langle\mathbf{p}_1|V_\nu(\hat{\boldsymbol{r}})|\mathbf{p}_2\rangle &= \int \mathrm{d}^3\mathbf{r}\langle\mathbf{p}_1|\,\mathbf{r}\rangle V_\nu(\mathbf{r})\langle\mathbf{r}|\,\mathbf{p}_2\rangle \\ &= \int \frac{\mathrm{d}^3\mathbf{r}}{(2\pi)^3}\mathrm{e}^{\mathrm{i}(\mathbf{p}_2-\mathbf{p}_1)\cdot\mathbf{r}}V_\nu(\mathbf{r}) = V_\nu(\mathbf{p}_1-\mathbf{p}_2)\,.\end{aligned} \tag{7.220}$$

From Eqs. (7.214), (7.219), and (7.220), we then find a simple expression of the Fourier transform of the four-current density vacuum expectation value

$$\int \frac{\mathrm{d}^3\mathbf{r}}{(2\pi)^3}\langle j_\mu(\mathbf{r})\rangle\,\mathrm{e}^{-\mathrm{i}\mathbf{k}\cdot\mathbf{r}} \simeq -e\Pi_{\mu\nu}(\mathbf{k})V_\nu(\mathbf{k}), \tag{7.221}$$

where $\Pi_{\mu\nu}(\mathbf{k})$ is a *vacuum polarization tensor*

$$\Pi_{\mu\nu}(k) = \frac{\mathrm{i}}{2\pi}\int_\Gamma \mathrm{d}E\int\frac{\mathrm{d}^3\mathbf{p}_1}{(2\pi)^3}\mathrm{Tr}\left[\gamma_\mu\frac{1}{\boldsymbol{\gamma}\cdot\boldsymbol{p}_1-m}\gamma_\nu\frac{1}{\boldsymbol{\gamma}\cdot(\boldsymbol{p}_1-\boldsymbol{k})-m}\right]$$

and corresponds to the Feynman diagram in Fig. 7.8. Here, we used the aforeintroduced notation

$$\boldsymbol{k} = (0,\mathbf{k}), \qquad \boldsymbol{k}^2 = -\omega^2. \tag{7.79}$$

Having introduced the vacuum polarization tensor, the energy shift (7.77) can now be written as

$$(\Delta E)^{\mathrm{VP}} = e^2\int\frac{\mathrm{d}^3\mathbf{k}}{\omega^2}\langle\overline{\Psi}_{\mathrm{at}}|\gamma_\mu\mathrm{e}^{\mathrm{i}\mathbf{k}\cdot\hat{\boldsymbol{r}}}|\Psi_{\mathrm{at}}\rangle\Pi_{\nu\sigma}(\mathbf{k})V_\sigma(\mathbf{k})P_{\mu\nu}\,. \tag{7.222}$$

Let us now calculate the tensor $\Pi_{\mu\nu}$[37]:

$$\Pi_{\mu\nu}(\boldsymbol{k}) = \mathrm{i}\int \frac{\mathrm{d}^4 p}{(2\pi)^4} \frac{\mathrm{Tr}\, \gamma_\mu(\boldsymbol{\gamma}\cdot\boldsymbol{p}+m)\gamma_\nu(\boldsymbol{\gamma}\cdot(\boldsymbol{p}-\boldsymbol{k})+m)}{(\boldsymbol{p}^2-m^2)((\boldsymbol{p}-\boldsymbol{k})^2-m^2)} \,. \tag{7.223}$$

With the use of the traces of the γ-matrices listed above, the numerator equals

$$\mathrm{Tr}\, \gamma_\mu(\boldsymbol{\gamma}\cdot\boldsymbol{p}+m)\gamma_\nu(\boldsymbol{\gamma}\cdot(\boldsymbol{p}-\boldsymbol{k})+m) = 4[p_\mu(\boldsymbol{p}-\boldsymbol{k})_\nu + p_\nu(\boldsymbol{p}-\boldsymbol{k})_\mu - \eta_{\mu\nu}\boldsymbol{p}\cdot(\boldsymbol{p}-\boldsymbol{k})] + 4m^2\eta_{\mu\nu} \,.$$

We combine the denominators by means of the Feynman parameter

$$\begin{aligned} \frac{1}{(\boldsymbol{p}^2-m^2)((\boldsymbol{p}-\boldsymbol{k})^2-m^2)} &= \int_0^1 \frac{\mathrm{d}y}{[\boldsymbol{p}^2-m^2+(-2\boldsymbol{p}\cdot\boldsymbol{k}+\boldsymbol{k}^2)y]^2} \\ &= \int_0^1 \frac{\mathrm{d}y}{[(\boldsymbol{p}-\boldsymbol{k}y)^2-m^2+\boldsymbol{k}^2 y(1-y)]^2} \,. \end{aligned}$$

After substituting these rearrangements into Eq. (7.223), we make the substitution $\boldsymbol{p} \to \boldsymbol{p}+\boldsymbol{k}y$.[38] The vacuum polarization tensor then can be manipulated into the form

$$\Pi_{\mu\nu} = (\boldsymbol{k}^2\eta_{\mu\nu} - k_\mu k_\nu)\Pi(\boldsymbol{k}^2) + \Delta_{\mu\nu} \,,$$

where

$$\Pi(\boldsymbol{k}^2) = 8\mathrm{i}\int \frac{\mathrm{d}^4\boldsymbol{p}}{(2\pi)^4}\int_0^1 \mathrm{d}y \frac{y(1-y)}{(\boldsymbol{p}^2-m^2+\boldsymbol{k}^2 y(1-y))^2} \,, \tag{7.224}$$

$$\Delta_{\mu\nu} = 4\mathrm{i}\int \frac{\mathrm{d}^4\boldsymbol{p}}{(2\pi)^4}\int_0^1 \mathrm{d}y \left(\frac{2p_\mu p_\nu}{(\boldsymbol{p}^2-m^2+\boldsymbol{k}^2 y(1-y))^2} - \frac{\eta_{\mu\nu}}{\boldsymbol{p}^2-m^2+\boldsymbol{k}^2 y(1-y)} \right)$$

and the terms odd in $\boldsymbol{p}$ vanish.

The integrals over the four-momenta $\boldsymbol{p}$ in the last two equations are divergent. To regularize them, we differentiate them with respect to the parameter m^2. By means of Eq. (7.168) we obtain

---

[37] Calculation of the vacuum polarization tensor can be found in any quantum field theory textbook, see, e.g., [2, 4, 15, 30, 33, 42].

[38] Here we are somewhat careless as the substitution is allowed only for a convergent integral. We should regularize the expression (7.223) first and only then make the substitution. Although the final result is the same in this particular case, the order of the regularization and substitution *could* generally matter and proceeding with care is appropriate. The noncommutativity of the regularization and substitution leads to the so-called *axial anomaly*, see e.g. [13, 30, 42], which is an indeed *interesting* effect. It shows, among others, that infinities appearing in the calculations are not a mere trouble we wish to dispose of, but rather very useful features! For further elaboration of this statement see, e.g., [14].

$$\frac{\partial \Pi(\boldsymbol{k}^2)}{\partial m^2} = 16\mathrm{i}\int \frac{\mathrm{d}^4\boldsymbol{p}}{(2\pi)^4}\int_0^1 \mathrm{d}y \frac{y(1-y)}{(\boldsymbol{p}^2 - m^2 + \boldsymbol{k}^2 y(1-y))^3}$$

$$= \frac{2}{(2\pi)^2}\int_0^1 \mathrm{d}y \frac{y(1-y)}{m^2 - \boldsymbol{k}^2 y(1-y)} \tag{7.225}$$

and

$$\frac{\partial^2 \Delta_{\mu\nu}}{\partial (m^2)^2} = 4\mathrm{i}\int \frac{\mathrm{d}^4\boldsymbol{p}}{(2\pi)^4}\int_0^1 \mathrm{d}y \left(\frac{2p_\mu p_\nu 3!}{(\boldsymbol{p}^2 - m^2 + \boldsymbol{k}^2 y(1-y))^4} - \frac{\eta_{\mu\nu} 2}{(\boldsymbol{p}^2 - m^2 + \boldsymbol{k}^2 y(1-y))^3}\right) = 0\,, \tag{7.226}$$

where the last equality follows from Eq. (7.168) and

$$\int \frac{p_\nu p_\mu \mathrm{d}^4\boldsymbol{p}_F}{(\boldsymbol{p}^2 - L)^4} = \frac{\eta_{\mu\nu}}{48L}\,. \tag{7.227}$$

This equality is obtained by differentiating Eq. (7.168)

$$\int \frac{p_\mu \mathrm{d}^4\boldsymbol{p}_F}{((\boldsymbol{p}-\boldsymbol{k})^2 - L)^3} = \frac{k_\mu}{8L}$$

with respect to k_ν and setting $k_\nu = 0$.

When integrating over p_0 in Eqs. (7.225) and (7.226), we considered the note following Eq. (7.168). Namely, when evaluating integrals of the type (7.168), it plays no role whether the integration path in the complex k_0 or p_0 plane lies in the upper or lower half of the complex plane. This means we once again avoid the separation of the contributions from the positive and negative energy modes. This separation was present in Eq. (7.80); however, we transformed it in Eq. (7.149) to different integration paths in the complex plane. By the extension (7.216), used in Eq. (7.223), the integrals over these different paths yield the same result.

7.6.2 *Gauge Invariance and Degree of Divergence*

Integrating back the expression (7.225) with respect to m^2, where we insert the square of the electron mass m^2 for the upper bound and Λ^2 for lower bound, we obtain

$$\Pi(\boldsymbol{k}^2) = \frac{2}{(2\pi)^2}\int_0^1 \mathrm{d}y\, y(1-y)\ln\frac{m^2 - \boldsymbol{k}^2 y(1-y)}{\Lambda^2} + O(1/\Lambda^2)\,. \tag{7.228}$$

Here, we assumed $\Lambda^2 \gg \boldsymbol{k}^2$, hence $\ln(\Lambda^2 - \boldsymbol{k}^2 y(1-y)) \simeq \ln \Lambda^2$. Setting the lower bound of the integration to Λ^2 is equivalent to the replacement

$$\frac{1}{\gamma \cdot \boldsymbol{p} - m} = \lim_{\Lambda \to \infty} \left(\frac{1}{\gamma \cdot \boldsymbol{p} - m} - \frac{1}{\gamma \cdot \boldsymbol{p} - \Lambda} \right)$$

of the electron propagator in the loop. Thus, one can readily see that from a physical point of view, the introduction of the finite Λ represents virtual creation and annihilation of another fermion pair with mass Λ which is much greater than the electron mass. The sign of the coupling of this heavy fermion to the EM field is opposite to that of electron. However, as in the case of self-energy regularization, nobody takes this interpretation seriously; the regulator Λ is viewed as a formal parameter which only purpose is to define well the pertinent integrals.

In the case of the electron self-energy, we showed that the dependence on the regulating parameter Λ^2 disappears once we realize what the measurable electron mass is. Likewise, we will show in a moment that in the case of vacuum polarization, the dependence on Λ^2 disappears once we realize what the measurable electron charge is. For now, however, we consider only the fact that the remaining dependence on Λ^2 in Eq. (7.228) is only logarithmic and that the quadratic dependence in the integral (7.223) is removed by the regularization, see Eq. (7.226). We thus arrive at

$$\Pi_{\mu\nu}(\boldsymbol{k}) = (\boldsymbol{k}^2 \eta_{\mu\nu} - k_\mu k_\nu) \Pi(\boldsymbol{k}^2) . \tag{7.229}$$

Is it a pure coincidence, or is there a deeper explanation? When we return to the expressions (7.221) and (7.222), we find that we could have expected this result. The expression (7.221) was derived for a static external field. For a time-dependent external field, the result is the same with the only exception that there is a four-dimensional Fourier transform of the vacuum expectation value on the lhs of Eq. (7.221) and the four-vector $\boldsymbol{k}$ on the rhs has a generally nonzero temporal component. Furthermore, $V(\boldsymbol{k})$ is a four-dimensional Fourier transform of the potential. Let us consider this more general case for a moment. The potential of the external EM field is determined up to the gauge transformation

$$V_\nu(\boldsymbol{x}) \to V_\nu(\boldsymbol{x}) + \partial_\nu \chi(\boldsymbol{x}) .$$

In terms of the Fourier transform, it reads

$$V_\nu(\boldsymbol{k}) \to V_\nu(\boldsymbol{k}) - \mathrm{i} k_\nu \chi(\boldsymbol{k}) .$$

The tensor part of the rhs of Eq. (7.229) is proportional to the projector $P_{\mu\nu}$, Eq. (7.78), to the (hyper)plane perpendicular to the vector $\boldsymbol{k}$. This guarantees that the expression (7.222) is invariant with respect to a gauge transformation. There we have to find the cause leading to the cancellation of the worst divergences: the gauge invariance!

The connection between the gauge invariance and the renormalizability is remarkable. Its universality was for the first time proved by Netherlander physicist Gerardus 't Hooft: a theory based on a very general notion of gauge invariance is renormalizable. This general "lightness" of a gauge theory bears a generally undesired consequence: the bosons mediating the interaction have to be massless; see for instance the following note about a massive vector field. In the framework of electrodynamics, this drawback causes no problems at all as the EM interaction has an infinite range as we have known for a long time. However, the range of the weak interaction is very short, about 10^{-18} m. As we show in the next section, the short range of a field is related to very large masses of the one-particle excitations of the field. Hence bosons $W^{\pm}$ and Z^0, which mediate weak interactions, are very massive, about 80 and 91 GeV, respectively. Roughly speaking, to preserve the gauge invariance it is necessary to add to them a massive scalar field whose one-particle excitations are called Higgs bosons. We will not explore this part of physics any further in this book and we refer the reader to a number of excellent textbooks, e.g., [13, 24, 30, 42].

7.6.3 Note on a Massive Vector Field

If we add a mass term to the Maxwell equations (7.12), we obtain

$$(\partial_\mu \partial_\mu + m^2)A_\nu - \partial_\nu \partial_\mu A_\mu = j_\nu \,. \tag{7.230}$$

It is not difficult to see that the *gauge invariance* of the theory, i.e., the invariance of the field equations with respect to the gauge transformation (7.10), $A_\mu = A'_\mu - \partial_\mu \chi$, is lost.

Acting on both sides of Eq. (7.230) with the operator of four-divergence and recalling the charge conservation law, Eq. (7.11), $\partial_\mu j_\mu = 0$, one obtains

$$\partial_\mu A_\mu = 0 \,. \tag{7.231}$$

Equation (7.230) consequently simplifies to

$$(\partial_\mu \partial_\mu + m^2)A_\nu = j_\nu \,. \tag{7.232}$$

We consider the case of a free field, $j_\nu = 0$, first. If we search for a solution to Eq. (7.232) in the form of a plane wave

$$A_\mu = \varepsilon_\mu \mathrm{e}^{\mathrm{i}\boldsymbol{k}\cdot\boldsymbol{x}} \,, \tag{7.233}$$

we obtain

$$\boldsymbol{k}^2 = m^2 \,. \tag{7.234}$$

This implies that, in contrast to the massless case, we get a nontrivial solution if we change to a frame connected with the particle, i.e., $\mathbf{k} = 0$. There are 4 equations (7.232) and one constraint (7.231) for the 4 components of the potential A. Therefore, the massive vector field features three independent components. It follows from Eqs. (7.231) and (7.233) that

$$k_\mu \varepsilon_\mu = 0\,. \tag{7.235}$$

In the frame connected with the particle, one finds from Eqs. (7.234) and (7.235)

$$\varepsilon_0 = 0\,.$$

In this frame, the three independent components are the space components and correspond to the three possible projections of spin of a particle with spin 1.

Now we consider the case of a static field, that is $\mathbf{j} = 0, \rho = \rho(\mathbf{r})$. Consequently, $\mathbf{A} = 0$ and the temporal component of the four-potential $A_0 = \varphi$ obeys the equation

$$\left(-\nabla^2 + m^2\right)\varphi = \rho\,.$$

In case of a point charge, $\rho = e\delta(\mathbf{r})$, the solution takes the form

$$\varphi = \frac{e}{(2\pi)^3}\int \mathrm{d}^3\mathbf{k}\frac{\mathrm{e}^{\mathrm{i}\mathbf{k}\cdot\mathbf{r}}}{|\mathbf{k}|^2 + m^2} = \frac{e}{4\pi}\frac{\mathrm{e}^{-mr}}{r}\,. \tag{7.236}$$

This form is widely know as the *Yukawa potential* in honor of Japanese physicist Hideki Yukawa who was the first one to realize that there is a connection between the mass and range of a field, as one can easily see from the expression. The shorter the range of the field, the more massive are its quanta and vice versa. The second equality in the last equation follows either from direct integration and use of the residue theorem, or from showing that both expressions of φ satisfy the same differential equation.

7.6.4 Charge Renormalization

Choosing the gauge $k_\rho V_\rho(\boldsymbol{k}) = 0$, Eqs. (7.78), (7.222) and (7.229) yield

$$(\Delta E)^{\mathrm{VP}} = \int \mathrm{d}^3\mathbf{k}\langle\overline{\psi}_{\mathrm{at}}|\gamma_\mu \mathrm{e}^{\mathrm{i}\mathbf{k}\cdot\hat{\boldsymbol{r}}}|\psi_{\mathrm{at}}\rangle V_\mu(\mathbf{k})e^2\Pi(\boldsymbol{k}^2)\,. \tag{7.237}$$

Since the expression for $\Pi(\boldsymbol{k}^2)$ is infinite in the limit $\Lambda \to \infty$, see Eq. (7.228), the last expression is infinite in this limit as well. As in the case of the electron self-energy, we now need to determine what is the measurable electron charge and what is the measurable energy shift. In the case of the self-energy, we succeeded in localizing the divergence into the electron mass. In a similar manner, we now attempt to localize the divergence into the electron charge.

The potential of an external EM field A appearing in the Dirac equation (7.19) is proportional to the elementary charge e creating the potential,

$$\boldsymbol{A} = e\boldsymbol{a}\,.$$

For instance, we have $V_0 = eA_0 = -Z\alpha/r$ for the Coulomb external field, hence $a_0 = -Z/(4\pi r)$. The Dirac equation (7.19) for an electron in an external EM field can be rewritten as

$$\left[\gamma \cdot (\hat{\boldsymbol{p}} - e^2\boldsymbol{a}(\hat{\boldsymbol{r}})) - m\right] \psi_{\text{at}} = 0\,. \tag{7.238}$$

So far, we always identified the bare electron charge e, which characterizes the coupling between the electron and the "bare" EM field, with the measured charge e_{exp}. However, this matching is not correct.

Photon exchange between particles, which leads to an observable interaction between them that is proportional to e^2_{exp}, constitutes an infinite sum of the processes depicted in Fig. 7.9. The resulting square of the electric charge e^2_{exp} differs from the square of the "bare" charge e^2. The latter is given by the exchange of a "bare" photon, i.e., a photon which does not undergo virtual decay to an electron-positron pair. It corresponds to the first diagram in Fig. 7.9. The "bare" photon is a mere theoretical construction; we observe effects of the EM field which are caused by a so-called "dressed" field which comprises an infinite sum of the processes showed in Fig. 7.9. We thus write

$$e^2 = e^2_{\text{exp}} + (e^2 - e^2_{\text{exp}})\,.$$

We insert this decomposition into the Dirac equation (7.238) and treat the term $(e^2 - e^2_{\text{exp}})\boldsymbol{\gamma} \cdot \boldsymbol{a}$ as a perturbation. A measurable correction to the atomic energy levels due to the vacuum polarization effect reads then

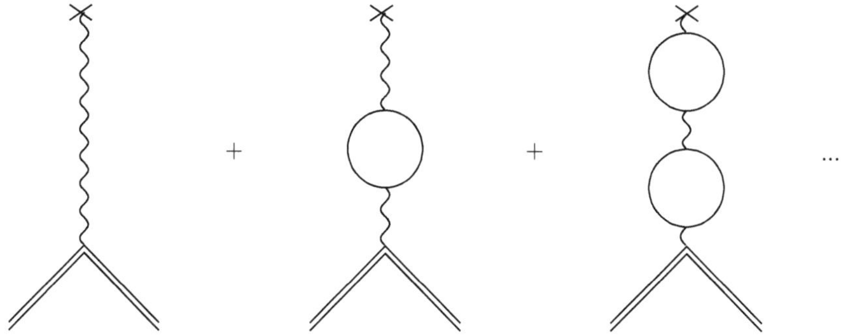

Fig. 7.9 The first three terms of the infinite series for the "dressed," observable charge. The *cross* denotes a nucleus and the *double line* an electron bound to the nucleus

$$(\Delta E)^{\mathrm{VP}}_{\mathrm{exp}} = (\Delta E)^{\mathrm{VP}} + (e^2 - e^2_{\mathrm{exp}})\langle\overline{\Psi}_{\mathrm{at}}|\boldsymbol{\gamma}\cdot\boldsymbol{a}(\hat{\boldsymbol{r}})|\psi_{\mathrm{at}}\rangle$$
$$= \int \mathrm{d}^3\mathbf{k}\langle\overline{\Psi}_{\mathrm{at}}|\gamma_\mu \mathrm{e}^{\mathrm{i}\mathbf{k}\cdot\hat{\boldsymbol{r}}}|\psi_{\mathrm{at}}\rangle a_\mu(\mathbf{k})\left[e^2(e^2\Pi(\boldsymbol{k}^2)+1) - e^2_{\mathrm{exp}}\right], \qquad (7.239)$$

where in the second equality we inserted from Eq. (7.237). The last equation states that the strength of the EM interaction depends on the exchanged momentum. The quantity

$$e^2(\boldsymbol{k}^2) = e^2(e^2\Pi(\boldsymbol{k}^2)+1) \qquad (7.240)$$

can be thus understood as the square of an effective charge. In that case, what is the value of the exchanged momentum for the fine structure constant α to equal $e^2_{\mathrm{exp}}/(4\pi) = 1/137.036\ldots$? It is the one measured for processes where the exchanged momentum approaches zero:

$$4\pi\alpha = e^2_{\mathrm{exp}} = e^2[1+e^2\Pi(0)] \Rightarrow e^2 \simeq e^2_{\mathrm{exp}}(1 - e^2_{\mathrm{exp}}\Pi(0)). \qquad (7.241)$$

We insert the last equation into Eq. (7.239) and neglect higher order terms in α to obtain

$$(\Delta E)^{\mathrm{VP}}_{\mathrm{exp}} = \langle\overline{\Psi}_{\mathrm{at}}|\int \mathrm{d}^3\mathbf{k}\,\mathrm{e}^{\mathrm{i}\mathbf{k}\cdot\hat{\boldsymbol{r}}}\boldsymbol{\gamma}\cdot\boldsymbol{V}(\mathbf{k})4\pi\alpha(\Pi(\boldsymbol{k}^2) - \Pi(0))|\psi_{\mathrm{at}}\rangle, \qquad (7.242)$$

where we substituted $e^2\boldsymbol{a} = \boldsymbol{V}$. It follows from Eq. (7.228) that the expression on the rhs of the last equation

$$\Pi(\boldsymbol{k}^2) - \Pi(0) = \frac{2}{(2\pi)^2}\int_0^1 \mathrm{d}y\,y(1-y)\ln\left(1 - \frac{\boldsymbol{k}^2}{m^2}y(1-y)\right) \qquad (7.243)$$

is now a finite number! The above-outlined procedure is called the *charge renormalization*. In an analogy to the self-energy case, we discovered that if we carefully distinguish between a theoretical construction and a measurable quantity, the resulting expression is finite and independent of the value of the regulating parameter Λ. Note that Eqs. (7.224) and (7.240) are manifestly invariant with respect to Lorentz transformations. Hence, the assignment (7.241) is invariant as well.

In summary, the vacuum polarization effect changes the strength of the EM interaction between two charged particles as follows:

$$4\pi\alpha \to 4\pi\alpha\left\{1 + 4\pi\alpha[\Pi(\boldsymbol{k}^2) - \Pi(0)]\right\}, \qquad (7.244)$$

where $[\Pi(\boldsymbol{k}^2) - \Pi(0)]$ is given by Eq. (7.243). One can readily see from this equation that with the increasing square of the exchanged momentum $-\boldsymbol{k}^2$, the square of the effective charge increases as well.

7.6.5 Calculation of the Observable Part of the Effect

Integrating over y by parts and recalling Eq. (7.79), we can transform Eq. (7.243) into the form

$$\Pi(\boldsymbol{k}^2) - \Pi(0) = -\frac{2}{(2\pi)^2}\frac{\omega^2}{m^2}\int_0^1 \mathrm{d}y \left(\frac{y^2}{2} - \frac{y^3}{3}\right)(1-2y)\frac{1}{1+\frac{\omega^2}{m^2}y(1-y)} . \tag{7.245}$$

Going back to Eq. (7.242) we see that we need to calculate the expression, called the *Uehling potential* [37][39]

$$V_\mu^U(\mathbf{r}) = \int \mathrm{d}^3\mathbf{k}\, \mathrm{e}^{\mathrm{i}\mathbf{k}\cdot\mathbf{r}} V_\mu(\mathbf{k}) 4\pi\alpha(\Pi(\boldsymbol{k}^2) - \Pi(0)) . \tag{7.246}$$

The Fourier transform of the Coulomb potential, $V_\mu(\mathbf{r}) = \eta_{\mu 0}\left(-\frac{Z\alpha}{r}\right)$, which appears on the rhs of Eq. (7.246), reads, see Eqs. (7.220) and (3.114),

$$V_0(\mathbf{k}) = -\frac{Z\alpha}{2\pi^2}\frac{1}{\omega^2} . \tag{7.247}$$

Substituting Eqs. (7.245) and (7.247) into Eq. (7.246) we find

$$V_0^U(\mathbf{r}) = \frac{\alpha(Z\alpha)}{\pi^3}\int_0^1 \mathrm{d}y \frac{\left(\frac{y^2}{2} - \frac{y^3}{3}\right)(1-2y)}{y(1-y)} \int \mathrm{d}^3\mathbf{k} \frac{\mathrm{e}^{\mathrm{i}\mathbf{k}\cdot\mathbf{r}}}{\frac{m^2}{y(1-y)} + \omega^2} \tag{7.248}$$
$$= \frac{2\alpha(Z\alpha)}{\pi}\int_0^1 \mathrm{d}y \frac{\left(\frac{y^2}{2} - \frac{y^3}{3}\right)(1-2y)}{y(1-y)} \frac{\mathrm{e}^{-\frac{m_V}{\sqrt{y(1-y)}}r}}{r} ,$$

where in the last equality we used Eq. (7.236) and replaced m by m_V to emphasize that it is the mass of the "virtual" particle. The virtual particle "in the loop" does not have to be identical with the particle bound in the atom. Next, we make transition to atomic units

$$r = \frac{r_A}{m_R Z\alpha} ,$$

where we replaced m by m_R to emphasize that it is the mass of the "real" particle, i.e., the one bound in an atom. After substituting Eq. (7.248) into Eq. (7.242), we obtain the following expression for the atomic energy shift due to vacuum polarization in the Coulomb field of the nucleus

[39]This potential was derived for the first time in the framework of the ordinary perturbation method.

$$(\Delta E)^{\rm VP}_{\rm exp} = \frac{2\alpha(Z\alpha)^2 m_R}{\pi} \int_0^1 {\rm d}y \frac{\left(\frac{y^2}{2} - \frac{y^3}{3}\right)(1-2y)}{y(1-y)} \int {\rm d}^3\mathbf{r}\, |\psi_{\rm at}(\mathbf{r})|^2 \frac{{\rm e}^{-\frac{m_V}{m_R Z\alpha\sqrt{y(1-y)}} r}}{r} . \tag{7.249}$$

For the purpose of further calculation, it is convenient to substitute

$$r \to r\sqrt{y(1-y)} Z\alpha \frac{m_R}{m_V} ;$$

thus we have

$$(\Delta E)^{\rm VP}_{\rm exp} = m_R \left(\frac{m_R}{m_V}\right)^2 \frac{2\alpha(Z\alpha)^4}{\pi} \int_0^1 {\rm d}y \left(\frac{y^2}{2} - \frac{y^3}{3}\right)(1-2y)$$
$$\times \int {\rm d}^3\mathbf{r} \left| \psi_{\rm at}\left(\frac{\mathbf{r}\sqrt{y(1-y)} Z\alpha m_R}{m_V}\right)\right|^2 \frac{{\rm e}^{-r}}{r} .$$

In the last equation, we can replace the Dirac wave function $\psi_{\rm at}$ by the Schrödinger one, $\psi^0_{\rm at}$. If the inequality $Z\alpha \frac{m_R}{m_V} \ll 1$ holds, the dependence of the wave function $\psi^0_{\rm at}$ on the distance r can be obviously neglected. Consequently, the wave function can be replaced by its value at the origin

$$(\Delta E)^{\rm VP}_{\rm exp} \simeq m_R \left(\frac{m_R}{m_V}\right)^2 \frac{2\alpha(Z\alpha)^4}{\pi} \int_0^1 {\rm d}y \left(\frac{y^2}{2} - \frac{y^3}{3}\right)(1-2y)|\psi^0_{\rm at}(0)|^2 \int {\rm d}^3\mathbf{r}\, \frac{{\rm e}^{-r}}{r} \tag{7.250}$$
$$= m_R \frac{\alpha}{\pi n^3} \delta_{l0} \left(\frac{m_R}{m_V}\right)^2 (Z\alpha)^4 \left(-\frac{4}{15}\right) ,$$

where in the last equality we used Eq. (4.101). Had we made the approximation $\ln\left(1 - \frac{k^2}{m^2} y(1-y)\right) \simeq -\frac{k^2}{m^2} y(1-y)$ in Eq. (7.243), we would have arrived at the last formula somewhat faster.

7.6.6 Comparison with Experiment

> How could one seriously believe that the electron really cared about my calculation, one way or the other? And yet the experiments (...) showed that it did care. Somehow or other, all this complicated mathematics I was scribbling established rules that the electron (...) was bound to follow.[5]

First, somewhat trivial, observation is that vacuum polarization at the leading order of $Z\alpha$ influences only s-states. For "ordinary" hydrogen-like atoms, we have $m_R = m_r$ and $m_V = m_e$. Adding contributions from Eqs. (6.243), (7.211), and (7.250) yields for $Z = 1$

$$(\Delta E)_{2s} = \frac{m_e \alpha^5}{\pi 2^3} \frac{1}{\left(1+\frac{m_e}{m_p}\right)^3} \left(10.8658433 - \frac{4}{3} \ln\left(1+\frac{m_e}{m_p}\right) - \frac{1}{3} - \frac{4}{15}\right)$$

and

$$(\Delta E)_{2p_{1/2}} = \frac{m_e \alpha^5}{\pi 2^3} \frac{1}{\left(1+\frac{m_e}{m_p}\right)^3} \left(0.0400340 - \frac{1+\frac{m_e}{m_p}}{6}\right),$$

where in the last equation we substituted from Eqs. (3.108) and (3.111). After the replacement according to Eq. (3.12), $m_e \to \frac{2R_\infty c}{\alpha^2}$, we obtain for Lamb shift in the hydrogen atom

$$\nu_{\text{theo}}(2s - 2p_{1/2}) = 1055.61\,\text{MHz}$$

which is in a very good agreement with the experimental value [26]

$$\nu_{\text{exp}}(2s - 2p_{1/2}) = 1057.845(9)\,\text{MHz}.$$

Figure 7.10 summarizes the structure of the lowest hydrogen levels. For muonium one obtains

$$\nu_{\text{theo}}(2s - 2p_{1/2}) = 1042.79\,\text{MHz}$$

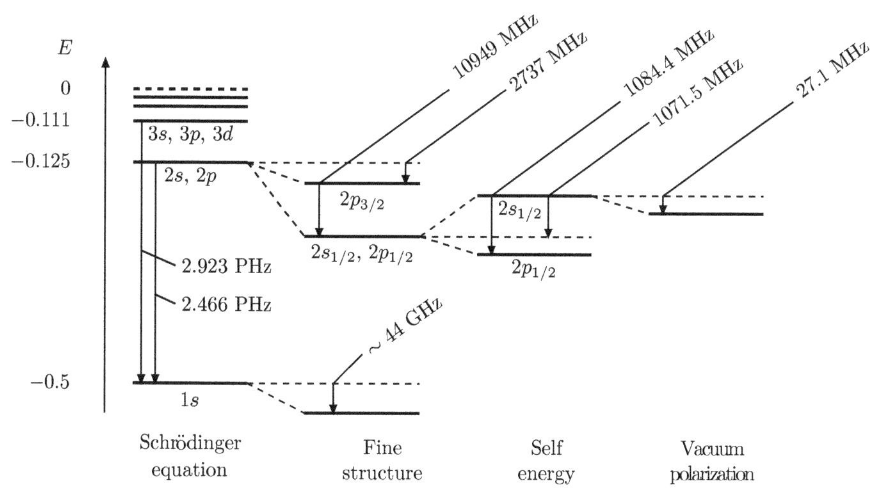

Fig. 7.10 The structure of the lowest hydrogen levels after taking into account relativistic and QED corrections. The displayed numbers are calculated in infinite nuclear mass limit. The columns, except for the first one, do not scale accordingly to the energy scale on the lhs. The symbol "PHz" stands for *petaherz*, 10^{15} Hz

which is in an even better agreement with the last experimental value [40]

$$\nu_{\mathrm{exp}}(2s - 2p_{1/2}) = 1042(23)\,\mathrm{MHz}\,.$$

However, measurement of this interval in muonium is significantly less accurate than in case of the hydrogen atom; the previous muonium experiment [28] established 1070(15) MHz.

Needless to say that our calculations remain still approximate. The most significant neglected contributions are: higher orders in α^2, i.e., two-loop corrections, higher orders in $Z\alpha$, one- and two-photon exchanges between the electron and the nucleus, nonzero proton radius and so on. The relative error of our calculation caused by these omissions does not exceed approximately $(1 - 5) \times 10^{-3}$. Contribution of photon exchanges, which we started discussing in Sects. 6.5.6 and 6.5.7 and we will continue this discussion in the next Sect. 7.7 as well, is suppressed with respect to the electron and photon self-energies by the ratio of the electron to nuclear masses. This ratio is smaller than α for both hydrogen and muonium; thus there is no point considering it unless higher-order corrections in α are taken into account as well.

As one can see from Eq. (7.250), the mass of the virtual particles appears in the denominator; therefore light particles, namely the electron-positron pair, affect the energy levels due to vacuum polarization the most. On the other hand, the mass of the bound particle appears in the numerator. Thus muonic hydrogen, where an electron is replaced by a muon, serves better for the study of the discussed quantum electrodynamic effects rather than the ordinary hydrogen atom.

Exercise 28: Corrections to the Gross Structure II

Continue Exercise 8. That is, calculate the theoretical prediction for the $2s - 1s$ interval in hydrogen-like atoms with accuracy up to the order of α^5. Compare the result with the experimental values listed in Table 3.1.

Exercise 29: Lamb Shift in Muonic Hydrogen

In case of muonic hydrogen, we insert the electron mass, $m_V = m_e$, and the reduced mass of the muon-proton complex, $m_R = \frac{m_\mu m_p}{m_\mu + m_p}$, for the masses of the virtual and real particles in Eq. (7.249), respectively. Thus we find $\frac{m_R}{m_V} = \frac{\frac{m_\mu}{m_e}}{1+\frac{m_\mu}{m_p}} = 185.840834$, where the mass ratios were taken from Sect. 3.2.6. The ratio $\frac{m_R}{m_V} Z\alpha = 1.356146$ is clearly not significantly smaller than 1 and the integrals in Eq. (7.249) have to be calculated exactly. Determine the theoretical prediction for the $2s - 2p_{1/2}$ splitting *and compare it with the experimental value [31]*

$$\nu_{\rm exp}(2s - 2p_{1/2}) = -49{,}881.88(70)\,\text{GHz}.$$

The virtual electron-positron pair bears very real consequences!
Hint: For the atomic wave function, take successively $\psi_{\rm at}(\mathbf{r}) \simeq R_{20}(r)Y_{00}(\mathbf{n})$ *and* $\psi_{\rm at}(\mathbf{r}) \simeq R_{21}(r)Y_{10}(\mathbf{n})$*, where you insert for* R_{20} *and* R_{21} *from Eqs. (3.95) and (3.96), respectively. Perform the analytic integration over r first and then the numerical integration over y.*

7.7 Two-Photon Exchange at High Energies

We now turn our attention to the two-photon exchange between charged particles. We have already dealt with this process in Sect. 6.5.7 where we described the low-energy region. Now we focus on the high-energy region where antiparticles appear in virtual states.[40]

As has been already shown in Sects. 6.5.6 and 6.5.7, one- and two-photon exchanges between charged particles shift the atomic energy levels at the order of α^5. At this order, the "radiative corrections," that is, electron self-energy and vacuum polarization, contribute as well. As has already been pointed out, in case of the "ordinary" hydrogen-like atoms, photon exchanges are suppressed with respect to the radiative corrections by the electron to nucleus mass ratio m_e/m_p. In case of more complex atoms such as helium, photon exchanges between two electrons are of the same order of magnitude as the radiative corrections. Thus we see that the "ordinary" hydrogen-like atoms are somewhat special in this respect.

For the description of a two-photon exchange with accuracy to α^5, it suffices to invoke the same approximations as in Sect. 7.3.2: we write the atomic state in a similar manner as in Eq. (7.89)

$$|\psi^0_{\rm at}\rangle \simeq \sum_{\mathbf{p}} \psi^0_{\rm at}(\mathbf{p})|U_0(1)\rangle|U_0(2)\rangle\,, \tag{7.251}$$

where $|U_0\rangle$ describes the state of the particle at rest, see Eq. (7.97). We will henceforth focus on the high-energy part of the effect, the low-energy part has already been analyzed in Sect. 6.5.7. Hence, it suffices to treat particles in virtual states as free.

[40] The results derived in this section were for the first time derived in [10, 34]; see also [12, 29].

7.7.1 Longitudinal Photons

For the purpose of the following consideration, it is advantageous to introduce "quantized" scalar potential:

$$\hat{\varphi}(\mathbf{r}) = \int \frac{d^3\mathbf{k}}{(2\pi)^{3/2}\omega} \left(\hat{a}(\mathbf{k}) e^{i\mathbf{k}\cdot\mathbf{r}} + \hat{a}^+(\mathbf{k}) e^{-i\mathbf{k}\cdot\mathbf{r}} \right) , \tag{7.252}$$

where we stipulate the commutation relations

$$\hat{a}(\mathbf{k})|0\rangle = 0 , \quad [\hat{a}(\mathbf{k}), \hat{a}^+(\mathbf{k}')] = \delta(\mathbf{k} - \mathbf{k}') .$$

Vacuum expectation value of the product of two such operators is the Coulomb potential

$$\langle 0|\hat{\varphi}(\mathbf{r}_1)\hat{\varphi}(\mathbf{r}_2)|0\rangle = \int \frac{d^3\mathbf{k}}{(2\pi)^3\omega^2} e^{i\mathbf{k}\cdot(\mathbf{r}_1-\mathbf{r}_2)} = \frac{1}{4\pi}\frac{1}{r_{12}} . \tag{7.253}$$

Needless to say that this is a purely formal invention, its only purpose being to simplify our following considerations. We have already emphasized that the Coulomb interaction between particles is instantaneous and carries no momentum. All experimentally detectable quanta of an EM field are polarized perpendicularly to the direction of propagation. Our above steps do not contradict it: the quantized scalar potential will appear only in the interaction Hamiltonian, see Eq. (7.254) below. This leads to the same results as if we treated the Coulomb interaction classically.

The advantage of introducing a quantized scalar potential is that we obtain the longitudinal part of the interaction between two particles in the same order of e as the transverse part. When we considered the interaction of an electron with fluctuations of the EM field, we found the contribution of the longitudinal and transverse parts of the interaction in the first and second orders of the perturbation method, see Eqs. (7.73) and (7.74), respectively. When we deal with the two-photon exchange, we would need to consider the second, third, and fourth order of the perturbation method to find everything we need. However, with the introduction of the quantized scalar potential, we find all of the desired parts at the fourth order of the method.

7.7.2 Two-Photon Exchange in Feynman Approach

For the transition amplitude that a particle finds itself in the same states after interacting twice with a quantized EM field, we derived Eq. (7.130),

$$e^2 \int_{t_0}^{\infty} \int_{t_0}^{\infty} \langle 0|\langle \psi_{at}(t)|\hat{G}_0^{(F)}(t,t'')\hat{H}_1(t'')\hat{G}_0^{(F)}(t'',t')\hat{H}_1(t')\hat{U}_0(t',t_0)|\psi_{at}(t_0)\rangle|0\rangle dt' dt'' . \tag{7.130}$$

Recall that we considered only the interaction with the transverse part of the EM field, Eq. (7.131). Generalization to the case when the particle interacts with the longitudinal part of the field as well is straightforward

$$\hat{\mathsf{H}}_1(t) = e\gamma_0 \int \mathrm{d}^3\mathbf{r}|\mathbf{r}\rangle\langle\mathbf{r}|[\gamma_0\hat{\varphi}(\mathbf{r}) - \boldsymbol{\gamma}\cdot\hat{\boldsymbol{A}}(t,\mathbf{r})]. \tag{7.254}$$

This form of the interaction Hamiltonian is obtained when one makes the usual replacement $\hat{\boldsymbol{p}} \to \hat{\boldsymbol{p}} - e\hat{\boldsymbol{A}}$ in the Dirac equation $\gamma_0(\boldsymbol{\gamma}\cdot\hat{\boldsymbol{p}} - m)\psi = 0$, see Eqs. (7.19) and (7.34), and separates the time differentiation $\mathrm{i}\frac{\partial}{\partial t}\psi = \gamma_0\left[\boldsymbol{\gamma}\cdot\hat{\boldsymbol{p}} + m + e\hat{\boldsymbol{A}}\cdot\boldsymbol{\gamma}\right]\psi$. The transition amplitudes for the individual particles are independent of each other. Hence, according to the basic principles of quantum mechanics, we need to multiply them with each other. Thus, the transition amplitude that two particles find themselves in the same state and each of them interacts twice with the EM field is the product of the one-particle amplitudes:

$$\begin{aligned}\mathscr{A} = e_1^2 e_2^2 \int\int\int\int &\langle 0|\langle\psi_1(t)|\hat{\mathsf{G}}_0^{(F)}(t,t_2)\hat{\mathsf{H}}_1(t_2)\hat{\mathsf{G}}_0^{(F)}(t_2,t_1)\hat{\mathsf{H}}_1(t_1)\hat{\mathsf{U}}_0(t_1,t_0)|\psi_1(t_0)\rangle \\ &\times \langle\psi_2(t)|\hat{\mathsf{G}}_0^{(F)}(t,t_4)\hat{\mathsf{H}}_1(t_4)\hat{\mathsf{G}}_0^{(F)}(t_4,t_3)\hat{\mathsf{H}}_1(t_3)\hat{\mathsf{U}}_0(t_3,t_0)|\psi_2(t_0)\rangle|0\rangle \mathrm{d}t_1\mathrm{d}t_2\mathrm{d}t_3\mathrm{d}t_4\,.\end{aligned}$$

After substituting from Eqs. (7.136) and (7.137),

$$\langle\psi_{\mathrm{at}}(t)|\hat{\mathsf{G}}_0^{(F)}(t,t'') = -\mathrm{i}\theta(t-t'')\langle\psi_{\mathrm{at}}(t'')| \tag{7.136}$$

and

$$\hat{\mathsf{U}}_0(t',t_0)|\psi_{\mathrm{at}}(t_0)\rangle = |\psi_{\mathrm{at}}(t')\rangle \tag{7.137}$$

respectively, we arrive at

$$\begin{aligned}\mathscr{A} = -e_1^2 e_2^2 \int\int\int\int &\langle 0|\langle\psi_1(t_2)|\hat{\mathsf{H}}_1(t_2)\hat{\mathsf{G}}_0^{(F)}(t_2,t_1)\hat{\mathsf{H}}_1(t_1)|\psi_1(t_1)\rangle \\ &\times\langle\psi_2(t_4)|\hat{\mathsf{H}}_1(t_4)\hat{\mathsf{G}}_0^{(F)}(t_4,t_3)\hat{\mathsf{H}}_1(t_3)|\psi_2(t_3)\rangle|0\rangle \mathrm{d}t_1\mathrm{d}t_2\mathrm{d}t_3\mathrm{d}t_4\,.\end{aligned} \tag{7.255}$$

7.7.3 Photon Propagator and Time Ordered Operator Product

We integrate over all possible times t_1, t_2, t_3, and t_4 in Eq. (7.255). When evaluating the products of the quantized EM potentials, we need to consider all of the possible time orderings,

$$\begin{aligned}&\langle 0|\mathscr{T}\left[\hat{\boldsymbol{A}}(\boldsymbol{x}_4)\hat{\boldsymbol{A}}(\boldsymbol{x}_3)\hat{\boldsymbol{A}}(\boldsymbol{x}_2)\hat{\boldsymbol{A}}(\boldsymbol{x}_1)\right]|0\rangle \\ &= \theta(t_4-t_3)\theta(t_3-t_2)\theta(t_2-t_1)\langle 0|\hat{\boldsymbol{A}}(\boldsymbol{x}_4)\hat{\boldsymbol{A}}(\boldsymbol{x}_3)\hat{\boldsymbol{A}}(\boldsymbol{x}_2)\hat{\boldsymbol{A}}(\boldsymbol{x}_1)|0\rangle + \ldots,\end{aligned}$$

where the dots stand for the remaining 23 possible orderings.[41] The vacuum expectation value of the product of four operators is a product of the vacuum expectation values of two operators where we sum over all pairs

$$\langle 0|\hat{\boldsymbol{A}}(\boldsymbol{x}_4)\hat{\boldsymbol{A}}(\boldsymbol{x}_3)\hat{\boldsymbol{A}}(\boldsymbol{x}_2)\hat{\boldsymbol{A}}(\boldsymbol{x}_1)|0\rangle = \langle 0|\hat{\boldsymbol{A}}(\boldsymbol{x}_4)\hat{\boldsymbol{A}}(\boldsymbol{x}_3)|0\rangle\langle 0|\hat{\boldsymbol{A}}(\boldsymbol{x}_2)\hat{\boldsymbol{A}}(\boldsymbol{x}_1)|0\rangle$$
$$+ \langle 0|\hat{\boldsymbol{A}}(\boldsymbol{x}_4)\hat{\boldsymbol{A}}(\boldsymbol{x}_2)|0\rangle\langle 0|\hat{\boldsymbol{A}}(\boldsymbol{x}_3)\hat{\boldsymbol{A}}(\boldsymbol{x}_1)|0\rangle + \langle 0|\hat{\boldsymbol{A}}(\boldsymbol{x}_4)\hat{\boldsymbol{A}}(\boldsymbol{x}_1)|0\rangle\langle 0|\hat{\boldsymbol{A}}(\boldsymbol{x}_3)\hat{\boldsymbol{A}}(\boldsymbol{x}_2)|0\rangle\,,$$

as the reader can easily verify from the expansion of the potentials in terms of creation and annihilation operators, see Eqs. (6.51) and (7.252). The reader can convince himself as well that the same statement holds also for the vacuum expectation value of the time ordered product of four operators

$$\begin{aligned}&\langle 0|\mathscr{T}\left[\hat{\boldsymbol{A}}(\boldsymbol{x}_4)\hat{\boldsymbol{A}}(\boldsymbol{x}_3)\hat{\boldsymbol{A}}(\boldsymbol{x}_2)\hat{\boldsymbol{A}}(\boldsymbol{x}_1)\right]|0\rangle\\ &=\langle 0|\mathscr{T}\left[\hat{\boldsymbol{A}}(\boldsymbol{x}_4)\hat{\boldsymbol{A}}(\boldsymbol{x}_3)\right]|0\rangle\langle 0|\mathscr{T}\left[\hat{\boldsymbol{A}}(\boldsymbol{x}_2)\hat{\boldsymbol{A}}(\boldsymbol{x}_1)\right]|0\rangle\\ &\quad+\langle 0|\mathscr{T}\left[\hat{\boldsymbol{A}}(\boldsymbol{x}_4)\hat{\boldsymbol{A}}(\boldsymbol{x}_2)\right]|0\rangle\langle 0|\mathscr{T}\left[\hat{\boldsymbol{A}}(\boldsymbol{x}_3)\hat{\boldsymbol{A}}(\boldsymbol{x}_1)\right]|0\rangle\\ &\quad+\langle 0|\mathscr{T}\left[\hat{\boldsymbol{A}}(\boldsymbol{x}_4)\hat{\boldsymbol{A}}(\boldsymbol{x}_1)\right]|0\rangle\langle 0|\mathscr{T}\left[\hat{\boldsymbol{A}}(\boldsymbol{x}_3)\hat{\boldsymbol{A}}(\boldsymbol{x}_2)\right]|0\rangle\,.\end{aligned}$$

We define the vacuum expectation value of a time ordered product of two time-independent operators as

$$D_{00}(\boldsymbol{x}''-\boldsymbol{x}') = \langle 0|\mathscr{T}\left(\hat{\varphi}(\mathbf{r}'')\hat{\varphi}(\mathbf{r}')\right)|0\rangle = \mathrm{i}\delta(t''-t')\langle 0|\hat{\varphi}(\mathbf{r}'')\hat{\varphi}(\mathbf{r}')|0\rangle\,, \tag{7.256}$$

and use Eq. (7.253) to evaluate the last expression. The factor i appears therein so that the longitudinal and transverse parts of the EM field combine to a Lorentz invariant photon propagator, see Sect. 7.7.4 below. Bearing in mind the two above equations, it is not necessary to discuss the instantaneous and delayed parts of the interaction separately. The advantage of the introduction of the quantized scalar potential thus becomes obvious.

Furthermore, one discovers the advantage of the Feynman space-time approach in its full extent when he considers a two-photon exchange. The reader can easily find out that the use of the ordinary perturbation method leads nowhere in this case.

After substituting the interaction Hamiltonian (7.254) and the last above equation into Eq. (7.255), we find

$$\begin{aligned}\mathscr{A} &= -e_1^2e_2^2\int \mathrm{d}^4\boldsymbol{x}_1\int \mathrm{d}^4\boldsymbol{x}_2\int \mathrm{d}^4\boldsymbol{x}_3\int \mathrm{d}^4\boldsymbol{x}_4\\ &\quad\times\langle\psi_1(t_2)|\,\mathbf{r}_2\rangle\gamma_0\gamma_\mu\hat{\mathsf{G}}_0^{(F)}(\boldsymbol{x}_2,\boldsymbol{x}_1)\gamma_0\gamma_\nu\langle\mathbf{r}_1|\,\psi_1(t_1)\rangle\end{aligned}$$

[41] The reader surely excuses us for not explicitly writing down all of them.

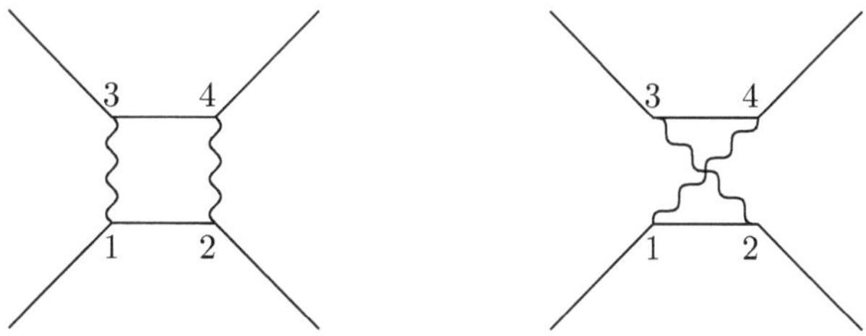

Fig. 7.11 Two-photon exchanges

$$\times \ \langle \psi_2(t_4)|\,\mathbf{r}_4\rangle \gamma_0 \gamma_\rho \hat{\mathsf{G}}_0^{(F)}(\boldsymbol{x}_4, \boldsymbol{x}_3)\gamma_0\gamma_\sigma \langle \mathbf{r}_3|\, \psi_2(t_3)\rangle$$
$$\times \ \left[D_{\mu\rho}(\boldsymbol{x}_2 - \boldsymbol{x}_4)D_{\nu\sigma}(\boldsymbol{x}_1 - \boldsymbol{x}_3) + D_{\mu\sigma}(\boldsymbol{x}_2 - \boldsymbol{x}_3)D_{\nu\rho}(\boldsymbol{x}_1 - \boldsymbol{x}_4)\right]. \qquad (7.257)$$

The two possibilities corresponding to the two terms in the last equation are illustrated in Fig. 7.11. Note that had we assigned the same expressions to the elements of the Feynman diagram in Fig. 7.11 as to the elements of the Feynman diagram in Fig. 7.6, the last equation could have be obtained somewhat faster. The first and second diagrams in Fig. 7.11 are usually referred to as the *ladder* and *crossed diagrams*, respectively. We omitted the term $D_{\mu\nu}(\boldsymbol{x}_2 - \boldsymbol{x}_1)D_{\rho\sigma}(\boldsymbol{x}_4 - \boldsymbol{x}_3)$ as it corresponds to self-energy effect for each of the particles separately and hence is of no interest to us presently. The expression for the photon propagator is given by a generalization of Eq. (7.134)

$$D_{\mu\rho}(\boldsymbol{x}_2 - \boldsymbol{x}_1) = \langle 0|\mathscr{T}\left(A_\mu(\boldsymbol{x}_2)A_\rho(\boldsymbol{x}_1)\right)|0\rangle = -\mathrm{i}\int \frac{\mathrm{d}^4\boldsymbol{k}}{(2\pi)^4}\mathrm{e}^{-\mathrm{i}\boldsymbol{k}\cdot(\boldsymbol{x}_2-\boldsymbol{x}_1)}D_{\mu\rho}(\boldsymbol{k}), \qquad (7.258)$$

where the Fourier transform of the photon propagator in the Coulomb gauge reads, see Eqs. (7.134), (7.253), and (7.256),

$$D_{00} = -\frac{1}{\omega^2}, \quad D_{ij} = -\frac{\delta_{ij}}{\boldsymbol{k}^2} + \frac{k_i k_j}{\boldsymbol{k}^2\omega^2}, \quad D_{i0} = D_{0i} = 0. \qquad (7.259)$$

The electron propagator is given by Eq. (7.126)

$$\hat{\mathsf{G}}_0^{(F)}(\boldsymbol{x}_2, \boldsymbol{x}_1)\gamma_0 = \langle \mathbf{r}_2|\frac{1}{2\pi}\int \mathrm{d}E\frac{\mathrm{e}^{-\mathrm{i}E(t_2-t_1)}}{E - \hat{\mathsf{h}}_\mathrm{D}}|\mathbf{r}_1\rangle\gamma_0 = \int \frac{\mathrm{d}^4\boldsymbol{p}}{(2\pi)^4}\frac{\mathrm{e}^{-\mathrm{i}\boldsymbol{p}\cdot(\boldsymbol{x}_2-\boldsymbol{x}_1)}}{\boldsymbol{\gamma}\cdot\boldsymbol{p} - m}. \qquad (7.260)$$

We consider the free-particle Hamiltonian for the one-particle Dirac Hamiltonian,

$$\frac{1}{E - \hat{\mathsf{h}}_\mathrm{D}} = \int \mathrm{d}^3\mathbf{p}\frac{|\mathbf{p}\rangle\langle\mathbf{p}|}{E - \gamma_0\boldsymbol{\gamma}\cdot\mathbf{p} - \gamma_0 m},$$

and the states (7.97) for the initial and final particle states,

$$|\psi(t)\rangle = \mathrm{e}^{-\mathrm{i}mt}|U_0\rangle. \qquad (7.261)$$

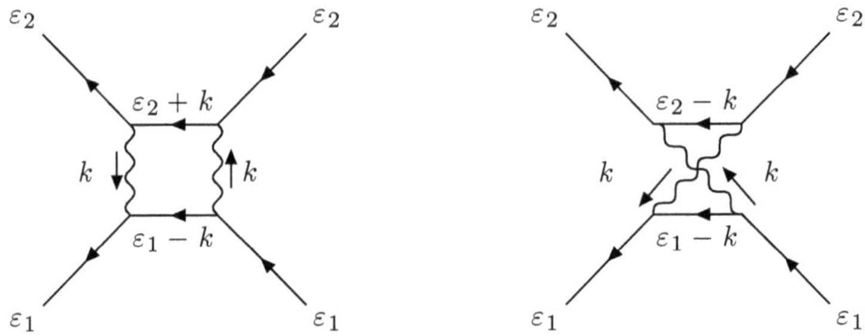

Fig. 7.12 Feynman diagrams for two-photon exchange in momentum space

After substituting the photon propagator (7.258), electron propagator (7.260) and initial and final states (7.261) into Eq. (7.257), the four integrations over four space-time variables $\boldsymbol{x}_1$-$\boldsymbol{x}_4$ yield four four-dimensional δ-functions. From a physical point of view, it corresponds to four-momentum conservations in the interaction vertices, see Fig. 7.12. The three of these four δ-functions cancel the three integrations over the four-momentum. The fourth one yields no new result, though. We are thus left with one integration over the four-momenta and the expression $(2\pi)^4\delta^{(4)}(0)$ which we replace by L^3T:

$$\mathscr{A} = e_1^2 e_2^2 L^3 T \int \frac{\mathrm{d}^4\boldsymbol{k}}{(2\pi)^4} \langle \overline{U}_0(1)| \gamma_\mu \frac{1}{\boldsymbol{\gamma}\cdot(\boldsymbol{\varepsilon}_1 - \boldsymbol{k}) - m_1} \gamma_\nu |U_0(1)\rangle \tag{7.262}$$

$$\times \langle \overline{U}_0(2)| \left[\gamma_\rho \frac{1}{\boldsymbol{\gamma}\cdot(\boldsymbol{\varepsilon}_2 + \boldsymbol{k}) - m_2} \gamma_\sigma + \gamma_\sigma \frac{1}{\boldsymbol{\gamma}\cdot(\boldsymbol{\varepsilon}_2 - \boldsymbol{k}) - m_2} \gamma_\rho \right]$$

$$\times |U_0(2)\rangle D_{\mu\rho}(\boldsymbol{k}) D_{\nu\sigma}(\boldsymbol{k}) ,$$

where

$$\boldsymbol{\varepsilon}_1 = (m_1, 0), \quad \boldsymbol{\varepsilon}_2 = (m_2, 0) .$$

We find for the shift of atomic energy level due to two-photon exchange from Eqs. (7.142) and (7.251)

$$\Delta E = \mathrm{i}\frac{\mathscr{A}}{T} \left| \sum_{\mathbf{p}} \psi^0_{\mathrm{at}}(\mathbf{p}) \right|^2 , \tag{7.263}$$

where the amplitude $\mathscr{A}$ is given by Eq. (7.262).

7.7.4 Note on Gauge Invariance

In an analogy to the expression for the electron self-energy, Eq. (7.148), the expression for the two-photon exchange, Eq. (7.262), is gauge invariant. Although we derived it in the Coulomb gauge, it is equivalent to the expressions one would derive in any other gauge. In this case, one means by the gauge invariance

$$\begin{aligned} k_\rho \langle \overline{U}_0(2)| &\left[\gamma_\rho \frac{1}{\boldsymbol{\gamma} \cdot (\boldsymbol{\varepsilon}_2 + \boldsymbol{k}) - m_2} \gamma_\sigma + \gamma_\sigma \frac{1}{\boldsymbol{\gamma} \cdot (\boldsymbol{\varepsilon}_2 - \boldsymbol{k}) - m_2} \gamma_\rho \right] |U_0(2)\rangle \\ = \langle \overline{U}_0(2)| &\left[(\boldsymbol{\gamma} \cdot (\boldsymbol{k} + \boldsymbol{\varepsilon}_2) - m_2) \frac{1}{\boldsymbol{\gamma} \cdot (\boldsymbol{\varepsilon}_2 + \boldsymbol{k}) - m_2} \gamma_\sigma \right. \\ &\left. + \gamma_\sigma \frac{1}{\boldsymbol{\gamma} \cdot (\boldsymbol{\varepsilon}_2 - \boldsymbol{k}) - m_2} (\boldsymbol{\gamma} \cdot (\boldsymbol{k} - \boldsymbol{\varepsilon}_2) + m_2) \right] |U_0(2)\rangle = 0 \,, \end{aligned} \tag{7.264}$$

where we used the Dirac equation $(\boldsymbol{\gamma} \cdot \boldsymbol{\varepsilon}_2 - m_2)|U_0(2)\rangle = 0$. It follows that we can add an arbitrary multiple of $\boldsymbol{k}$ to the photon propagators. If we subtract the expression $\frac{k_\mu k_\rho}{\boldsymbol{k}^2 \omega^2}$, from $D_{\mu\rho}(\boldsymbol{k})$, we get from the form of the photon propagator in the Coulomb gauge, Eqs. (7.258) and (7.259), to the form of the photon propagator (7.159) in the Lorenz gauge. We could elaborate a bit on the last statement. Writing Eq. (7.262) symbolically as

$$\mathscr{A} = G_{\mu\rho} D_{\mu\rho}(\boldsymbol{k})$$

we are free to change it to

$$\mathscr{A} = G_{\mu\rho} \left[D_{\mu\rho}(\boldsymbol{k}) - \frac{k_\mu k_\rho}{\boldsymbol{k}^2 \omega^2} \right].$$

Writing it explicitly in the space-time components, we have

$$\begin{aligned} \mathscr{A} = G_{00} &\left[D_{00}(\boldsymbol{k}) - \frac{k_0 k_0}{\boldsymbol{k}^2 \omega^2} \right] - G_{0i} \left[D_{0i}(\boldsymbol{k}) - \frac{k_0 k_i}{\boldsymbol{k}^2 \omega^2} \right] \\ - G_{i0} &\left[D_{i0}(\boldsymbol{k}) - \frac{k_i k_0}{\boldsymbol{k}^2 \omega^2} \right] + G_{ij} \left[D_{ij}(\boldsymbol{k}) - \frac{k_i k_j}{\boldsymbol{k}^2 \omega^2} \right]. \end{aligned}$$

Now, Eq. (7.264) yields

$$G_{\mu\rho} k_\rho = 0 \Rightarrow G_{0i} k_i = G_{00} k_0 \,, \quad G_{i0} k_i = G_{00} k_0 \,.$$

Taking into account the explicit form of $D_{\mu\rho}(\boldsymbol{k})$, Eq. (7.259), we finally arrive at

$$\mathscr{A} = G_{00} \left[-\frac{1}{\omega^2} + \frac{k_0 k_0}{\boldsymbol{k}^2 \omega^2} \right] - G_{ij} \frac{\delta_{ij}}{\boldsymbol{k}^2} = G_{\mu\rho} \frac{\eta_{\mu\rho}}{\boldsymbol{k}^2} \,.$$

Clearly, $\eta_{\mu\rho}\boldsymbol{k}^{-2}$ is the Fourier transform of the photon propagator in the Lorenz gauge. In fact, we already went through similar exercise when discussing the self-energy problem, see Sect. 7.4.6.

In the following text, we will retain the Coulomb gauge. The reason is, the integration over the four-momentum $\boldsymbol{k}$ of virtual photons in Eq. (7.262) leads to a divergent integral at *low* frequencies of virtual photons, while the integration over high frequencies causes no problems at all in this case. In contrast to the so-called ultraviolet divergences, i.e., divergences at the upper bound of the integration which are inherent to the theory, the so-called infrared divergences, i.e., divergences at the lower bound of the integration, are a mere consequence of the invalidity of the chosen approximation at low energies of virtual particles. While the manifest Lorentz invariance is of greatest importance when dealing with ultraviolet divergences, it is inessential or even leads to undesirable problems when tackling the infrared divergences. We need to distinguish between the longitudinal and transverse parts of the interaction. The Coulomb part of the interaction in Eq. (7.262) has to be "renormalized" as it contains Coulomb interaction in the nonrelativistic limit and this interaction must be removed, see the next section. The transverse part of the interaction has to be regularized. Regularization, in this case, means only to stop the integration over the frequencies at the lower bound Λ. As we will see further, the dependence on the regulating parameter Λ vanishes if we add Eqs. (7.263) and (6.256) together. In the Lorenz gauge, the distinction between the longitudinal and transverse parts of the interaction is obscured.

7.7.5 Longitudinal Part of the Interaction

After substituting Eq. (7.262) into Eq. (7.263), setting $\mu = \nu = \rho = \sigma = 0$ and substituting $\boldsymbol{k} \to -\boldsymbol{k}$, the atomic energy shift due to the exchange of two longitudinal photons reads

$$\Delta E_{||} = \mathrm{i}\left|\sum_{\mathbf{p}} \psi^0_{\mathrm{at}}(\mathbf{p})\right|^2 L^3 e_1^2 e_2^2 \int \frac{\mathrm{d}^4\boldsymbol{k}}{(2\pi)^4\omega^4} \langle U_0(1)| \frac{1}{k_0 + m_1 - \mathsf{h}_1} |U_0(1)\rangle \tag{7.265}$$

$$\times \langle U_0(2)| \left[-\frac{1}{k_0 - m_2 + \mathsf{h}_2^-} + \frac{1}{k_0 + m_2 - \mathsf{h}_2^+} \right] |U_0(2)\rangle ,$$

where

$$\mathsf{h}_1 = \gamma_0 \boldsymbol{\gamma} \cdot \mathbf{k} + \gamma_0 m_1 \quad \mathsf{h}_2^{\pm} = \pm\gamma_0 \boldsymbol{\gamma} \cdot \mathbf{k} + \gamma_0 m_2 . \tag{7.266}$$

Spectral decomposition of the free particle propagator reads

$$\frac{1}{z+\mathsf{h}} = \frac{\mathsf{P}^{+}}{z+E} + \frac{\mathsf{P}^{-}}{z-E}, \quad E = \sqrt{\omega^2 + m^2} \tag{7.267}$$

where $\mathsf{P}^{\pm}$ represent positive and negative energy projectors for a free particle and are special cases of the general case introduced in Eq. (7.81). The projectors obey equations

$$\mathsf{P}^{+} + \mathsf{P}^{-} = 1$$

and

$$(+E)\mathsf{P}^{+} + (-E)\mathsf{P}^{-} = \mathsf{h},$$

where the last equation is a spectral decomposition of the Hamiltonian h. From these two equations we can easily find

$$\mathsf{P}^{\pm} = \frac{E \pm \mathsf{h}}{2E}. \tag{7.268}$$

Recalling that the integration contour is closed in the lower half of the complex plane, the integration over k_0 by means of the residue theorem yields important equalities

$$\int \frac{\mathrm{d}k_0}{2\pi\mathrm{i}} \frac{1}{(k_0 - a \pm \mathrm{i}\varepsilon)(k_0 - b \pm \mathrm{i}\varepsilon)} = 0, \quad \int \frac{\mathrm{d}k_0}{2\pi\mathrm{i}} \frac{1}{(k_0 - a \pm \mathrm{i}\varepsilon)(k_0 - b \mp \mathrm{i}\varepsilon)} = \frac{\pm 1}{b-a}.$$

The integration over k_0 in Eq. (7.265) then yields

$$\begin{aligned}
&\int \frac{\mathrm{d}k_0}{2\pi\mathrm{i}} \frac{1}{(k_0 + m_1 - \mathsf{h}_1)(k_0 - m_2 + \mathsf{h}_2)} \\
&\quad = \int \frac{\mathrm{d}k_0}{2\pi\mathrm{i}} \left(\frac{\mathsf{P}_1^{+}}{k_0 + m_1 - E_1 + \mathrm{i}\varepsilon} + \frac{\mathsf{P}_1^{-}}{k_0 + m_1 + E_1 - \mathrm{i}\varepsilon} \right) \\
&\qquad\qquad \times \left(\frac{\mathsf{P}_2^{+}}{k_0 - m_2 + E_2 - \mathrm{i}\varepsilon} + \frac{\mathsf{P}_2^{-}}{k_0 - m_2 - E_2 + \mathrm{i}\varepsilon} \right) \\
&\quad = \frac{\mathsf{P}_1^{+}\mathsf{P}_2^{+}}{m_1 + m_2 - E_1 - E_2} - \frac{\mathsf{P}_1^{-}\mathsf{P}_2^{-}}{m_1 + m_2 + E_1 + E_2}
\end{aligned} \tag{7.269}$$

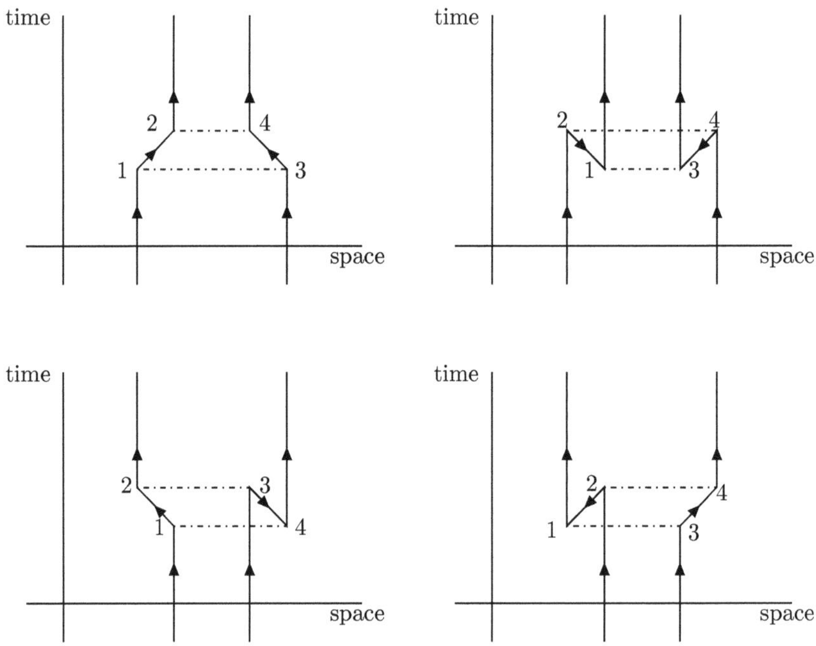

Fig. 7.13 Spacetime diagram of Coulomb interaction between charged particles

and

$$-\int \frac{dk_0}{2\pi i} \frac{1}{(k_0 + m_1 - \mathsf{h}_1)(k_0 + m_2 - \mathsf{h}_2)}$$

$$= \frac{\mathsf{P}_1^+ \mathsf{P}_2^-}{-m_1 + m_2 + E_1 + E_2} - \frac{\mathsf{P}_1^- \mathsf{P}_2^+}{m_2 - m_1 - E_1 - E_2} . \tag{7.270}$$

The last two equations are equations for a product of two one-particle propagators in the momentum space. As should be clear now, the positive and negative energy projectors correspond to virtual particles and antiparticles, respectively, see Fig. 7.13. Hence, Eq. (7.265) contains four types of virtual states. If the particles are an electron and a antimuon, we find the following particles in virtual states: an electron and an antimuon, a positron and a muon (both possibilities are contained in Eq. (7.269)) and a positron and an antimuon, an electron and a muon (both possibilities are contained in Eq. (7.270)). Comparison of Figs. 7.11 and 7.13 reveals that real spacetime ordering of events can be very different from the impression one may get from the Feynman diagrams.

As already mentioned in previous section, one has to subtract the superfluous part from Eq. (7.265). However, which one is it? We have already considered the Coulomb interaction between two particles before, see, e.g., Eq. (3.9). We have thus clearly taken into account a part of the Coulomb interaction in the Schrödinger equation to which we are now trying to calculate a correction!

After substituting from Eqs. (7.267), (7.268), and

$$\langle U_0|\mathsf{h}|U_0\rangle = \frac{m}{L^3} \tag{7.271}$$

we obtain for the average of the first term on the rhs of Eq. (7.269)

$$\begin{aligned}
&\langle U_0(1)|\langle U_0(2)|\frac{\mathsf{P}_1^+\mathsf{P}_2^+}{m_1+m_2-E_1-E_2}|U_0(1)\rangle|U_0(2)\rangle \\
&= \frac{1}{4L^6}\frac{\left(1+\frac{m_1}{\sqrt{m_1^2+\omega^2}}\right)\left(1+\frac{m_2}{\sqrt{m_2^2+\omega^2}}\right)}{m_1+m_2-\sqrt{m_1^2+\omega^2}-\sqrt{m_2^2+\omega^2}}.
\end{aligned} \tag{7.272}$$

We now substitute the expansion

$$\sqrt{m^2+\omega^2} \simeq m + \frac{\omega^2}{2m} - \frac{\omega^4}{8m^3}$$

into the last above equation, and expand further all the terms in Eq. (7.272) in the powers of ω. The first two terms of this expansion read

$$-\frac{1}{L^6}\left(\frac{2m_r}{\omega^2} - \frac{1}{2(m_1+m_2)}\right),$$

where m_r is the reduced mass. The first term in this expression is, but for an irrelevant normalization constant, the eigenvalue of Green operator $(E_0 - \hat{\mathsf{H}}_0)^{-1}$, where $E_0 = 0$ and $\hat{\mathsf{H}}_0 = \hat{\mathsf{p}}^2/(2m_r)$. The first term in the expression

$$-\left|\sum_{\mathbf{p}} \psi_{\text{at}}^0(\mathbf{p})\right|^2 L^{-3} e_1^2 e_2^2 \int \frac{\mathrm{d}^3\mathbf{k}}{(2\pi)^3\omega^4}\left(\frac{2m_r}{\omega^2} - \frac{1}{2(m_1+m_2)}\right), \tag{7.273}$$

thus represents the part of the expression (7.265) corresponding to the double count of the Coulomb interaction between two particles which are initially and finally at rest and even in virtual states move slowly. If we count this interaction infinitely many times, we obtain bound state energy. The second term of the last expression represents a relativistic correction due to the dependence of the inertial mass on velocity. However, we have already calculated both of these contributions before. We obtained the former by directly solving the Schrödinger equation (3.9) and we considered the latter back in Sect. 3.6.2, see Eq. (3.122). Therefore, we need to subtract both of them from Eq. (7.265). Using Eqs. (7.269) and (7.270), where we insert from Eq. (7.268) for the projectors $\mathsf{P}_{1,2}^{\pm}$ and subsequently use Eq. (7.271), we derive the following expression for the energy shift due to Coulomb interaction between charged particles present neither in the Schrödinger equation nor in the Breit correction given in Sect. 3.6.2,

$$\begin{aligned}\Delta E_{||,\mathrm{exp}} &= (7.265)\text{–}(7.273)\\ &= -\frac{\left|\sum_{\mathbf{p}} \psi^0_{\mathrm{at}}(\mathbf{p})\right|^2}{L^3} e_1^2 e_2^2 \int \frac{\mathrm{d}^3\boldsymbol{k}}{(2\pi)^3\omega^4} \frac{1}{4E_1E_2}\left[-\frac{(E_1+m_1)(E_2-m_2)}{E_1+E_2-m_1+m_2}\right.\\ &\quad + \frac{(E_1-m_1)(E_2+m_2)}{-E_1-E_2-m_1+m_2} + \frac{(E_1-m_1)(E_2-m_2)}{E_1+E_2+m_1+m_2}\\ &\quad \left. - \frac{(E_1+m_1)(E_2+m_2)}{m_1+m_2-E_1-E_2} - \frac{4E_1E_2}{m_2+m_1}\left(\frac{2m_1m_2}{\omega^2}-\frac{1}{2}\right)\right].\end{aligned}$$

Recall that, see Eq. (7.267), $E_{1,2} = \sqrt{\omega^2 + m_{1,2}^2}$. The integral over the frequencies in the last equation is now finite and the integral over the angles trivial. If we evaluate these integrals and insert from Eq. (7.102), we arrive at the final result

$$\Delta E_{||,\mathrm{exp}} = -\frac{e_1^2 e_2^2}{(4\pi)^2}\left|\psi^{(0)}_{\mathrm{at}}(\mathbf{r}_{12}=0)\right|^2 \frac{4}{3m_1m_2}. \tag{7.274}$$

7.7.6 The Remaining Part of the Interaction

Calculation of the remaining part of the interaction contained in Eq. (7.262) requires calculational prowess, but no new ideas. After substituting Eq. (7.262) into Eq. (7.263) and setting $\mu = i$, $\nu = k$, $\rho = j$, and $\sigma = l$ one obtains

$$\begin{aligned}(\Delta E)_\perp &= \mathrm{i}\left|\sum_{\mathbf{p}} \psi^0_{\mathrm{at}}(\mathbf{p})\right|^2 L^3 e_1^2 e_2^2 \int \frac{\mathrm{d}^4\boldsymbol{k}}{(2\pi)^4(\boldsymbol{k}^2)^2}\left(\delta_{ij}-\eta_i\eta_j\right)\left(\delta_{kl}-\eta_k\eta_l\right)\\ &\quad \times \left\langle \overline{U}_0(1)\right|\gamma_i \frac{\gamma_0(m_1-k_0)+m_1}{(m_1-k_0)^2-E_1^2}\gamma_k\left|U_0(1)\right\rangle\\ &\quad \times \left\langle \overline{U}_0(2)\right|\left[\gamma_j\frac{\gamma_0(m_2+k_0)+m_2}{(m_2+k_0)^2-E_2^2}\gamma_l + \gamma_l\frac{\gamma_0(m_2-k_0)+m_2}{(m_2-k_0)^2-E_2^2}\gamma_j\right]\left|U_0(2)\right\rangle\\ &= \frac{e_1^2e_2^2}{(4\pi)^2}\left|\psi^{(0)}_{\mathrm{at}}(\mathbf{r}_{12}=0)\right|^2 \frac{2}{m_1m_2}\left[\ln\left(\frac{2\Lambda}{m_1m_2}\right) + \frac{m_1^2\ln m_1 - m_2^2\ln m_2}{m_1^2-m_2^2}\right.\\ &\quad \left. + \frac{4m_1m_2}{3(m_1^2-m_2^2)}\ln\left(\frac{m_1}{m_2}\right)\left\langle\hat{\boldsymbol{S}}_1\cdot\hat{\boldsymbol{S}}_2\right\rangle\right].\end{aligned} \tag{7.275}$$

The integration over k_0 was performed via partial fraction decomposition,[42] summation of residua in poles and multiplication by $(-2\pi \mathrm{i})$.[43] When integrating over angles, we used Eqs. (3.47) and (6.106). Calculation of the spinorial part was performed using the properties of the Pauli matrices, see Eq. (1.25). As alluded to above, one of the integrations over the frequencies was stopped at the lower bound Λ.

In a similar way, we find for the mixed contribution of the longitudinal and transverse parts

$$
\begin{aligned}
(\Delta E)_{\perp,\|} &= \mathrm{i}\left|\sum_{\mathbf{p}} \psi^0_{\mathrm{at}}(\mathbf{p})\right|^2 L^3 e_1^2 e_2^2 \int \frac{\mathrm{d}^4 k}{(2\pi)^4 \boldsymbol{k}^2 \omega^2}\left(\delta_{ij} - \eta_i \eta_j\right) \\
&\times \Bigg\{ \langle \overline{U}_0(1)| \gamma_0 \frac{\boldsymbol{\gamma}\cdot\mathbf{k}}{(m_1 - k_0)^2 - E_1^2} \gamma_i |U_0(1)\rangle \\
&\times \langle \overline{U}_0(2)| \left[\gamma_0 \frac{-\boldsymbol{\gamma}\cdot\mathbf{k}}{(m_2 + k_0)^2 - E_2^2} \gamma_j + \gamma_j \frac{\boldsymbol{\gamma}\cdot\mathbf{k}}{(m_2 - k_0)^2 - E_2^2} \gamma_0 |U_0(2)\rangle \right] \\
&+ \langle \overline{U}_0(1)| \gamma_i \frac{\boldsymbol{\gamma}\cdot\mathbf{k}}{(m_1 - k_0)^2 - E_1^2} \gamma_0 |U_0(1)\rangle \\
&\times \langle \overline{U}_0(2)| \left[\gamma_j \frac{-\boldsymbol{\gamma}\cdot\mathbf{k}}{(m_2 + k_0)^2 - E_2^2} \gamma_0 + \gamma_0 \frac{\boldsymbol{\gamma}\cdot\mathbf{k}}{(m_2 - k_0)^2 - E_2^2} \gamma_j |U_0(2)\rangle \right] \Bigg\} \\
&= -\frac{e_1^2 e_2^2}{(4\pi)^2} \left|\psi^{(0)}_{\mathrm{at}}(\mathbf{r}_{12} = 0)\right|^2 \frac{32}{3(m_1^2 - m_2^2)} \ln\left(\frac{m_1}{m_2}\right) \left\langle \hat{\boldsymbol{S}}_1 \cdot \hat{\boldsymbol{S}}_2 \right\rangle . \qquad (7.276)
\end{aligned}
$$

7.7.7 *Comparison with Experiment*

The atomic energy shift $\Delta E_{12\mathrm{p}}$ due to the one- and two-photon exchanges is for the state-independent part of the s-states given by the sum of Eqs. (6.252), (6.256), (7.274), (7.275), and (7.276):

$$
\begin{aligned}
\frac{\Delta E_{12\mathrm{p}}(1s)}{n^3} &= \frac{m_r^3 (Z\alpha)^5}{m_1 m_2 \pi n^3} \Bigg\{ -\frac{4}{3} \langle 1s| (\hat{\mathrm{p}}_A)_i (\hat{\mathrm{h}}_0 - \varepsilon_0) \ln\left[2(\hat{\mathrm{h}}_0 - \varepsilon_0)\right] (\hat{\mathrm{p}}_A)_i |1s\rangle \\
&+ \frac{62}{9} - \frac{2}{3}\ln Z\alpha + \frac{8}{3}\ln 2 + 2\left[\ln\frac{2}{m_1 + m_2} + \frac{m_1^2 \ln m_1 - m_2^2 \ln m_2}{m_1^2 - m_2^2}\right] \\
&-8\frac{m_1 m_2}{m_1^2 - m_2^2} \ln\frac{m_1}{m_2} \left\langle \hat{\boldsymbol{S}}_1 \cdot \hat{\boldsymbol{S}}_2 \right\rangle \Bigg\} . \qquad (7.277)
\end{aligned}
$$

---

[42] It was accomplished by means of Maple.

[43] Recall that the integration over the time components of the four-vectors is always closed in the lower half of the complex plane.

Note that the dependence on the regulating parameter Λ disappears. Consequent generalization of the results for photon exchanges and radiative corrections derived here to helium-like atoms is relatively straightforward, see [29].[44] One can readily see from the last equation that the photons exchange is suppressed with respect to radiative correction by the electron to nucleus mass ratio m_e/m_p for the spin-spin interaction, but it is also slightly amplified by the logarithm of the ratio. If we add the spin dependent terms from the last equation to Eq. (7.206), our final result for the muonium hyperfine splitting reads

$$\nu_{\text{theo}}(1^3s - 1^1s) = \frac{8}{3}\frac{m_e}{m_p}\frac{2R_\infty c\alpha^2}{\left(1+\frac{m_e}{m_p}\right)^3}\left[1+\frac{\alpha}{\pi}\left(1+\frac{3\frac{m_e}{m_p}}{1-\left(\frac{m_e}{m_p}\right)^2}\ln\left(\frac{m_e}{m_p}\right)\right)\right]$$
$$\simeq 4463.388\,\text{MHz}\,.$$

In comparison with the theoretical prediction 4464.2 MHz made in Sect. 7.5.7, our estimate agrees with the experimental value 4463.303 MHz, Eq. (3.60) by another order of magnitude. We will use the entire equation (7.277) in Sect. 7.8.5 for a precise determination of positronium spectral lines.

7.8 Positronium II

As has been repeatedly pointed out, in case of "ordinary" hydrogen-like atoms, the photon exchanges are suppressed with respect to the radiative corrections by the electron to nucleus mass ratio m_e/m_p. In case of positronium, this ratio equals 1 but the situation is complicated by virtual annihilation, as discussed in Sect. 7.3.2 where we showed that it contributes at the order α^4. Therefore at the order α^5, one needs to consider corrections to it, as we will do in this section. There is another interesting aspect of these corrections: all one-loop QED corrections considered so far were for space-like momentum exchanges, while now we enter the realm of time-like momentum exchanges which brings a few new features. We commence by returning to the leading part in the virtual annihilation effect in the Feynman approach. We then successively discuss three corrections to this process: vacuum polarization, photon exchange between an electron and a positron in initial and final states (which is the most complicated and the most interesting part), and finally virtual two-photon annihilation. To conclude this section we summarize all the contributions to the ground state positronium fine splitting up to the order α^5 and compare their sum with the experimental value.[45]

[44] To be more precise, one can easily see what expressions are to be evaluated.

[45] The theoretical results were first derived in [16].

7.8.1 Virtual Positronium Annihilation in Feynman Approach

For the purpose of our further considerations, it is now very useful (and enjoyable) to return to the virtual positronium annihilation. We analyzed it in Sect. 7.3.2 in the framework of the time-independent perturbation method; now we will adopt the Feynman approach.

The transition amplitude for the annihilation of an electron-positron pair and the creation of a photon succeeded by photon annihilation and creation of "another" electron-positron pair is in the Feynman view a product of two transition amplitudes. The first one corresponds to an electron evolving from an initial positive energy state at initial time t_0 to a later time t_1 when it emits or absorbs a photon, and then evolves back in time to a negative energy state. The second one corresponds to "another" electron evolving from an initial negative energy state at initial time t_3 to an earlier time t_2 when it emits or absorbs a photon, and then evolves forward in time to a positive energy state. The amplitude reads

$$\mathscr{A} = \int_{-\infty}^{\infty} \mathrm{d}t_1 \int_{-\infty}^{\infty} \mathrm{d}t_2 \langle 0|\langle \psi_c(t_0)|\hat{\mathsf{G}}_0^{(F)}(t_0,t_1)\hat{\mathsf{H}}_1(t_1)\hat{\mathsf{U}}_0(t_1,t_0)|\psi(t_0)\rangle$$

$$\times \langle \psi(t_3)|\hat{\mathsf{G}}_0^{(F)}(t_3,t_2)\hat{\mathsf{H}}_1(t_2)\hat{\mathsf{U}}_0(t_2,t_3)|\psi_c(t_3)\rangle|0\rangle\,,$$

where we integrate over all possible interaction times t_1 and t_2. The interaction Hamiltonian is given by Eq. (7.131) and the action of the propagators on the wave functions by Eqs. (7.136), (7.137), and

$$\langle \psi_c(t_0)|\hat{\mathsf{G}}_0^{(F)}(t_0,t_1) = \mathrm{i}\theta(t_1-t_0)\langle \psi_c(t_1)| = \mathrm{i}\theta(t_1-t_0)\mathrm{e}^{-\mathrm{i}mt_1}\langle V_0|\,, \qquad (7.278)$$

$$\hat{\mathsf{U}}_0(t_2,t_3)|\psi_c(t_3)\rangle = |\psi_c(t_2)\rangle\,.$$

The states U_0 and V_0 are given by Eqs. (7.97) and (7.98), respectively. After substituting all of these expressions into the previous equation for the transition amplitude $\mathscr{A}$, we obtain

$$\mathscr{A} = e^2 \int \mathrm{d}^4x_1 \int \mathrm{d}^4x_2 D_{ij}(\boldsymbol{x}_2-\boldsymbol{x}_1)\mathrm{e}^{-2\mathrm{i}m(t_1-t_2)} U_0^+\gamma_0\gamma_i V_0 V_0^+\gamma_0\gamma_j U_0\,, \qquad (7.279)$$

where we again introduced the time ordered product of the operators, Eq. (7.132). From a physical point of view, it corresponds to the necessity to consider both of the possible time orderings. Either $t_2 > t_1$: the "original" pair is annihilated and a photon created first and then the photon is annihilated and "another" pair created, or $t_1 > t_2$: "another" pair and a photon are created first and then the photon and the "original" pair annihilated. It is worth noticing the connection between the left and right diagrams in Fig. 7.14 and the first and second terms in Eqs. (7.91), or (7.92). The time-independent perturbation method operates on the (hyper)plane of constant

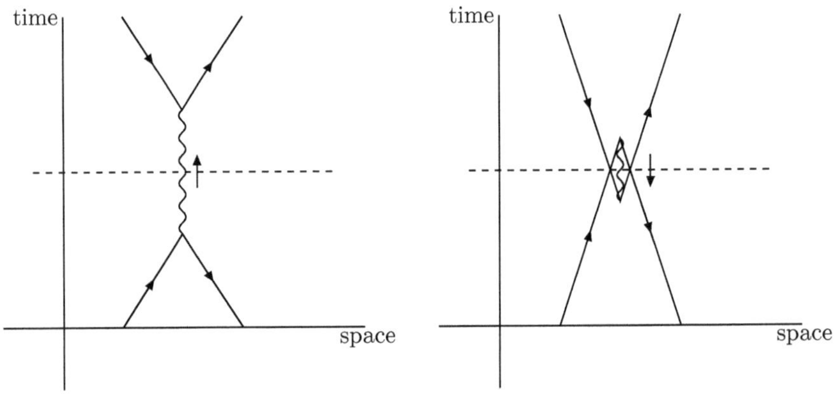

Fig. 7.14 A virtual photon "rests at one place" in space and moves forward (*left picture*) or backward (*right picture*) in time. An electron and a positron at initial and final states are also at rest. The *dashed line* denotes the (hyper)plane of constant time, that is the "operating place" of the time-independent perturbation method. The reader shall be warned that strictly speaking all space-time pictures drawn in this book are somewhat misleading as they originate from the classical view. A particle with a definite, albeit zero, momentum is completely delocalized and drawing its trace in space-time bears a very restricted meaning

time. Hence, "it sees" only one photon in the virtual state in the first case, whereas "it sees" one photon and two electron-positron pairs in the virtual state in the second case. As one can readily see, the two denominators in the first form of Eq. (7.92) corresponds to it. Recall that within the time-independent perturbation method, we subtract the energy of the initial state from the energy of the virtual state. Hence, the first denominator reads $\omega - 2m$ and the second $\omega + 4m - 2m = \omega + 2m$. In the Feynman space-time approach, these two possibilities correspond to a virtual photon evolving either forward or backward in time. If we now insert for the photon propagator from Eq. (7.134) and perform integrations over the space-time coordinates, we obtain

$$\mathscr{A} = \mathrm{i}L^3Te^2 \int \frac{\mathrm{d}^4\boldsymbol{k}}{(2\pi)^4\boldsymbol{k}^2} 2\pi\delta(k_0 - 2m)(2\pi)^3\delta(\mathbf{k})U_0^+\gamma_0\gamma_i V_0 V_0^+\gamma_0\gamma_i U_0 \,, \tag{7.280}$$

where we again replaced $(2\pi)^4\delta^{(4)}(0)$ by L^3T. After performing trivial integration over the photon four-momentum, multiplying the result by $\left|\sum_{\mathbf{p}} \psi^0_{\mathrm{at}}(\mathbf{p})\right|^2$, and dividing it by $\mathrm{i}T$, we arrive at Eq. (7.92),

$$E_2 = \frac{e^2}{4m^2}\left|\sum_{\mathbf{p}} \psi^0_{\mathrm{at}}(\mathbf{p})\right|^2 L^3 U_0^+\gamma_0\gamma_i V_0 V_0^+\gamma_0\gamma_i U_0 \,. \tag{7.92}$$

The relation between the amplitude and energy has, in this case, the opposite sign to that in Eq. (7.142). This minor discrepancy can be traced back to the opposite sign in Eq. (7.278) with respect to Eq. (7.136) and is a consequence of the electron moving now in the "final" state backward in time.

Henceforth, it should be clear that the time-like momentum exchanges are associated with pair creation and annihilation. As one can notice from Fig. 7.14, a photon in a virtual state whose only nonvanishing component of the four-momentum is the time component "rests at one place," see Eq. (7.280).

7.8.2 Vacuum Polarization Correction

We begin our calculations of the corrections to the virtual annihilation by the simplest of them, namely the one stemming from virtual decay of a virtual photon into an electron-positron pair, see Fig. 7.15.

We derived back in Sect. 7.6 that in case of a photon with four-momentum $\boldsymbol{k}$, the vacuum polarization effect changes the strength of the interaction between the particles exchanging the photon according to the equation

$$4\pi\alpha \to 4\pi\alpha\left\{1 + 4\pi\alpha[\Pi(\boldsymbol{k}^2) - \Pi(0)]\right\} . \tag{7.244}$$

Although we derived this replacement for vacuum polarization in a static EM field, it is clearly Lorentz invariant. Hence, we can extend its validity to an arbitrary four-momentum $\boldsymbol{k}$. In a static EM field, the four-momentum of the Fourier transform of a four-potential is purely spatial, see Eq. (7.79). For virtual photon considered in the previous section, the four-momentum is purely temporal, $\boldsymbol{k} = (2m, 0)$. Thus vacuum polarization of a virtual photon with a zero three-momentum leads to the multiplication of Eq. (7.105) by the factor

$$\left(1 + 4\pi\alpha\left[\Pi(4m^2) - \Pi(0)\right]\right) = \left(1 - \frac{8\alpha}{9\pi}\right) . \tag{7.281}$$

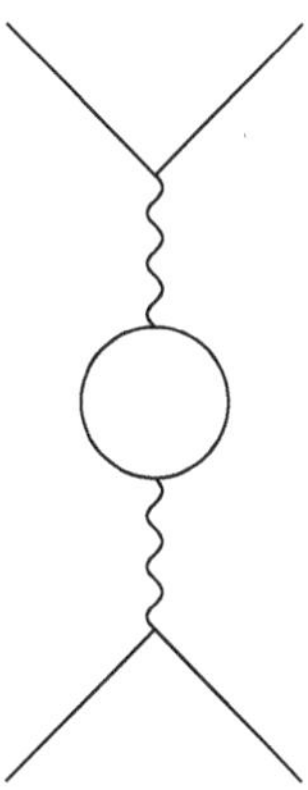

Fig. 7.15 Feynman diagram for virtual decay of a virtual photon into an electron-positron pair

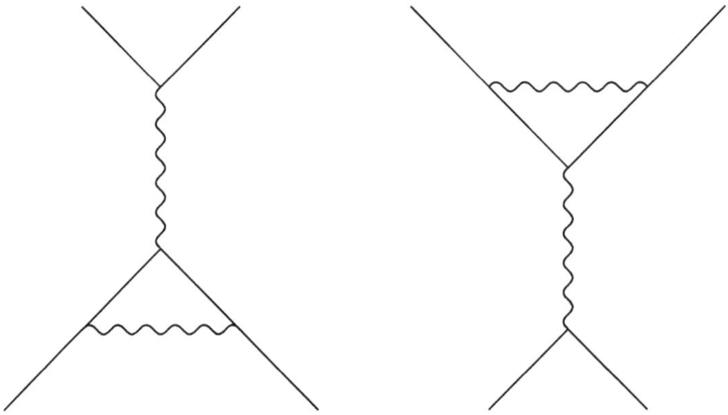

Fig. 7.16 Feynman diagrams for photon exchange between "initial" and "final" electron-positron pairs

7.8.3 *Photon Exchange Correction*

Another correction to the virtual annihilation originates in the photon exchange between either "initial" or "final" electron-positron pairs, see Fig. 7.16.

We have shown above that the electron-positron annihilation can be viewed as an electron going forward and backward in time. Hence, the photon exchange between an electron-positron pair is nothing but an interaction of an electron with its own EM field already thoroughly analyzed in Sect. 7.5. The situation under consideration now differs from that considered back in Sect. 7.5, nevertheless. In case of an electron described in its initial and final states by a bound state function, it suffices to consider the region where the momentum imparted on the electron by an external EM field is small in comparison with the electron rest mass m, i.e., of the order $mZ\alpha$. On the other hand, in case of an electron moving in the final state backward in time, the exchanged momentum between the initial and final states is $2m$, which is certainly not small in comparison with m. However, a crucial simplification appears now: in contrast to the "ordinary" self-energy effect, the photon exchange between an electron-positron pair is a "high-energy" effect. Meaning, the contribution of the long-wavelength photons is in this case completely negligible. Hence, it suffices to consider the influence of the "external" EM field in the first order. In other words, it suffices to consider the correction due to emission and absorption of a photon in electron-photon scattering, where the electron is treated as free. We assume that the electron in the initial and final states carries four-momentum $\boldsymbol{p}_1$ and $\boldsymbol{p}_2$, respectively, and $\boldsymbol{p}_1^2 = \boldsymbol{p}_2^2 = m^2$. The difference of these four-momenta equals the four-momentum $\boldsymbol{q}$ imparted on the electron by an "external" EM field

$$\boldsymbol{q} = \boldsymbol{p}_2 - \boldsymbol{p}_1\,, \qquad \boldsymbol{q}^2 = 2(m^2 - \boldsymbol{p}_1 \cdot \boldsymbol{p}_2)\,. \tag{7.282}$$

In the case of the electron-positron annihilation, this "external" EM field is constituted by a virtual photon with purely temporal four-momentum. At the end of calculation we set, see Eqs. (7.89) and (7.251),

$$\boldsymbol{p}_1 = (m, 0)\,, \qquad \boldsymbol{p}_2 = (-m, 0)\,.$$

1. *Expansion and rearrangement of the self-energy operator*
Back in Sect. 7.4, we derived an expression for the Lorentz invariant self-energy operator, Eq. (7.148), which takes the following form after regularization (7.162)

$$\hat{\mathsf{M}} = \frac{\alpha}{\pi}\int_{\mu^2}^{\Lambda^2} \mathrm{d}\lambda \int \frac{\mathrm{d}^4 \boldsymbol{k}_F}{(\boldsymbol{k}^2-\lambda)^2}\gamma_\mu \frac{1}{\boldsymbol{\gamma}\cdot(\hat{\boldsymbol{\Pi}}-\boldsymbol{k})-m}\gamma_\mu\,,$$

where we introduced the mass of long-wavelength photons μ. This is, once again, a purely formal step to avoid infrared divergence in our following calculation and does not contradict our above considerations. While the contribution of the long-wavelength photons is negligible for time-like photon exchanges of the order $2m$, one cannot neglect it for small space-like exchanges. As we are now considering all possible exchanges, the long-wavelength region needs to be cut off. The physical cause of the infrared divergence will be discussed later.

We derived the self-energy operator for an electron moving in a static EM field. However, owing to the manifest Lorentz transformation of the operator $\hat{\mathsf{M}}$, it holds for an electron moving in an arbitrary EM field.

We use the rearrangement (7.160), now in a symmetric form

$$\frac{1}{\boldsymbol{\gamma}\cdot(\hat{\boldsymbol{\Pi}}-\boldsymbol{k})-m} = \frac{1}{2}\left\{\frac{1}{\boldsymbol{k}^2-2\boldsymbol{k}\cdot\hat{\boldsymbol{\Pi}}+\mathscr{H}}, (\boldsymbol{\gamma}\cdot(\hat{\boldsymbol{\Pi}}-\boldsymbol{k})+m)\right\}\,.$$

Using the explicit form of the second order Dirac Hamiltonian, Eq. (7.24), we expand the self-energy operator to the first order in the potential of "external" EM field using Eq. (6.198), see Fig. 7.17, $\hat{\mathsf{M}} \simeq \hat{\mathsf{M}}_0 + \hat{\mathsf{M}}_1 + \ldots$.

Here, $\hat{\mathsf{M}}_0$ is the self-energy operator of a free particle, which is of no interest for now, and

$$\hat{\mathsf{M}}_1 = \frac{\alpha}{2\pi}\int_{\mu^2}^{\Lambda^2} \mathrm{d}\lambda \int \frac{\mathrm{d}^4 \boldsymbol{k}_F}{(\boldsymbol{k}^2-\lambda)^2}\gamma_\mu \left(-\left\{\boldsymbol{\gamma}\cdot\boldsymbol{V}(\hat{\boldsymbol{x}}), \frac{1}{\hat{\mathsf{D}}_0}\right\}\right.$$
$$\left. + \left\{\boldsymbol{\gamma}\cdot(\hat{\boldsymbol{p}}-\boldsymbol{k})+m, \frac{1}{\hat{\mathsf{D}}_0}\hat{\mathsf{D}}_1\frac{1}{\hat{\mathsf{D}}_0}\right\}\right)\gamma_\mu\,, \tag{7.283}$$

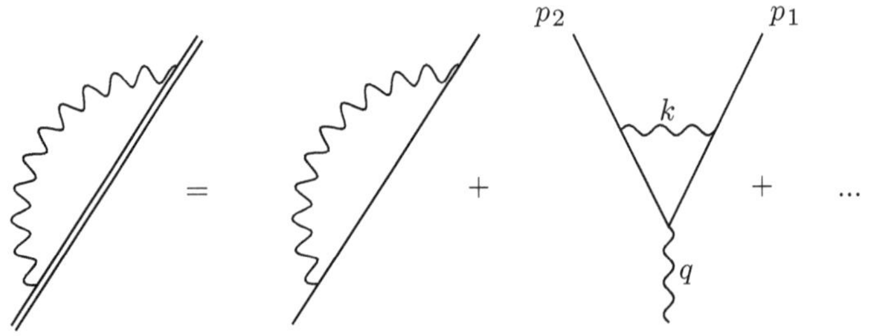

Fig. 7.17 The first two terms of the expansion of the self-energy operator for an electron moving in an external EM field in number of the interactions of the electron with the field. The first term corresponds to no scattering at all, the second term is referred to as one-potential scattering

where $\boldsymbol{V} = e\boldsymbol{A}$, see Eq. (7.215), and we introduced the notation

$$\frac{1}{\hat{\mathsf{D}}_0} = \frac{1}{\boldsymbol{k}^2 - 2\boldsymbol{k}\cdot\hat{\boldsymbol{p}} + \hat{\boldsymbol{p}}^2 - m^2}\,, \qquad \hat{\mathsf{D}}_1 = -2\boldsymbol{k}\cdot\boldsymbol{V}(\hat{\boldsymbol{x}}) + \{\boldsymbol{\gamma}\cdot\hat{\boldsymbol{p}}, \boldsymbol{\gamma}\cdot\boldsymbol{V}(\hat{\boldsymbol{x}})\}\,,$$

$$\{\boldsymbol{\gamma}\cdot\hat{\boldsymbol{p}}, \boldsymbol{\gamma}\cdot\boldsymbol{V}(\hat{\boldsymbol{x}})\} = \hat{\boldsymbol{p}}\cdot\boldsymbol{V}(\hat{\boldsymbol{x}}) + \boldsymbol{V}(\hat{\boldsymbol{x}})\cdot\hat{\boldsymbol{p}} - \hat{\mathscr{H}}_S\,, \quad \hat{\mathscr{H}}_S = \frac{1}{4}[\hat{\Pi}_\mu, \hat{\Pi}_\nu][\gamma_\mu, \gamma_\nu]\,. \tag{7.284}$$

In case of positronium, the photon acting on the electron at the event $\boldsymbol{x}_1$ is created by the current $\mathrm{e}^{2\mathrm{i}mt_2} V_0^+ \gamma_0\gamma_j U_0$ at the event $\boldsymbol{x}_2$, see Eq. (7.279),

$$\begin{aligned}
&(\boldsymbol{V})_0(\boldsymbol{x}_1) = 0\\
&(\boldsymbol{V})_i(\boldsymbol{x}_1) = e^2 \int \mathrm{d}^4x_2 D_{ij}(\boldsymbol{x}_2 - \boldsymbol{x}_1)\mathrm{e}^{2\mathrm{i}mt_2} V_0^+ \gamma_0\gamma_j U_0 = e^2\mathrm{i}\frac{\mathrm{e}^{\mathrm{i}2mt_1}}{(2m)^2} V_0^+ \gamma_0\gamma_i U_0\,,
\end{aligned} \tag{7.285}$$

where we used Eq. (7.134) in the last step. Let us consider the matrix element of $\hat{\mathsf{M}}_1$ between two states of a free particle with definite values of its four-momenta $\boldsymbol{p}_1$ and $\boldsymbol{p}_2$. The states obey the Dirac equation

$$(\boldsymbol{\gamma}\cdot\hat{\boldsymbol{p}} - m)\,|\boldsymbol{p}_{1,2}\rangle = (\boldsymbol{\gamma}\cdot\boldsymbol{p}_{1,2} - m)\,|\boldsymbol{p}_{1,2}\rangle = 0\,. \tag{7.286}$$

By means of Eq. (7.173) we find for the first term in the parenthesis on the rhs of Eq. (7.283)

$$-\frac{1}{2}\gamma_\mu \left\{\boldsymbol{\gamma}\cdot\boldsymbol{V}(\hat{\boldsymbol{x}}), \frac{1}{\hat{\mathsf{D}}_0}\right\}\gamma_\mu = \left\{\boldsymbol{\gamma}\cdot\boldsymbol{V}(\hat{\boldsymbol{x}}), \frac{1}{\hat{\mathsf{D}}_0}\right\}\,.$$

Furthermore, Eqs. (7.17) and (7.286) yield, compare with Eq. (7.161),

$$\frac{1}{2}\langle\overline{\boldsymbol{p}}_2|\gamma_\mu\left\{(\boldsymbol{\gamma}\cdot\hat{\boldsymbol{p}}+m),\frac{1}{\hat{\mathsf{D}}_0}\hat{\mathsf{D}}_1\frac{1}{\hat{\mathsf{D}}_0}\right\}\gamma_\mu|\boldsymbol{p}_1\rangle=\langle\overline{\boldsymbol{p}}_2|\hat{\mathsf{p}}_\mu\frac{1}{\hat{\mathsf{D}}_0}\hat{\mathsf{D}}_1\frac{1}{\hat{\mathsf{D}}_0}\gamma_\mu+\gamma_\mu\frac{1}{\hat{\mathsf{D}}_0}\hat{\mathsf{D}}_1\frac{1}{\hat{\mathsf{D}}_0}\hat{\mathsf{p}}_\mu|\boldsymbol{p}_1\rangle.$$

Using the last two equations and Eqs. (7.190) and (7.191), we arrive at

$$\begin{aligned}\langle\overline{\boldsymbol{p}}_2|\hat{\mathsf{M}}_1|\boldsymbol{p}_1\rangle &= \frac{2\alpha}{\pi}\int_{\mu^2}^{\Lambda^2}\mathrm{d}\lambda\int\frac{\mathrm{d}^4\boldsymbol{k}_F}{(\boldsymbol{k}^2-\lambda)^2}\langle\overline{\boldsymbol{p}}_2|\left(\frac{1}{2}\left\{\boldsymbol{\gamma}\cdot\boldsymbol{V}(\hat{\boldsymbol{x}}),\frac{1}{\hat{\mathsf{D}}_0}\right\}\right.\\ &\quad+\frac{1}{2}\left\{m+\boldsymbol{\gamma}\cdot\boldsymbol{k},\frac{1}{\hat{\mathsf{D}}_0}\hat{\mathsf{D}}_1\frac{1}{\hat{\mathsf{D}}_0}\right\}-\frac{1}{2}\left[\hat{\mathsf{p}}_\mu,\left[\gamma_\mu,\frac{1}{\hat{\mathsf{D}}_0}\hat{\mathsf{D}}_1\frac{1}{\hat{\mathsf{D}}_0}\right]\right]\\ &\quad\left.-\frac{1}{8}\left\{\boldsymbol{\gamma}\cdot\boldsymbol{k},\left[\gamma_\mu,\left[\gamma_\mu,\frac{1}{\hat{\mathsf{D}}_0}\hat{\mathsf{D}}_1\frac{1}{\hat{\mathsf{D}}_0}\right]\right]\right\}\right)|\boldsymbol{p}_1\rangle\,.\end{aligned}$$

Proceeding with rearrangements, Eq. (7.193) yields

$$-\frac{1}{2}[\hat{\mathsf{p}}_\mu,[\gamma_\mu,\hat{\mathsf{D}}_1]]=[\hat{\mathsf{p}}_\mu,[\hat{\Pi}_\mu,\hat{\Pi}_\lambda]]\gamma_\lambda\quad\text{and}\quad-\frac{1}{8}[\gamma_\mu,[\gamma_\mu,\hat{\mathsf{D}}_1]]=\hat{\mathscr{H}}_S\,.$$

Furthermore, Eq. (7.286) yields

$$\frac{1}{\hat{\mathsf{D}}_0}|\boldsymbol{p}_{1,2}\rangle=\frac{1}{\boldsymbol{k}^2-2\boldsymbol{k}\cdot\boldsymbol{p}_{1,2}}|\boldsymbol{p}_{1,2}\rangle$$

and

$$\langle\overline{\boldsymbol{p}}_2|[\hat{\mathsf{p}}_\mu,[\hat{\Pi}_\mu,\hat{\Pi}_\lambda]]\gamma_\lambda|\boldsymbol{p}_1\rangle=-\boldsymbol{q}^2\langle\overline{\boldsymbol{p}}_2|\boldsymbol{\gamma}\cdot\boldsymbol{V}(\hat{\boldsymbol{x}})|\boldsymbol{p}_1\rangle\,,$$

where the four-momentum $\boldsymbol{q}$ imparted on the electron by the EM field described by the four-potential $\boldsymbol{V}$ obeys Eq. (7.282). These rearrangements bring the matrix element into the form

$$\begin{aligned}\langle\overline{\boldsymbol{p}}_2|\hat{\mathsf{M}}_1|\boldsymbol{p}_1\rangle &= \frac{2\alpha}{\pi}\int_{\mu^2}^{\Lambda^2}\mathrm{d}\lambda\int\frac{\mathrm{d}^4\boldsymbol{k}_F}{(\boldsymbol{k}^2-\lambda)^2}\langle\overline{\boldsymbol{p}}_2|\left[\frac{1}{2}\boldsymbol{\gamma}\cdot\boldsymbol{V}(\hat{\boldsymbol{x}})\right.\\ &\quad\times\left(\frac{1}{\boldsymbol{k}^2-2\boldsymbol{k}\cdot\boldsymbol{p}_1}+\frac{1}{\boldsymbol{k}^2-2\boldsymbol{k}\cdot\boldsymbol{p}_2}\right)\\ &\quad+\Big(-2m\boldsymbol{k}\cdot\boldsymbol{V}(\hat{\boldsymbol{x}})+(2m^2-\boldsymbol{q}^2)\boldsymbol{\gamma}\cdot\boldsymbol{V}(\hat{\boldsymbol{x}})\\ &\quad+\frac{1}{2}\left\{\boldsymbol{\gamma}\cdot\boldsymbol{k},-2\boldsymbol{k}\cdot\boldsymbol{V}(\hat{\boldsymbol{x}})+\hat{\boldsymbol{p}}\cdot\boldsymbol{V}(\hat{\boldsymbol{x}})+\boldsymbol{V}(\hat{\boldsymbol{x}})\cdot\hat{\boldsymbol{p}}+\hat{\mathscr{H}}_S\right\}\Big)\\ &\quad\left.\times\frac{1}{\boldsymbol{k}^2-2\boldsymbol{k}\cdot\boldsymbol{p}_1}\frac{1}{\boldsymbol{k}^2-2\boldsymbol{k}\cdot\boldsymbol{p}_2}\right]|\boldsymbol{p}_1\rangle\,.\end{aligned}$$

2. *Introduction of form-factors*
   Our next calculations, once again, involve somewhat extended algebraic manipulations. Therefore, prior to tackling them, we reflect where they will lead us.

   On the basis of now familiar considerations, we know that the transition amplitude for an electron with the initial four-momentum $\boldsymbol{p}_1$ to be scattered by a photon with the four-momentum $\boldsymbol{q}$ and thus acquiring the final momentum $\boldsymbol{p}_2$, reads

$$\mathscr{A}_1 = \mathrm{i}(2\pi)^4\delta(\boldsymbol{p}_2 - \boldsymbol{p}_1 - \boldsymbol{q})\langle\overline{\boldsymbol{p}}_2|\boldsymbol{\gamma}\cdot\boldsymbol{V}(\hat{\boldsymbol{x}})|\boldsymbol{p}_1\rangle\,.$$

   Likewise, the transition amplitude of the same process but now accompanied by emission and subsequent absorption of a virtual photon reads

$$\mathscr{A}_3 = \mathrm{i}(2\pi)^4\delta(\boldsymbol{p}_2 - \boldsymbol{p}_1 - \boldsymbol{q})\langle\overline{\boldsymbol{p}}_2|\hat{\mathsf{M}}_1|\boldsymbol{p}_1\rangle\,.$$

   We can express the sum of these amplitudes as

$$\langle\overline{\boldsymbol{p}}_2|\hat{\mathsf{M}}_1 + \boldsymbol{\gamma}\cdot\boldsymbol{V}(\hat{\boldsymbol{x}})|\boldsymbol{p}_1\rangle = \langle\overline{\boldsymbol{p}}_2|\boldsymbol{\gamma}\cdot\boldsymbol{V}(\hat{\boldsymbol{x}})F_1(\boldsymbol{q}^2) - \frac{\hat{\mathscr{H}}_S}{2m}F_2(\boldsymbol{q}^2)|\boldsymbol{p}_1\rangle\,, \tag{7.287}$$

   where F_1 and F_2 measure the electron responses to the external EM field and are called *electric and magnetic electron form-factors*, respectively. If the emission and absorption of the virtual photon are not considered, then clearly $F_1 = 1$ and $F_2 = 0$. This corresponds to the fact, mentioned back in Chaps. 1 and 3, that a "bare" electron behaves as an electric monopole and a magnetic dipole with the gyromagnetic ratio $g_e = 1$. We have already seen back in Sect. 7.5 that the self-energy effect changes this simple picture.
3. *Renormalization*
   Consider the four-potential $\boldsymbol{V}(\hat{\boldsymbol{x}})$ in the last equation to have nonvanishing only the temporal component, the momentum exchange $\boldsymbol{q}$ to be purely spatial, $\boldsymbol{q}^2 = -\mathbf{q}^2$, and take the nonrelativistic limit of the Dirac wave function, Eq. (7.22) with $\mathbf{A} = 0$; one then finds

$$(7.287) \simeq \left\langle\mathbf{p}_2^{(0)}\middle|V_0(\hat{\boldsymbol{r}})\middle|\mathbf{p}_1^{(0)}\right\rangle\left[F_1(\mathbf{q}^2) - \frac{\mathbf{q}^2}{4m^2}F_2(\mathbf{q}^2)\right] + \ldots\,,$$

   where the states $\left\langle\mathbf{r}\middle|\mathbf{p}_{1,2}^{(0)}\right\rangle = \mathrm{e}^{\mathrm{i}\mathbf{p}_{1,2}\cdot\mathbf{r}}/(2\pi)^{3/2}$ are the nonrelativistic limits of the states $|\boldsymbol{p}_{1,2}\rangle$. Hence, e times the square bracket on the rhs of the last equation corresponds to the Fourier transform of the charge density, i.e.[46]

---

[46]Notice that the form-factor in the nonrelativistic limit appeared already in Eq. (6.168). The angular part is already integrated out there, though.

$$e\left[F_1(\mathbf{q}^2)-\frac{\mathbf{q}^2}{4m^2}F_2(\mathbf{q}^2)\right]=\int \mathrm{d}^3\mathbf{r}\rho(\mathbf{r})\mathrm{e}^{\mathrm{i}\mathbf{q}\cdot\mathbf{r}}.$$

For a point charge, $\rho(\mathbf{r})=e\delta(\mathbf{r})$, clearly $F_1=1$ and $F_2=0$. Measurement of the proton form-factor in low energy electron-proton scattering experiments was one of the very first indications that the proton is not a point particle.

An electron, on the other hand, *is* indeed a point particle. Is it not in disagreement with the fact that the self-energy effect leads to the electron form-factor different from 1? It is not. We will see further that the self-energy effect does not lead to an unambiguous determination of the function $F_1(\boldsymbol{q}^2)$. As we have already discussed in Sect. 7.6, the electric charge value yielding $e^2/(4\pi)\simeq 1/137.036$ is measured in processes when the exchanged squared four-momentum $\boldsymbol{q}^2$ approaches zero:

$$F_1(0)=1. \tag{7.288}$$

As we will see further, this requirement determines the function $F_1(\boldsymbol{q}^2)$ uniquely. It will also be clear that the electric form-factor differs from 1 for nonzero values of the square of the exchanged four-momentum.

4. *Integration over the four-momentum of a virtual photon*

In order to convert the needed integrals over the four-momentum $\boldsymbol{k}$ of a virtual photon to known integrals, Eq. (7.169), we use the Feynman parameters and differentiate with respect to a parameter

$$\frac{1}{\boldsymbol{k}^2-2\boldsymbol{k}\cdot\boldsymbol{p}_1}\frac{1}{\boldsymbol{k}^2-2\boldsymbol{k}\cdot\boldsymbol{p}_2}=-\frac{\partial}{\partial\Delta}\int_0^1\mathrm{d}x\frac{1}{\boldsymbol{k}^2-2\boldsymbol{k}\cdot\boldsymbol{p}_x+\Delta}\bigg|_{\Delta=0}$$

and

$$\frac{2k_\sigma}{\boldsymbol{k}^2-2\boldsymbol{k}\cdot\boldsymbol{p}_1}\frac{1}{\boldsymbol{k}^2-2\boldsymbol{k}\cdot\boldsymbol{p}_2}=\frac{\partial}{\partial(p_x)_\sigma}\int_0^1\mathrm{d}x\frac{1}{\boldsymbol{k}^2-2\boldsymbol{k}\cdot\boldsymbol{p}_x+\Delta}\bigg|_{\Delta=0},$$

where

$$\boldsymbol{p}_x=\boldsymbol{p}_1x+\boldsymbol{p}_2(1-x). \tag{7.289}$$

Consequently, the sought integrals over $\boldsymbol{k}$ can be again reduced to a basic integral, see Eq. (7.169),

$$\int\frac{\mathrm{d}^4\boldsymbol{k}_F(1,k_\nu)}{(\boldsymbol{k}^2-\lambda)^2(\boldsymbol{k}^2-2\boldsymbol{k}\cdot\boldsymbol{p}_x+\Delta)}=\int_0^1\mathrm{d}y\frac{1}{4}(1-y)\frac{(1,(p_x)_\nu y)}{(\boldsymbol{p}_x)^2y^2-\Delta y+\lambda(1-y)},$$

The last four equations yield

$$\langle \overline{\boldsymbol{p}}_2|\hat{\mathsf{M}}_1|\boldsymbol{p}_1\rangle = \frac{\alpha}{2\pi}\int_{\mu^2}^{\Lambda^2} \mathrm{d}\lambda \int_0^1 \mathrm{d}y(1-y)\int_0^1 \mathrm{d}x\langle \overline{\boldsymbol{p}}_2| \left[\frac{\boldsymbol{\gamma}\cdot\boldsymbol{V}(\hat{\boldsymbol{x}})}{m^2y^2+\lambda(1-y)} \right.$$
$$- \left(m\hat{\mathsf{V}}_\sigma \frac{\partial}{\partial (p_x)_\sigma} + (2m^2 - \boldsymbol{q}^2)\boldsymbol{\gamma}\cdot\boldsymbol{V}(\hat{\boldsymbol{x}})\frac{\partial}{\partial \Delta}\right.$$
$$\left. + \frac{y}{2}\left\{ \gamma_\nu, \hat{\mathsf{V}}_\sigma \frac{\partial}{\partial (p_x)_\sigma} + (\hat{\boldsymbol{p}}\cdot\boldsymbol{V}(\hat{\boldsymbol{x}}) + \boldsymbol{V}(\hat{\boldsymbol{x}})\cdot\hat{\boldsymbol{p}} + \hat{\mathscr{H}}_S)\frac{\partial}{\partial \Delta}\right\}(p_x)_\nu \right)$$
$$\left.\times \left. \frac{1}{(\boldsymbol{p}_x)^2y^2 - \Delta y + \lambda(1-y)}\right|_{\Delta=0}\right]|\boldsymbol{p}_1\rangle .$$

Next, we convert the parametric differentiations with respect to Δ and $(p_x)_\sigma$ to parametric differentiations with respect to λ:

$$-\frac{\partial}{\partial \Delta}\left.\frac{(1,(p_x)_\nu)}{(\boldsymbol{p}_x)^2y^2 - \Delta y + \lambda(1-y)}\right|_{\Delta=0} = \frac{y}{1-y}\frac{\partial}{\partial\lambda}\frac{(1,(p_x)_\nu)}{(\boldsymbol{p}_x)^2y^2 + \lambda(1-y)}$$

and

$$\frac{\partial}{\partial (p_x)_\sigma}\frac{(1,(p_x)_\nu)}{(\boldsymbol{p}_x)^2y^2 + \lambda(1-y)} = \frac{(0,\eta_{\nu\sigma})}{(\boldsymbol{p}_x)^2y^2 + \lambda(1-y)}$$
$$+ \frac{2y^2}{1-y}\frac{\partial}{\partial\lambda}\frac{(p_x)_\sigma(1,(p_x)_\nu)}{(\boldsymbol{p}_x)^2y^2 + \lambda(1-y)} .$$

This brings the matrix element to the form

$$\langle \overline{\boldsymbol{p}}_2|\hat{\mathsf{M}}_1|\boldsymbol{p}_1\rangle = \frac{\alpha}{2\pi}\int_0^1 \mathrm{d}x\int_0^1 \mathrm{d}y\langle \overline{\boldsymbol{p}}_2|\boldsymbol{\gamma}\cdot\boldsymbol{V}(\hat{\boldsymbol{x}})$$
$$\times\left[(1-y)\ln\left(\frac{\Lambda^2(1-y)}{m^2y^2}\right) + y\ln\left(\frac{(\boldsymbol{p}_x)^2}{m^2}\right)\right]$$
$$+ \left(-2my\boldsymbol{p}_x\cdot\boldsymbol{V}(\hat{\boldsymbol{x}}) + (2m^2-\boldsymbol{q}^2)\boldsymbol{\gamma}\cdot\boldsymbol{V}(\hat{\boldsymbol{x}})\right.$$
$$\left. + \frac{1}{2}\left\{\boldsymbol{\gamma}\cdot\boldsymbol{p}_x y, -2y\boldsymbol{p}_x\cdot\boldsymbol{V}(\hat{\boldsymbol{x}}) + \hat{\boldsymbol{p}}\cdot\boldsymbol{V}(\hat{\boldsymbol{x}}) + \boldsymbol{V}(\hat{\boldsymbol{x}})\cdot\hat{\boldsymbol{p}} + \hat{\mathscr{H}}_S\right\}\right)$$
$$\times\frac{-y}{(\boldsymbol{p}_x)^2y^2 + \mu^2(1-y)}|\boldsymbol{p}_1\rangle . \tag{7.290}$$

5. *Further rearrangements*
   As we have already discussed in Sect. 7.4.10, the expressions corresponding to the Feynman diagrams are independent of the chosen gauge. Thus, we can assume that the four-potential of the EM field obeys the condition

$$\frac{\partial\hat{\mathsf{V}}_\mu}{\partial x_\mu} = 0 \Rightarrow \boldsymbol{q}\cdot\langle \overline{\boldsymbol{p}}_2|\boldsymbol{V}(\hat{\boldsymbol{x}})|\boldsymbol{p}_1\rangle = 0 , \tag{7.291}$$

known as the Lorenz gauge. For the case which is presently of our greatest interest, namely Eq. (7.285), this condition is clearly satisfied. Owing to the relation, see Eq. (7.289),

$$\boldsymbol{p}_x = \boldsymbol{p}_1 + \boldsymbol{q}(1-x) = \boldsymbol{p}_2 - \boldsymbol{q}x\,,$$

one obtains, using Eqs. (7.284), (7.286) and (7.291)

$$\langle\overline{\boldsymbol{p}}_2|2\boldsymbol{p}_x\cdot\boldsymbol{V}(\hat{\boldsymbol{x}})|\boldsymbol{p}_1\rangle = \langle\overline{\boldsymbol{p}}_2|\boldsymbol{p}\cdot\boldsymbol{V}(\hat{\boldsymbol{x}}) + \boldsymbol{V}\cdot\hat{\boldsymbol{p}}|\boldsymbol{p}_1\rangle = \langle\overline{\boldsymbol{p}}_2|2m\boldsymbol{\gamma}\cdot\boldsymbol{V}(\hat{\boldsymbol{x}}) + \hat{\mathscr{H}}_S|\boldsymbol{p}_1\rangle\,.$$

Furthermore, it follows from Eq. (7.286)

$$\langle\overline{\boldsymbol{p}}_2|\boldsymbol{\gamma}\cdot\boldsymbol{q}|\boldsymbol{p}_1\rangle = 0\,;$$

hence

$$\begin{aligned}
&\frac{1}{2}\langle\overline{\boldsymbol{p}}_2|\left\{\boldsymbol{\gamma}\cdot\boldsymbol{p}_x y, -2y\boldsymbol{p}_x\cdot\boldsymbol{V}(\hat{\boldsymbol{x}}) + \hat{\boldsymbol{p}}\cdot\boldsymbol{V}(\hat{\boldsymbol{x}}) + \boldsymbol{V}(\hat{\boldsymbol{x}})\cdot\hat{\boldsymbol{p}} + \hat{\mathscr{H}}_S\right\}|\boldsymbol{p}_1\rangle \\
&= \frac{y}{2}\langle\overline{\boldsymbol{p}}_2|2m(1-y)(\hat{\boldsymbol{p}}\cdot\boldsymbol{V}(\hat{\boldsymbol{x}}) + \boldsymbol{V}(\hat{\boldsymbol{x}})\cdot\hat{\boldsymbol{p}}) + 2m\hat{\mathscr{H}}_S - x[\boldsymbol{q}\cdot\boldsymbol{\gamma},\hat{\mathscr{H}}_S]|\boldsymbol{p}_1\rangle \\
&= y\langle\overline{\boldsymbol{p}}_2|(2m^2(1-y) + x\boldsymbol{q}^2)\boldsymbol{\gamma}\cdot\boldsymbol{V}(\hat{\boldsymbol{x}}) + m(2-y)\hat{\mathscr{H}}_S|\boldsymbol{p}_1\rangle\,.
\end{aligned}$$

In the first equality we used the symmetry of the integrand with respect to the replacement $x \to 1-x$. In the second equality we used the identity

$$\langle\overline{\boldsymbol{p}}_2|[\boldsymbol{q}\cdot\boldsymbol{\gamma},\hat{\mathscr{H}}_S]|\boldsymbol{p}_1\rangle = -2\boldsymbol{q}^2\langle\overline{\boldsymbol{p}}_2|\boldsymbol{\gamma}\cdot\boldsymbol{V}(\hat{\boldsymbol{x}})|\boldsymbol{p}_1\rangle\,.$$

6. *The resulting formula for the form-factors*
After substituting the above rearrangements into Eq. (7.290) and comparing the resulting equation with Eq. (7.287), we obtain for the electric and magnetic form-factors

$$\begin{aligned}
F_1(\boldsymbol{q}^2) &= 1 + \frac{\alpha}{2\pi}\int_0^1 \mathrm{d}x\int_0^1 \mathrm{d}y\left\{(1-y)\ln\left(\frac{\Lambda^2(1-y)}{m^2y^2}\right) + y\ln\left(\frac{(\boldsymbol{p}_x)^2}{m^2}\right)\right. \\
&\quad\left. - \frac{y[2m^2(1-y^2) + \boldsymbol{q}^2(-1+yx)]}{(\boldsymbol{p}_x)^2y^2 + \mu^2(1-y)}\right\} \\
&= 1 - \frac{\alpha}{2\pi}\left\{\int_0^1 \mathrm{d}x\left[-\frac{1}{2}\ln\left(\frac{(\boldsymbol{p}_x)^2}{m^2}\right) + \frac{(2m^2-\boldsymbol{q}^2)\frac{1}{2}\ln\left(\frac{(\boldsymbol{p}_x)^2}{\mu^2}\right) + x\boldsymbol{q}^2 - m^2}{(\boldsymbol{p}_x)^2}\right]\right. \\
&\quad\left. - \int_0^1 \mathrm{d}y(1-y)\ln\left(\frac{\Lambda^2(1-y)}{m^2y^2}\right)\right\}
\end{aligned}$$

and

$$F_2(\boldsymbol{q}^2) = m^2 \frac{\alpha}{2\pi} \int_0^1 \frac{\mathrm{d}x}{(\boldsymbol{p}_x)^2} .$$

The electric form-factor F_1 was (re)normalized by Eq. (7.288). By virtue of the identity, see Eqs. (7.282) and (7.289),

$$(\boldsymbol{p}_x)^2 = m^2 - \boldsymbol{q}^2 x(1-x) \tag{7.292}$$

we have for the nonzero values of the exchanged four-momentum

$$F_1(\boldsymbol{q}^2) - F_1(0) = -\frac{\alpha}{2\pi} \int_0^1 \mathrm{d}x \left\{ -\frac{1}{2} \ln\left(\frac{(\boldsymbol{p}_x)^2}{m^2}\right) + \frac{m^2}{(\boldsymbol{p}_x)^2} \left[\ln\left(\frac{(\boldsymbol{p}_x)^2}{\mu^2}\right) - 1 \right] \right.$$
$$\left. - \left[\ln\left(\frac{m^2}{\mu^2}\right) - 1 \right] - \frac{\boldsymbol{q}^2}{(\boldsymbol{p}_x)^2} \left[\frac{1}{2} \ln\left(\frac{(\boldsymbol{p}_x)^2}{\mu^2}\right) - x \right] \right\} .$$

By means of the identity

$$\frac{m^2}{(\boldsymbol{p}_x)^2} = 1 + \frac{\boldsymbol{q}^2 x(1-x)}{(\boldsymbol{p}_x)^2}$$

following from Eq. (7.292), and integration by parts

$$\int_0^1 \mathrm{d}x \ln\left(\frac{(\boldsymbol{p}_x)^2}{m^2}\right) = \boldsymbol{q}^2 \int_0^1 \mathrm{d}x \frac{x(1-2x)}{(\boldsymbol{p}_x)^2}$$

we finally arrive at

$$F_1(\boldsymbol{q}^2) - F_1(0) = -\frac{\alpha}{4\pi} \boldsymbol{q}^2 \int_0^1 \mathrm{d}x \frac{x + [2x(1-x) - 1] \ln\left(\frac{(\boldsymbol{p}_x)^2}{\mu^2}\right)}{(\boldsymbol{p}_x)^2} .$$

Substitution $x = (1-t)/2$ converts the form-factors into a form convenient for our purposes

$$F_1(\boldsymbol{q}^2) - F_1(0) = -\frac{\alpha}{2\pi} \frac{\boldsymbol{q}^2}{4m^2} \int_0^1 \mathrm{d}t \frac{1 - (1+t^2) \ln\left(\frac{m^2}{\mu^2}\left[1 - \frac{\boldsymbol{q}^2}{4m^2}(1-t^2)\right]\right)}{1 - \frac{\boldsymbol{q}^2}{4m^2}(1-t^2)}$$

and [47]

$$F_2(q^2) = \frac{\alpha}{2\pi} \int_0^1 \mathrm{d}t \frac{1}{1 - \frac{q^2}{4m^2}(1-t^2)} .$$

7. *Application to positronium*
The last two equations yield the electric and magnetic form-factors as functions of the exchanged four-momentum. We now express them as functions of relative electron-positron velocity. To do so, we insert into Eq. (7.282) for the initial and final electron four-momentum

$$p_1 = (E, \mathbf{p}) \quad p_2 = (-E, \mathbf{p}) .$$

By means of the relativistic relations between energy and velocity, $E^2 = m^2/(1-v^2)$, we express the square of the exchanged momentum as

$$q^2 = (p_2 - p_1)^2 = 4E^2 = \frac{4m^2}{1-v^2} .$$

An electron with the four-momentum $p_2 = (-E, \mathbf{p})$ is a positron with the four-momentum $p_2 = (E, -\mathbf{p})$. To see this, it suffices to insert $\psi^M \sim \mathrm{e}^{\mathrm{i}p\cdot x}$ into Eq. (7.58); then $(\psi^M)_C \sim \mathrm{e}^{-\mathrm{i}p\cdot x}$. Hence, $2\mathbf{p}$ is the relative electron-positron momentum in the center of mass frame. Recalling the relativistic relation between energy and momentum $\mathbf{p}\cdot\mathbf{p} = E^2 - m^2 = m^2v^2/(1-v^2)$, clearly $2v$ is the relative electron-positron velocity in the center of mass frame. The last equation expresses the square of the exchanged momentum q^2 through the relative velocity.

We are interested in the limit where the relative velocity approaches zero. However, as will be clear later, it is better to take the limit at the very end. By means of the last equation we have for the form-factors

[47]No quantum field theory textbook fails to note that in the limit of zero momentum transfer we have $F_2(0) = \frac{\alpha}{2\pi}$. For purely spatial components of $V(\hat{x})$ and purely spatial momentum exchanges $q^2 = -\mathbf{q}^2$, we obtain in the limit $\mathbf{q}^2 \to 0$

$$(7.287) \simeq -g_e \frac{e}{m} \hat{\mathbf{S}} \cdot \mathbf{B} + \dots ,$$

where $\boldsymbol{B}$ is the Fourier transform of magnetic induction, $\mathbf{B} = \left\langle \mathbf{p}_2^{(0)} \middle| \nabla \times \mathbf{A}(\hat{\boldsymbol{r}}) \middle| \mathbf{p}_1^{(0)} \right\rangle$, and the gyromagnetic ratio reads $g_e = [F_1(0) + F_2(0)] = 1 + \frac{\alpha}{2\pi}$. This corresponds to the Schwinger correction to the electron gyromagnetic ratio which we derived already in Sect. 7.5. Thus it certainly suffices to comment on it in a footnote. As should be clear now, the magnetic form-factor is not affected by any divergence. Therefore, it cannot—at least at the leading order of α—confirm the validity of the renormalization procedure.

$$F_1\left(\frac{4m^2}{1-v^2}\right) - F_1(0) = -\frac{\alpha}{2\pi}\int_0^1 dt \frac{1-(1+t^2)\ln\left(\frac{m^2}{\mu^2}\frac{t^2-v^2}{1-v^2}\right)}{t^2-v^2}$$

and

$$F_2\left(\frac{4m^2}{1-v^2}\right) = \frac{\alpha}{2\pi}(1-v^2)\int_0^1 dt \frac{1}{t^2-v^2}.$$

For real $v \in (0, 1)$, the integrals exist only in the sense of the Cauchy principal value[48] and have nonvanishing imaginary part. We saw back in Sect. 6.5.3 that this imaginary part is associated with the possibility that one of the particles "detaches" from the rest and becomes real. We consequently calculate the decay probability, instead of the energy shift. This is presently of no interest to us as we wish to calculate the energy shift. In the limit $v \to 0$ one has

$$\Re \int_0^1 \frac{dt}{t^2-v^2} \to -1 \Rightarrow \Re \int_0^1 dt \frac{(1+t^2)}{t^2-v^2} \ln\left(\frac{m^2}{\mu^2(1-v^2)}\right) \to 0$$

and

$$\Re \int_0^1 dt \frac{(1+t^2)\ln(t^2-v^2)}{t^2-v^2} \to -4 + \frac{\pi^2}{2v}.$$

[48]For instance, consider the integral

$$\int_0^1 \frac{dt}{t^2-v^2} = \frac{1}{2v}\int_0^1 dt \left[\frac{1}{t-v} - \frac{1}{t+v}\right].$$

The second integral is well defined. To define the first one, we move the pole, as usually, into the lower half of the complex plane

$$\frac{1}{t-v} \to \frac{1}{t-v+i\varepsilon} = \frac{t-v}{(t-v)^2+\varepsilon^2} - i\frac{\varepsilon}{(t-v)^2+\varepsilon^2}.$$

Recalling now Eq. (1.72), clearly

$$\lim_{\varepsilon\to 0+} \Im\left[\frac{1}{t-v+i\varepsilon}\right] = -\pi\delta(t-v).$$

The real part of the integral reads

$$\lim_{\varepsilon\to 0+}\int_0^1 dt \frac{t-v}{(t-v)^2+\varepsilon^2} = \lim_{\varepsilon\to 0+}\frac{1}{2}\ln\left(\frac{\varepsilon^2+(1-v)^2}{\varepsilon^2+v^2}\right) = \ln\left(\frac{1-v}{v}\right).$$

The integral defined in this way is what is usually called the Cauchy principal value. Returning now back to the original integral we have for the real part

$$\Re\left[\int_0^1 \frac{dt}{t^2-v^2}\right] = \frac{1}{2v}\ln\left(\frac{1-v}{1+v}\right).$$

Hence, it follows that

$$F_1\left(4m^2\right) - F_1(0) = \lim_{v\to 0} -\frac{\alpha}{2\pi}\left(3 - \frac{\pi^2}{2v}\right)$$

and

$$F_2(4m^2) = -\frac{\alpha}{2\pi}\,.$$

For the case in question, the four-potential $\boldsymbol{V}$ of a virtual photon is purely spatial and the exchanged momentum $\boldsymbol{q}$ purely temporal, see Eq. (7.285). Then we have $\mu \neq \nu$ in the expression for the spin part of the Hamiltonian, $\hat{\mathscr{H}}_S$, Eq. (7.284), and γ-matrices anticommute:

$$\langle\overline{\boldsymbol{p}}_2|\hat{\mathscr{H}}_S|\boldsymbol{p}_1\rangle = -\langle\overline{\boldsymbol{p}}_2|(\boldsymbol{p}_2-\boldsymbol{p}_1)\cdot\boldsymbol{\gamma}\boldsymbol{\gamma}\cdot\boldsymbol{V}(\hat{\boldsymbol{x}})|\boldsymbol{p}_1\rangle = -2m\langle\overline{\boldsymbol{p}}_2|\boldsymbol{\gamma}\cdot\boldsymbol{V}(\hat{\boldsymbol{x}})|\boldsymbol{p}_1\rangle\,. \qquad (7.293)$$

After substitution of the last three equations into Eq. (7.287), we find that the photon exchange between an electron and a positron in initial and final states leads to the multiplication of Eq. (7.105) by the factor

$$1 + 2\left[F_1(4m^2) - F_1(0) + F_2(4m^2)\right] = \lim_{v\to 0}\left[1 + \frac{\alpha\pi}{2v} - \frac{4\alpha}{\pi}\right]. \qquad (7.294)$$

8. *Yet another "renormalization"*
As mentioned above, the whole process of a photon exchange between an electron and a positron in initial and final states of virtual electron-positron annihilation is a "high-energy" process, i.e., involving short-wavelength photons. In the limit where the relative velocity of the electron and positron vanishes, the dependence on the regulating parameter μ vanishes as well. However, a glance over the last equation reveals another problem: the calculated correction diverges in the limit when the relative electron-positron velocity vanishes. We can guess where the problem stems from if we recognize that the photon exchange between an electron and a positron includes a complete electromagnetic interaction between the electron and the positron. However, owing to the Coulomb interaction, the electron and positron form a bound state at all. Clearly, this part of the interaction has already been taken into account in the Schrödinger equation! Obviously, we encounter here the same problem as in the two-photon exchange analyzed in the last section. We need to subtract the part of the photon exchange already contained in the Schrödinger equation from the last equation. To determine which part it is, one needs to recognize that the last equation matches that of the correction due to the photon exchange to the probability that the electron "touches down" on the positron when both of them are moving freely. Intuitively, it should be clear that should the electron and positron annihilate, the electron has to "touch down" on the positron. In other words, the transition amplitude for the electron-positron annihilation is

always proportional to the transition amplitude that the electron and positron world-lines meet somewhere in the space-time. Since we are considering a photon exchange between the both initial and final electron-positron pairs, the amplitude for pair annihilation and subsequent pair creation is proportional to the absolute value of the square of the transition amplitude. However, that is the transition probability. The probability that the electron "touches down" on the positron, if we take into account the Coulomb interaction between the electron and positron, relative to the probability that this happens if the electron and positron are moving freely equals the ratio between the square of the absolute value of the Coulomb function, Eq. (6.146), at the origin and the square of the absolute value of the plane wave at the origin

$$\frac{|\psi_{\mathbf{k}_e}(\mathbf{r}=0)|^2}{\frac{1}{(2\pi)^3}} = |e^{\pi/(2k_e)}\Gamma(1+i/k_e)|^2 = \frac{\frac{2\pi}{k_e}}{1-e^{-\frac{2\pi}{k_e}}}, \tag{7.295}$$

where in the last equality we used the relations for the Γ-function, Eq. (6.155). Recall that $\mathbf{r}$ denotes the difference between the electron and positron radius vectors in the center of mass frame. We then substitute

$$k_e = \frac{2v}{\alpha},$$

which corresponds to the transformation of the electron-positron relative velocity from the atomic to natural units. If we now expand the expression (7.295) into the first order in α (we can do that irrespective of the magnitude of v), we obtain

$$\frac{|\psi_{\mathbf{k}_e}(\mathbf{r}=0)|^2}{\frac{1}{(2\pi)^3}} \simeq 1 + \frac{\pi\alpha}{2v} + \ldots .$$

The second term on the rhs clearly matches the troublesome term in Eq. (7.294). As follows from the above considerations, this term should be subtracted from Eq. (7.294).[49]

9. *The final result*

The final expression for the correction to the virtual annihilation, Eq. (7.105), stemming from the photon exchange between an electron and a positron in the initial and final states reads

$$\left(1-\frac{4\alpha}{\pi}\right). \tag{7.296}$$

[49] As far as we know this resolution of divergence at small velocities is due to J. Schwinger [36].

10. *Note on the infrared divergence*
The divergence of the form-factor $F_1(\boldsymbol{q}^2)$ for a general value of the square of the exchanged momentum $\boldsymbol{q}^2$ originates in the region of long wavelengths of the virtual photon. The physical reason behind it is the following. When the free electron is scattered by the EM field, its initial and final momenta differ. Obviously, the electron is accelerated or, for that matter, decelerated. However, an accelerated charged particle radiates EM waves. If we add the amplitude of the process when the electron is scattered by the EM field and radiates and absorbs back a virtual photon, and the process when the electron is scattered by the EM field and radiates the photons away (they can be registered in a detector), then the dependence on the regulating parameter μ disappears and the whole expression is finite. For an explicit calculation see, e.g., [30].

7.8.4 Virtual Two-Photon Annihilation

Finally, we consider the possibility that the pair annihilates into two virtual photons which subsequently annihilate into "another" pair, see Fig. 7.18.

The positronium energy shift due to this process equals, compare with Eqs. (7.262) and (7.263),[50]

$$\Delta E = \left|\sum_{\mathbf{p}} \psi^0_{\text{at}}(\mathbf{p})\right|^2 L^3 \frac{e^4}{(2\pi)^2} \int \frac{\mathrm{d}^4 \boldsymbol{k}_F}{\boldsymbol{k}^2(\boldsymbol{k}-2\boldsymbol{\varepsilon})^2} \langle \overline{V}_0 | \gamma_\nu \frac{1}{\boldsymbol{\gamma}\cdot(\boldsymbol{\varepsilon}-\boldsymbol{k})-m} \gamma_\mu | U_0\rangle$$
$$\times \langle \overline{U}_0 | \left[\gamma_\nu \frac{1}{\boldsymbol{\gamma}\cdot(-\boldsymbol{\varepsilon}+\boldsymbol{k})-m}\gamma_\mu + \gamma_\mu \frac{1}{\boldsymbol{\gamma}\cdot(\boldsymbol{\varepsilon}-\boldsymbol{k})-m}\gamma_\nu\right] | V_0\rangle . \tag{7.297}$$

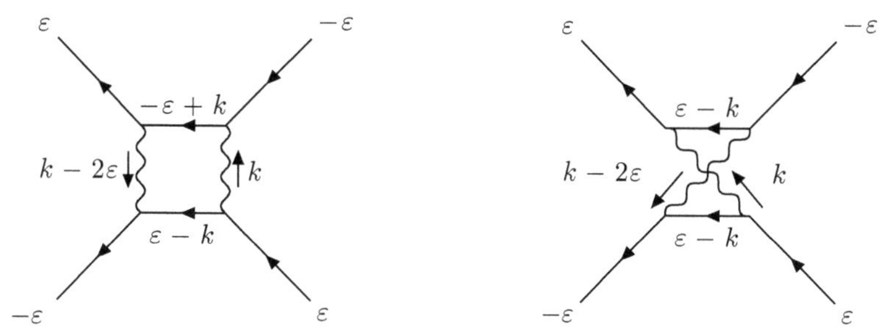

Fig. 7.18 Feynman diagram for two-photon virtual annihilation in momentum space

[50]The reader surely recalls that in the case the electron is moving backward in the final state. The relation between the amplitude and energy is the opposite to the one in Eq. (7.263), see the note below Eq. (7.280).

The reader can easily verify the validity of the equations[51]

$$
\begin{aligned}
&\langle\overline{V}_0|\gamma_\nu[\boldsymbol{\gamma}\cdot(\boldsymbol{\varepsilon}-\boldsymbol{k})+m]\gamma_\mu|U_0\rangle \\
&\qquad\times\langle\overline{U}_0|\left[\gamma_\nu[\boldsymbol{\gamma}\cdot(-\boldsymbol{\varepsilon}+\boldsymbol{k})+m]\gamma_\mu+\gamma_\mu[\boldsymbol{\gamma}\cdot(\boldsymbol{\varepsilon}-\boldsymbol{k})-m]\gamma_\nu\right]|V_0\rangle \\
&=-\frac{k_m k_l}{8}\langle\overline{V}_0|\{\gamma_l,[\gamma_j,\gamma_i]\}|U_0\rangle\langle\overline{U}_0|\{\gamma_m,[\gamma_j,\gamma_i]\}|V_0\rangle=\frac{1}{L^6}4k_m k_l\delta_{ml}|\langle S|_p\sigma_2|S\rangle_e|^2
\end{aligned}
$$

and, see Eq. (7.227),

$$
\begin{aligned}
\int\frac{\mathrm{d}^4k k_m k_l}{\boldsymbol{k}^2(\boldsymbol{k}-2\boldsymbol{\varepsilon})^2(\boldsymbol{k}^2-2\boldsymbol{k}\cdot\boldsymbol{\varepsilon})^2}&=-\frac{\delta_{ml}}{8m^2}\int_0^1\mathrm{d}x\int_0^1\mathrm{d}y\frac{y(1-y)}{1-2y+y^2(1-2x)^2} \\
&=-\frac{\delta_{ml}}{8m^2}\frac{1-\ln 2}{3}.
\end{aligned}
$$

One has to consider the last integral again in the sense of the Cauchy principal value and take only the real part. After substituting the last three equations into Eq. (7.297) we obtain for the energy shift

$$
\begin{aligned}
\Delta E&=-\frac{e^4}{(2\pi)^2m^2}\left|\psi_{\mathrm{at}}^0(\mathbf{r}=0)\right|^2|\langle S|_p\sigma_2|S\rangle_e|^2\frac{1-\ln 2}{2} \\
&=-\frac{\alpha^2}{m^2}\left|\psi_{\mathrm{at}}^0(\mathbf{r}=0)\right|^2 2(1-\ln 2)\left\langle 2-\hat{\mathsf{S}}^2\right\rangle, \qquad (7.298)
\end{aligned}
$$

where in the second equality we used Eq. (7.111).

7.8.5 Comparison with Experiment

Summarizing now all the terms contributing to the singlet-triplet splitting of the positronium s-states with accuracy up to the order of α^5, one obtains

$$
\begin{aligned}
\Delta E_{\mathrm{spin}}&=\frac{\alpha\pi}{m^2}\left|\psi_{\mathrm{at}}^0(\mathbf{r}=0)\right|^2\left\{\frac{8}{3}\left\langle\hat{\boldsymbol{S}}_e\cdot\hat{\boldsymbol{S}}_p\right\rangle\left[1+\frac{\alpha}{\pi}\left(1-\frac{3}{2}\right)\right]\right. \\
&\left.+\left\langle\hat{\mathsf{S}}^2\right\rangle\left[1+\frac{\alpha}{\pi}\left(-4-\frac{8}{9}\right)\right]-\left\langle 2-\hat{\mathsf{S}}^2\right\rangle\frac{\alpha}{\pi}2(1-\ln 2)\right\}. \qquad (7.299)
\end{aligned}
$$

The first term comes from the magnetic interaction between two dipoles, Eq. (3.124). Corrections to this term originate either from the sum of the Schwinger corrections to the gyromagnetic ratio, Eq. (7.204), for both the electron and positron, or from the

[51] In the second equality use Eq. (7.203) and $\gamma_0\{\gamma_l,\Sigma_k\}=2\delta_{l,k}\gamma_5$, see Eq. (7.218).

two-photon exchange between the electron and positron, Eqs. (7.275) and (7.276). Here, we take the limit

$$\lim_{m_2 \to m_1} \frac{m_1 m_2}{m_1^2 - m_2^2} \ln \frac{m_1}{m_2} = \frac{1}{2}. \tag{7.300}$$

The second term stems from the virtual one-photon annihilation of positronium, Eq. (7.105), and corrections to it come either from the photon exchange between the electron and positron in the initial and final states, Eq. (7.296), or from the self-energy of the virtual photon, Eq. (7.281). The third term appears from the virtual two-photon annihilation of positronium, Eq. (7.298).

By inserting Eqs. (7.106) and (3.12) into Eq. (7.299), we find for the splitting between the ground singlet and triplet states of positronium

$$\begin{aligned}\nu_{\text{theo}}(1^3s - 1^1s) = \frac{2R_\infty c\alpha^2}{8} &\left\{ \frac{8}{3}\left[1 + \frac{\alpha}{\pi}\left(1 - \frac{3}{2}\right)\right] \right. \\ &\left. + 2\left[1 + \frac{\alpha}{\pi}\left(-4 - \frac{8}{9}\right)\right] + 2\frac{\alpha}{\pi}2(1 - \ln 2)\right\} \\ = R_\infty c\alpha^2 &\left\{\frac{7}{6} - \frac{\alpha}{\pi}\left(\frac{16}{9} + \ln 2\right)\right\} = 203.38113\ \text{GHz}.\end{aligned}$$

Back in Sect. 7.3.2, when we considered only the leading term, we obtained the value 204.4 GHz, see Eq. (7.107). One thus sees that all of the above considered corrections of the order α to this result significantly improve the agreement with the experimental value 203.3875(16) GHz, see Eq. (7.108). Clearly, the singlet-triplet splitting of the positronium s-states is a "high-energy" effect as the atomic structure enters here only through the absolute value of the square of the wave function at the origin.

To conclude, we summarize all of the terms contributing to the differences between arbitrary two positronium energy levels at the order of α^5.

1. *Electron and positron self-energies*: From Eqs. (7.211), (7.212), (7.210), (3.132), and (7.204) we find for the energy shift

$$\Delta E_{\text{exp}}^{\text{SE}} = \frac{m_e \alpha^5}{\pi (2n)^3}(F_{\text{low}} + F_{\text{high}} + F_{\text{anom}}),$$

where

$$F_{\text{low}} = 2\left[(6.239) + \delta_{l,0}\frac{4}{3}\ln 2\right],$$

$$F_{\text{high}} = -\frac{2}{3}\delta_{l,0} + n^3\left\langle \frac{\hat{\boldsymbol{S}} \cdot \hat{\boldsymbol{L}}}{2r_A^3} \right\rangle$$

and

$$F_{\text{anom}} = n^3 \left\langle \delta(\mathbf{r}_A)\pi \frac{8}{3} \hat{\boldsymbol{S}}_e \cdot \hat{\boldsymbol{S}}_p + \frac{1}{r_A^3} \left[\frac{\hat{\boldsymbol{S}} \cdot \hat{\boldsymbol{L}}}{2} - \left(\hat{\boldsymbol{S}}_e \cdot \hat{\boldsymbol{S}}_p - 3\mathbf{n} \cdot \hat{\boldsymbol{S}}_e \mathbf{n} \cdot \hat{\boldsymbol{S}}_p \right) \right] \right\rangle ,$$

where $\hat{\boldsymbol{S}} = \hat{\boldsymbol{S}}_e + \hat{\boldsymbol{S}}_p$ is the total spin operator. In contrast to it, in Eq. (7.210) $\hat{\boldsymbol{S}}$ stands for a one-particle operator, either $\hat{\boldsymbol{S}}_e$ or $\hat{\boldsymbol{S}}_p$. The average is taken with respect to Pauli wave function. The last term F_{anom} was obtained by substituting Eq. (7.204) for g into Eq. (3.132).

2. *Vacuum polarization*: Equation (7.250) where we substitute $m_R = m_e/2$ and $m_V = m_e$ yields

$$\Delta E_{\text{exp}}^{\text{VP}} = \frac{m_e \alpha^5}{\pi (2n)^3} \delta_{l,0} \left(-\frac{4}{15} \right) ,$$

3. *One- and two-photon exchanges*: The energy shift is given by Eqs. (6.245), (6.246), (6.247), (6.254), (6.258), and (7.277), where one takes the limit (7.300) and

$$\lim_{m_2 \to m_1} \frac{m_1^2 \ln m_1 - m_2^2 \ln m_2}{m_1^2 - m_2^2} = \frac{1}{2} + \ln m_1 .$$

4. *Correction to virtual annihilation*: Here, the situation for an arbitrary positronium state is the same as for the s-states, see Eq. (7.299),

$$\Delta E_{\text{an}} = \frac{m_e \alpha^5}{\pi (2n)^3} \delta_{l,0} \left\langle \hat{\mathrm{S}}^2 \left(-4 - \frac{8}{9} \right) - \left(2 - \hat{\mathrm{S}}^2 \right) 2(1 - \ln 2) \right\rangle$$

Exercise 30: Positronium Gross Structure

Determine the magnitude of the $2^3s - 1^3s$ interval in positronium with accuracy up to α^5 and compare your result with the experimental value [6]

$$\nu_{\text{exp}}(2^3s - 1^3s) = 1{,}233{,}607{,}216.4\,(3.2)\,\text{MHz} .$$

7.9 Final Notes

In this and previous chapter we analyzed merely the basic quantum electrodynamic processes. There is, of course, a great number of other indeed interesting processes which we did not discuss at all. We refer an interested reader to the classics [2, 15] for a more complete picture.

The view of the renormalization adopted herein suffices well for low-energy QED processes. However, when dealing with second-order phase transitions as well as strong interactions at high energies, one needs a somewhat deeper view based on the renormalization group, see, e.g., [23, 30].

For an in-depth treatment of spinors see, e.g., [4, 42, 44].

Throughout this book, we treated nuclei as point-like. For the calculation of finite-size corrections to the hyperfine and fine structure see [9, 45].

As repeatedly pointed out, QED needs some input, namely we need to take the mass of the electron or better the Rydberg constant, the charge of the electron or better the fine-structure constant, electron-muon, electron-proton, etc. ratios from experiment. For an up-to-date review of how this is done see [26]. For a reflection on and survey of the tests of QED see, e.g., [17, 18].

References

1. A.H. Al-Ramadhan, D.W Gidley, Phys. Rev. Lett. **72**, 1632 (1994)
2. V.B. Beresteckij, E.M. Lifshitz, L.P. Pitaevskij, *Quantum Electrodynamics* (Butterworth-Heinemann, Oxford, 1982)
3. J.D. Bjorken, S.D. Drell, *Relativistic Quantum Mechanics* (McGraw-Hill, New York, 1965)
4. L.S. Brown, *Quantum Field Theory* (Cambridge University Press, Cambridge, 1994)
5. F. Dyson, *Disturbing the Universe* (Harper and Row Publishers, New York, 1979)
6. M.S. Fee et al., Phys. Rev. Lett. **70**, 1397 (1993)
7. R.P. Feynman, *Quantum Electrodynamics* (Westview Press, Boulder, 1998)
8. R.P. Feynman, R.B. Leighton, M. Sands, *Feynman Lectures on Physics 1, 2* (Addison-Wesley, Reading, 1977)
9. J.L. Friar, Ann. Phys. (NY) **122**, 151 (1979)
10. T. Fulton, P. Martin, Phys. Rev. **95**, 811 (1954)
11. W. Greiner, J. Reinhardt, *Quantum Electrodynamics* (Springer, Berlin, 2008)
12. H. Grotch, D.R. Yennie, Rev. Mod. Phys. **41**, 350 (1969)
13. J. Hořejší, *Fundamentals of Electroweak Theory* (Karolinum, Praha, 2002); J. Hořejší, *Introduction to Electroweak Unification: Standard Model from Tree Unitarity* (World Scientific, Singapore, 1994)
14. R. Jackiw, What good are quantum field theory infinities? arXiv:hep-th/9911071 (1999)
15. J.M. Jauch, F. Rohrlich, *The Theory of Photons and Electrons: The Relativistic Quantum Field Theory of Charged Particles with Spin One-Half* (Addisson Wesley, Reading, 1955)
16. R. Karplus, A. Klein, Phys. Rev. **87**, 848 (1952)
17. S.G. Karshenboim, Phys. Rep. **422**, 1 (1995)
18. T. Kinoshita, *Quantum Electrodynamics* (World Scientific, Singapore, 1990)
19. T. Kinoshita, Rep. Prog. Phys. **59**, 1459 (1996)
20. N.M. Kroll, W.E. Lamb, Phys. Rev. **75**, 388 (1949); J.B. French, V.F. Weisskopf, Phys. Rev. **75**, 1240 (1949)

21. G.P. Lepage, What is Renormalization? arXiv:hep-ph/0506330 (2005); G.P. Lepage, How to renormalize Schrödinger equation. arXiv:nucl-th/9706029 (1997)
22. I. Lindgren, J. Morrison, *Atomic Many-Body Theory* (Springer, Berlin, 1982)
23. S.-K. Ma, *Modern Theory of Critical Phenomena* (Westview Press, Boulder, 2000)
24. F. Mandl, G. Shaw, *Quantum Field Theory* (Wiley, New York, 1993)
25. A.P. Mils Jr., G.H. Bearman, Phys. Rev. Lett. **34**, 246 (1975)
26. P.J. Mohr, B.N. Taylor, D.B. Newell, Rev. Mod. Phys. **84**, 1527 (2012)
27. B. Odom et al., Phys. Rev. Lett. **97**, 030801 (2006)
28. C.J. Oram et al., Phys. Rev. Lett. **52**, 910 (1984)
29. K. Pachucki, J. Phys. B **31**, 5123 (1998)
30. M. Peskin, D. Schroeder, *An Introduction to Quantum Field Theory* (Westview Press, Boulder, 1995)
31. R. Pohl et al., Nature **466**, 213 (2010)
32. F. Rohrlich, *Classical Charged Particles*, 3rd edn. (World Scientific Publishing, 2007)
33. J.J. Sakurai, *Advanced Quantum Mechanics* (Addison Wesley, Reading, 1967)
34. E.E. Salpeter, Phys. Rev. **87**, 328 (1952)
35. S.S. Schweber, *QED and the Man Who Made It: Dyson, Feynman, Schwinger Tomonaga* (Princeton University Press, Princeton, 1994)
36. J. Schwinger, *Particles, Sources and Fields, Volume II* (Addison-Wesley, New York, 1973)
37. E.A. Uehling, Phys. Rev. **48**, 55 (1935)
38. M. Veltman, *Diagrammatica: The Path to Feynman Diagrams* (Cambridge University Press, Cambridge, 1994)
39. S. Weinberg, *Gravitation and Cosmology: Principles and Applications of the General Theory of Relativity* (Wiley, New York, 1972)
40. K.A. Woodle et al., Phys. Rev. A **41**, 93 (1990)
41. J. Zamastil, V. Patkóš, Phys. Rev. A **88**, 032501 (2013); J. Zamastil, V. Patkóš, Phys. Rev. A **86**, 042514 (2012)
42. A. Zee, *Quantum Field Theory in a Nutshell*, 2nd edn. (Princeton University Press, Princeton, 2010)
43. A. Zee, *Einstein Gravity in a Nutshell* (Princeton University Press, Princeton, 2013)
44. A. Zee, *Group Theory in a Nutshell for Physicists* (Princeton University Press, Princeton, 2016)
45. A.C. Zeemach, Phys. Rev. **104**, 1771 (1956)

Closing Remarks

> It is as if, in order to carry the quantum theory to its logical conclusion, a new breed of people had to be created: the theory was so great that it could stand the strain going farther, but its creators could not. If Albert Einstein could not be happy with the quantum theory, who could expect any mortal to do better? [2]

From its very inception, quantum mechanics (QM) was looked upon with suspicion and this somehow still persists to the present day. All the worse for QED, or for the quantum field theory (QFT) in general. It is quite amazing to observe how often and forcefully QFT was regarded as something to be discarded and replaced with something better: in the 1930s QFT was thought not to be able to describe EM interactions, in the 1950s and 1960s to describe strong interactions, and so on and so forth. And yet to this day, QED and its generalization, the electroweak theory, passed every decisive test with flying colors. Ingenious thought experiments devised by Einstein, Schrödinger, and other great minds to show that QM cannot be taken seriously to its full extent have been eventually technically realized and showed that at least Nature herself follows QM to the letter. Although one always has to keep open mind to the possibility that there could indeed be something better that would replace QM and QFT one day, one should acknowledge at the same time how these schemes survived the stringent scientific scrutiny to this day.

Time to time, one can hear about scientific revolutions—there is even an entire book devoted to their structure—but between us friends, there has been solely one revolution in the history of modern physics (since the Galileo times), namely the discovery of QM. It is quite remarkable that from that time on, all successful solutions of new problems were conservative: of all the more or less revolutionary ideas, so far only those preserving the original structure of QM and QFT proved to be the correct ones. Is there a lesson to be taken here?

What makes the physics so fascinating is that it combines very precise thinking and very precise experiments. One has to acknowledge that unless the both are present, much of the fascination disappears. Only ingeniously devised and carefully

J. Zamastil, J. Benda, *Quantum Mechanics and Electrodynamics*,
DOI 10.1007/978-3-319-65780-6

executed experiments enforced theorists to come up with such a strange scheme as QM. It is then quite remarkable how this scheme is able to deal with new phenomena, completely unknown at the time of its invention. Look how the basic principle of quantum mechanics, i.e., the principle of superposition, enters the electroweak theory, Eq. (E.26) of Epilogue. It enforces us to extend the notion of a particle to an object without definite mass, see Eq. (E.52) of Epilogue. The strangest features of QM are the most useful ones. Also, as peripherally mentioned in Sects. 7.5 and 7.6, the infinities appearing in successful QFTs are not something to dispose of as early as possible. In a way, at least in our present state of understanding, they should be there.

At the same time, one must acknowledge that there is no such a thing as an absolutely accurate experiment or an absolutely accurate calculation. We have to think in terms of clearly defined concepts; yet again there is no such thing as an absolute clarity of thinking. In other words, we make approximations all the time. Thus in the future we will know more, and there is still a lot to be learned, for sure. Maybe some of the readers of this book will take us further one day.

Epilogue: Electrodynamics as a Part of a Greater Structure

This Epilogue lies somewhat out of the scope of this book and also the style of the exposition differs from the rest: we do not work out everything as in detail as previously. We rather want to wet the reader's appetite to study further and show him or her one of many possible ways one can follow further. Last but not least, we want to demonstrate how atomic physics manages after all these years to stay at the frontier of our exploration of the fundamental physical laws.

We will give an exposition of those elements of the theory of electroweak interactions that are of relevance at low energies. That is, we will focus on β-decay and electron-nucleus electroweak interactions in atoms. We restrict ourselves to a world (in fact not much different from the one we inhabit) comprising only one generation of leptons and quarks: electron, its neutrino, and up and down quarks, and consider neutrinos strictly massless. Clearly, what follows is only a tiny bit of the whole theory. Our aim is to show the reader why the calculation and measurement of atomic electroweak interactions form such an interesting topic, and at the same time spare him or her the necessity to learn the whole theory with all its complications. This final part of the text requires no more knowledge of elementary particles than the quark composition of nucleons, i.e., of protons or neutrons.

β-decay and Its Problems

In his experiments, Rutherford subjected radioactively emitted particles to an external magnetic field. From the results, he deduced the electric charge and mass of the emitted particles and distinguished three types of radioactivity:

In the original version of this book, W superscripts were inadvertently attached to the sigma symbol appearing in the second pair of equations on page 437 and equation E.7. This has now been updated.

J. Zamastil, J. Benda, *Quantum Mechanics and Electrodynamics*,
DOI 10.1007/978-3-319-65780-6

1. α-decay: The nucleus emits α-particles (helium nuclei). This decay was explained by Russian physicist Georgij Gamow as a consequence of the quantum-mechanical tunneling effect.
2. β-decay: The nucleus emits electrons.
3. γ-decay: The nucleus emits photons. It is a spontaneous emission, but the initial and final states are nuclear energy levels.

In case of β-decay the nucleus emits electrons, hence it seems reasonable to assume that nuclei comprise protons and electrons. However, experiments contradict this assumption. Let us elucidate using deuteron as an example. Its mass is about twice the proton's and its charge is the same as proton's. Therefore in order to yield the correct charge and a reasonable value of mass, the deuteron would have to constitute two protons and one electron. These three particles of spin $1/2$ would then combine to a half-integer spin, yet we know from the experiment that deuteron has an integer spin.

These problems were resolved after the discovery of neutron by British physicist James Chadwick in 1932. Only then did it become clear that nuclei are constituted by neutrons and protons. Comparing the mass and charge of nuclei prior and after β-decay, one can deduce that the masses are about the same and that the charge is increased by one elementary unit. Gradually, but by no means easily, it was established that the energy spectrum of the emitted electrons is continuous.[1] This means that either β-decay does not conserve energy or that there must be another neutral and very light particle that is emitted along with the electron but escapes the detection. The latter, first proposed by Pauli, turned out to be the correct explanation. Thus it gradually became clear that β-decay is nothing but disintegration of neutron into proton, electron, and antineutrino. There is also inverse β-decay when an electron in the atomic s-state is captured by the proton in the nucleus and produces a neutron and a neutrino. But how are these reactions possible at all?

This question was answered by Italian physicist Enrico Fermi who noticed a similarity between β-decay and spontaneous emission. A photon is not present in the atom prior its radiation and it is not a particle, but a quantum of EM field instead. The interaction Hamiltonian (6.65) includes the vector potential of the quantized EM field, which in turn comprises photon creation and annihilation operators, see Eq. (6.39). This Hamiltonian capturing the interaction of an atom and EM field then allows for transition between the vacuum and one-photon states. Likewise, neutron, proton, electron, and antineutrino are not particles, but quanta of neutron, proton, electron, and neutrino fields. The problem thus reduces to the construction of the interaction Hamiltonian that allows for a transition between the initial and final states of β-decay.

[1] If the reader is interested in the fascinating history of scientific endeavor that lead to this picture, we refer him to [6].

Fermi Theory

> He [Oppenheimer] often presented physics in rather abstract terms which contrasted, at least in my mind, with the simple, direct approach to which Fermi had accustomed me. I remember a remark that Fermi made in 1940 at the time of his visit to Berkley for the Hitchcock lecture. After attending a seminar given by one of Oppenheimer's pupils on Fermi's beta-ray theory, Fermi met me and said: "Emilio, I am getting rusty and old. I cannot follow the highbrow theory developed by Oppenheimer's pupils any more. I went to their seminar and was depressed by my inability to understand them. Only the last sentence cheered me up; it was: 'And this is Fermi's theory of beta decay.'." [13]

Weak interactions are indeed weak; thus within our focus on low-energy (the so-called "table-top") experiments it perfectly suffices to consider only the first-order of the perturbation method. Fermi was well aware of the Fermi golden rule, Eq. (6.79), which for β-decay takes the form

$$\mathrm{d}w_{I\to F} = 2\pi\delta(\omega_{FI})\,|(\hat{\mathsf{H}}_\beta)_{IF}|^2\mathrm{d}^3\mathbf{p}_p\mathrm{d}^3\mathbf{p}_e\mathrm{d}^3\mathbf{p}_\nu\,, \tag{E.1}$$

where

$$(\hat{\mathsf{H}}_\beta)_{IF} = \langle 0|\hat{\mathsf{b}}_n\hat{\mathsf{H}}_\beta\hat{\mathsf{b}}_p^+\hat{\mathsf{b}}_e^+\hat{\mathsf{d}}_\nu^+|0\rangle\,.$$

Clearly, in order for $(\hat{\mathsf{H}}_\beta)_{IF}$ to be nonzero, $\hat{\mathsf{H}}_\beta$ must constitute a neutron creation operator, a proton annihilation operator, an electron annihilation operator, and an antineutrino annihilation operator. Recalling the decomposition of the Dirac field into modes, Eq. (7.41),

$$\hat{\psi} = \sum_\sigma\left(\hat{\mathsf{b}}_\sigma U_\sigma + \hat{\mathsf{d}}_\sigma^+ V_\sigma\right)\,, \tag{7.41}$$

we see that the simplest possible choice of the interaction Hamiltonian is

$$\hat{\mathsf{H}}_\beta = \frac{G_\beta}{\sqrt{2}}\int \mathrm{d}^3\mathbf{r}(\hat{\psi}_n^+\hat{\psi}_p\hat{\psi}_\nu^+\hat{\psi}_e)(\mathbf{r}) \Rightarrow (\hat{\mathsf{H}}_\beta)_{IF} = \frac{G_\beta}{\sqrt{2}}\int \mathrm{d}^3\mathbf{r}(U_n^+U_pV_\nu^+U_e)(\mathbf{r})\,. \tag{E.2}$$

The constant G_β is called the Fermi coupling constant for β-decay and the factor $2^{-1/2}$ appears there for historical reasons. Believe it or not, this is the correct form of the interaction Hamiltonian in the limit when the proton recoil can be neglected and the transition preserves the nucleus spin (i.e., the spins of the neutron and proton are the same). To find the correct Hamiltonian also for other cases we will follow the similarity between the EM and weak interactions.

Recall Eq. (7.230) for a massive vector field:

$$(\partial_\mu\partial_\mu + m^2)A_\nu - \partial_\nu\partial_\mu A_\mu = j_\nu\,. \tag{7.230}$$

On the scales $r \gg m^{-1}$ and $t \gg m^{-1}$, the spatial and temporal changes of the field can be safely neglected and we can write

$$\boldsymbol{A} \simeq \frac{\boldsymbol{j}}{m^2}$$

Inserting this expression into the interaction Hamiltonian we find the effective current-current interaction

$$\hat{\mathsf{H}}_{\text{int}} = \int \mathrm{d}^3\mathbf{r}\boldsymbol{j} \cdot \boldsymbol{A} \simeq \frac{1}{m^2} \int \mathrm{d}^3\mathbf{r}\boldsymbol{j} \cdot \boldsymbol{j} \,. \tag{E.3}$$

Assuming now that weak interactions are mediated by very massive vector bosons we can write

$$\hat{\mathsf{H}}_\beta = \frac{G_\beta}{2^{1/2}} \int \mathrm{d}^3\mathbf{r}\hat{\boldsymbol{j}}^N \cdot \hat{\boldsymbol{j}}^l \,, \tag{E.4}$$

where

$$\hat{\boldsymbol{j}}^N = \hat{\bar{\psi}}_n \gamma_\mu \hat{\psi}_p \,, \quad \hat{\boldsymbol{j}}^l = \hat{\bar{\psi}}_\nu \gamma_\mu \hat{\psi}_e \,.$$

The interaction (E.2) is then the temporal component of the complete interaction (E.4), proposed by Fermi as early as in 1934, and $G_\beta 2^{-1/2} = m^{-2}$. The form (E.4) is almost, but not completely, correct. The correct one was determined 24 years later only after it became clear that weak interactions do not preserve parity. To elucidate the connection we have to turn our attention to the solution of the Dirac equation in the massless limit.

Weyl Representation

To investigate such a limit, it is useful to introduce yet another representation of the γ-matrices. We once again perform a unitary transformation, confront with Eq. (7.54),

$$\psi = \mathsf{U}\psi^W$$

this time choosing the matrix

$$\mathsf{U} = \frac{1}{\sqrt{2}} \begin{pmatrix} 1 & 1 \\ -1 & 1 \end{pmatrix} .$$

Substitution of the last expression into $\gamma_\mu^W = \mathsf{U}^+\gamma_\mu \mathsf{U}$ leads to the *Weyl* or *chiral realization* of the γ-matrices

$$\boldsymbol{\gamma}^W = \begin{pmatrix} 0 & \boldsymbol{\sigma}^W \\ \overline{\boldsymbol{\sigma}}^W & 0 \end{pmatrix}, \quad \gamma_5^W = \begin{pmatrix} -1 & 0 \\ 0 & 1 \end{pmatrix},$$

where

$$\boldsymbol{\sigma}^W = (1, \boldsymbol{\sigma}) \quad \overline{\boldsymbol{\sigma}}^W = (1, -\boldsymbol{\sigma}) \, .$$

Defining now the left- and right-handed components of the fermion fields

$$\psi^L = \frac{1-\gamma_5}{2}\psi \quad \psi^R = \frac{1+\gamma_5}{2}\psi \, , \tag{E.5}$$

where the nomenclature becomes clear in a moment, the Dirac wave function of a free particle can be written as

$$\psi^W = \begin{pmatrix} \psi^L \\ \psi^R \end{pmatrix} .$$

Taking the Hermitian conjugate of Eq. (E.5) and multiplying it by γ_0^W we obtain

$$\overline{\psi}^L = \overline{\psi}\frac{1+\gamma_5}{2} \quad \overline{\psi}^R = \overline{\psi}\frac{1-\gamma_5}{2} \, . \tag{E.6}$$

We consider only the states with a definite value of *helicity*, i.e., the states with a given projection of spin along the direction of motion. Writing the momentum of the particle as

$$\mathbf{p} = p\boldsymbol{\eta}$$

the helicity spinors are defined as

$$\hat{\mathbf{S}} \cdot \boldsymbol{\eta}|h\rangle = \frac{h}{2}|h\rangle \, , \quad \hat{\mathbf{S}} = \frac{\boldsymbol{\sigma}}{2} \, , \quad h = \pm 1 \, . \tag{E.7}$$

These states are the same states as those encountered in Exercise 1: The states denoted here by $|h\rangle$ correspond to the states denoted as $|h\boldsymbol{\eta}\rangle$ in the exercise. Helicity is an integral of motion. To see this, it suffices to solve the Dirac equation $\left[\boldsymbol{\gamma}^W \cdot \boldsymbol{p} - m\right]\psi^W = 0$ in the form

$$\left|\psi_h^W\right\rangle = |\chi_h\rangle|h\rangle \tag{E.8}$$

and use Eq. (E.7); one thus obtains

$$\begin{pmatrix} -m & E-hp \\ E+hp & -m \end{pmatrix}\begin{pmatrix} \chi_h^L \\ \chi_h^R \end{pmatrix} = 0\,. \tag{E.9}$$

Writing now

$$E = m\cosh\chi\,, \quad p = m\sinh\chi\,, \tag{E.10}$$

where the parameter χ is called the *rapidity*, Eq. (E.9) takes the form

$$\begin{pmatrix} -1 & \mathrm{e}^{-h\chi} \\ \mathrm{e}^{h\chi} & -1 \end{pmatrix}\begin{pmatrix} \chi_h^L \\ \chi_h^R \end{pmatrix} = 0\,.$$

The solution to this equation reads

$$|\chi_h\rangle = N\begin{pmatrix} \mathrm{e}^{-h\chi/2} \\ \mathrm{e}^{h\chi/2} \end{pmatrix}, \quad N^2 = \frac{1}{\mathrm{e}^{-\chi}+\mathrm{e}^{\chi}}\,. \tag{E.11}$$

The reader can easily verify[2] that the charge conjugated solution is

$$\left|\psi^W\right\rangle_c = \gamma_2|\chi_h\rangle|h\rangle^* = -\mathrm{i}hN\begin{pmatrix} \mathrm{e}^{h\chi/2} \\ -\mathrm{e}^{-h\chi/2} \end{pmatrix}|-h\rangle^* \tag{E.12}$$

It follows from Eq. (E.10) that in the massless limit the rapidity approaches infinity. A general solution, Eqs. (E.8) and (E.11), consequently reduces to

$$\chi\to\infty:\ \left|\psi_+^W\right\rangle = \begin{pmatrix} 0 \\ 1 \end{pmatrix}|+\rangle\,, \quad \left|\psi_-^W\right\rangle = \begin{pmatrix} 1 \\ 0 \end{pmatrix}|-\rangle\,. \tag{E.13}$$

It should be clear from the last equation that in the massless limit the left- and right-handed fermions have negative and positive values of the spin projection along the direction of motion, respectively. The charge conjugated solution, Eq. (E.12), then reduces to

$$\chi\to\infty:\ \left|\psi_+^W\right\rangle_c = -\mathrm{i}\begin{pmatrix} 1 \\ 0 \end{pmatrix}|-\rangle^* \quad \left|\psi_-^W\right\rangle_c = -\mathrm{i}\begin{pmatrix} 0 \\ 1 \end{pmatrix}|+\rangle^*\,. \tag{E.14}$$

In the limit where the particle is at rest, the general solution (E.11) reduces to

$$\chi\to 0:\ |\chi_\pm\rangle = \frac{1}{\sqrt{2}}\begin{pmatrix} 1 \\ 1 \end{pmatrix}\,. \tag{E.15}$$

[2]One only needs to work out the transition between the Weyl and Majorana realizations and use the result of Exercise 1.

Feynman – Gell-Mann Theory

In the massless limit, the Dirac equation for the four-component bispinor decouples into two equations for two-component spinors. It then suggests that neutrino, a massless 1/2-spin particle, should be described by only one of the two equations, i.e., neutrino would be either left-handed or right-handed. This idea was put forward first by German mathematician Hermann Weyl as early as in 1929. However, it was rejected immediately by physicists on the grounds that the two-component equation for neutrino violates the parity. Indeed, upon inversion of the direction of coordinate axes, the direction of the particle motion reverses, $\boldsymbol{\eta} \to -\boldsymbol{\eta}$, but not the particle spin, $\hat{\boldsymbol{S}} \to \hat{\boldsymbol{S}}$. In common jargon it is said that the momentum is a polar vector, while angular momentum is an axial vector, or a pseudovector. Thus by inversion of the direction of the coordinate axes, the state of the positive helicity is transformed into a state of negative helicity and vice versa. In addition, the two-component equation for neutrino itself violates the charge parity. It follows from the comparison of Eqs. (E.13) and (E.14) that upon charge conjugation the left-handed neutrino transforms into the right-handed antineutrino and vice versa. Nevertheless, the idea of the two-component neutrino was rescued immediately after it had become clear that weak interactions violate parity. Then, in a beautiful experiment, see, e.g., [8], M. Goldhaber et al. found out that neutrino produced in an inverse β-decay is left-handed. This discovery led American physicists Richard Feynman and Murray Gell-Mann (and also George Sudarshan and Robert Marshak) to a bold suggestion (which turned out to be correct) that the weak interaction acts only on the left-handed parts of the fermions. Thus, the only part that needs to be modified in the Fermi form of the interaction Hamiltonian (E.4) is the definition of the hadronic and leptonic currents to[3]

$$\hat{j}^N = 2(\hat{\bar{\psi}}_n)_L\gamma_\mu(\hat{\psi}_p)_L = \hat{\bar{\psi}}_n\gamma_\mu(1-\gamma_5)\hat{\psi}_p \tag{E.16}$$

and

$$\hat{j}^l = 2(\hat{\bar{\psi}}_\nu)_L\gamma_\mu(\hat{\psi}_e)_L = \hat{\bar{\psi}}_\nu\gamma_\mu(1-\gamma_5)\hat{\psi}_e\,, \tag{E.17}$$

where in the second equalities we used Eqs. (E.5) and (E.6). In fact, the form of the hadronic current (E.16) is not entirely correct. The correct one reads

$$\hat{j}^N = \hat{\bar{\psi}}_n\gamma_\mu(1-g_A\gamma_5)\hat{\psi}_p\,, \tag{E.18}$$

[3]In fact, Lorentz invariance together with the assumption that only the left-handed parts of the fermion fields enter the interaction Hamiltonian determines uniquely the form (E.4) with the currents given by Eqs. (E.16) and (E.17), see, e.g., [4].

where $g_A \simeq 1.27$ [7]. This value is determined from nucleus-spin-changing β-decay. It originates in the fact that proton and neutron are not elementary particles, but bound states of quarks. On the quark level, a down quark changes into an up quark in the β-decay and the corresponding current is given by

$$\hat{j}^q = \hat{\overline{\psi}}_d \gamma_\mu (1 - \gamma_5) \hat{\psi}_u .$$

Thus it is not surprising that the strong interaction enters the scene. What surprising is that it enters the scene in an indeed innocent way: the fact that the vector part $\hat{\overline{\psi}}_n \gamma_\mu \hat{\psi}_p$ is not at all influenced by the strong interaction will come in hand later.

Exercise 31: Neutron Lifetime

From the known value of the lifetime of a free neutron, $\tau_n = 880\,\mathrm{s}$ *[7], determine the Fermi coupling constant for β-decay. Compare your result with the official value* $G_\beta \simeq 1.0 \times 10^{-5}\, m_p^{-2}$*, where m_p stands for the proton mass* $m_p \simeq 0.938\,\mathrm{GeV}$*. Hints:*

- *Kinematics: Write* $U = L^{-3/2} e^{i\mathbf{p}\cdot\mathbf{r}} u$*,* $V = L^{-3/2} e^{i\mathbf{p}\cdot\mathbf{r}} v$*; then*

$$(\hat{\mathsf{H}}_\beta)_{IF} = \frac{G_\beta}{\sqrt{2}} \frac{1}{L^3} \int \mathrm{d}^3\mathbf{r} e^{-i(\mathbf{p}_n - \mathbf{p}_p - \mathbf{p}_\nu - \mathbf{p}_e)\cdot\mathbf{r}} (h_\beta)_{IF} ,$$

where

$$(h_\beta)_{IF} = \langle \overline{u}_n | \gamma_\mu (1 - g_A \gamma_5) \big| u_p \big\rangle \langle \overline{v}_\nu | \gamma_\mu (1 - \gamma_5) | u_e \rangle .$$

Integration over the space yields momentum conservation

$$\int \mathrm{d}^3\mathbf{r} e^{-i(\mathbf{p}_n - \mathbf{p}_p - \mathbf{p}_\nu - \mathbf{p}_e)\cdot\mathbf{r}} = \delta_{\mathbf{p}_n, \mathbf{p}_p + \mathbf{p}_\nu + \mathbf{p}_e} .$$

This cancels the integration over the proton momentum. Next change to the neutron rest frame and neglect the proton recoil, i.e., set $E_p \simeq m_p$*. After squaring the matrix element take a continuous limit; then* $L^{-6} \to (2\pi)^{-6}$*. Equation (E.1) then takes the form*

$$\mathrm{d}w_{I\to F} = 2\pi \delta(\Delta - E_e - p_\nu) \frac{G_\beta^2}{2} |(h_\beta)_{IF}|^2 \frac{\mathrm{d}^3 p_e}{(2\pi)^3} \frac{\mathrm{d}^3 p_\nu}{(2\pi)^3} ,$$

where [7]

$$\Delta = m_n - m_p \simeq 1.29\,\text{MeV}\,.$$

Write for the differentials $\mathrm{d}^3p = \mathrm{d}pp^2\mathrm{d}\Omega$, $\mathrm{d}p_e p_e = \mathrm{d}E_e E_e$. *The* δ-*function cancels the integration over the magnitude of the neutrino momentum, thus one should end up with*

$$\mathrm{d}w_{I\to F} = \frac{G_\beta^2}{4\pi^3}|(h_\beta)_{IF}|^2 \frac{\mathrm{d}\Omega_e}{4\pi}\frac{\mathrm{d}\Omega_\nu}{4\pi}\sqrt{E_e^2 - m_e^2}(\Delta - E_e)^2 E_e \mathrm{d}E_e\,.$$

- *Reduction to spinors: Insert into* $(h_\beta)_{IF}$ *for electron, antineutrino, and nucleon wave functions from Eqs. (E.11), (E.14), and (E.15), respectively. You should arrive at*

$$\langle \overline{u}_n|\gamma_\mu(1 - g_A\gamma_5)\big|u_p\rangle = \langle h_n|(\delta_{\mu,0} - g_A\delta_{\mu,i}\sigma_i\big|h_p\rangle$$

and

$$\langle \overline{v}_\nu|\gamma_\mu(1-\gamma_5)|u_e\rangle = 2N_e \mathrm{e}^{-h_e\chi_e/2}\mathrm{i}\langle -_\nu|^*\overline{\sigma}_\mu|h_e\rangle\,.$$

- *Averaging over the directions of the emitted electrons and neutrinos and over nucleons polarizations: Use the result of Exercise 1 to show*

$$\frac{1}{2}\sum_{h_n=\pm 1}\sum_{h_p=\pm 1}\int\frac{\mathrm{d}\Omega_e}{4\pi}\int\frac{\mathrm{d}\Omega_\nu}{4\pi}|(h_\beta)_{IF}|^2 = 2N_e^2\mathrm{e}^{-h_e\chi_e}\,.$$

It follows from the last equation and Eq. (E.10) that

$$\frac{N_- - N_+}{N_- + N_+} = \tanh\chi_e = \frac{p_e}{E_e}$$

where N_- *and* N_+ *are the total numbers of emitted electrons with helicities* -1 *and* $+1$, *respectively. This is the experimental measure of the parity violation.*

- *Summing now over the electron helicities we finally obtain*

$$w_{I\to F} = \frac{G_\beta^2}{\pi^3}\frac{1+3g_A^2}{2}\int_{m_e}^{\Delta}\sqrt{E_e^2 - m_e^2}(\Delta - E_e)^2 E_e\mathrm{d}E_e\,,$$

where the last integral has to be calculated numerically.

Conserved Lepton Number and Generalization of Electrodynamics

We now assume the weak interaction is indeed mediated by vector bosons and try to develop this idea further. We begin by considering a world where masses of all the particles vanish, which is clearly a very high-energy limit of our world. The reason is that at sufficiently high energies all the masses become (at least in the first approximation) unimportant. Despite it not being true even for the energies reached at LHC, in theory one can consider such a limit very easily.

Experiments show that weak interactions preserve electron lepton number l. That is, if we assign $l = 1$ to the neutrino, $l = -1$ to the antineutrino, $l = 1$ to the electron, and $l = -1$ to the positron, the values of l prior and after any reaction have to be the same. This suggests that from the point of view of weak interactions, electron and neutrino are merely different quanta of the same underlying field

$$\psi_l = \begin{pmatrix} \psi_\nu^L \\ \psi_e^L \end{pmatrix} . \tag{E.19}$$

Henceforth, ψ will generally stand for the classical Dirac field. This field is to be eventually quantized applying the procedure outlined in Sect. 7.2.1. In an analogy to the spin states, we may regard neutrino and electron as the "spin up" and "spin down" states, respectively. These "spin" states are usually referred to as the *weak isospin* states; neutrino and electron have weak isospin $1/2$ and $-1/2$, respectively.

Even if we knew nothing about classical electrodynamics, we could stipulate the Hamiltonian describing the interaction of the spin with an external magnetic field to be of the form $\hat{\mathsf{H}} = -K\hat{\boldsymbol{S}} \cdot \mathbf{B}$, see Eq. (1.41), on the grounds that this is the only reasonable scalar constructed of the two available vectors $\hat{\boldsymbol{S}}$ and $\mathbf{B}$.[4] In a similar manner, we stipulate that the interaction Hamiltonian of the isospin field interacting with the vector bosons behaves as a scalar, but this time as a scalar with respect to rotations in the isospin space

$$[\boldsymbol{\gamma} \cdot (\mathrm{i}\boldsymbol{\partial} + g\boldsymbol{W}^a \mathsf{S}_a)]\, \psi_l = 0 \tag{E.20}$$

where

$$\mathsf{S}_a = \frac{\sigma_a}{2}, \quad \boldsymbol{W}^a \mathsf{S}_a = \frac{1}{\sqrt{2}}\left(\boldsymbol{W}^+ \mathsf{S}_- + \boldsymbol{W}^- \mathsf{S}_+\right) + \boldsymbol{W}^3 \mathsf{S}_3$$

and

$$\boldsymbol{W}^{\pm} = \frac{1}{\sqrt{2}}(\boldsymbol{W}^1 \pm \mathrm{i}\boldsymbol{W}^2) .$$

[4] A scalar of the type $\hat{\boldsymbol{S}} \cdot \hat{\boldsymbol{S}}$ does not seem reasonable as it clearly does not describe the interaction between the spin and the magnetic field.

Here σ_a, $a = 1, 2, 3$, are the Pauli spin matrices, but now acting in the isospin space, and we used Eq. (1.35).

First we verify that Eq. (E.20) preserves the electron lepton number. Multiplying this equation by the Dirac conjugated wave function from the left

$$\overline{\psi}_l \left[\boldsymbol{\gamma} \cdot (\mathrm{i}\partial^{\rightarrow} + g\boldsymbol{W}^a \mathsf{S}_a)\right] \psi_l = 0$$

and considering the equation for the Dirac conjugated wave function and multiplying it by the wave function from the right

$$\overline{\psi}_l \left[\boldsymbol{\gamma} \cdot (\mathrm{i}\partial^{\leftarrow} + g\boldsymbol{W}^a \mathsf{S}_a)\right] \psi_l = 0$$

and subtracting the two equations, we arrive at the continuity equation

$$\partial_\mu j_\mu = 0, \; j_\mu = \overline{\psi}_l \gamma_\mu \psi_l \,.$$

Integrating now over the whole space we obtain

$$\frac{d}{dt} Q_l = \oint \mathbf{j}_l \cdot \mathbf{dS} \,, \quad Q_l = \int dV |\psi_l|^2 = \int dV (|\psi_\nu^L|^2 + |\psi_e^L|^2) \,,$$

where clearly Q_l is the total electron lepton number.

From Eq. (E.20) we can deduce the form of the interaction Hamiltonian

$$H_{\text{int}} = \int \mathrm{d}^3\mathbf{r} \mathscr{H}_{\text{int}} \,, \quad \mathscr{H}_{\text{int}} = g\boldsymbol{j}^a \cdot \boldsymbol{W}^a \,, \quad \boldsymbol{j}^a = \psi^+ \mathsf{S}_a \boldsymbol{\gamma} \psi \,, \tag{E.21}$$

where $\mathscr{H}_{\text{int}}$ is called the *Hamiltonian interaction density* and $\boldsymbol{j}^a$ the *isotopic currents*. The last equation is a straightforward generalization of the interaction Hamiltonian of quantum electrodynamics $H_{\text{int}} = \int \mathrm{d}^3\mathbf{r}\boldsymbol{j} \cdot \boldsymbol{A}$, confront with Eq. (7.46).

What form does the law of conservation of isotopic currents take on? To find it, we multiply Eq. (E.20) by $\overline{\psi}_l \mathsf{S}_a$ from the left

$$\overline{\psi}_l \mathsf{S}_a \left[\boldsymbol{\gamma} \cdot (\mathrm{i}\partial^{\rightarrow} + g\boldsymbol{W}^b \mathsf{S}_b)\right] \psi_l = 0 \,.$$

We consider once again the Dirac conjugate of Eq. (E.20) and multiply it this time by $\mathsf{S}_a \psi_l$ from the right

$$\overline{\psi}_l \left[\boldsymbol{\gamma} \cdot (-\mathrm{i}\partial^{\leftarrow} + g\boldsymbol{W}^b \mathsf{S}_b)\right] \mathsf{S}_a \psi_l = 0 \,.$$

Subtracting the last two equations we find

$$\overline{\psi}_l \boldsymbol{\gamma} \cdot \left\{\mathsf{S}_a \mathrm{i}(\partial^{\rightarrow} + \partial^{\leftarrow}) + g\boldsymbol{W}^b [\mathsf{S}_a, \mathsf{S}_b]\right\} \psi_l = 0 \,.$$

Recalling now the definition of the isotopic currents, Eq. (E.21), and the commutation relation for the Pauli spin matrices, $[\mathsf{S}_a, \mathsf{S}_b] = \mathrm{i}\varepsilon_{abc}\mathsf{S}_c$, we arrive at

$$\partial_\mu j^a_\mu + g\varepsilon_{abc} W^b_\mu j^c_\mu = 0 \,. \tag{E.22}$$

One can immediately see that Eq. (7.230) has to be modified to

$$\partial_\mu \partial_\mu W^a_\nu - \partial_\nu \partial_\mu W^a_\mu = g j^a_\nu + g^2 N^a_\nu \,, \tag{E.23}$$

where N^a_ν is a nonlinear function of the field W^a_ν determined from the requirement Eq. (E.22) be satisfied. We will not try to find its form and restrict ourselves to a few comments. First, the nonlinearity of the classical equation for the field W means that after quantization the quanta of W-field interact among themselves. Although the existence of this self-interaction will be important in our next considerations, its precise form will not since this self-interaction is perfectly negligible at low energies.[5] One way to deduce the precise form of the classical equation obeyed by the W-field is to notice that Eq. (E.22) follows from the requirement the W-field be determined up to a *generalized gauge transformation*

$$W^a_\nu \to W^a_\nu + \partial_\nu \chi^a + g\varepsilon_{abc} W^b_\nu \chi^c \,. \tag{E.24}$$

Indeed, if we require the integral of the interaction Hamiltonian density over the whole space-time be not changed by this gauge transformation

$$\int \mathrm{d}^4\boldsymbol{x}\,\mathscr{H}_{int} = g \int \mathrm{d}^4\boldsymbol{x}\, W^a_\mu j^a_\mu = g \int \mathrm{d}^4\boldsymbol{x} (W^a_\mu + \partial_\mu \chi^a + g\varepsilon_{abc} W^b_\mu \chi^c) j^a_\mu$$

and integrate by parts, we obtain Eq. (E.22). Thus, the usual way of determining the nonlinear terms in Eq. (E.23) starts from the requirement the theory be invariant with respect to a generalized gauge transformation (E.24).

Glashow Theory of Electroweak Interactions

Bosons $W^\pm$, "carriers" of the weak force, are *charged*. Thus they *must* exchange photons between one another. Hence we stand no chance of creating a consistent theory of the weak interaction with intermediate carriers of the weak force unless we consider EM interactions as well.[6] One can easily trace the origin of this problem

[5] However, this is not the case of high energies. In fact, there is now an experimental proof of this self-interaction, see Figs. 14 and 15 in [11].

[6] However, note that we do not face this problem the other way around! Quantum electrodynamics is "sufficiently" consistent without taking into account the weak interaction.

back to the fact that the currents (E.16) and (E.17) are charged (neutron changes into proton, antineutrino into electron).

The boson W^3 is electrically neutral. Is it a photon? Were that the case, it would be really beautiful as the above-mentioned self-interactions of the W-field, necessary for the inner consistency of the theory, would be the so needed EM interactions of bosons $W^{\pm}$! In such a case, we would really have a unified theory of EM and weak interactions, as was the original theory of J. Schwinger from 1957. Unfortunately, it does not work. The reason is simple: the weak interaction acts only on the left-handed components of the fermions while the EM interaction on both. In other words, the weak interaction violates the parity, while the EM interaction preserves it. The correct theory was found in 1961 by Schwinger's student Sheldon Glashow: we need to introduce an additional carrier of force. Instead of Eq. (E.20), we need to consider the equations

$$\left[\boldsymbol{\gamma}\cdot(\mathrm{i}\boldsymbol{\partial}+g\boldsymbol{W}^a\mathsf{S}_a+g'\boldsymbol{X}Y_L)\right]\psi_l=0$$

and

$$\left[\boldsymbol{\gamma}\cdot(\mathrm{i}\boldsymbol{\partial}+g'\boldsymbol{X}Y_R)\right]\psi_e^R=0\,,$$

where obviously ψ_e^R stands for the right-handed component of the electron field and the number Y, to be determined later, is called the *hypercharge*. The interaction Hamiltonian density then has the form

$$\begin{aligned}\mathscr{H}_{\text{int}} &= +g\boldsymbol{W}^a\cdot\overline{\psi}_l\mathsf{S}_a\boldsymbol{\gamma}\psi_l+g'\boldsymbol{X}\cdot\left(\overline{\psi}_l\boldsymbol{\gamma}\psi_lY_L+\overline{\psi}{}_e^R\boldsymbol{\gamma}\psi_e^RY_R\right) \\ &= \left(\overline{\psi}{}_\nu^L\overline{\psi}{}_e^L\right)\boldsymbol{\gamma}\cdot\begin{pmatrix}\frac{gW^3}{2}+g'\boldsymbol{X}Y_L, & g\frac{W^-}{\sqrt{2}}\\ \frac{gW^+}{\sqrt{2}}, & -g\frac{W^3}{2}+g'\boldsymbol{X}Y_L\end{pmatrix}\begin{pmatrix}\psi_\nu^L\\ \psi_e^L\end{pmatrix} \\ &\quad+g'\overline{\psi}{}_e^R\boldsymbol{\gamma}\cdot\boldsymbol{X}\psi_e^RY_R\,.\end{aligned}\tag{E.25}$$

Now we come to the key step: the electrically neutral fields $\boldsymbol{X}$ and $\boldsymbol{W}^3$ are a linear combination of the fields $\boldsymbol{A}$ and $\boldsymbol{Z}$, where $\boldsymbol{A}$ is the vector potential of EM field well known to the reader from the previous chapters:

$$\begin{pmatrix}\boldsymbol{X}\\ \boldsymbol{W}^3\end{pmatrix}=\begin{pmatrix}\cos\vartheta, & -\sin\vartheta\\ \sin\vartheta, & \cos\vartheta\end{pmatrix}\begin{pmatrix}\boldsymbol{A}\\ \boldsymbol{Z}\end{pmatrix}.\tag{E.26}$$

The angle ϑ is called the *weak mixing angle*.[7] After substituting Eq. (E.26) into Eq. (E.25) we obtain for the interaction Hamiltonian density

[7] It is very often called the Weinberg angle, but this makes little sense since Mr. Glashow introduced it 6 years before Mr. Weinberg.

$$\mathscr{H}_{\text{int}} = +\left(\overline{\psi}^L_\nu \overline{\psi}^L_e\right) \boldsymbol{\gamma} \cdot \begin{pmatrix} \boldsymbol{Z}\left(\frac{g}{2}\cos\vartheta - g'\sin\vartheta Y_L\right), g\frac{W^-}{\sqrt{2}} \\ g\frac{W^+}{\sqrt{2}}, \boldsymbol{Z}\left(-\frac{g}{2}\cos\vartheta - g'\sin\vartheta Y_L\right) \end{pmatrix} \begin{pmatrix} \psi^L_\nu \\ \psi^L_e \end{pmatrix} \tag{E.27}$$

$$+ g'(-\sin\vartheta)\overline{\psi}^R_e \boldsymbol{\gamma} \cdot \boldsymbol{Z}\psi^R_e Y_R + \mathscr{H}_{\text{EM}}\,.$$

Thus, the Glashow theory leads to the following interactions:

$$\mathscr{H}_{\text{int}} = \mathscr{H}_{\text{EM}} + \mathscr{H}_{\text{CC}} + \mathscr{H}_{\text{NC}}\,, \tag{E.28}$$

where

$$\mathscr{H}_{\text{EM}} = \boldsymbol{J}_{EM} \cdot \boldsymbol{A},$$

stands for EM interactions,

$$\mathscr{H}_{\text{CC}} = \boldsymbol{J}_+ \cdot \boldsymbol{W}^- + \boldsymbol{J}_- \cdot \boldsymbol{W}^+\,, \tag{E.29}$$

stands for weak interactions, and finally

$$\mathscr{H}_{\text{NC}} = \boldsymbol{Z} \cdot \boldsymbol{J}_Z\,, \tag{E.30}$$

stands for electroweak interactions. The currents participating in these interactions can be found from Eq. (E.27):

1. Weak charged currents:

$$\boldsymbol{J}^l_+ = \frac{g}{2\sqrt{2}}\overline{\psi}_\nu \boldsymbol{\gamma}(1-\gamma_5)\psi_e,\ \boldsymbol{J}^l_- = \frac{g}{2\sqrt{2}}\overline{\psi}_e \boldsymbol{\gamma}(1-\gamma_5)\psi_\nu, \tag{E.31}$$

2. EM currents:

$$\boldsymbol{J}^l_{EM} = \left(\overline{\psi}^L_\nu \overline{\psi}^L_e\right) \boldsymbol{\gamma} \begin{pmatrix} \frac{g\sin\vartheta}{2} + g'\cos\vartheta Y_L, 0 \\ 0, -\frac{g\sin\vartheta}{2} + g'\cos\vartheta Y_L \end{pmatrix} \begin{pmatrix} \psi^L_\nu \\ \psi^L_e \end{pmatrix} + \overline{\psi}^R_e \boldsymbol{\gamma}\psi^R_e g'\cos\vartheta Y_R \tag{E.32}$$

$$= -e\overline{\psi}_e \boldsymbol{\gamma}\psi_e\,,$$

where e is the elementary electric charge. Recall that neutrino is electrically neutral, hence does not couple with the EM field. It follows from the last equation that

$$\frac{g\sin\vartheta}{2} + g'\cos\vartheta Y_L = 0$$

$$-\frac{g\sin\vartheta}{2} + g'\cos\vartheta Y_L = -e$$

$$g'\cos\vartheta Y_R = -e\,.$$

These equations determine the product $Y_L g'$. Once we choose a specific value of Y_L, by convention $Y_L = -1/2$, the quantities Y_R, g, and g' are determined unambiguously

$$Y_L = -\frac{1}{2},\ Y_R = -1, \tag{E.33}$$

$$g = \frac{e}{\sin\vartheta},\ g' = \frac{e}{\cos\vartheta}. \tag{E.34}$$

3. Electroweak neutral currents:

$$\boldsymbol{J}_Z^l = \frac{g}{2\cos\vartheta}\left\{\overline{\psi}_\nu \boldsymbol{\gamma}\left[\frac{1}{2} - \frac{\gamma_5}{2}\right]\psi_\nu + \overline{\psi}_e \boldsymbol{\gamma}\left[-\frac{1}{2} + 2\sin^2\vartheta + \frac{\gamma_5}{2}\right]\psi_e\right\}, \tag{E.35}$$

where we used Eqs. (E.33) and (E.34). The last equation may be rewritten into a concise form

$$\boldsymbol{J}_Z^l = \frac{g}{2\cos\vartheta}\overline{\psi}_l \boldsymbol{\gamma}\sigma_3\psi_l - \tan\vartheta\, \boldsymbol{J}_{EM}^l. \tag{E.36}$$

We thus see that for the charged currents we obtain what we should: $\boldsymbol{J}_+^l$ has the form of the current (E.17) entering the Fermi interaction Hamiltonian for β-decay and $\boldsymbol{J}_-^l$ is necessary for the description of the inverse β-decay. We know very well from the previous chapters what we should obtain for EM currents. This thus allows us to express the coupling constants g and g' in terms of the elementary electric charge and weak mixing angle. What is novel in the Glashow theory is the prediction of the electroweak neutral current $\boldsymbol{J}_Z$. It does so by introducing a single new parameter, namely the weak mixing angle.

Extension to Quarks

Now we extend the Glashow theory to hadrons. We do so by appealing to an apparently correct common belief that weak and EM interactions for lepton and quarks have the same form. Instead of the lepton doublet, Eq. (E.19), we consider now the quark doublet

$$\psi_q = \begin{pmatrix} \psi_u^L \\ \psi_d^L \end{pmatrix} \tag{E.37}$$

and instead of the Hamilton interaction density (E.25) we consider

$$\mathscr{H}_{\text{int}} = g\boldsymbol{W}^a \cdot \overline{\psi}_q \mathsf{S}_a \boldsymbol{\gamma}\psi_q + g'\boldsymbol{X}\cdot\left(\overline{\psi}_q \boldsymbol{\gamma}\psi_q Y_L^q + \overline{\psi}_u^R \boldsymbol{\gamma}\psi_u^R Y_R^u + \overline{\psi}_d^R \boldsymbol{\gamma}\psi_d^R Y_R^d\right). \tag{E.38}$$

The difference is brought by the fact that we have to consider the right-handed fields for both quarks. In the real world, none of the quarks is massless. Insertion of the transformation (E.26) into the last equation leads to, instead of Eq. (E.27),

$$\mathscr{H}_{\text{int}} = +\left(\overline{\psi}_u^L \overline{\psi}_d^L\right) \boldsymbol{\gamma} \cdot \begin{pmatrix} \boldsymbol{Z}\left(\frac{g}{2}\cos\vartheta - g'\sin\vartheta Y_L^q\right),\ g\frac{\boldsymbol{W}^-}{\sqrt{2}} \\ g\frac{\boldsymbol{W}^+}{\sqrt{2}},\ \boldsymbol{Z}\left(-\frac{g}{2}\cos\vartheta - g'\sin\vartheta Y_L^q\right) \end{pmatrix} \begin{pmatrix} \psi_u^L \\ \psi_d^L \end{pmatrix} \tag{E.39}$$
$$+ (-g'\sin\vartheta)\boldsymbol{Z}\cdot\left(\overline{\psi}_u^R \boldsymbol{\gamma}\psi_u^R Y_R^u + \overline{\psi}_d^R \boldsymbol{\gamma}\psi_d^R Y_R^d\right) + \mathscr{H}_{\text{EM}}\,.$$

We can again write the same form of the interaction Hamiltonian density, Eq. (E.28), as for leptons where now the currents entering the interaction take the following forms:

1. Weak charged currents:
   Instead of Eq. (E.31) we now obtain

$$\boldsymbol{J}_+^q = \frac{g}{2\sqrt{2}}\overline{\psi}_u \boldsymbol{\gamma}(1-\gamma_5)\psi_d,\ \boldsymbol{J}_-^q = \frac{g}{2\sqrt{2}}\overline{\psi}_d \boldsymbol{\gamma}(1-\gamma_5)\psi_u\,, \tag{E.40}$$

   where $\boldsymbol{J}_-^q$ and $\boldsymbol{J}_+^q$ enter the Fermi interaction Hamiltonian for the description of β-decay and inverse β-decay at the quark level.
2. EM currents:
   Instead of Eq. (E.32) we now have

$$\boldsymbol{J}_{EM}^q = \left(\overline{\psi}_u^L \overline{\psi}_d^L\right) \boldsymbol{\gamma} \begin{pmatrix} \frac{g\sin\vartheta}{2} + g'\cos\vartheta Y_L^q,\ 0 \\ 0,\ -\frac{g\sin\vartheta}{2} + g'\cos\vartheta Y_L^q \end{pmatrix} \begin{pmatrix} \psi_u^L \\ \psi_d^L \end{pmatrix} \tag{E.41}$$
$$+ g'\cos\vartheta\left(\overline{\psi}_u^R \boldsymbol{\gamma}\psi_u^R Y_R^u + \overline{\psi}_d^R \boldsymbol{\gamma}\psi_d^R Y_R^d\right) = \frac{2e}{3}\overline{\psi}_u \boldsymbol{\gamma}\psi_u - \frac{e}{3}\overline{\psi}_d \boldsymbol{\gamma}\psi_d\,.$$

   The last equality holds because we know the charges of up and down quarks are $2/3$ and $-1/3$ multiples of the elementary charge, respectively. It follows from the last equation, using Eq. (E.34), that the values of the quark hypercharges are

$$Y_L^q = \frac{1}{6},\quad Y_R^u = \frac{2}{3},\quad Y_R^d = -\frac{1}{3}\,. \tag{E.42}$$

3. Electroweak neutral currents:
   Instead of Eq. (E.35) we now have

$$\boldsymbol{J}_Z^q = \frac{g}{2\cos\vartheta}\left\{\overline{\psi}_u \boldsymbol{\gamma}\left[\frac{1}{2} - \frac{4}{3}\sin^2\vartheta - \frac{\gamma_5}{2}\right]\psi_u + \overline{\psi}_d \boldsymbol{\gamma}\left[-\frac{1}{2} + \frac{2}{3}\sin^2\vartheta + \frac{\gamma_5}{2}\right]\psi_d\right\}\,. \tag{E.43}$$

   One may write the last equation again in a concise form, confront with Eq. (E.36),

$$\boldsymbol{J}_Z^q = \frac{g}{2\cos\vartheta}\overline{\psi}_q \boldsymbol{\gamma}\sigma_3\psi_q - \tan\vartheta \boldsymbol{J}_{EM}^q\,.$$

Extension to Nucleons

So far so good, but for the fact that at low energies we do not observe free quarks but bound states of quarks instead. However, the precise form of the above currents for nucleons at low energies is a very difficult and still unresolved problem. We know only for sure that these currents depend on the exchanged momentum. In view of the foregoing discussion, see Sect. 7.8, this should not be surprising. We can therefore appeal to our knowledge of weak and EM interactions for vanishing momentum transfers to deduce the form of the electroweak interactions at this zero-momentum limit.

1. Weak charged currents:
   Instead of Eq. (E.40) we now have, confront with Eq. (E.18),

$$\boldsymbol{J}_+^N = \frac{g}{2\sqrt{2}}\overline{\psi}_p\boldsymbol{\gamma}(1-g_A\gamma_5)\psi_n,\ \boldsymbol{J}_-^N = \frac{g}{2\sqrt{2}}\overline{\psi}_n\boldsymbol{\gamma}(1-g_A\gamma_5)\psi_p\,. \tag{E.44}$$

2. EM currents:
   We know that proton and neutron are constituted by two up and one down quarks and two down and one up quarks, respectively. Therefore using Eq. (E.41) and assuming the strong interaction does not enter, we write

$$\boldsymbol{J}_{EM}^N = \overline{\psi}_p\boldsymbol{\gamma}\psi_p\left(2\frac{2e}{3}-\frac{e}{3}\right)+\overline{\psi}_n\boldsymbol{\gamma}\psi_n\left(\frac{2e}{3}-2\frac{e}{3}\right) = e\overline{\psi}_p\boldsymbol{\gamma}\psi_p\,. \tag{E.45}$$

   However, this expression is not completely correct. If it were, the gyromagnetic ratio of proton would be $1+\alpha/(2\pi)$ and that of neutron 0, see Sects. 7.5.7 and 7.8. Yet we know from experiment that these ratios equal approximately 2.79 and -1.86, respectively. The strong interaction causes the EM current of nucleons not to be purely vectorial, even at zero momentum transfer. Nevertheless, the strong interactions leave the vector part of the EM current intact, similar to the weak charged currents (E.44).
3. Electroweak neutral currents:
   Given the structure of the electroweak theory and knowing that strong interactions leave purely vector part of the currents intact, recalling the composition of nucleons in terms of quarks and using Eq. (E.43), it is reasonable to guess that the electroweak neutral current for nucleons takes the form

$$\begin{aligned}\boldsymbol{J}_Z^N =& \frac{g}{2\cos\vartheta}\left\{\overline{\psi}_p\boldsymbol{\gamma}\left[2\left(\frac{1}{2}-\frac{4}{3}\sin^2\vartheta-\frac{\gamma_5}{2}g_A\right)+\left(-\frac{1}{2}+\frac{2}{3}\sin^2\vartheta+\frac{\gamma_5}{2}g_A\right)\right]\psi_p\right.\\ &\left.+\overline{\psi}_n\boldsymbol{\gamma}\left[\left(\frac{1}{2}-\frac{4}{3}\sin^2\vartheta-\frac{\gamma_5}{2}g_A\right)+2\left(-\frac{1}{2}+\frac{2}{3}\sin^2\vartheta+\frac{\gamma_5}{2}g_A\right)\right]\psi_n\right\}\\ =& \frac{g}{2\cos\vartheta}\left\{\overline{\psi}_p\boldsymbol{\gamma}\left[\frac{1}{2}-2\sin^2\vartheta-\frac{\gamma_5}{2}g_A\right]\psi_p+\overline{\psi}_n\boldsymbol{\gamma}\left[-\frac{1}{2}+\frac{\gamma_5}{2}g_A\right]\psi_n\right\}\,.\end{aligned} \tag{E.46}$$

Effective Interactions at Low Energies

In the real world, the masses of the vector bosons are not zero; bosons are in fact very heavy. Due to the very high masses of the W and Z bosons, the interactions (E.29) and (E.30) reduce to, confront with Eq. (E.3),

$$\mathcal{H}_{\mathrm{CC}} \to \frac{1}{m_W^2} \boldsymbol{J}_+ \cdot \boldsymbol{J}_- = \frac{g^2}{8m_W^2} \boldsymbol{j}_+ \cdot \boldsymbol{j}_- , \tag{E.47}$$

where

$$\boldsymbol{J}_\pm = \frac{g}{2^{3/2}} \boldsymbol{j}_\pm$$

and

$$\mathcal{H}_{\mathrm{NC}} \to \frac{1}{m_Z^2} \boldsymbol{J}_Z \cdot \boldsymbol{J}_Z = \frac{g^2}{16 m_Z^2 \cos^2 \vartheta} \boldsymbol{j}_Z \cdot \boldsymbol{j}_Z , \tag{E.48}$$

where

$$\boldsymbol{J}_Z = \frac{g}{4\cos\vartheta} \boldsymbol{j}_Z .$$

Here the currents are sums of leptonic and nucleonic currents

$$\boldsymbol{j} = \boldsymbol{j}^l + \boldsymbol{j}^N .$$

If we insert the last equation into Eqs. (E.47) and (E.48), we find that Eq. (E.47) includes the Fermi Hamiltonian for β-decay, (E.4),

$$\mathcal{H}_\beta = \frac{g^2}{8m_W^2} \boldsymbol{j}_+^N \cdot \boldsymbol{j}_-^l \tag{E.49}$$

and Eq. (E.48) features parity violating interaction arising from a Z-boson exchange between the nucleus and electrons

$$\mathcal{H}_{\mathrm{ANC}} = \frac{g^2}{16 m_Z^2 \cos^2 \vartheta} \boldsymbol{j}_Z^N \cdot \boldsymbol{j}_Z^e . \tag{E.50}$$

The interaction arising from the Z-boson exchange between electrons is too small to be detectable. The Z-boson exchange between nucleons leads to the parity violating nuclear forces.

Masses of Intermediate Bosons

Comparison of Eqs. (E.4) and (E.49) yields an important relation

$$\frac{G_\beta}{\sqrt{2}} = \frac{g^2}{8m_W^2}\,. \tag{E.51}$$

Recalling further the relations (E.34) we see that the mass of the W bosons can be expressed by means of the Fermi coupling constant, fine structure constant, and weak mixing angle; that is, no new parameter is introduced by giving mass to the W bosons. However, we know nothing about the mass of Z boson so far. This absence of knowledge hinders the predictive power of the Glashow theory as the strength of the electroweak interaction, Eq. (E.48), depends on it.

Abdus Salam and Steven Weinberg substantially improved the predictive power of the Glashow theory by assuming that the three W bosons, the charged $W^\pm$ and the neutral W^3 possess the same mass.[8] However, the W^3 boson is not a particle with a definite mass. Its mass term in the Hamiltonian density has to be considered together with the X boson and is represented by the mass matrix

$$\left(W_\mu^3, X_\mu\right)\begin{pmatrix} m_W^2, & m_{XW}^2 \\ m_{XW}^2, & m_X^2 \end{pmatrix}\begin{pmatrix} W_\mu^3 \\ X_\mu \end{pmatrix}. \tag{E.52}$$

This means that, for instance, the W^3 boson can be converted by a mere time evolution into the X boson and vice versa. That is, these states are not stationary even when the particles are free. On the other hand, the Z boson and photon are particles of definite mass, hence their mass matrix takes the form

$$\left(Z_\mu, A_\mu\right)\begin{pmatrix} m_Z^2, & 0 \\ 0, & 0 \end{pmatrix}\begin{pmatrix} Z_\mu \\ A_\mu \end{pmatrix}.$$

Inserting now the relation (E.26) between W^3, X and Z, A fields into Eq. (E.52) and comparing the last two equations we arrive at a very important relation connecting the masses of the $W^\pm$ and Z bosons

$$m_W^2 = m_Z^2\cos^2\vartheta\,. \tag{E.53}$$

---

[8]This is not the way Salam and Weinberg contribution is usually presented. The argument given here is taken from [8].

Using now Eqs. (E.51) and (E.53) we find that the strength of the electroweak interactions is governed by the Fermi coupling constant for β-decay

$$\frac{g^2}{16m_Z^2\cos^2\vartheta} = \frac{G_\beta}{2^{3/2}}\,. \tag{E.54}$$

To conclude, the Glashow-Salam-Weinberg theory predicts electroweak interactions of neutral currents, Eqs. (E.48) and (E.54). The measurement of this interaction, e.g., in atomic experiments, yields the value of the weak mixing angle. With the Salam-Weinberg improvement, this weak mixing angle remains the only new free parameter of the theory even after finite masses are assigned to the W and Z bosons.

Electroweak Neutral Currents in Atoms

In atoms, electroweak interaction (E.50) is dominated by the temporal component of the vector part of the nucleonic current

$$\mathscr{H}_{\text{ANC,NSI}} = \frac{G_\beta}{2^{3/2}}\rho_W\psi_e^+\gamma_5\psi_e\,, \quad \rho_W = \psi_p^+\psi_p(1-4\sin^2\vartheta) - \psi_n^+\psi_n\,.$$

The nucleus is placed at rest, with utterly negligible intrinsic dynamics. The nucleon fields may be thus considered classically. The "weak charge density" ρ_W is then given in terms of the proton and neutron densities, ρ_p and ρ_n, respectively, as

$$\rho_W(\mathbf{r}) = \rho_p(\mathbf{r})(1-4\sin^2\vartheta) - \rho_n(\mathbf{r})\,.$$

What remains is the one-particle operator written in terms of the electron field ψ_e. However, this is merely a complicated way of saying that what remains is the following addition to the potential energy of the one-particle Dirac Hamiltonian

$$\hat{\mathsf{h}}_{\text{ANC,NSI}} = \frac{G_\beta}{2^{3/2}}\rho_W(\mathbf{r})\gamma_5\,. \tag{E.55}$$

In contrast to the nuclear spin-dependent part which is further complicated by the parity-violating part of the nuclear forces, the nuclear spin-independent part arising from the electroweak force is nearly free of uncertainities of nuclear physics. In the first approximation, one can consider the nucleus to be point-like

$$\rho(\mathbf{r}) = Q_W\delta(\mathbf{r})\,, \quad Q_W = Z(1-4\sin^2\vartheta) - N\,,$$

where clearly Z is the charge of the nucleus and N the number of the neutrons. Q_W is commonly termed the *weak charge* of the nucleus.

Let us summarize the main features of the interaction (E.55). First, it is parity violating (also called parity-odd). That is, it mixes states of different parity. To see the mixing, it suffices to consider the matrix elements of (E.55) between an s-state and a $p_{1/2}$ state

$$\langle n' p_{1/2} | \gamma_5 \delta(\mathbf{r}_A) | ns \rangle \simeq \frac{Z\alpha}{2} \langle n' p_{1/2} |_0 \left\{ \delta(\mathbf{r}), \boldsymbol{\sigma} \cdot \hat{\boldsymbol{p}} \right\} | ns \rangle_0 ,$$

where we inserted Eq. (7.22) with $\mathbf{A} = 0$ and (7.218), and the state vectors with the subscripts 0 are the Pauli spinors, Eq. (4.58).

Second, the interaction (E.55) is really weak. Changing to atomic units, $r = r_A/(m_e Z\alpha)$, we find that the interaction (E.55) is proportional to

$$Q_W G_\beta m_e^3 (Z\alpha)^4 \simeq m_e (Z\alpha)^2 \left[10^{-5} \left(\frac{m_e}{m_p} \right)^2 \alpha^2 \right] Q_W Z^2 ,$$

where we substituted $G_\beta \simeq 1 \times 10^{-5} m_p^{-2}$. The factor in the square brackets on the rhs is of the order 10^{-15}. However, given that $Q_W \approx Z$, we see that for heavy atoms there is an enhancement of the interaction strength by the factor Z^3.

The interaction (E.55) was measured on several atomic transitions in heavy atoms, the most accurate being the $6S_{1/2} - 7S_{1/2}$ transition in ^{133}Cs (the superscript denotes the total number of nucleons in nucleus). Due to the weakness of the electroweak interaction, the measurements rely on the interference of parity violating transition amplitude with forbidden EM transition amplitude. For a recent review of the current status of both the theory and experiment see [12]. The value cited there obtained from the comparison of the theory and experiment for parity violating transition amplitude in the cesium atom reads

$$\sin^2 \vartheta = 0.2356(20) .$$

This is the "weighting" of the W and Z bosons. From Eqs. (E.34) and (E.51) we obtain

$$m_W = \sqrt{\frac{\alpha \pi}{\sqrt{2} G_\beta \sin^2 \vartheta}} \simeq 79\,\text{GeV} .$$

The experimentally determined mass of the W boson is $m_W \simeq 80\,\text{GeV}$. If we substitute the experimentally determined masses of the W and Z bosons, $m_Z \simeq 91\,\text{GeV}$, into the Salam-Weinberg formula (E.53), we find $\sin^2 \vartheta \simeq 0.23$. The theory thus seems to be on the right track. The masses of the W and Z bosons are determined from the position of resonances in high-energy electron-positron annihilation [11]. In fact, the above calculation of the masses of the W and Z bosons produces values in better agreement with the experiment than one would (rightfully) expect. The reason is that various effects arising from the presence of additional two fermion generations (which we neglected) cancel each other.

Exercise 32: Parity Violating Interaction

Show that for the hydrogenic wave functions

$$\langle np_{1/2}|_0 \left\{\delta(\mathbf{r}_A), \boldsymbol{\sigma}\cdot\hat{\boldsymbol{p}}_A\right\} |ns\rangle_0 = -\mathrm{i}\sqrt{\frac{n^2-1}{n^2}\frac{4}{n^3}}\,.$$

Hint: Use the results of Ex. 11 and Eqs. (4.76) and (4.101).

Final Notes

In our discussion of electroweak interactions we completely avoided the problem associated with particle masses at high energies and its solution via the Higgs mechanism. We barely mentioned the generalization of the gauge invariance and associated nonlinear interactions between the quanta of the gauge fields. For a first encounter with these topics we refer the reader to, e.g., [1, 4, 5, 10], for a more advanced treatment see, e.g., [9], and for a thorough discussion of both the theory and experiment see [3].

References

1. I.J.R. Aitchinson, A.J.G. Hey, *Gauge Theories in Particle Physics* (IOP Publishing Ltd, Bristol, 1989)
2. P.W. Anderson, *More and Different: Notes From a Thoughtful Curmudgeon* (World Scientific, Singapore, 2011)
3. E.D. Commins, P.H. Bucksbaum, *Weak Interactions of Leptons and Quarks* (Cambridge University Press, Cambridge, 1983)
4. J. Hořejší, *Fundamentals of Electroweak Theory* (Karolinum, Prague, 2002); J. Hořejší, *Introduction to Electroweak Unification: Standard Model From Tree Unitarity* (World Scientific, Singapore, 1994)
5. F. Mandl, G. Shaw, *Quantum Field Theory* (Wiley, New York, 1993)
6. A. Pais, *Inward Bound: of Matter and Forces in the Physical World* (Oxford University Press, Oxford, 1986)
7. Particle Data Group, Phys. Rev. D **86**, 010001 (2012)
8. D.H. Perkins, *Introduction to High Energy Physics* (Addison-Wesley, Reading, 1987)
9. M. Peskin, D. Schroeder, *An Introduction to Quantum Field Theory* (Westview Press, Boulder, 1995)

10. C. Quigg, *Gauge Theories of the Strong, Weak and Electromagnetic Interactions* (Benjamin/Cummings, Menlo Park, 1984)
11. S. Riemann, Rep. Prog. Phys. **73**, 126201 (2010)
12. B.M. Roberts, V.A. Dzuba, V.V. Flambaum, Annu. Rev. Nucl. Part. Sci. **65**, 63 (2015)
13. E. Segre, *Enrico Fermi Physicist* (The University of Chicago Press, Chicago, 1970)

Index

J. Zamastil, J. Benda, *Quantum Mechanics and Electrodynamics*,
DOI 10.1007/978-3-319-65780-6

MIX
Papier aus verantwortungsvollen Quellen
Paper from responsible sources
FSC® C105338

If you have any concerns about our products,
you can contact us on
ProductSafety@springernature.com

In case Publisher is established outside the EU,
the EU authorized representative is:
Springer Nature Customer Service Center GmbH
Europaplatz 3, 69115 Heidelberg, Germany

Printed by Libri Plureos GmbH
in Hamburg, Germany